STUDENT SOLUTIONS MANUAL

&

STUDY GUIDE

To Accompany

W9-CUQ-598

INTERMEDIATE ALGEBRA
Third Edition

Dennis Weltman
North Harris County College

Gilbert Perez
San Antonio College

Richard C. Spangler
Tacoma Community College

PWS Publishing Company
Boston, MA
A Division of Wadsworth, Inc.

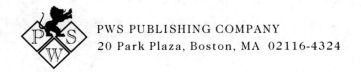

PWS PUBLISHING COMPANY
20 Park Plaza, Boston, MA 02116-4324

PWS Publishing Company is a division of Wadsworth, Inc.

I(T)P ™
International Thomson Publishing
The trademark ITP is used under license

PREFACE

This Solutions Manual and Study Guide accompanies *Intermediate Algebra, Third Edition*, by Dennis Weltman and Gilbert Perez. The booklet contains worked-out solutions to odd-numbered exercises and a Study Guide for all sections in each chapter.

To use this manual:

1. For each section in the main text, read the explanations and work through the examples using pencil and paper.

2. Next, work as many exercises as assigned by your instructor.

3. Compare your answers with those at the end of your text, and rework the ones that were not correct.

4. If further help is needed, check your work against the detailed solutions provided in this manual.

5. The worked-out solutions also may be used as examples to help find solutions to other problems.

6. For additional practice try the Study Guide self-tests at the end of each chapter. Self-test solutions are also included.

CONTENTS

Chapter 1   Fundamental Concepts **1**
          Study Guide **11**

Chapter 2   Linear Equations and Inequalities **13**
          Study Guide **44**

Chapter 3   Polynomials and Factorable Quadratic Equations **48**
          Study Guide **83**

Chapter 4   Rational Expressions **86**
          Study Guide **134**

Chapter 5   Exponential and Radical Expressions **138**
          Study Guide **166**

Chapter 6   Linear Relations and Functions **169**
          Study Guide **184**

Chapter 7   Quadratic and Higher Degree Equations and Inequalities **187**
          Study Guide **226**

Chapter 8   Conic Sections **230**
          Study Guide **236**

Chapter 9   Systems of Equations **240**
          Study Guide **269**

Chapter 10   Exponential and Logarithmic Functions **275**
          Study Guide **302**

Chapter 11   Sequences and Series **306**
          Study Guide **332**

CHAPTER 1 FUNDAMENTAL CONCEPTS

Solutions to Text Exercises

1.-19.
The exercises do not require worked-out solutions.

21.
$M \cap N = \{1,2,3,...,10\} \cap \{1,2,3,4,...\}$
$= \{1,2,3,...,10\} = M$

23.
$M \cap P = \{1,2,3,...\} \cap \{2,4,6,8,10\} = \{2,4,6,8,10\}$

25.
$M \cup P = \{1,2,3,...,10\} \cup \{2,4,6,8,10\}$
$= \{1,2,3,...,10\} = M$

27.
$N \cup D = \{1,2,3,...\} \cup \{1,3,5,7,...\} = \{1,2,3,...\} = N$

29.
$D \cap P = \{1,3,5,7,...\} \cap \{2,4,6,8,10\} = \varnothing$

31.
$N \cup P = \{1,2,3,...\} \cup \{2,4,6,8,10\} = \{1,2,3,...\} = N$

33.
$M \cap M = M$

35.
$M \cup \varnothing = M$

37.
$(M \cap D) \cup P = M \cup P = M$
Combine #21 and #25.

39.
$(M \cup D) \cap P = \{1,2,3,...,10\} \cup \{1,3,5,7,...\}$
$\cap \{2,4,6,8,10\}$
$= \{1,2,3,...,10\} \cap \{2,4,6,8,10\}$
$= \{2,4,6,8,10\} = P$

41.
The exercise does not require a worked-out solution.

1.-9.
These exercises do not require worked-out solutions.

11.
$-30 - (-52) = -30 + (52)$ Add opposites
$= 22$

13.-25.
The exercises do not require worked-out solutions.

27.
$(-2)^3 = (-2)(-2)(-2) = 4(-2) = -8$

29.
$-5^4 = -1 \cdot 5 \cdot 5 \cdot 5 \cdot 5 = -1 \cdot 625 = -625$

31.
The exercise does not require a worked-out solution.

33.
$(-2^4) = -1 \cdot 2 \cdot 2 \cdot 2 \cdot 2 = -1 \cdot 16 = -16$

35.
$-57 + 17 - (-23) - 7 = -57 + 17 + 23 - 7$
$= -40 + 23 - 7$
$= -17 - 7 = -24$

37.
$59 + 41 - (-10) - 9 = 59 + 41 + 10 - 9$
$= 100 + 10 - 9 = 110 - 9 = 101$

39.
$(-7)(2)(10)(-3) = -14(10)(-3) = -140(-3) = 420$

1

41.-43.
The exercises do not require worked-out solutions.

45.
$$(-2)(2)(-3)(-4)(8) = -4(-3)(-4)(8) = 12(-4)(8)$$
$$= -48(8) = -384$$

47.-49.
The exercises do not require worked-out solutions.

Exercises 1.3

1.
$$\frac{64}{48} = \frac{\cancel{16} \cdot 4}{\cancel{16} \cdot 3} = \frac{4}{3}$$

3.
$$\frac{35}{18} = \frac{5 \cdot 7}{2 \cdot 3 \cdot 3} = \frac{35}{18} \quad \text{Already reduced}$$

5.
$$\frac{56}{63} = \frac{\cancel{7} \cdot 8}{\cancel{7} \cdot 9} = \frac{8}{9}$$

7.
$$\frac{\cancel{2} \cdot \cancel{7}}{\cancel{7} \cdot \cancel{2} \cdot 5} = \frac{1}{5}$$

9.
$$\frac{243}{162} = \frac{3 \cdot \cancel{9} \cdot \cancel{9}}{2 \cdot \cancel{9} \cdot \cancel{9}} = \frac{3}{2}$$

11.
The exercise does not require a worked-out solution.

13.
$$\frac{7}{8} - \frac{-11}{20} \qquad \text{LCD} = 40$$
$$= \frac{5}{5} \cdot \frac{7}{8} - \frac{-11}{20} \cdot \frac{2}{2} = \frac{35}{40} - \frac{-22}{40}$$
$$= \frac{35}{40} + \frac{22}{40} \qquad \text{Add opposites.}$$
$$= \frac{57}{40}$$

15.
$$\frac{8}{21} \cdot \frac{35}{6} = \frac{\cancel{2} \cdot 4}{3 \cdot \cancel{7}} \cdot \frac{5 \cdot \cancel{7}}{\cancel{2} \cdot 3} = \frac{20}{9}$$

17.
$$\frac{2}{15} + \frac{10}{27} = \frac{2}{15} \cdot \frac{27}{10} = \frac{\cancel{2}}{\cancel{5} \cdot 5} \cdot \frac{\cancel{3} \cdot 9}{\cancel{2} \cdot 5} = \frac{9}{25}$$

19.
The exercise does not require a worked-out solution.

21.
$$\frac{18}{\frac{-4}{9}} = 18 \div \frac{-4}{9} = \frac{18}{1} \cdot \frac{9}{-4} = \frac{\cancel{2} \cdot 9}{1} \cdot \frac{9}{-2 \cdot \cancel{2}} = -\frac{81}{2}$$

23.
$$2 - \left(\frac{-11}{39}\right) - \frac{25}{26} = \frac{156}{78} - \left(\frac{-22}{78}\right) - \frac{75}{78}$$
$$= \frac{156}{78} + \frac{22}{78} - \frac{75}{78} = \frac{178}{78} - \frac{75}{78} = \frac{103}{78}$$

25.
$$\left(3\frac{1}{4}\right)\left(\frac{-5}{-8}\right) = \frac{13}{4} \cdot \frac{5}{8} = \frac{65}{32}$$

27.
$$5\frac{1}{2} - \left(-1\frac{1}{6}\right) - 8\frac{3}{4} = \frac{11}{2} + \frac{7}{6} - \frac{35}{4} = \frac{66}{12} + \frac{14}{12} - \frac{105}{12}$$
$$= \frac{80}{12} - \frac{105}{12} = -\frac{25}{12}$$

29.
$$\left(\frac{25}{-9}\right)(-6)\left(-\frac{21}{40}\right)\left(\frac{2}{35}\right)$$
$$= \left(\frac{\cancel{5} \cdot \cancel{5}}{-\cancel{3} \cdot \cancel{3}}\right)\left(\frac{-1 \cdot \cancel{2} \cdot \cancel{3}}{1}\right)\left(\frac{-\cancel{3} \cdot \cancel{7}}{\cancel{5} \cdot \cancel{2} \cdot \cancel{2} \cdot 2}\right)\left(\frac{\cancel{2}}{\cancel{5} \cdot \cancel{7}}\right) = -\frac{1}{2}$$

31.

$\dfrac{3}{8} = 0.375$ Work

$$8\overline{)3.000}$$

$$\begin{array}{r}.375\\\hline\underline{24}\\60\\\underline{56}\\40\\\underline{40}\end{array}$$

33.

$-\dfrac{2}{9} = -0.2\overline{2}$ Work

$$9\overline{)2.00}$$

$$\begin{array}{r}.22\cdots\\\hline\underline{18}\\20\\\underline{18}\\2\end{array}$$

35.

$\dfrac{1}{7} = 0.\overline{142857}$ Work

$$7\overline{)1.0000000}$$

$$\begin{array}{r}.1428571\\\hline\underline{7}\\30\\\underline{28}\\20\\\underline{14}\\60\\\underline{56}\\40\\\underline{35}\\50\\\underline{49}\\10\\\underline{7}\end{array}$$

37.

$0.65 = \dfrac{65}{100} = \dfrac{\cancel{5}\cdot 13}{\cancel{5}\cdot 20} = \dfrac{13}{20}$

39.

$3.875 = \dfrac{3875}{1000} = \dfrac{\cancel{125}\cdot 31}{\cancel{125}\cdot 8} = \dfrac{31}{8}$

41.

$0.0025 = \dfrac{25}{10,000} = \dfrac{\cancel{25}\cdot 1}{\cancel{25}\cdot 400} = \dfrac{1}{400}$

43.

$-16.45 + (-8.403) = -24.853$ Work

$$\begin{array}{r}-16.450\\-8.403\\\hline-24.853\end{array}$$

45.

$(-13.356) \div (-6.3) = 2.12$ Work

$$6.3\overline{)13.356}$$

$$\begin{array}{r}2.12\\\hline\underline{126}\\75\\\underline{63}\\126\\\underline{126}\end{array}$$

47..

$7.81 + 16.095 - 9.96 = 23.905 - 9.96 = 13.945$

49.

$(-2.5)(-40.8)(-7.25) = 102(-7.25) = -739.5$

51.

$14.5 + 8.91 - 3.407 = 23.41 - 3.407 = 20.003$

53.

$(-152.2) \div (6.15) = -24.8$ Work

$$6.15\overline{)-152.52\,0}$$

$$\begin{array}{r}24.8\\\hline\underline{1230}\\2952\\\underline{2460}\\4920\\\underline{4920}\end{array}$$

55.

a. $P = 2W + 2L$

$$P = 2\left(1\dfrac{5}{6}\right) + 2\left(3\dfrac{2}{3}\right) = 2\left(\dfrac{11}{6}\right) + 2\left(\dfrac{11}{3}\right)$$

$$= \dfrac{22}{6} + \dfrac{22}{3} \qquad \text{LCD} = 6$$

#55. continued

$$= \frac{22}{6} + \frac{44}{6} = \frac{66}{6} \text{ or } 11$$

Answer: 11 in

b. $A = a + b + c$; $A = 2.4 + 4.45 + 3.82 = 10.67$

Answer: 10.67 m

57.

She worked $16\frac{1}{4}$ or 16.25 hours overtime. For time and one - half she earned $6.75 per hour.

a. Thus, gross income $= 40(4.50) + 16.25(6.75)$

$$= 180 + 109.69 = 289.69$$

Answer: $289.69

b. $\dfrac{164.25}{4.50} = 36.5$ Answer: 36.5 hr

59.

$$16\frac{3}{4} + 1\frac{1}{2} + \frac{3}{8} - 2\frac{1}{4} = \frac{67}{4} + \frac{3}{2} + \frac{3}{8} - \frac{9}{4} \qquad LCD = 8$$

$$\frac{134}{8} + \frac{12}{8} + \frac{3}{8} - \frac{18}{8} = \frac{149}{8} - \frac{18}{8} = \frac{131}{8} = 16\frac{3}{8}$$

61.

a. $n = 0.555...$

$10n = 5.555...$

$9n = 5$ Subtract

So, $n = \dfrac{5}{9}$

b. $n = 0.4545...$

$100n = 45.4545...$

$99n = 45$ Subtract

So, $n = \dfrac{45}{99}$ or $\dfrac{5}{11}$

Exercises 1.4

1.-11.
The exercises do not require worked - out solutions.

13.

$D^2 = 3^2 + 3^2 = 9 + 9 = 18$

$D = \sqrt{18} \doteq 4.24 mm$

15.

$D^2 = 5^2 + 2^2 = 25 + 4 = 29$ $D = \sqrt{29} \doteq 5.39 m$

17.

$D^2 = 5^2 + 7^2 = 25 + 49 = 74$

$D = \sqrt{74} \doteq 8.60$ in or $8\frac{3}{5}$ in

19.

$D^2 = 17^2 + 17^2 = 289 + 289 = 578$

$D = \sqrt{578} \doteq 24.04$ in or $24\frac{1}{25}$ in

21.

$C = 2(0.6)(3.14) = 3.768 \doteq 3.77$ m

23.

$A = (3.14)(3.5)^2 = (3.14)(12.25) = 38.465$

$\doteq 38.47$ cm^2

25.

$A = \dfrac{22}{7}\left(\dfrac{7}{4}\right)^2 = \dfrac{22}{7}\left(\dfrac{49}{16}\right) = \dfrac{77}{8} = 9\dfrac{5}{8}$ in^2

27.-31.
The exercises do not require worked - out solutions.

Exercises 1.5

1.-83.
The exercises do not require worked - out solutions.

Exercises 1.6

1.

$17 - 8 + 7 + (-6) = 17 + (-8) + 7 + (-6) = 9 + 7 + (-6)$

$$= 16 + (-6)$$
$$= 10$$

3.

$24 \div 4 \cdot 3 = 6 \cdot 3 = 18$

5.

$5 \cdot 3^2 = 5 \cdot 9 = 45$

7.

$$-8 + 30 \div 2^2 - \frac{1}{4} = -8 + 30 \div 4 - \frac{1}{4} = -8 + \frac{30}{4} - \frac{1}{4}$$

$$= -8 + \frac{29}{4} = \frac{-32}{4} + \frac{29}{4} = -\frac{3}{4}$$

9.

$$1.8 \div (1.5)^2 + 0.2 = 1.8 \div 2.25 + 0.2 = 0.8 + 0.2 = 1$$

11.

$$11(7 - 15) \div 66 = 11(-8) \div 66 = -88 \div 66 = -\frac{88}{66}$$

$$= -\frac{8}{6} = -\frac{4}{3}$$

13.

$$12 - 2(6 + 3 \cdot 4) + 15 = 12 - 2(6 + 12) + 15$$

$$= 12 - 2(18) + 15 = 12 - 36 + 15$$

$$= -24 + 15 = -9$$

15.

$$-3 - \sqrt{6 \cdot 2 - 8} = -3 - \sqrt{12 - 8} = -3 - \sqrt{4} = -3 - 2$$

$$= -5$$

17.

$$4 \cdot 7 \cdot \sqrt{16} = 4 \cdot 7 \cdot 4 = 112$$

19.

$$7 + 8(-11) \div 4 = 7 + (-88) \div 4 = 7 + (-22) = -15$$

21.

$$\frac{2.1 + 5(0.7)}{(0.2)^2 + 0.1} = \frac{2.1 + 3.5}{0.04 + 0.1} = \frac{5.6}{0.14} = 40$$

23.

$$4 \cdot \left| 6 - 4 \div \frac{1}{2} \right| = 4 \cdot |6 - 8| = 4|-2| = 4 \cdot 2 = 8$$

25.

$$\left[7 + 3\left(2^3 - 1 \right) \right] \div 21 = [7 + 3(8 - 1)] \div 21$$

$$= [7 + 3(7)] \div 21$$

$$= [7 + 21] \div 21$$

#25. continued

$$= 28 \div 21 = \frac{28}{21} = \frac{4}{3}$$

27.

$$3 - [32 \div (2 \cdot 7 - 6)] = 3 - [32 \div (14 - 6)]$$

$$= 3 - [32 \div 8] = 3 - 4 = -1$$

29.-39.
The graphs are in the answer section of the main text.

41.-55.
The exercises do not require worked - out solutions.

57.

$$x^3 + 4x^2 - x + 3 \text{ for } x = -2$$

$$(-2)^3 + 4(-2)^2 - (-2) + 3 = -8 + 4 \cdot 4 + 2 + 3$$

$$= -8 + 16 + 2 + 3 = 13$$

59.

$$-2(x + 3)^2 - 4 \text{ for } x = 1$$

$$-2(1 + 3)^2 - 4 = -2(4)^2 - 4$$

$$= -2(16) - 4 = -32 - 4 = -36$$

61.

$$\frac{1}{2}(x - 1)^2 + 3 \text{ for } x = -3$$

$$\frac{1}{2}(-3 - 1)^2 + 3 = \frac{1}{2}(-4)^2 + 3 = \frac{1}{2}(16) + 3 = 8 + 3 = 11$$

63.

$$\frac{2}{3}\sqrt{13 - x^2} \text{ for } x = 2$$

$$\frac{2}{3}\sqrt{13 - 2^2} = \frac{2}{3}\sqrt{13 - 4} = \frac{2}{3}\sqrt{9} = \frac{2}{3} \cdot 3 = 2$$

65.

$$|2x + 3| - 4 \text{ for } x = -4$$

$$|2(-4) + 3| - 4 = |-8 + 3| - 4 = |-5| - 4 = 5 - 4 = 1$$

67.

$$4x^2 - xy + 3y^2 \text{ for } x = 3, \ y = -1$$

$$4(3)^2 - 3(-1) + 3(-1)^2 = 4 \cdot 9 + 3 + 3(1)$$

#67. continued

$$= 36 + 3 + 3 = 42$$

69.

$$\frac{2x+1}{x} \quad \text{for} \quad x = \frac{1}{4}$$

$$\frac{2\left(\frac{1}{4}\right)+1}{\frac{1}{4}} = \frac{\frac{1}{2}+1}{\frac{1}{4}} = \frac{\frac{3}{2}}{\frac{1}{4}} = \frac{3}{2} \div \frac{1}{4} = \frac{3}{2} \cdot \frac{4}{1} = 6$$

71.

$$\frac{x}{y} + 3y \quad \text{for} \quad x = \frac{1}{3}, \ y = \frac{1}{2}$$

$$\frac{\frac{1}{3}}{\frac{1}{2}} + 3\left(\frac{1}{2}\right) = \frac{2}{3} + \frac{3}{2} = \frac{4}{6} + \frac{9}{6} = \frac{13}{6}$$

73.

$$\frac{10x^2 + 11xy - 6y^2}{5x - 2y} \quad \text{for} \quad x = \frac{1}{2}, \ y = 2$$

$$\frac{10\left(\frac{1}{2}\right)^2 + 11\left(\frac{1}{2}\right)(2) - 6(2)^2}{5\left(\frac{1}{2}\right) - 2(2)} = \frac{10\left(\frac{1}{4}\right) + 11 - 6(4)}{\frac{5}{2} - 4}$$

$$= \frac{\frac{10}{4} + 11 - 24}{\frac{5}{2} - \frac{8}{2}}$$

$$= \frac{\frac{10}{4} - 13}{-\frac{3}{2}} = \frac{\frac{10}{4} - \frac{52}{4}}{-\frac{3}{2}}$$

$$= \frac{-\frac{42}{4}}{-\frac{3}{2}} = -\frac{21}{2} \div -\frac{3}{2}$$

$$= -\frac{21}{2} \cdot \left(-\frac{2}{3}\right) = 7$$

75.

$$2x\left[x^2 + y\left(3x - z^2\right)\right] \quad \text{for} \quad x = 5, \ y = -2, \ z = 3$$

$$2(5)\left[5^2 + (-2)\left(3 \cdot 5 - 3^2\right)\right] = 10\left[25 + (-2)(15 - 9)\right]$$

$$= 10\left[25 + (-2)(6)\right]$$

#75. continued

$$= 10[25 - 12]$$
$$= 10[13] = 130$$

Review Exercises

1.-21.
The exercises do not require worked - out solutions.

23.
$$-14 - (-2) = -14 + 2 = -12$$

25.-27.
The exercises do not require worked - out solutions.

29.
$$17 - (-5) - 14 = 17 + 5 - 14 = 22 - 14 = 8$$

31.
$$22(-6) = -132$$

33.
$$(-4)(2)(-5)(-3) = -8(-5)(-3) = 40(-3) = -120$$

35.-43.
The exercises do not require worked - out solutions.

45.
$$\frac{54}{81} = \frac{2 \cdot 27}{3 \cdot 27} = \frac{2}{3}$$

47.
$$\frac{19}{38} = \frac{1 \cdot 19}{2 \cdot 19} = \frac{1}{2}$$

49.
$$\frac{60}{35} = \frac{5 \cdot 12}{5 \cdot 7} = \frac{12}{7}$$

51.
$$\frac{3}{4} \cdot \frac{5}{6} = \frac{3}{4} \cdot \frac{5}{2 \cdot 3} = \frac{5}{8}$$

53.

$$\frac{3}{4} + \frac{5}{6} = \frac{9}{12} + \frac{10}{12} = \frac{19}{12} \qquad \text{LCD} = 12$$

55.

$$\frac{7}{8} \div 14 = \frac{7}{8} \cdot \frac{1}{14} = \frac{\cancel{7}}{8} \cdot \frac{1}{2 \cdot \cancel{7}} = \frac{1}{16}$$

57.

$$\left(\frac{-21}{40}\right)\left(\frac{-88}{-49}\right) = \frac{-3 \cdot \cancel{7}}{5 \cdot \cancel{8}} \cdot \frac{-11 \cdot \cancel{8}}{-7 \cdot \cancel{7}} = -\frac{33}{35}$$

59.

$$\frac{2}{15} - \frac{4}{9} = \frac{6}{45} - \frac{20}{45} = -\frac{14}{45} \qquad \text{LCD} = 45$$

61.

$$\left(-\frac{7}{8}\right)\left(\frac{-4}{21}\right)(9) = \frac{-1 \cdot \cancel{7}}{2 \cdot \cancel{4}} \cdot \frac{-1 \cdot \cancel{4}}{\cancel{7} \cdot \cancel{8}} \cdot \frac{\cancel{8} \cdot 3}{1} = \frac{3}{2}$$

63.

$$\frac{5}{12} - \frac{11}{42} + \frac{3}{7} = \frac{35}{84} - \frac{22}{84} + \frac{36}{84} = \frac{35 - 22 + 36}{84}$$
$$= \frac{49}{84} \text{ or } \frac{7}{12} \qquad \text{LCD} = 84$$

65.

$$\frac{-28}{45} \cdot \frac{60}{77} = \frac{-4 \cdot \cancel{7}}{3 \cdot \cancel{15}} \cdot \frac{\cancel{15} \cdot 4}{\cancel{7} \cdot 11} = -\frac{16}{33}$$

67.

$$\frac{9}{2} + \frac{13}{4} - \frac{21}{8} = \frac{36}{8} + \frac{26}{8} - \frac{21}{8} = \frac{36 + 26 - 21}{8} = \frac{41}{8}$$

69.

$$\frac{5}{2} \cdot \frac{-14}{15} \cdot \frac{8}{7} \cdot \frac{-1}{4} = \frac{\cancel{5}}{\cancel{2}} \cdot \frac{-1 \cdot \cancel{2} \cdot \cancel{7}}{3 \cdot \cancel{5}} \cdot \frac{2 \cdot \cancel{4}}{\cancel{7}} \cdot \frac{-1}{\cancel{4}} = \frac{2}{3}$$

71.

$$\frac{5}{16} = 0.3125 \qquad \text{Work } 16\overline{)5.0000}$$

```
         .3125
  16)5.0000
      48
      20
      16
       40
       32
       80
       80
```

73.

$$\frac{4}{11} = 0.\overline{36} \qquad \text{Work } 11\overline{)4.000}$$

```
        .363 ....
  11)4.000
     33
     70
     66
      40
      33
```

75.

$$0.24 = \frac{24}{100} = \frac{\cancel{4} \cdot 6}{\cancel{4} \cdot 25} = \frac{6}{25}$$

77.

$$2.6 = \frac{26}{10} = \frac{\cancel{2} \cdot 13}{\cancel{2} \cdot 5} = \frac{13}{5}$$

79.-81.
These exercises do not require worked-out solutions.

83.

$$-43.1 - 8.905 + 17.46 = -43.1 + (-8.905) + 17.46$$
$$= -52.005 + 17.46$$
$$= -34.545$$

85.

$$(7.08)(-5.5)(-31.25) = (-38.94)(-31.25)$$
$$= 1216.875$$

87.

a. $A = s^2$; $\frac{7}{4} \cdot \frac{7}{4} = \frac{49}{16}$ or $3\frac{1}{16}$ sq in

b. $A = \frac{b \cdot h}{2}$; $\frac{1.8(4.15)}{2} = \frac{7.47}{2} = 3.735$ cm^2

#87. continued

c. $A = b \cdot h; \quad 5.01(2.4) = 12.024 \text{ m}^2$

d.

$A = \left(1\frac{1}{3}\right)\left(2\frac{1}{2}\right) + \left(1\frac{1}{4}\right)\left(1\frac{5}{6}\right)$

$= \frac{4}{3} \cdot \frac{5}{2} + \frac{5}{4} \cdot \frac{11}{6} = \frac{10}{3} + \frac{55}{24} = \frac{80}{24} + \frac{55}{24}$

$= \frac{135}{24} = \frac{45}{8} \quad \text{or} \quad 5\frac{5}{8} \text{ ft}^2$

89.

8 oz bowl; $\frac{1.95}{8} = \$0.24375$ per oz

18 oz pat; $\frac{4.45}{18} = \$0.247\overline{2}$ per oz

Answer: 8 oz bowl.

91.-95.

The exercises do not require worked-out solutions.

97.

$D^2 = 2^2 + 2^2 = 4 + 4 = 8$

$D = \sqrt{8} \quad \text{or} \quad 2.83\text{cm}$

99.

$D^2 = 5^2 + 8^2 \qquad D^2 = 25 + 64 \qquad D^2 = 89$

$D = \sqrt{89} \doteq 9.43 = 9\frac{43}{100}\text{in}$

101.

$A = \frac{22}{7} \cdot \left(\frac{7}{4}\right)^2 = \frac{22}{7} \cdot \frac{49}{16} = \frac{\cancel{2} \cdot 11}{\cancel{7}} \cdot \frac{\cancel{7} \cdot 7}{\cancel{2} \cdot 8} = \frac{77}{8} \text{ in}^2$

103.

$C = 2(3.14)(2.1) = 13.188 \text{ cm}$

105.-113.

The exercises do not require worked-out solutions.

115.

The graph is in the answer section of the main text.

117.-147.

The exercises do not require worked-out solutions.

149.

$8 - 24 + \sqrt{16} = 8 - 24 + 4 = -12$

151.

$5 \cdot 3^2 + 6 \cdot 3 - 17 = 5 \cdot 9 + 18 - 17$

$= 45 + 18 - 17 = 46$

153.

$3 - 2\left(\frac{1}{2} + 4\right) + 6 = 3 - 2\left(\frac{9}{2}\right) + 6$

$= 3 - 9 + 6 = 3 - \frac{3}{2} = \frac{3}{2}$

155.

$4 - \sqrt{15 - 2 \cdot 7} = 4 - \sqrt{15 - 14} = 4 - \sqrt{1} = 4 - 1 = 3$

157.

$\frac{35.6 - 3(5.01)}{2.6 - 6} = \frac{35.6 - 15.03}{-3.4} = \frac{20.57}{-3.4} = -6.05$

159.

$2 - |7 \cdot 5 - 46| = 2 - |35 - 46|$

$= 2 - |-11| = 2 - 11 = -9$

161.

$8 - 5[-4 - 6(4 \cdot 3 - 7)] = 8 - 5[-4 - 6(12 - 7)]$

$= 8 - 5[-4 - 6(5)]$

$= 8 - 5[-4 - 30]$

$= 8 - 5[-34] = 8 + 170 = 178$

163.-165.

The graphs are in the answer section of the main text.

167.-171.

The exercises do not require worked out solutions.

173.

$x^5 - 8x^3 - x + 2$ for $x = -2$

$(-2)^5 - 8(-2)^3 - (-2) + 2 = -32 - 8(-8) + 2 + 2$

$= -32 + 64 + 4$

$= 32 + 4 = 36$

175.

$\left(-\dfrac{1}{5}\right)(x-3)^2 - 2$ for $x = -2$

$\left(-\dfrac{1}{5}\right)(-2-3)^2 - 2 = \left(-\dfrac{1}{5}\right)(-5)^2 - 2 = \left(-\dfrac{1}{5}\right)(25) - 2$

$= -5 - 2 = -7$

177.

$x - \left|x^2 + 3y\right|$ for $x = 5$, $y = -9$

$5 - \left|5^2 + 3(-9)\right| = 5 - |25 - 27| = 5 - |-2| = 5 - 2 = 3$

179.

$\dfrac{3x}{x^2+1}$ for $x = \dfrac{2}{3}$ $\dfrac{3\left(\dfrac{2}{3}\right)}{\left(\dfrac{2}{3}\right)^2 + 1} = \dfrac{2}{\dfrac{4}{9}+1} = \dfrac{2}{\dfrac{13}{9}} = \dfrac{18}{13}$

181.

$\dfrac{x}{2a} + y^2 - \dfrac{2}{3}$ for $x = -3$, $a = 3$, $y = \dfrac{1}{2}$

$\dfrac{-3}{2 \cdot 3} + \left(\dfrac{1}{2}\right)^2 - \dfrac{2}{3} = \dfrac{-3}{6} + \dfrac{1}{4} - \dfrac{2}{3} = -\dfrac{1}{2} + \dfrac{1}{4} - \dfrac{2}{3}$

$= -\dfrac{1}{4} - \dfrac{2}{3}$

$= -\dfrac{3}{12} - \dfrac{8}{12} = -\dfrac{11}{12}$

Chapter 1 Test Solutions.

1.-15.

The answers do not require worked-out solutions.

17.

The graph is in the answer section of the main text.

19.-25.

The answers do not require worked-out solutions.

27.

$-14 - 8 - (-21) - 5 = -14 - 8 + 21 - 5$

$= -22 + 21 - 5$

$= -1 - 5 = -6$

29.

$-8^2 = -1 \cdot 8 \cdot 8 = -1 \cdot 64 = -64$

31.

$\dfrac{\dfrac{-8}{15}}{\dfrac{20}{1}} = \dfrac{-8}{15} \div \dfrac{20}{1} = \dfrac{-8}{15} \cdot \dfrac{1}{20} = \dfrac{-2 \cdot \cancel{4}}{15} \cdot \dfrac{1}{\cancel{4} \cdot 5} = -\dfrac{2}{75}$

33.

$16.7 + (-8.94) - 4.506 = 7.76 - 4.506 = 3.254$

35.

$\left(\dfrac{2}{3} + 2 \cdot 3\right) \div \dfrac{5}{6} - 7 = \left(\dfrac{2}{3} + 6\right) \div \dfrac{5}{6} - 7 = \dfrac{20}{3} \div \dfrac{5}{6} - 7$

$= 8 - 7 = 1$

37.

$\dfrac{(-2)^4 - 3^2}{(-7-9) - (3-5)} = \dfrac{16 - 9}{-16 - (-2)}$

$= \dfrac{7}{-16 + 2} = \dfrac{7}{-14} = -\dfrac{1}{2}$

39.

$\left(\dfrac{-5}{16}\right)(-8)\left(\dfrac{-26}{65}\right) = \dfrac{-1 \cdot \cancel{5}}{\cancel{2} \cdot \cancel{8}} \cdot \dfrac{-1 \cdot \cancel{8}}{1} \cdot \dfrac{-1 \cdot \cancel{13} \cdot \cancel{2}}{\cancel{5} \cdot \cancel{13}} = -1$

41.

$96 \div 2^2 \cdot 8 = 96 \div 4 \cdot 8 = 24 \cdot 8 = 192$

43.

$2\left|4^2 - 3\right| = 2|16 - 3| = 2|13| = 2 \cdot 13 = 26$

45.

$\dfrac{3}{25} \cdot \dfrac{4}{4} = \dfrac{12}{100} = 0.12$

47.

$2x^4 - x^2 + 5x$ for $x = 2$

$2(2)^4 - (2)^2 + 5(2) = 2(16) - 4 + 10$
$= 32 - 4 + 10 = 38$

49.

$x - |2x - 9|$ for $x = 3$

$3 - |2 \cdot 3 - 9| = 3 - |6 - 9| = 3 - |-3| = 3 - 3 = 0$

Chapter 1 Study Guide

Self-Test Exercises

I. Use the sets A={1,2,3,4}, B={2,4,6}, E={2,4,6,...}, N, I, Q, H, and R for problems 1-6. Are the following true or false?

1. $28 \in E$ 2. $N \subset E$ 3. $\sqrt{15} \in R$ 4. $\emptyset \subseteq N$

   Find the following.

5. $A \cap E$ 6. $E \cup N$

II. Place the appropriate symbol, <, >, or =, between each pair of numbers.

7. $|-9|$ _____ -5 8. $|-29|$ _____ 29

III. Graph the following set.   **IV.** Find the additive and multiplicative inverses.

9. $\{x : x < -3.5\}$ 10. -8.2

V. Rewrite using the distributive property.   **VI.** State the property illustrated.

11. $23z + 19z$.   12. $(2x - 5) + 16x = 16x + (2x - 5)$

VII. Name the property that justifies each step in the following.

13. $-7 + (x + 7) = -7 + (7 + x)$
$= (-7 + 7) + x$
$= 0 + x$

VIII. Perform the indicated operations, if possible.

14. (-5)(-3)(-9)   15. -30 + (-25)   16. $\dfrac{70}{-35}$ 17. $\dfrac{7}{0}$

18. $(-0.85)(7.04)(-300)$ 19. $0.2091 \div 0.17$ 20. $|9 - 23|$ 21. $\dfrac{2}{15} + \dfrac{13}{45} - 3$

22. $\sqrt{3 \cdot 9 + 6^2 + 1}$ 23. $-\dfrac{3}{4} - \dfrac{1}{3} \div \dfrac{5}{12} + 2\left(\dfrac{7}{24}\right)$

IX. Make the conversion.

24. Decimal to fraction in lowest terms:

     0.18

X. Evaluate the algebraic expression for the given values of the variable.

25. $(x^2 + 4y^2) - 3xy$ for $x = -2$, $y = 3$

The worked-out solutions begin on the next page.

11

Self-Test Solutions

1. True   2. False   3. True   4. True   5. {2,4}
6. {1,2,3,...}   7. >   8. =

9.

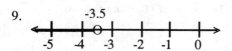

-3.5

$\begin{array}{ccccccc} | & | & \circ & | & | & | & \\ -5 & -4 & -3 & -2 & -1 & 0 \end{array}$

10. Additive: 8.2

Multiplicative: $-\dfrac{1}{8.2}$ or $-\dfrac{5}{41}$

11. $(23+19)z$ 12. Commutative property

13. Commutative property, associative property, additive identity.

---

14. $(-5)(-3)(-9) = 15(-9) = -135$

---

15. $-30 + (-25) = -55$

---

16. $\dfrac{70}{-35} = -2$ 17. $\dfrac{7}{0}$ is undefined

---

18. $(-0.85)(7.04)(-300) = -5.984(-300) = 1795.2$

19.

$0.2091 \div 0.17 = 1.23$ Work $0.17\overline{)0.2091}$

$$
\begin{array}{r}
1.23 \\
0.17\overline{)0.2091} \\
\underline{17} \\
39 \\
\underline{34} \\
51 \\
\underline{51}
\end{array}
$$

---

20. $|9 - 23| = |-14| = 14$

---

21. $\dfrac{2}{15} + \dfrac{13}{45} - 3 = \dfrac{6}{45} + \dfrac{13}{45} - \dfrac{135}{45} = \dfrac{19}{45} - \dfrac{135}{45} = -\dfrac{116}{45}$

---

22. $\sqrt{3 \cdot 9 + 6^2 + 1} = \sqrt{27 + 36 + 1} = \sqrt{64} = 8$

23.

$-\dfrac{3}{4} - \dfrac{1}{3} \div \dfrac{5}{12} + 2\left(\dfrac{7}{24}\right) = -\dfrac{3}{4} - \dfrac{4}{5} + \dfrac{7}{12}$

$= -\dfrac{45}{60} - \dfrac{48}{60} + \dfrac{35}{60}$

$= -\dfrac{93}{60} + \dfrac{35}{60} = -\dfrac{58}{60} = -\dfrac{29}{30}$

---

24. $0.18 = \dfrac{18}{100} = \dfrac{9}{50}$

25.

$\left(x^2 + 4y^2\right) - 3xy$ for $x = -2,\ y = 3$

$\left((-2)^2 + 4(3)^2\right) - 3(-2)(3) = (4 + 4 \cdot 9) - 3(-2)(3)$

$= (4 + 36) + 18$

$= 40 + 18 = 58$

CHAPTER 2   LINEAR EQUATIONS AND INEQUALITIES

Solutions to Text Exercises

Exercises 2.1

1.
$$5x = 100$$
$$x = 20 \qquad \text{Multiply by } \frac{1}{5}$$
Check: $5 \cdot 20 = 100$
$$100 = 100$$

3.
$$-9x = 39$$
$$x = -\frac{39}{9} \text{ or } -\frac{13}{3} \qquad \text{Multiply by } -\frac{1}{9}$$
Check: $-9\left(-\frac{13}{3}\right) = 39$
$$39 = 39$$

5.
$$0.3x = 51$$
$$x = 170 \qquad \text{Multiply by } \frac{1}{0.3}$$
Check: $\quad 0.3x = 51$
$$0.3 \cdot 170 = 51$$
$$51 = 51$$

7.
$$\frac{2}{3}x = 24$$
$$\frac{3}{2} \cdot \frac{2}{3}x = \frac{3}{2} \cdot 24$$
$$x = 36 \qquad \text{Multiply by } \frac{3}{2}$$
Check: $\frac{2}{3} \cdot 36 = 24$
$$24 = 24$$

9.
$$-\frac{1}{4}x = \frac{3}{8}$$
$$-4\left(-\frac{1}{4}x\right) = -4\left(\frac{3}{8}\right) \qquad \text{Multiply by } -4$$
$$x = -\frac{3}{2}$$

#9. continued
Check: $-\frac{1}{4}\left(-\frac{3}{2}\right) = \frac{3}{8}$
$$\frac{3}{8} = \frac{3}{8}$$

11.
$$3x + 7 = 15$$
$$3x + 7 - 7 = 15 - 7 \qquad \text{Subtract } 7$$
$$3x = 8$$
$$\frac{1}{3} \cdot 3x = \frac{1}{3} \cdot 8 \qquad \text{Multiply by } \frac{1}{3}$$
$$x = \frac{8}{3}$$
Check: $3 \cdot \frac{8}{3} + 7 = 15$
$$8 + 7 = 15$$
$$15 = 15$$

13.
$$-x + 3 = -8$$
$$-x + 3 - 3 = -8 - 3 \qquad \text{Subtract } 3$$
$$-x = -11$$
$$-1(-x) = -1(-11) \qquad \text{Multiply by } -1$$
$$x = 11$$
Check: $-(11) + 3 = -8$
$$-11 + 3 = -8$$
$$-8 = -8$$

15.
$$-6x - 5 = 11$$
$$-6x - 5 + 5 = 11 + 5 \qquad \text{Add } 5$$
$$-6x = 16$$
$$-\frac{1}{6}(-6x) = -\frac{1}{6} \cdot 16 \qquad \text{Multiply by } -\frac{1}{6}$$
$$x = -\frac{16}{6} \text{ or } -\frac{8}{3}$$
Check: $-6\left(-\frac{8}{3}\right) - 5 = 11$
$$16 - 5 = 11$$
$$11 = 11$$

17.
$$8x + 7 = 7$$
$$8x + 7 - 7 = 7 - 7 \qquad \text{Subtract } 7$$
$$8x = 0$$
$$\frac{1}{8} \cdot 8x = \frac{1}{8} \cdot 0 \qquad \text{Multiply by } \frac{1}{8}$$
$$x = 0$$
$$\text{Check: } 8 \cdot 0 + 7 = 7$$
$$0 + 7 = 7$$
$$7 = 7$$

19.
$$\frac{3}{7}x + 4 = -\frac{1}{2}$$
$$\frac{3}{7}x + 4 - 4 = -\frac{1}{2} - 4 \qquad \text{Subtract } 4$$
$$\frac{3}{7}x = -\frac{9}{2}$$
$$\frac{7}{3} \cdot \frac{3}{7}x = \frac{7}{3}\left(-\frac{9}{2}\right) \qquad \text{Multiply by } \frac{7}{3}$$
$$x = -\frac{21}{2}$$
$$\text{Check: } \frac{3}{7}\left(-\frac{21}{2}\right) + 4 = -\frac{1}{2}$$
$$-\frac{9}{2} + 4 = -\frac{1}{2}$$
$$-\frac{9}{2} + \frac{8}{2} = -\frac{1}{2}$$
$$-\frac{1}{2} = -\frac{1}{2}$$

21.
$$-\frac{2}{3}x + 1 = 5$$
$$-\frac{2}{3}x + 1 - 1 = 5 - 1 \qquad \text{Subtract } 1$$
$$-\frac{2}{3}x = 4$$
$$-\frac{3}{2}\left(-\frac{2}{3}x\right) = -\frac{3}{2} \cdot 4 \qquad \text{Multiply by } -\frac{3}{2}$$
$$x = -6$$
$$\text{Check: } -\frac{2}{3}(-6) + 1 = 5$$
$$4 + 1 = 5$$
$$5 = 5$$

23.
$$0.2x - 1.31 = 2.39$$
$$0.2x - 1.31 + 1.31 = 2.39 + 1.31 \qquad \text{Add } 1.31$$
$$0.2x = 3.70$$
$$\frac{1}{0.2} \cdot 0.2x = \frac{1}{0.2} \cdot 3.70$$
$$x = 18.5$$
$$\text{Check: } 0.2 \cdot 18.5 - 1.31 = 2.39$$
$$3.70 - 1.31 = 2.39$$
$$2.39 = 2.39$$

25.
$$12 - 4x = 5x + 3$$
$$12 - 3 - 4x = 5x + 3 - 3 \qquad \text{Subtract } 3$$
$$9 - 4x = 5x$$
$$9 - 4x + 4x = 5x + 4x \qquad \text{Add } 4x$$
$$9 = 9x$$
$$\frac{1}{9} \cdot 9 = \frac{1}{9} \cdot 9x \qquad \text{Multiply by } \frac{1}{9}$$
$$1 = x$$
$$\text{Check: } 12 - 4 \cdot 1 = 5 \cdot 1 + 3$$
$$12 - 4 = 5 + 3$$
$$8 = 8$$

27.
$$-4x - 1 = -2x + 15$$
$$-4x - 1 - 15 = -2x + 15 - 15 \qquad \text{Subtract } 15$$
$$-4x - 16 = -2x$$
$$-4x + 4x - 16 = -2x + 4x \qquad \text{Add } 4x$$
$$-16 = 2x$$
$$\frac{1}{2}(-16) = \frac{1}{2} \cdot 2x \qquad \text{Multiply by } \frac{1}{2}$$
$$-8 = x$$
$$\text{Check: } -4(-8) - 1 = -2(-8) + 15$$
$$32 - 1 = 16 + 15$$
$$31 = 31$$

29.
$$-5x + 7 = 10x + 11$$
$$-5x + 7 - 11 = 10x + 11 - 11 \qquad \text{Subtract } 11$$
$$-5x - 4 = 10x$$
$$-5x + 5x - 4 = 10x + 5x \qquad \text{Add } 5x$$
$$-4 = 15x$$

#29. continued

$$\frac{1}{15}(-4) = \frac{1}{15} \cdot 15x \qquad \text{Multiply by } \frac{1}{15}$$

$$-\frac{4}{15} = x$$

Check: $-5\left(-\frac{4}{15}\right) + 7 = 10\left(-\frac{4}{15}\right) + 11$

$$\frac{4}{3} + 7 = -\frac{8}{3} + 11$$

$$\frac{4}{3} + \frac{21}{3} = -\frac{8}{3} + \frac{33}{3}$$

$$\frac{25}{3} = \frac{25}{3}$$

31.

$$5x - 7 = 5x + 2$$

$$5x - 5x - 7 = 5x - 5x + 2 \qquad \text{Subtract } 5x$$

$$-7 = 2 \qquad \text{No solution}$$

33.

$$-6x - 4 = -9x - 3$$

$$-6x - 4 + 4 = -9x - 3 + 4 \qquad \text{Add } 4$$

$$-6x = -9x + 1$$

$$-6x + 9x = -9x + 9x + 1 \qquad \text{Add } 9x$$

$$3x = 1$$

$$\frac{1}{3} \cdot 3x = \frac{1}{3} \cdot 1 \qquad \text{Multiply by } \frac{1}{3}$$

$$x = \frac{1}{3}$$

Check: $-6 \cdot \frac{1}{3} - 4 = -9 \cdot \frac{1}{3} - 3$

$$-2 - 4 = -3 - 3$$

$$-6 = -6$$

35.

$$\frac{2}{3}x - \frac{4}{9} = x - \frac{1}{6} \qquad \text{LCD} = 18$$

$$18\left(\frac{2}{3}x - \frac{4}{9}\right) = 18\left(x - \frac{1}{6}\right) \qquad \text{Multiply by } 18$$

$$12x - 8 = 18x - 3 \qquad \text{Distributive}$$

$$12x - 8 + 3 = 18x - 3 + 3 \qquad \text{Add } 3$$

$$12x - 5 = 18x$$

$$12x - 12x - 5 = 18x - 12x \qquad \text{Subtract } 12x$$

$$-5 = 6x$$

#35. continued

$$\frac{1}{6}(-5) = \frac{1}{6} \cdot 6x \qquad \text{Multiply by } \frac{1}{6}$$

$$-\frac{5}{6} = x$$

Check: $\frac{2}{3}\left(-\frac{5}{6}\right) - \frac{4}{9} = -\frac{5}{6} - \frac{1}{6}$

$$-\frac{5}{9} - \frac{4}{9} = -\frac{5}{6} - \frac{1}{6}$$

$$-\frac{9}{9} = -\frac{6}{6}$$

$$1 = 1$$

37.

$$\frac{3}{4}x + 9 = \frac{1}{3}x + 4 \qquad \text{LCD} = 12$$

$$12\left(\frac{3}{4}x + 9\right) = 12\left(\frac{1}{3}x + 4\right) \qquad \text{Multiply by } 12$$

$$9x + 108 = 4x + 48 \qquad \text{Distributive}$$

$$9x + 108 - 108 = 4x + 48 - 108 \qquad \text{Subtract } 108$$

$$9x = 4x - 60$$

$$9x - 4x = 4x - 4x - 60 \qquad \text{Subtract } 4x$$

$$5x = -60$$

$$\frac{1}{5} \cdot 5x = \frac{1}{5}(-60)$$

$$x = -12$$

Check: $\frac{3}{4}(-12) + 9 = \frac{1}{3}(-12) + 4$

$$-9 + 9 = -4 + 4$$

$$0 = 0$$

39.

$$-7(x + 2) = -14$$

$$-7x - 14 = -14 \qquad \text{Distributive}$$

$$-7x - 14 + 14 = -14 + 14 \qquad \text{Add } 14$$

$$-7x = 0$$

$$-\frac{1}{7}(-7x) = -\frac{1}{7} \cdot 0 \qquad \text{Multiply by } -\frac{1}{7}$$

$$x = 0$$

Check: $-7(0 + 2) = -14$

$$-7(2) = -14$$

$$-14 = -14$$

41.

$$-5(3x+6) = x+82$$
$$-15x-30 = x+82 \qquad \text{Distributive}$$
$$-15x-30-82 = x+82-82 \qquad \text{Subtract 82}$$
$$-15x-112 = x$$
$$-15x+15x-112 = x+15x \qquad \text{Add 15x}$$
$$-112 = 16x$$
$$\frac{1}{16}(-112) = \frac{1}{16} \cdot 16x \qquad \text{Multiply by } \frac{1}{16}$$
$$-7 = x$$

Check: $-5(3(-7)+6) = -7+82$
$$-5(-21+6) = 75$$
$$-5(-15) = 75$$
$$75 = 75$$

43.

$$8(2x-6) = 4x-35$$
$$16x-48 = 4x-35 \qquad \text{Distributive}$$
$$16x-48+48 = 4x-35+48 \qquad \text{Add 48}$$
$$16x = 4x+13$$
$$16x-4x = 4x-4x+13 \qquad \text{Subtract 4x}$$
$$12x = 13$$
$$\frac{1}{12} \cdot 12x = \frac{1}{12} \cdot 13 \qquad \text{Multiply by } \frac{1}{12}$$
$$x = \frac{13}{12}$$

Check: $8\left(2 \cdot \frac{13}{12} - 6\right) = 4 \cdot \frac{13}{12} - 35$
$$8\left(\frac{13}{6} - 6\right) = \frac{13}{3} - 35$$
$$\frac{52}{3} - 48 = \frac{13}{3} - 35$$
$$\frac{52}{3} - \frac{144}{3} = \frac{13}{3} - \frac{105}{3}$$
$$-\frac{92}{3} = -\frac{92}{3}$$

45.

$$4x-6 = 3(x-2)+x$$
$$4x-6 = 3x-6+x \qquad \text{Distributive}$$
$$4x-6 = 4x-6 \qquad \text{Simplify}$$
$$4x-4x-6 = 4x-4x-6 \qquad \text{Subtract 4x}$$
$$-6 = -6 \qquad \text{All real numbers}$$

47.

$$\frac{2(2x-9)}{3} = x-8$$
$$3 \cdot \frac{2(2x-9)}{3} = 3(x-8) \qquad \text{Multiply by 3}$$
$$2(2x-9) = 3(x-8)$$
$$4x-18 = 3x-24 \qquad \text{Distributive}$$
$$4x-18+18 = 3x-24+18 \qquad \text{Add 18}$$
$$4x = 3x-6$$
$$4x-3x = 3x-3x-6 \qquad \text{Subtract 3x}$$
$$x = -6$$

Check: $\dfrac{2(2(-6)-9)}{3} = -6-8$
$$\frac{2(-12-9)}{3} = -14$$
$$\frac{2(-21)}{3} = -14$$
$$\frac{-42}{3} = -14$$
$$-14 = -14$$

49.

$$\frac{1}{2}(10-3x) = \frac{1}{3}x+4$$
$$5-\frac{3}{2}x = \frac{1}{3}x+4 \qquad \text{LCD} = 6$$
$$6\left(5-\frac{3}{2}x\right) = 6\left(\frac{1}{3}x+4\right) \qquad \text{Multiply by 6}$$
$$30-9x = 2x+24$$
$$30-24-9x = 2x+24-24 \qquad \text{Subtract 24}$$
$$6-9x = 2x$$
$$6-9x+9x = 2x+9x \qquad \text{Add 9x}$$
$$6 = 11x$$
$$\frac{1}{11} \cdot 6 = \frac{1}{11} \cdot 11x \qquad \text{Multiply by } \frac{1}{11}$$
$$\frac{6}{11} = x$$

Check: $\dfrac{1}{2}\left(10-3 \cdot \dfrac{6}{11}\right) = \dfrac{1}{3} \cdot \dfrac{6}{11} + 4$
$$\frac{1}{2}\left(10-\frac{18}{11}\right) = \frac{6}{33}+4$$
$$\frac{1}{2}\left(\frac{110}{11} - \frac{18}{11}\right) = \frac{6}{33} + \frac{132}{33}$$

#49. continued

$$\frac{1}{2}\left(\frac{92}{11}\right) = \frac{138}{33}$$

$$\frac{92}{22} = \frac{138}{33}$$

$$\frac{46}{11} = \frac{46}{11}$$

51.

$$-0.5x + 3.2 = 1.3x + 4.4$$

$-0.5x + 3.2 - 4.4 = 1.3x + 4.4 - 4.4$ Subtract 4.4

$$-0.5x - 1.2 = 1.3x$$

$-0.5x + 0.5x - 1.2 = 1.3x + 0.5x$ Add 0.5x

$$-1.2 = 1.8x$$

$$\frac{1}{1.8}(-1.2) = \frac{1}{1.8}\cdot 1.8x$$

$$-\frac{1.2}{1.8} = x$$

$$-\frac{2}{3} = x$$

Check: $-0.5\left(-\frac{2}{3}\right) + 3.2 = 1.3\left(-\frac{2}{3}\right) + 4.4$

$$\frac{1}{3} + 3.2 = -\frac{2.6}{3} + 4.4$$

$$\frac{1}{3} + 3\frac{1}{5} = -\frac{13}{15} + 4\frac{2}{5}$$

$$3\frac{8}{15} = 3\frac{8}{15}$$

53.

$$\frac{1}{2}x + \frac{3}{4} = \frac{1}{4}(2x - 3) + 1$$

$\frac{1}{2}x + \frac{3}{4} = \frac{1}{2}x - \frac{3}{4} + 1$ LCD = 4

$$4\left(\frac{1}{2}x + \frac{3}{4}\right) = 4\left(\frac{1}{2}x - \frac{3}{4} + 1\right)$$

$2x + 3 = 2x - 3 + 4$ Multiply by 4

$$2x + 3 = 2x + 1$$

$3 = 1$ Subtract 2x

       No Solution

55.

$$3(2x - 5) + 7 = 2(4x - 1) + 12$$

$6x - 15 + 7 = 8x - 2 + 12$ Distributive

$6x - 8 = 8x + 10$ Simplify

$6x - 8 - 10 = 8x + 10 - 10$ Subtract 10

#55. continued

$$6x - 18 = 8x$$

$6x - 6x - 18 = 8x - 6x$ Subtract 6x

$$-18 = 2x$$

$\frac{1}{2}(-18) = \frac{1}{2}\cdot 2x$ Multiply by $\frac{1}{2}$

$$-9 = x$$

Check: $3\big(2(-9) - 5\big) + 7 = 2\big(4(-9) - 1\big) + 12$

$$3(-18 - 5) + 7 = 2(-36 - 1) + 12$$

$$3(-23) + 7 = 2(-37) + 12$$

$$-69 + 7 = -74 + 12$$

$$-62 = -62$$

57.

$$2(4x - 8) + 9 = 5(2x + 1) - 12$$

$$8x - 16 + 9 = 10x + 5 - 12$$

$8x - 7 = 10x - 7$ Simplify

$8x - 7 + 7 = 10x - 7 + 7$ Add 7

$$8x = 10x$$

$8x - 8x = 10x - 8x$ Subtract 8x

$$0 = 2x$$

$0 = x$ Multiply by $\frac{1}{2}$

Check: $2(4\cdot 0 - 8) + 9 = 5(2\cdot 0 + 1) - 12$

$$2(-8) + 9 = 5(1) - 12$$

$$-16 + 9 = 5 - 12$$

$$-7 = -7$$

59.

$$-3(x + 7) - 4x = 3(1 - 3x) - x$$

$-3x - 21 - 4x = 3 - 9x - x$ Distributive

$-7x - 21 = 3 - 10x$ Simplify

$-7x - 21 + 21 = 3 + 21 - 10x$ Add 21

$$-7x = 24 - 10x$$

$-7x + 10x = 24 - 10x + 10x$ Add 10x

$$3x = 24$$

$$x = 8$$

Check: $-3(8 + 7) - 4\cdot 8 = 3(1 - 3\cdot 8) - 8$

$$-3(15) - 32 = 3(1 - 24) - 8$$

$$-45 - 32 = 3(-23) - 8$$

$$-77 = -69 - 8$$

$$-77 = -77$$

61.

$$3(x-7)-4(2x+1)=2(x+5)$$

$3x-21-8x-4=2x+10$	Distributive
$-5x-25=2x+10$	Simplify
$-5x-25-10=2x+10-10$	Subtract 10
$-5x-35=2x$	
$-5x+5x-35=2x+5x$	Add 5x
$-35=7x$	
$\frac{1}{7}(-35)=\frac{1}{7}\cdot 7x$	Multiply by $\frac{1}{7}$
$-5=x$	

Check: $3(-5-7)-4(2(-5)+1)=2(-5+5)$

$$3(-12)-4(-10+1)=2\cdot 0$$
$$-36-4(-9)=0$$
$$-36+36=0$$
$$0=0$$

63.

$$5(3x-4)+2(4x+1)=3(7x-2)$$

$15x-20+8x+2=21x-6$	Distributive
$23x-18=21x-6$	Simplify
$23x-18+18=21x-6+18$	Add 18
$23x=21x+12$	
$23x-21x=21x-21x+12$	Subtract 21x
$2x=12$	
$\frac{1}{2}\cdot 2x=\frac{1}{2}\cdot 12$	
$x=6$	

Check: $5(3\cdot 6-4)+2(4\cdot 6+1)=3(7\cdot 6-2)$

$$5(18-4)+2(24+1)=3(42-2)$$
$$5(14)+2(25)=3(40)$$
$$70+50=120$$
$$120=120$$

65.

$$10(3x-5)-7(3x-5)=4(2x+6)$$

$30x-50-21x+35=8x+24$	Distributive
$9x-15=8x+24$	Simplify
$9x-15-24=8x+24-24$	Subtract 24
$9x-39=8x$	
$9x-9x-39=8x-9x$	Subtract 9x
$-39=-x$	

#65. continued

$-39=-x$	
$39=x$	Multiply by -1

Check: $10(3\cdot 39-5)-7(3\cdot 39-5)=4(2\cdot 39+6)$

$$10(117-5)-7(117-5)=4(78+6)$$
$$10(112)-7(112)=4(84)$$
$$1120-784=336$$
$$336=336$$

67.

$$\frac{5(x+1)}{6}+\frac{1}{3}(5x-6)=\frac{1}{6}(7x-2) \qquad \text{LCD}=6$$

$$6\left[\frac{5(x+1)}{6}+\frac{1}{3}(5x-6)\right]=\left[\frac{1}{6}(7x-2)\right]6$$

$5(x+1)+2(5x-6)=7x-2$	
$5x+5+10x-12=7x-2$	Distributive
$15x-7=7x-2$	Simplify
$15x-7+7=7x-2+7$	Add 7
$15x=7x+5$	
$15x-7x=7x-7x+5$	Subtract 7x
$8x=5$	
$\frac{1}{8}\cdot 8x=\frac{1}{8}\cdot 5$	Multiply by $\frac{1}{8}$
$x=\frac{5}{8}$	

Check: $\dfrac{5\left(\frac{5}{8}+1\right)}{6}+\dfrac{1}{3}\left(5\cdot\frac{5}{8}-6\right)=\dfrac{1}{6}\left(7\cdot\frac{5}{8}-2\right)$

$$\frac{5\left(\frac{13}{8}\right)}{6}+\frac{1}{3}\left(\frac{25}{8}-6\right)=\frac{1}{6}\left(\frac{35}{8}-2\right)$$

$$\frac{\frac{65}{8}}{6}+\frac{1}{3}\left(\frac{25}{8}-\frac{48}{8}\right)=\frac{1}{6}\left(\frac{35}{8}-\frac{16}{8}\right)$$

$$\frac{65}{48}+\frac{1}{3}\left(\frac{-23}{8}\right)=\frac{1}{6}\left(\frac{19}{8}\right)$$

$$\frac{65}{48}-\frac{46}{48}=\frac{19}{48}$$

$$\frac{19}{48}=\frac{19}{48}$$

69.

$$2.8(3x-5.063)-11.2(x+3.75)=0.24(6.5x-486.04)$$
$$8.4x-14.1764-11.2x-42=1.56x-116.6496$$
$$-2.8x-56.1764=1.56x-116.6496$$
$$60.4732=4.36x$$
$$13.87=x$$

The check is left to the student.

Exercises 2.2

> The graphs of the solution sets are found in the answer section of the main text book.

1.

$$x+5\geq 10$$
$$x+5-5\geq 10-5 \qquad \text{Subtract } 5$$
$$x\geq 5$$

3.

$$x-8\leq 0$$
$$x-8+8\leq 0+8 \qquad \text{Add } 8$$
$$x\leq 8$$

5.

$$3x\leq 27$$
$$\frac{1}{3}\cdot 3x\leq \frac{1}{3}\cdot 27 \qquad \text{Multiply by } \frac{1}{3}$$
$$x\leq 9$$

7.

$$7x>0$$
$$\frac{1}{7}\cdot 7x>\frac{1}{7}\cdot 0 \qquad \text{Multiply by } \frac{1}{7}$$
$$x>0$$

9.

$$-x\leq \frac{2}{3}$$
$$x\geq -\frac{2}{3} \qquad \text{Multiply by } -1$$

11.

$$\frac{3}{4}x>-9$$
$$\frac{4}{3}\cdot\frac{3}{4}x>\frac{4}{3}(-9)$$
$$x>-12$$

> The graphs of the solution sets are found in the answer section of the main text book.

13.

$$2x+1\leq -7$$
$$2x+1-1\leq -7-1 \qquad \text{Subtract } 1$$
$$2x\leq -8$$
$$\frac{1}{2}\cdot 2x\leq \frac{1}{2}(-8) \qquad \text{Multiply by } \frac{1}{2}$$
$$x\leq -4$$

15.

$$4x-2>-10$$
$$4x-2+2>-10+2 \qquad \text{Add } 2$$
$$4x>-8$$
$$\frac{1}{4}\cdot 4x>\frac{1}{4}(-8) \qquad \text{Multiply by } \frac{1}{4}$$
$$x>-2$$

17.

$$\frac{1}{3}x-2<1$$
$$3\left(\frac{1}{3}x-2\right)<3(1) \qquad \text{Multiply by } 3$$
$$x-6<3 \qquad \text{Distributive}$$
$$x-6+6<3+6 \qquad \text{Add } 6$$
$$x<9$$

19.

$$\frac{1}{5}x-\frac{1}{3}\geq -\frac{1}{2} \qquad \text{LCD}=30$$
$$30\left(\frac{1}{5}x-\frac{1}{3}\right)\geq 30\left(-\frac{1}{2}\right) \qquad \text{Multiply by } 30$$
$$6x-10\geq -15 \qquad \text{Distributive}$$
$$6x-10+10\geq -15+10 \qquad \text{Add } 10$$
$$6x\geq -5$$
$$\frac{1}{6}\cdot 6x\geq \frac{1}{6}(-5)$$
$$x\geq -\frac{5}{6}$$

21.

$$0.2x+0.07<0.13$$
$$0.2x+0.07-0.07<0.13-0.07 \qquad \text{Subtract } 0.07$$
$$0.2x<0.06$$

| The graphs of the solution sets are found in the answer section of the main text book. | The graphs of the solution sets are found in the answer section of the main text book. |

#21. continued

$$\frac{1}{0.2} \cdot 0.2x < \frac{1}{0.2} \cdot 0.06$$

$$x < 0.3$$

23.

$$5x - 12 \le 2x$$

$$5x - 12 + 12 \le 2x + 12 \qquad \text{Add } 12$$

$$5x \le 2x + 12$$

$$5x - 2x \le 2x - 2x + 12$$

$$3x \le 12$$

$$\frac{1}{3} \cdot 3x \le \frac{1}{3} \cdot 12$$

$$x \le 4$$

25.

$$3x + 10 > 7x$$

$$3x - 3x + 10 > 7x - 3x \qquad \text{Subtract } 3x$$

$$10 > 4x$$

$$\frac{1}{4} \cdot 10 > \frac{1}{4} \cdot 4x \qquad \text{Multiply by } \frac{1}{4}$$

$$\frac{10}{4} > x \text{ or } x < \frac{5}{2} \qquad \text{Reduce}$$

27.

$$-6 - 2x > -x$$

$$-6 - 2x + 2x > -x + 2x \qquad \text{Add } 2x$$

$$-6 > x \text{ or } x < -6$$

29.

$$\frac{2}{3}x - 1 \le \frac{3}{4}x \qquad \text{LCD } = 12$$

$$12\left(\frac{2}{3}x - 1\right) \le 12 \cdot \frac{3}{4}x$$

$$8x - 12 \le 9x \qquad \text{Distributive}$$

$$8x - 8x - 12 \le 9x - 8x \qquad \text{Subtract } 8x.$$

$$-12 \le x \text{ or } x \ge -12$$

31.

$$2x + 5 < 4x - 9$$

$$2x + 5 + 9 < 4x - 9 + 9 \qquad \text{Add } 9$$

$$2x + 14 < 4x$$

$$2x - 2x + 14 < 4x - 2x \qquad \text{Subtract } 2x$$

$$14 < 2x$$

#31. continued

$$14 < 2x$$

$$\frac{1}{2} \cdot 14 < \frac{1}{2} \cdot 2x$$

$$7 < x \text{ or } x > 7$$

33.

$$8 - 2x \ge 2x + 7$$

$$8 - 7 - 2x \ge 2x + 7 - 7 \qquad \text{Subtract } 7$$

$$1 - 2x \ge 2x$$

$$1 - 2x + 2x \ge 2x + 2x \qquad \text{Add } 2x$$

$$1 \ge 4x$$

$$\frac{1}{4} \cdot 1 \ge \frac{1}{4} \cdot 4x \qquad \text{Multiply by } \frac{1}{4}$$

$$\frac{1}{4} \ge x \text{ or } x \le \frac{1}{4}$$

35.

$$-3x - 2 > -x + 14$$

$$-3x - 2 - 14 > -x + 14 - 14 \qquad \text{Subtract } 14$$

$$-3x - 16 > -x$$

$$-3x + 3x - 16 > -x + 3x \qquad \text{Add } 3x$$

$$-16 > 2x$$

$$\frac{1}{2}(-16) > \frac{1}{2} \cdot 2x$$

$$-8 > x \text{ or } x < -8$$

37.

$$\frac{1}{5}x - 3 \le \frac{5}{4} - \frac{3}{2}x \qquad \text{LCD } = 20$$

$$20\left(\frac{1}{5}x - 3\right) \le \left(\frac{5}{4} - \frac{3}{2}x\right)20$$

$$4x - 60 \le 25 - 30x$$

$$4x - 60 + 60 \le 25 + 60 - 30x \qquad \text{Add } 60$$

$$4x \le 85 - 30x$$

$$4x + 30x \le 85 - 30x + 30x \qquad \text{Add } 30x$$

$$34x \le 85$$

$$\frac{1}{34} \cdot 34x \le \frac{1}{34} \cdot 85$$

$$x \le \frac{85}{34} \text{ or } x \le \frac{5}{2} \qquad \text{Reduce}$$

The graphs of the solution sets are found in the answer section of the main text book.

39.

$$0.4x - 1.6 > 1.5x + 0.6$$

$$0.4x - 1.6 - 0.6 > 1.5x + 0.6 - 0.6 \quad \text{Subtract } 0.6$$

$$0.4x - 2.2 > 1.5x$$

$$0.4x - 0.4x - 2.2 > 1.5x - 0.4x \quad \text{Subtract } 0.4x$$

$$-2.2 > 1.1x$$

$$-2 > x \text{ or } x < -2$$

41.

$$3(2x + 1) < 4(x - 3)$$

$$6x + 3 < 4x - 12$$

$$6x + 3 - 3 < 4x - 12 - 3 \quad \text{Subtract } 3$$

$$6x < 4x - 15$$

$$6x - 4x < 4x - 4x - 15 \quad \text{Subtract } 4x$$

$$2x < -15$$

$$\frac{1}{2} \cdot 2x < \frac{1}{2}(-15) \quad \text{Multiply by } \frac{1}{2}$$

$$x < -\frac{15}{2}$$

43.

$$4(2x - 3) - 7 > 3 + 6(1 - x)$$

$$8x - 12 - 7 > 3 + 6 - 6x \quad \text{Distributive}$$

$$8x - 19 > 9 - 6x \quad \text{Simplify}$$

$$8x - 19 + 19 > 9 + 19 - 6x \quad \text{Add } 19$$

$$8x > 28 - 6x$$

$$8x + 6x > 28 - 6x + 6x \quad \text{Add } 6x$$

$$14x > 28$$

$$\frac{1}{14} \cdot 14x > \frac{1}{14} \cdot 28 \quad \text{Multiply by } \frac{1}{14}$$

$$x > 2$$

45.

$$3(2x + 8) - 8x \geq 32$$

$$6x + 24 - 8x \geq 32 \quad \text{Distributive}$$

$$-2x + 24 \geq 32 \quad \text{Simplify}$$

$$-2x + 24 - 24 \geq 32 - 24 \quad \text{Subtract } 24$$

$$-2x \geq 8$$

$$-\frac{1}{2}(-2x) \leq -\frac{1}{2} \cdot 8 \quad \text{Reverse sign}$$

$$x \leq -4$$

The graphs of the solution sets are found in the answer section of the main text book.

47.

$$4(2x - 3) - 5 < 2(3x - 1) - 15$$

$$8x - 12 - 5 < 6x - 2 - 15 \quad \text{Distributive}$$

$$8x - 17 < 6x - 17 \quad \text{Simplify}$$

$$8x - 17 + 17 < 6x - 17 + 17 \quad \text{Add } 17$$

$$8x < 6x$$

$$8x - 6x < 6x - 6x \quad \text{Subtract } 6x$$

$$2x < 0$$

$$x < 0$$

49.

$$\frac{1}{5}(2x - 3) + 1 > \frac{1}{2}(2x - 1) \quad \text{LCD} = 10$$

$$10\left[\frac{1}{5}(2x - 3) + 1\right] > \left[\frac{1}{2}(2x - 1)\right]10 \quad \text{Multiply by } 10$$

$$2(2x - 3) + 10 > 5(2x - 1) \quad \text{Distributive}$$

$$4x - 6 + 10 > 10x - 5 \quad \text{Distributive}$$

$$4x + 4 > 10x - 5 \quad \text{Simplify}$$

$$4x + 4 + 5 > 10x - 5 + 5 \quad \text{Add } 5$$

$$4x + 9 > 10x$$

$$4x - 4x + 9 > 10x - 4x \quad \text{Subtract } 4x$$

$$9 > 6x$$

$$\frac{1}{6} \cdot 9 > \frac{1}{6} \cdot 6x \quad \text{Multiply by } \frac{1}{6}$$

$$\frac{9}{6} > x \text{ or } x < \frac{3}{2} \quad \text{Reduce}$$

51.

The exercise does not require a worked - out solution.

53.

$$-7 < x - 4 < 1$$

$$-7 + 4 < x - 4 + 4 < 1 + 4 \quad \text{Add } 4$$

$$-3 < x < 5$$

55.

$$2 \leq 6 - x \leq 8$$

$$2 - 6 \leq 6 - 6 - x \leq 8 - 6 \quad \text{Subtract } 6$$

$$-4 \leq -x \leq 2$$

$$4 \geq x \geq -2 \quad \text{Multiply by } -1$$

$$-2 \leq x \leq 4 \quad \text{Reverse signs}$$

The graphs of the solution sets are found in the answer section of the main text book.

57.
$$1 \le 2x - 3 \le 9$$
$$1 + 3 \le 2x - 3 + 3 \le 9 + 3 \qquad \text{Add } 3$$
$$4 \le 2x \le 12$$
$$2 \le x \le 6 \qquad \text{Multiply by } \frac{1}{2}$$

59.
$$7 < 5x + 2 \le 27$$
$$7 - 2 < 5x + 2 - 2 \le 27 - 2 \quad \text{Subtract } 2$$
$$5 < 5x \le 25$$
$$1 < x \le 5 \qquad \text{Multiply by } \frac{1}{5}$$

61.
$$5 < 2 - 3x < 11$$
$$5 - 2 < 2 - 2 - 3x < 11 - 2 \quad \text{Subtract } 2$$
$$3 < -3x < 9$$
$$-1 > x > -3 \qquad \text{Multiply by } -\frac{1}{3}$$
$$-3 < x < -1$$

63.
$$1 < \frac{1}{3}x < 4$$
$$3 \cdot 1 < 3 \cdot \frac{1}{3}x < 3 \cdot 4 \qquad \text{Multiply by } 3$$
$$3 < x < 12$$

65.
$$0 < \frac{3}{5}x + 2 < 8$$
$$5 \cdot 0 < 5\left(\frac{3}{5}x + 2\right) < 5 \cdot 8 \qquad \text{Multiply by } 5$$
$$0 < 3x + 10 < 40 \qquad \text{Distributive}$$
$$-10 < 3x + 10 - 10 < 40 - 10 \qquad \text{Subtract 10}$$
$$-10 < 3x < 30$$
$$-\frac{10}{3} < x < 10 \qquad \text{Multiply by } \frac{1}{3}$$

67.
$$3x - 1 < 2 \quad \text{and} \quad 5x + 4 > 9$$
$$3x < 3 \quad \text{and} \qquad 5x > 5$$
$$x < 1 \quad \text{and} \qquad x > 1 \quad \text{There is no intersection.}$$

69.
$$2x - 1 > -3 \quad \text{or} \quad x + 5 \le 3x$$

The graphs of the solution sets are found in the answer section of the main text book.

#69. continued
$$2x > -2 \quad \text{or} \qquad 5 \le 3x - x$$
$$x > -1 \quad \text{or} \qquad 5 \le 2x$$
$$x > -1 \quad \text{or} \qquad \frac{5}{2} \le x$$
$$x > -1 \quad \text{or} \qquad x \ge \frac{5}{2}$$
The union includes all numbers in both sets.

71.
$$6x \le 4 \quad \text{or} \quad 4x - 3 \ge 7$$
$$x \le \frac{4}{6} \quad \text{or} \qquad 4x \ge 10$$
$$x \le \frac{2}{3} \quad \text{or} \qquad x \ge \frac{10}{4}$$
$$x \le \frac{2}{3} \quad \text{or} \qquad x \ge \frac{5}{2}$$

73.
$$5x - 2 < 4 \quad \text{and} \quad 2x + 7 > 3$$
$$5x < 6 \quad \text{and} \qquad 2x > -4$$
$$x < \frac{6}{5} \quad \text{and} \qquad x > -2$$
$$\text{Thus,} \quad -2 < x < \frac{6}{5}$$

75.
$$x - 1 \le 4 \quad \text{or} \quad 1 - x \le 4$$
$$x \le 5 \quad \text{or} \quad -x \le 3$$
$$x \le 5 \quad \text{or} \qquad x \ge -3 \qquad \text{All real numbers}$$

Exercises 2.3

1.
$$D = R \cdot T \quad \text{for R}$$
$$\frac{1}{T} \cdot D = \frac{1}{T} \cdot RT \qquad \text{Multiply by } \frac{1}{T}$$
$$\frac{D}{T} = R$$

3.
$$A = LW \quad \text{for W}$$
$$\frac{1}{L} \cdot A = \frac{1}{L} \cdot LW \qquad \text{Multiply by } \frac{1}{L}$$
$$\frac{A}{L} = W$$

5.

$C = 2\pi r$ for r

$$\frac{1}{2\pi} \cdot C = \frac{1}{2\pi} \cdot 2\pi r$$

$$\frac{C}{2\pi} = r$$

7.

$V = \frac{1}{3}\pi r^2 h$ for h

$3V = 3 \cdot \frac{1}{3}\pi r^2 h$ Multiply by   3

$3V = \pi r^2 h$

$\frac{1}{\pi r^2} \cdot 3V = \frac{1}{\pi r^2} \cdot \pi r^2 h$ Multiply by $\frac{1}{\pi r^2}$

$$\frac{3V}{\pi r^2} = h$$

9.

$a = \frac{GM}{R^2}$ for G

$R^2 \cdot a = R^2 \cdot \frac{GM}{R^2}$ Multiply by R^2

$aR^2 = GM$

$\frac{1}{M} \cdot aR^2 = \frac{1}{M} \cdot GM$ Multiply by $\frac{1}{M}$

$$\frac{aR^2}{M} = G$$

11.

$P = 2L + 2W$ for L

$P - 2W = 2L$ Subtract   2W

$\frac{1}{2}(P - 2W) = \frac{1}{2} \cdot 2L$ Multiply by $\frac{1}{2}$

$$\frac{P - 2W}{2} = L \text{ or } L = \frac{P}{2} - W$$

13.

$180 = x + y + z$ for z

$180 - x - y = z$ Subtract x, y.

15.

$y = mx + b$ for m

$y - b = mx$ Subtract b

#15. continued

$\frac{1}{x}(y - b) = mx \cdot \frac{1}{x}$ Multiply by $\frac{1}{x}$

$$\frac{y - b}{x} = m$$

17.

$m(x - x_1) = y - y_1$ for m

$$\frac{1}{(x - x_1)} \cdot m(x - x_1) = \frac{1}{(x - x_1)} \cdot (y - y_1)$$

$$m = \frac{y - y_1}{x - x_1}$$

19.

$S = 2x^2 + 4xy$ for y

$S - 2x^2 = 4xy$ Subtract $2x^2$

$\frac{1}{4x} \cdot (S - 2x^2) = \frac{1}{4x} \cdot 4xy$ Multiply by $\frac{1}{4x}$

$$\frac{S - 2x^2}{4x} = y \text{ or } y = \frac{S}{4x} - \frac{x}{2}$$

21.

$x = \frac{x_1 + x_2}{2}$ for x_2.

$2x = x_1 + x_2$ Multiply by 2

$2x - x_1 = x_2$ Subtract x

23.

$P = 2L + \pi r + 2r$ for r

$P - 2L = \pi r + 2r$ Subtract 2L

$P - 2L = r(\pi + 2)$ Distributive

$\frac{1}{\pi + 2} \cdot (P - 2L) = \frac{1}{\pi + 2} \cdot r(\pi + 2)$

$$\frac{P - 2L}{\pi + 2} = r$$

25.

$S = 2LW + 2LH + 2WH$ for W

$S - 2LH = 2LW + 2WH$ Subtract 2LH

$S - 2LH = W(2L + 2H)$ Distributive

$\frac{1}{2L + 2H} \cdot (S - 2LH) = \frac{1}{2L + 2H} \cdot W(2L + 2H)$

$$\frac{S - 2LH}{2L + 2H} = W$$

27.

$t = \dfrac{1}{2}mv^2 + mad \quad \text{for } m$

$\qquad 2t = mv^2 + 2mad \qquad \text{Multiply by 2}$

$\qquad 2t = m\left(v^2 + 2ad\right) \qquad \text{Distributive}$

$\dfrac{1}{v^2 + 2ad} \cdot 2t = \dfrac{1}{v^2 + 2ad} \cdot m\left(v^2 + 2ad\right)$

$\qquad \dfrac{2t}{v^2 + 2ad} = m$

29.

$C = \dfrac{5}{9}(F - 32) = \dfrac{5}{9}(98.6 - 32) = \dfrac{5}{9}(66.6) = 37$

Answer: $37°$ C

31.

$T = \dfrac{D}{R} = \dfrac{372}{62} = 6 \qquad$ Answer: 6 hr.

33.

$r = \dfrac{I}{p \cdot t} = \dfrac{1683.50}{7800 \cdot \dfrac{7}{3}} = \dfrac{1683.50}{18200}$

$\qquad\qquad r = 0.0925 \quad \text{or} \quad 9\dfrac{1}{4}\%$

Note: 28 months $= 2\dfrac{1}{3}$ years

35.

$L = \dfrac{A}{W} = \dfrac{96}{6} = 16 \qquad$ Answer: 16 ft

37.

$h = \dfrac{2A}{b} = \dfrac{2 \cdot 68}{8} = \dfrac{136}{8} = 17 \qquad$ Answer: 17 m

39.

$r = \dfrac{C}{2\pi} = \dfrac{28\pi}{2\pi} = 14 \qquad$ Answer: 14 in

41.

$L = \sqrt{\dfrac{S}{6}} = \sqrt{\dfrac{96}{6}} = \sqrt{16} = 4 \qquad$ Answer: 4 in

---

Exercises 2.4

1.

Let x = the number.

Then, $2x + 1 =$ one more than twice the number.

Thus, $2x + 1 = 43$

$\qquad 2x + 1 - 1 = 43 - 1$

$\qquad\qquad 2x = 42$

$\qquad \dfrac{1}{2} \cdot 2x = \dfrac{1}{2} \cdot 42$

$\qquad\qquad x = 21 \qquad$ The number is 21.

3.

Let x = the number.

Then, $3x + 1 =$ one more than 3 times the number.

Thus, $3x + 1 = -14$

$\qquad 3x + 1 - 1 = -14 - 1$

$\qquad\qquad 3x = -15$

$\qquad \dfrac{1}{3} \cdot 3x = \dfrac{1}{3}(-15)$

$\qquad\qquad x = -5 \qquad$ The number is - 5.

5.

Let x = the number.

Then $3x - \dfrac{1}{2} =$ three times a number decreased by $\dfrac{1}{2}$.

Thus, $3x - \dfrac{1}{2} = 2$

$\qquad 6x - 1 = 4 \qquad$ Multiply by 2.

$\qquad 6x - 1 + 1 = 4 + 1$

$\qquad\qquad 6x = 5$

$\qquad\qquad x = \dfrac{5}{6} \qquad$ The number is $\dfrac{5}{6}$.

7.

Let x = the larger number.

Then 13 - x = the other number.

Thus, $x = 3(13 - x) + 1$

$\qquad x = 39 - 3x + 1$

$\qquad x = 40 - 3x$

$\qquad x + 3x = 40 - 3x + 3x$

$\qquad\qquad 4x = 40$

$\qquad \dfrac{1}{4} \cdot 4x = \dfrac{1}{4} \cdot 40$

#7. continued
$$x = 10$$
$$13 - x = 3 \qquad \text{The two numbers are 3 and 10.}$$

9.
Let x = the larger number.

Then 5 - x = the other number.

Thus, $x = 2(5 - x) + 17$
$$x = 10 - 2x + 17$$
$$x = 27 - 2x$$
$$x + 2x = 27 - 2x + 2x$$
$$3x = 27$$
$$\frac{1}{3} \cdot 3x = \frac{1}{3} \cdot 27$$
$$x = 9$$
$$5 - x = -4 \qquad \text{The numbers are -4 and 9.}$$

11.
Let x = the larger number.

Then, x - 3 = the other number.

Thus, $x = 33 - 2(x - 3)$
$$x = 33 - 2x + 6$$
$$x = 39 - 2x$$
$$x + 2x = 39 - 2x + 2x$$
$$3x = 39$$
$$\frac{1}{3} \cdot 3x = \frac{1}{3} \cdot 39$$
$$x = 13$$
$$x - 3 = 10 \qquad \text{The numbers are 10 and 13.}$$

13.
Three consecutive integers are x, x + 1, and x + 2.

Thus, $x + (x + 1) + (x + 2) = 78$
$$3x + 3 = 78$$
$$3x + 3 - 3 = 78 - 3$$
$$3x = 75$$
$$\frac{1}{3} \cdot 3x = \frac{1}{3} \cdot 75$$
$$x = 25$$

The integers are 25, 26, and 27.

15.
Three consecutive odd integers are x, x+2, and x+4.

Thus, $x + (x + 2) + (x + 4) = 111$
$$3x + 6 = 111$$
$$3x + 6 - 6 = 111 - 6$$
$$3x = 105$$
$$\frac{1}{3} \cdot 3x = \frac{1}{3} \cdot 105$$
$$x = 35$$

The integers are 35, 37, and 39.

17.
Three consecutive even integers are x, x+2, and x+4.

Thus, $(x + 2) + (x + 4) = x + 18$
$$2x + 6 = x + 18$$
$$2x - x + 6 = x - x + 18$$
$$x + 6 = 18$$
$$x + 6 - 6 = 18 - 6$$
$$x = 12$$

The integers are 12, 14, and 16.

19.
Let x = measurement of the width.

Then $2x + 9$ = measurement of the length.

$$\begin{array}{|c|} \hline \text{2x+9} \\ \text{x} \quad \\ \hline \end{array}$$

Also, $P = 2\ell + 2w$ Thus, $54 = 2(2x + 9) + 2x$
$$54 = 4x + 18 + 2x$$
$$54 = 6x + 18$$
$$54 - 18 = 6x + 18 - 18$$
$$36 = 6x$$
$$\frac{1}{6} \cdot 36 = \frac{1}{6} \cdot 6x$$
$$6 = x$$
$$21 = 2x + 9$$

Answer: 6 m by 21 m

21.
Let x = the length of the rectangle.

Then 13 - x = the width of the rectangle.

Also, $(13 - x) + 5$ or 18 - x = the side of the square.

Thus, $36 = 4(18 - x)$
$$36 = 72 - 4x$$
$$4x = 72 - 36$$
$$4x = 36$$

#21. continued

$$\frac{1}{4} \cdot 4x = \frac{1}{4} \cdot 36$$
$$x = 9$$
$$13 - x = 4 \qquad \text{Answer: 4 in by 9 in}$$

23.

Let x = measurement of the first side.

Then 2x = measurement of the second side.

And 3x - 1 = measurement of the third side.

Thus, $x + 2x + (3x - 1) = 17$
$$6x - 1 = 17$$
$$6x - 1 + 1 = 17 + 1$$
$$6x = 18$$
$$x = 3$$
$$2x = 6$$
$$3x - 1 = 8$$

Answer: 3 cm, 6 cm, 8 cm

25.

Let x = number of nickels.

Then 2x + 1 = number of quarters.

Let 5x = value of x numbers of nickels in cents.

Then 25(2x + 1) = value of 2x + 1 number of quarters in cents.

Thus, $5x + 25(2x + 1) = 465$
$$5x + 50x + 25 = 465$$
$$55x + 25 = 465$$
$$55x + 25 - 25 = 465 - 25$$
$$55x = 440$$
$$x = 8$$
$$2x + 1 = 17$$

Answer: 8 nickels, 17 quarters

27.

Let x = number of quarters.

Then 30 - x = number of dimes.

Let 25x = value of the quarters in cents.

Then 10(30 - x) = value of the dimes in cents.

Thus, $25x + 10(30 - x) = 480$
$$25x + 300 - 10x = 480$$
$$15x + 300 = 480$$
$$15x = 480 - 300$$
$$15x = 180$$

#27. continued

$$x = 12$$
$$30 - x = 18$$

Answer: 18 dimes, 12 quarters

29.

Let x = number of twenties.

Let x + 7 = number of fives.

Let 2x - 3 = number of tens.

Then 20x = value of the twenties.

Then 5(x + 7) = value of the fives.

Then 10(2x - 3) = value of the tens.

Thus, $20x + 5(x + 7) + 10(2x - 3) = 545$
$$20x + 5x + 35 + 20x - 30 = 545$$
$$45x + 5 = 545$$
$$45x = 540$$
$$x = 12$$
$$x + 7 = 19$$
$$2x - 3 = 21$$

Answer: 12 twenties, 19 fives, 21 tens

31.

Let x = selling price of house.

Thus, $x - 0.05x = 80,000$
$$0.95x = 80,000$$
$$\frac{1}{0.95} \cdot 0.95x = \frac{1}{0.95} \cdot 80,000$$
$$x = 84,210.52632$$

Answer: $84,210.53

33.

Let x = original price of the car.

Let (x + 0.2x) = marked up cost.

Thus, $(x + 0.2x) - 0.15(x + 0.2x) = 11,515.80$
$$x + 0.2x - 0.15x - 0.03x = 11,515.8$$
$$1.02x = 11,515.8$$
$$x = 11,290$$

Answer: $11,290

35.

Let x = number of hours of travel time for Henry.

Then x - 4 = number of hours of travel time for Sarah.

#35. continued

	Rate	· Time	= Distance
Henry	15	x	15x
Sarah	45	x-4	45(x-4)

The distances are the same.

$15x = 45(x - 4)$

$15x = 45x - 180$

$180 = 30x$

$6 = x$

$2 = x - 4$ Answer: 2 hr

37.

Let x = number of hours of travel time for both Dottie and Gilbert.

	Rate	· Time	= Distance
Dottie	45	x	45x
Gilbert	55	x	55x

Thus, $45x + 55x = 250$

$100x = 250$

$x = 2.5$ Answer: $2\frac{1}{2}$ hr

Exercises 2.5

1.
$|x| = 2$

$x = 2$ or $x = -2$ $\{2, -2\}$

3.
$|z| = 23$

Case I   Case II

$z = 23$ or $z = -23$ $\{23, -23\}$

5.
$|5| = -13$

$\varnothing$ Absolute value is never negative.

7.
$|-z| = -27$

$\varnothing$ Absolute value is never negative.

9.
$|-S| = 0$

$-S = 0$ or $-S = 0$ $\{0\}$

11.
$|-5x| = 35$

$-5x = 35$ or $-5x = -35$

$x = -7$ $x = 7$ $\{-7, 7\}$

13.
$|-4y| = -48$

$\varnothing$ Absolute value is never negative.

15.
$|5s| = 13$

$5s = 13$ or $5s = -13$

$s = \frac{13}{5}$ $s = -\frac{13}{5}$ $\left\{-\frac{13}{5}, \frac{13}{5}\right\}$

17.
$3|5p| = 0$

$|5p| = 0$ Isolate absolute value.

$5p = 0$ or $5p = 0$

$p = 0$ or $p = 0$ $\{0\}$

19.
$3|y| = -12$ or $|y| = -4$

$\varnothing$ Absolute value is never negative.

21.
$4|-s| = 24$

$|-s| = 6$ Isolate absolute value.

$-s = 6$ or $-s = -6$

$s = -6$ $s = 6$ $\{-6, 6\}$

23.
$-3|y| = -27$

$|y| = 9$ Isolate absolute value.

$y = 9$ or $y = -9$ $\{-9, 9\}$

25.

$-4|-y| = -36$

$\quad |-y| = 9 \qquad$ Isolate absolute value.

$-y = 9 \quad$ or $\; -y = -9$

$\quad y = -9 \qquad y = 9 \qquad \{-9, 9\}$

27.

$|x| + 5 = 9$

$\quad |x| = 4 \qquad$ Isolate absolute value.

$x = 4 \;$ or $\; x = -4 \qquad \{-4, 4\}$

29.

$|y| + 4 = 1$

$\quad |y| = -3 \qquad$ Isolate absolute value.

$\varnothing \qquad$ Absolute value is never negative.

31.

$|z| - 7 = -9$

$\quad |z| = -2 \qquad$ Isolate absolute value.

$\varnothing \qquad$ Absolute value is never negative.

33.

$\quad |x - 7| = 9$

$\quad x - 7 = 9 \qquad$ or $\qquad x - 7 = -9$

$x - 7 + 7 = 9 + 7 \qquad x - 7 + 7 = 7 - 9$

$\quad\quad x = 16 \qquad\qquad x = -2 \qquad \{-2, 16\}$

35.

$|3x + 4| = -8$

$\varnothing \qquad$ Absolute value is never negative.

37.

$2|x - 1| = 10$

$\quad |x - 1| = 5 \qquad$ Isolate absolute value.

$\quad x - 1 = 5 \quad$ or $\quad x - 1 = -5$

$x - 1 + 1 = 5 + 1 \quad x - 1 + 1 = -5 + 1$

$\quad\quad x = 6 \qquad\qquad x = -4 \qquad \{-4, 6\}$

39.

$2|3 - 4x| = 15$

$\quad |3 - 4x| = \dfrac{15}{2} \qquad$ Isolate absolute value.

#39. continued

$3 - 4x = \dfrac{15}{2} \qquad$ or $\qquad 3 - 4x = -\dfrac{15}{2}$

$6 - 8x = 15 \qquad\qquad 6 - 8x = -15$

$6 - 6 - 8x = 15 - 6 \qquad 6 - 6 - 8x = -15 - 6$

$\quad -8x = 9 \qquad\qquad -8x = -21$

$-\dfrac{1}{8}(-8x) = -\dfrac{1}{8} \cdot 9 \quad -\dfrac{1}{8}(-8x) = -\dfrac{1}{8}(-21)$

$\quad x = -\dfrac{9}{8} \qquad\qquad x = \dfrac{21}{8}$

$\left\{ -\dfrac{9}{8}, \dfrac{21}{8} \right\}$

41.

$-3|x + 4| = -21$

$\quad |x + 4| = 7 \qquad$ Isolate absolute value.

$\quad x + 4 = 7 \qquad$ or $\qquad x + 4 = -7$

$x + 4 - 4 = 7 - 4 \qquad x + 4 - 4 = -7 - 4$

$\quad\quad x = 3 \qquad\qquad x = -11$

$\{-11, 3\}$

43.

$-6|1 - 3x| = 0$

$\quad |1 - 3x| = 0 \qquad$ Isolate absolute value.

$\quad 1 - 3x = 0 \qquad$ or $\qquad 1 - 3x = 0$

$1 - 1 - 3x = 0 \qquad\qquad 1 - 1 - 3x = 0$

$\quad -3x = -1 \qquad\qquad -3x = -1$

$-\dfrac{1}{3}(-3x) = -\dfrac{1}{3}(-1) \qquad -\dfrac{1}{3}(-3x) = -\dfrac{1}{3}(-1)$

$\quad x = \dfrac{1}{3} \qquad\qquad x \;\; = \dfrac{1}{3}$

$\left\{ \dfrac{1}{3} \right\}$

45.

$|2x - 4| - 2 = 11$

$\quad |2x - 4| = 13 \qquad$ Isolate absolute value.

$\quad 2x - 4 = 13 \qquad$ or $\qquad 2x - 4 = -13$

$2x - 4 + 4 = 13 + 4 \qquad 2x - 4 + 4 = -13 + 4$

$\quad\quad 2x = 17 \qquad\qquad 2x = -9$

$\dfrac{1}{2} \cdot 2x = \dfrac{1}{2} \cdot 17 \qquad \dfrac{1}{2} \cdot 2x = \dfrac{1}{2}(-9)$

$\quad x = \dfrac{17}{2} \qquad\qquad x \;\; = -\dfrac{9}{2}$

#45. continued

$$\left\{-\frac{9}{2}, \frac{17}{2}\right\}$$

47.

$|4x + 3| + 7 = 2$

$\quad |4x + 3| = -5$ Isolate absolute value.

$\varnothing \qquad\qquad$ Absolute value is never negative.

49.

$|6x - 3| - 8 = -8$

$\quad |6x - 3| = 0$ Isolate absolute value.

$\quad 6x - 3 = 0 \qquad$ or $\qquad 6x - 3 = 0$

$6x - 3 + 3 = 0 + 3 \qquad 6x - 3 + 3 = 0 + 3$

$\qquad 6x = 3 \qquad\qquad\qquad 6x = 3$

$\quad \frac{1}{6} \cdot 6x = \frac{1}{6} \cdot 3 \qquad \frac{1}{6} \cdot 6x = \frac{1}{6} \cdot 3$

$\qquad x = \frac{1}{2} \qquad\qquad\quad x = \frac{1}{2}$

$\left\{\frac{1}{2}\right\}$

51.

$2|x + 3| - 5 = 7$

$\quad 2|x + 3| = 12$

$\quad |x + 3| = 6$ Isolate absolute value.

$\quad x + 3 = 6 \qquad$ or $\qquad x + 3 = -6$

$x + 3 - 3 = 6 - 3 \qquad x + 3 - 3 = -6 - 3$

$\qquad x = 3 \qquad\qquad\qquad x = -9$

$\{-9, 3\}$

53.

$5 - |3x - 4| = 1$

$\quad -|3x - 4| = -4$

$\quad |3x - 4| = 4$ Isolate absolute value.

$\quad 3x - 4 = 4 \qquad$ or $\qquad 3x - 4 = -4$

$3x - 4 + 4 = 4 + 4 \qquad 3x - 4 + 4 = -4 + 4$

$\qquad 3x = 8 \qquad\qquad\qquad 3x = 0$

$\quad \frac{1}{3} \cdot 3x = \frac{1}{3} \cdot 8 \qquad \frac{1}{3} \cdot 3x = \frac{1}{3} \cdot 0$

$\qquad x = \frac{8}{3} \qquad\qquad\quad x = 0$

$\left\{0, \frac{8}{3}\right\}$

55.

$11 - 2|2x - 3| = 5$

$\quad -2|2x - 3| = -6$

$\quad |2x - 3| = 3$ Isolate absolute value.

$\quad 2x - 3 = 3 \qquad$ or $\qquad 2x - 3 = -3$

$2x - 3 + 3 = 3 + 3 \qquad 2x - 3 + 3 = -3 + 3$

$\qquad 2x = 6 \qquad\qquad\qquad 2x = 0$

$\quad \frac{1}{2} \cdot 2x = \frac{1}{2} \cdot 6 \qquad \frac{1}{2} \cdot 2x = \frac{1}{2} \cdot 0$

$\qquad x = 3 \qquad\qquad\quad x = 0$

$\{0, 3\}$

57.

$10 - 5|6 - 4x| = 12$

$\quad -5|6 - 4x| = 2$

$\quad |6 - 4x| = -\frac{2}{5}$ Isolate absolute value.

$\varnothing \quad$ Absolute value is never negative.

59.

$1.2|5.67x - 8.5035| + 6.451 = 18.7$

$\qquad 1.2|5.67x - 8.5035| = 12.249$

$\qquad |5.67x - 8.5035| = 10.2075$

Isolate absolute value.

$\qquad 5.67x - 8.5035 = 10.2075$

$\qquad\qquad 5.67x = 18.711$

$\qquad\qquad\qquad x = 3.3$

or $\quad 5.67x - 8.5035 = -10.2075$

$\qquad\qquad 5.67x = -1.704$

$\qquad\qquad\qquad x = -0.300529$

$\{-0.300529, 3.3\}$

61.

$|2x + 1| = |x - 4|$

$\quad 2x + 1 = x - 4 \qquad$ or $\qquad 2x + 1 = -(x - 4)$

$2x + 1 - 1 = x - 4 - 1 \qquad\quad 2x + 1 = -x + 4$

$\qquad 2x = x - 5 \qquad\qquad 2x + 1 - 1 = -x + 4 - 1$

$2x - x = x - x - 5 \qquad\qquad 2x = -x + 3$

$\qquad x = -5 \qquad\qquad 2x + x = -x + x + 3$

$\qquad\qquad\qquad\qquad\qquad 3x = 3$

$\qquad\qquad\qquad\qquad\qquad x = 1$

$\{-5, 1\}$

63.

$$|6-x| = |4x+3|$$

$$6-x = 4x+3 \quad \text{or} \quad 6-x = -(4x+3)$$

$$6-3-x = 4x+3-3 \qquad\qquad 6\text{-}x = -4x\text{-}3$$

$$3\text{-}x = 4x \qquad\qquad 6\text{-}6\text{-}x = -4x\text{-}3\text{-}6$$

$$3\text{-}x+x = 4x+x \qquad\qquad -x = -4x+4x-9$$

$$3 = 5x \qquad\qquad -x+4x = -4x+4x-9$$

$$\frac{1}{5}\cdot 3 = \frac{1}{5}\cdot 5x \qquad\qquad 3x = \text{-}9$$

$$\frac{3}{5} = x \qquad\qquad \frac{1}{3}\cdot x = \frac{1}{3}(-9)$$

$$x = \text{-}3$$

$$\left\{-3, \frac{3}{5}\right\}$$

65.

$$|5x-1| = |4x-9|$$

$$5x-1 = 4x-9 \quad \text{or} \quad 5x\text{-}1 = -(4x\text{-}9)$$

$$5x-1+1 = 4x-9+1 \qquad 5x\text{-}1 = -4x+9$$

$$5x = 4x\text{-}8 \qquad 5x\text{-}1+1 = -4x+9+1$$

$$5x\text{-}4x = 4x\text{-}4x\text{-}8 \qquad 5x = -4x+10$$

$$x = \text{-}8 \qquad 5x+4x = -4x+4x+10$$

$$9x = 10$$

$$\frac{1}{9}\cdot 9x = \frac{1}{9}\cdot 10$$

$$x = \frac{10}{9}$$

$$\left\{-8, \frac{10}{9}\right\}$$

67.

$$|7-x| = |9-3x|$$

$$7-x = 9-3x \quad \text{or} \quad 7\text{-}x = -(9\text{-}3x)$$

$$7-7-x = 9-7-3x \qquad 7\text{-}x = -9+3x$$

$$-x = 2\text{-}3x \qquad 7+9\text{-}x = -9+9+3x$$

$$-x+3x = 2\text{-}3x+3x \qquad 16\text{-}x = 3x$$

$$2x = 2 \qquad 16\text{-}x+x = 3x+x$$

$$\frac{1}{2}\cdot 2x = \frac{1}{2}\cdot 2 \qquad 16 = 4x$$

$$x = 1 \qquad \frac{1}{4}\cdot 16 = \frac{1}{4}\cdot 4x$$

$$4 = x$$

$$\{1,4\}$$

69.

$$\left|\frac{1}{2}x+5\right| = |x-2|$$

$$\frac{1}{2}x+5 = x-2 \quad \text{or} \quad \frac{1}{2}x+5 = -(x-2)$$

$$\frac{1}{2}x+5+2 = x-2+2 \qquad \frac{1}{2}x+5 = -x+2$$

$$\frac{1}{2}x+7 = x \qquad \frac{1}{2}x+5-5 = -x+2-5$$

$$\frac{1}{2}x-\frac{1}{2}x+7 = x-\frac{1}{2}x \qquad \frac{1}{2}x = -x-3$$

$$7 = \frac{1}{2}x \qquad \frac{1}{2}x+x = -x+x-3$$

$$2\cdot 7 = 2\cdot\frac{1}{2}x \qquad \frac{3}{2}x = -3$$

$$14 = x \qquad \frac{2}{3}\cdot\frac{3}{2}x = \frac{2}{3}(-3)$$

$$x = \text{-}2$$

$$\{-2,14\}$$

71.

$$\left|\frac{3}{4}x+\frac{1}{3}\right| = \left|\frac{5}{2}x-7\right|$$

$$\frac{3}{4}x+\frac{1}{3} = \frac{5}{2}x-7$$

$$12\left(\frac{3}{4}x+\frac{1}{3}\right) = 12\left(\frac{5}{2}x-7\right)$$

$$9x+4 = 30x\text{-}84$$

$$9x+4+84 = 30x\text{-}84+84$$

$$9x+88 = 30x$$

$$9x\text{-}9x+88 = 30x\text{-}9x$$

$$88 = 21x$$

$$\frac{88}{21} = x$$

$$\text{or} \quad \frac{3}{4}x+\frac{1}{3} = -\left(\frac{5}{2}x-7\right)$$

$$12\left(\frac{3}{4}x+\frac{1}{3}\right) = \left(-\frac{5}{2}x+7\right)12$$

$$9x+4 = \text{-}30x+84$$

$$9x+4\text{-}4 = \text{-}30x+84\text{-}4$$

$$9x = \text{-}30x+80$$

$$9x+30x = \text{-}30x+30x+80$$

$$39x = 80$$

#71. continued

$$x = \frac{80}{39}$$

$$\left\{ \frac{88}{21}, \frac{80}{39} \right\}$$

73.

$$|5 - 2x| = |2x + 7|$$
$$5 - 2x = 2x + 7$$
$$5 - 7 - 2x = 2x + 7 - 7$$
$$-2 - 2x = 2x$$
$$-2 - 2x + 2x = 2x + 2x$$
$$-2 = 4x$$
$$\frac{1}{4}(-2) = \frac{1}{4} \cdot 4x$$
$$-\frac{1}{2} = x$$

or $\quad 5 - 2x = -(2x + 7)$
$$5 - 2x = -2x - 7$$
$$5 + 7 - 2x = -2x - 7 + 7$$
$$12 - 2x = -2x$$
$$12 - 2x + 2x = -2x + 2x$$
No solution.

$$\left\{ -\frac{1}{2} \right\}$$

75.

$$\left| \frac{1}{2}x - \frac{1}{3} \right| = \left| \frac{1}{2}x + \frac{1}{4} \right|$$
$$\frac{1}{2}x - \frac{1}{3} = \frac{1}{2}x + \frac{1}{4}$$
$$12\left(\frac{1}{2}x - \frac{1}{3} \right) = 12\left(\frac{1}{2}x + \frac{1}{4} \right)$$
$$6x - 4 = 6x + 3$$
$$-4 = 3$$

No solution.

or $\quad \frac{1}{2}x - \frac{1}{3} = -\left(\frac{1}{2}x + \frac{1}{4} \right)$
$$12\left(\frac{1}{2}x - \frac{1}{3} \right) = 12\left(-\frac{1}{2}x - \frac{1}{4} \right)$$
$$6x - 4 = -6x - 3$$
$$6x - 4 + 4 = -6x - 3 + 4$$
$$6x = -6x + 1$$
$$6x + 6x = 1$$
$$12x = 1$$

#75. continued

$$\frac{1}{12} \cdot 12x = \frac{1}{12} \cdot 1$$
$$x = \frac{1}{12}$$

$$\left\{ \frac{1}{12} \right\}$$

77.

$$|ax + b| = c$$
$$ax + b = c \quad \text{or} \quad ax + b = -c$$
$$ax = c - b \qquad\qquad ax = -c - b$$
$$x = \frac{c - b}{a} \qquad\qquad x = \frac{-c - b}{a}$$
$$\left\{ \frac{c - b}{a}, \frac{-b - c}{a} \right\}$$

Exercises 2.6

> The graphs of the solution sets are found in the answer section of the main text book.

1.

$$|x| < 1$$
$$-1 < x < 1 \qquad \text{Theorem 2.6.1}$$

3.

$$|x| > 2$$
$$x > 2 \text{ or } x < -2 \qquad \text{Theorem 2.6.2}$$
Thus, $x < -2$ or $x > 2$

5.

$$|x| \le \frac{3}{2}$$
$$-\frac{3}{2} \le x \le \frac{3}{2} \qquad \text{Theorem 2.6.1}$$

7.

$$|x| \ge -3$$
$$x \ge -3 \text{ or } x \le 3 \qquad \text{Theorem 2.6.2}$$
All real numbers.

9.

$$|x| + 5 < 10$$
$$|x| + 5 - 5 < 10 - 5$$
$$|x| < 5$$
$$-5 < x < 5 \qquad \text{Theorem 2.6.1}$$

| The graphs of the solution sets are found in the answer section of the main text book. | The graphs of the solution sets are found in the answer section of the main text book. |

11.

$$|x| + 1 \geq 2$$

$$|x| + 1 - 1 \geq 2 - 1$$

$$|x| \geq 1$$

$x \geq 1$ or $x \leq -1$ Theorem 2.6.2

Thus, $x \leq -1$ or $x \geq 1$

13.

$$|x| - 2 < 0$$

$$|x| - 2 + 2 < 2$$

$$|x| < 2$$

$$-2 < x < 2 \qquad \text{Theorem 2.6.1}$$

15.

$$|x| - 6 < -3$$

$$|x| - 6 + 6 < -3 + 6 \qquad \text{Theorem 6.2.1}$$

$$|x| < 3$$

$$-3 < x < 3$$

17.

$$|x| + 7 \geq 7$$

$$|x| + 7 - 7 \geq 7 - 7 \qquad \text{Theorem 6.2.2}$$

$$|x| \geq 0$$

$x \geq 0$ or $x \leq 0$ All real numbers.

19.

$$|x + 1| \leq 5$$

$$-5 \leq x + 1 \leq 5 \qquad \text{Theorem 6.2.1}$$

$$-5 - 1 \leq x + 1 - 1 \leq 5 - 1$$

$$-6 \leq x \leq 4$$

21.

$$|x - 1| \geq 5$$

$x - 1 \geq 5$ or $x - 1 \leq -5$ Theorem 6.2.2

$x - 1 + 1 \geq 5 + 1$ $x - 1 + 1 \leq -5 + 1$

$x \geq 6$ $x \leq -4$

Thus, $x \leq -4$ or $x \geq 6$

23.

$$|x - 3| < \frac{1}{2}$$

$$-\frac{1}{2} < x - 3 < \frac{1}{2} \qquad \text{Theorem 6.2.1}$$

$$2 \cdot \left(-\frac{1}{2}\right) < 2(x - 3) < 2 \cdot \frac{1}{2}$$

$$-1 < 2x - 6 < 1$$

$$-1 + 6 < 2x - 6 + 6 < 1 + 6$$

$$5 < 2x < 7$$

$$\frac{1}{2} \cdot 5 < \frac{1}{2} \cdot 2x < \frac{1}{2} \cdot 7$$

$$\frac{5}{2} < x < \frac{7}{2}$$

25.

$$|5x - 4| \leq 9$$

$$-9 \leq 5x - 4 \leq 9$$

$$-9 + 4 \leq 5x - 4 + 4 \leq 9 + 4$$

$$-5 \leq 5x \leq 13$$

$$\frac{1}{5}(-5) \leq \frac{1}{5} \cdot 5x \leq \frac{1}{5} \cdot 13$$

$$-1 \leq x \leq \frac{13}{5}$$

27.

$$|1 - x| > \frac{3}{2}$$

$1 - x > \frac{3}{2}$ or $1 - x < -\frac{3}{2}$ Theorem 6.2.2

$1 - 1 - x > \frac{3}{2} - 1$ $1 - 1 - x < -\frac{3}{2} - 1$

$-x > \frac{1}{2}$ $-x < -\frac{5}{2}$

$x < -\frac{1}{2}$ $x > \frac{5}{2}$ Reverse signs

Thus, $x < -\frac{1}{2}$ or $x > \frac{5}{2}$

29.

$$|3 - 2x| \leq 5$$

$$-5 \leq 3 - 2x \leq 5 \qquad \text{Theorem 6.2.1}$$

$$-5 - 3 \leq 3 - 3 - 2x \leq 5 - 3$$

#29. continued

$$-8 \le -2x \le 2$$

$$-\frac{1}{2}(-8) \ge -\frac{1}{2}(-2x) \ge -\frac{1}{2} \cdot 2 \qquad \text{Reverse signs}$$

$$4 \ge x \ge -1$$

Thus, $-1 \le x \le 4$

31.

$$|1 - 3x| < 1$$

$$-1 < 1 - 3x < 1 \qquad \text{Theorem } 6.2.1$$

$$-1 - 1 < 1 - 1 - 3x < 1 - 1$$

$$-2 < -3x < 0$$

$$-\frac{1}{3}(-2) > -\frac{1}{3}(-3x) > -\frac{1}{3} \cdot 0$$

$$\frac{2}{3} > x > 0 \qquad \text{Reverse signs}$$

Thus, $0 < x < \dfrac{2}{3}$

33.

$$|-2 - 3x| > 5$$

$$-2 - 3x > 5 \qquad \text{Theorem } 6.2.2$$

$$-2 + 2 - 3x > 5 + 2$$

$$-3x > 7$$

$$-\frac{1}{3}(-3x) < -\frac{1}{3} \cdot 7 \qquad \text{Reverse signs}$$

$$x < -\frac{7}{3}$$

or $\quad -2 - 3x < -5 \qquad \text{Theorem } 6.2.2$

$$-2 + 2 - 3x < -5 + 2$$

$$-3x < -3$$

$$-\frac{1}{3}(-3x) > -\frac{1}{3}(-3) \qquad \text{Reverse signs}$$

$$x > 1$$

Thus, $x < -\dfrac{7}{3}$ or $x > 1$

35.

$$|x + 2| - 1 < 3$$

$$|x + 2| - 1 + 1 < 3 + 1$$

$$|x + 2| < 4 \qquad \text{Isolate absolute value.}$$

#35. continued

$$-4 < x + 2 < 4 \qquad \text{Theorem } 6.2.1$$

$$-4 - 2 < x + 2 - 2 < 4 - 2$$

$$-6 < x < 2$$

37.

$$|2x + 5| - 3 \ge 0$$

$$|2x + 5| - 3 + 3 \ge 3$$

$$|2x + 5| \ge 3 \qquad \text{Isolate absolute value.}$$

$$2x + 5 \ge 3 \quad \text{or} \quad 2x + 5 \le -3$$

$$2x + 5 - 5 \ge 3 - 5 \qquad 2x + 5 - 5 \le -3 - 5$$

$$2x \ge -2 \qquad\qquad 2x \le -8$$

$$\frac{1}{2} \cdot 2x \ge \frac{1}{2}(-2) \qquad \frac{1}{2} \cdot 2x \le \frac{1}{2}(-8)$$

$$x \ge -1 \qquad\qquad x \le -4$$

Thus, $x \le -4$ or $x \ge -1$

39.

$$|4 - x| + 2 < 5$$

$$|4 - x| + 2 - 2 < 5 - 2$$

$$|4 - x| < 3 \qquad \text{Isolate absolute value.}$$

$$-3 < 4 - x < 3 \qquad \text{Theorem } 6.1.1$$

$$-3 - 4 < 4 - 4 - x < 3 - 4$$

$$-7 < -x < -1$$

$$7 > x > 1 \qquad \text{Multiply by } -1, \text{ reverse signs}$$

Thus, $1 < x < 7$

41.

$$|4x - 3| + 6 > 1$$

$$|4x - 3| + 6 - 6 > 1 - 6$$

$$|4x - 3| > -5 \qquad \text{Isolate absolute value.}$$

$$4x - 3 > -5 \quad \text{or} \quad 4x - 3 < 5$$

$$4x - 3 + 3 > -5 + 3 \qquad 4x - 3 + 3 < 5 + 3$$

$$4x > -2 \qquad\qquad 4x < 8$$

$$\frac{1}{4} \cdot 4x > \frac{1}{4}(-2) \qquad \frac{1}{4} \cdot 4x < \frac{1}{4} \cdot 8$$

$$x > -\frac{1}{2} \qquad\qquad x < 2$$

Thus, $x > -\dfrac{1}{2}$ or $x < 2$

All real numbers.

The graphs of the solution sets are found in the answer section of the main text book.

The graphs of the solution sets are found in the answer section of the main text book.

43.

$$2|x+7| < 8$$

$$\frac{1}{2} \cdot 2|x+7| < \frac{1}{2} \cdot 8$$

$$|x+7| < 4 \qquad \text{Isolate absolute value.}$$

$$-4 < x+7 < 4 \qquad \text{Theorem } 6.2.1$$

$$-4-7 < x+7-7 < 4-7$$

$$-11 < x < -3$$

45.

$$2|3x-1| \geq 9$$

$$\frac{1}{2} \cdot 2|3x-1| \geq \frac{1}{2} \cdot 9$$

$$|3x-1| \geq \frac{9}{2} \qquad \text{Isolate absolute value.}$$

$$3x-1 \geq \frac{9}{2}$$

$$2(3x-1) \geq 2 \cdot \frac{9}{2}$$

$$6x-2 \geq 9$$

$$6x-2+2 \geq 9+2$$

$$6x \geq 11$$

$$\frac{1}{6} \cdot 6x \geq \frac{1}{6} \cdot 11$$

$$x \geq \frac{11}{6}$$

$$\text{or} \quad 3x-1 \leq -\frac{9}{2} \qquad \text{Theorem } 6.2.2$$

$$2(3x-1) \leq 2\left(-\frac{9}{2}\right)$$

$$6x-2 \leq -9$$

$$6x-2+2 \leq -9+2$$

$$6x \leq -7$$

$$\frac{1}{6} \cdot 6x \leq \frac{1}{6}(-7)$$

$$x \leq -\frac{7}{6}$$

Thus, $x \leq -\dfrac{7}{6}$ or $x \geq \dfrac{11}{6}$

47.

$$2|x-1| + 4 \leq 8$$

$$2|x-1| + 4 - 4 \leq 8 - 4$$

#47. continued

$$2|x-1| \leq 4$$

$$\frac{1}{2} \cdot 2|x-1| \leq \frac{1}{2} \cdot 4$$

$$|x-1| \leq 2 \qquad \text{Isolate absolute value.}$$

$$-2 \leq x-1 \leq 2 \qquad \text{Theorem } 6.2.1$$

$$-2+1 \leq x-1+1 \leq 2+1$$

$$-1 \leq x \leq 3$$

49.

$$3|2x+7| + 1 > 10$$

$$3|2x+7| + 1 - 1 > 10 - 1$$

$$3|2x+7| > 9$$

$$\frac{1}{3} \cdot 3|2x+7| > \frac{1}{3} \cdot 9$$

$$|2x+7| > 3 \qquad \text{Isolate absolute value.}$$

$$2x+7 > 3 \quad \text{or} \quad 2x+7 < -3 \qquad \text{Theorem } 6.2.2$$

$$2x+7-7 > 3-7 \qquad 2x+7-7 < -3-7$$

$$2x > -4 \qquad\qquad 2x < -10$$

$$\frac{1}{2} \cdot 2x > \frac{1}{2}(-4) \qquad \frac{1}{2} \cdot 2x < \frac{1}{2}(-10)$$

$$x > -2 \qquad\qquad x < -5$$

Thus, $x < -5$ or $x > -2$

51.

$$|x+1| > 0$$

$$x+1 > 0 \quad \text{or} \quad x+1 < 0 \qquad \text{Theorem } 6.2.2$$

$$x+1-1 > 0-1 \qquad x+1-1 < 0-1$$

$$x > -1 \qquad\qquad x < -1$$

Thus, $x < -1$ or $x > -1$, $x \neq -1$

53.

$$|3x-4| \leq 0$$

$$0 \leq 3x-4 \leq 0 \qquad \text{Theorem } 6.2.1$$

$$0+4 \leq 3x-4+4 \leq 0+4$$

$$4 \leq 3x \leq 4$$

$$\frac{1}{3} \cdot 4 \leq \frac{1}{3} \cdot 3x \leq \frac{1}{3} \cdot 4$$

$$\frac{4}{3} \leq x \leq \frac{4}{3}, \ x = \frac{4}{3}$$

The graphs of the solution sets are found in the answer section of the main text book.

55.

$|2x - 6| > 0$

$2x - 6 > 0$ or $2x - 6 < 0$ Theorem 6.2.2

$2x - 6 + 6 > 0 + 6$ $2x - 6 + 6 < 0 + 6$

      $2x > 6$ $2x < 6$

$\frac{1}{2} \cdot 2x > \frac{1}{2} \cdot 6$ $\frac{1}{2} \cdot 2x < \frac{1}{2} \cdot 6$

     $x > 3$ $x < 3$

Thus, $x < 3$ or $x > 3$, $x \neq 3$

57.

$|x + 7| \geq -3$

    $x + 7 \geq -3$ or $x + 7 \leq 3$ Theorem 6.2.2

$x + 7 - 7 \geq -3 - 7$ $x + 7 - 7 \leq 3 - 7$

      $x \geq -10$ $x \leq -4$

Thus, $x \geq -10$ or $x \leq -4$.

All real numbers.

59.

     $|5x - 7| \leq -1$

   $1 \leq 5x - 7 \leq -1$ Theorem 6.2.1

$1 + 7 \leq 5x - 7 + 7 \leq -1 + 7$

    $8 \leq 5x \leq 6$

$\frac{1}{5} \cdot 8 \leq \frac{1}{5} \cdot 5x \leq \frac{1}{5} \cdot 6$

   $\frac{8}{5} \leq x \leq \frac{6}{5}$

$\emptyset$ since $x \leq \frac{6}{5}$ and $x \geq \frac{8}{5}$ is impossible.

Review Exercises

1.

    $7x = 56$

$\frac{1}{7} \cdot 7x = \frac{1}{7} \cdot 56$

    $x = 8$

Check: $7 \cdot 8 = 56$

          $56 = 56$

3.

    $3x - 1 = 15$

$3x - 1 + 1 = 15 + 1$

    $3x = 16$

$\frac{1}{3} \cdot 3x = \frac{1}{3} \cdot 16$

    $x = \frac{16}{3}$

Check: $3 \cdot \frac{16}{3} - 1 = 15$

         $16 - 1 = 15$

           $15 = 15$

5.

     $4x + 7 = 2x - 5$

$4x + 7 - 7 = 2x - 5 - 7$

      $4x = 2x - 12$

$4x - 2x = 2x - 2x - 12$

    $2x = -12$

$\frac{1}{2} \cdot 2x = \frac{1}{2}(-12)$

      $x = -6$

Check: $4(-6) + 7 = 2(-6) - 5$

      $-24 + 7 = -12 - 5$

         $-17 = -17$

7.

    $2(3x - 7) = 4$

    $6x - 14 = 4$

$6x - 14 + 14 = 4 + 14$

      $6x = 18$

$\frac{1}{6} \cdot 6x = \frac{1}{6} \cdot 18$

      $x = 3$

Check: $2(3 \cdot 3 - 7) = 4$

       $2(9 - 7) = 4$

         $2(2) = 4$

           $4 = 4$

9.

$3(5x - 2) + 3x = 12x + 4$

$15x - 6 + 3x = 12x + 4$

    $18x - 6 = 12x + 4$

$18x - 6 + 6 = 12x + 4 + 6$

#9. continued

$$18x = 12x + 10$$
$$18x - 12x = 12x - 12x + 10$$
$$6x = 10$$
$$\frac{1}{6} \cdot 6x = \frac{1}{6} \cdot 10$$
$$x = \frac{10}{6} \quad \text{or} \quad \frac{5}{3}$$

Check: $3\left(5 \cdot \frac{5}{3} - 2\right) + 3 \cdot \frac{5}{3} = 12 \cdot \frac{5}{3} + 4$

$$3\left(\frac{25}{3} - 2\right) + 5 = 20 + 4$$
$$3\left(\frac{25}{3} - \frac{6}{3}\right) + 5 = 24$$
$$3\left(\frac{19}{3}\right) + 5 = 24$$
$$19 + 5 = 24$$
$$24 = 24$$

11.

$$5(2x - 1) - 7(x - 2) = 9$$
$$10x - 5 - 7x + 14 = 9$$
$$3x + 9 = 9$$
$$3x + 9 - 9 = 9 - 9$$
$$3x = 0$$
$$x = 0$$

Check: $5(2 \cdot 0 - 1) - 7(0 - 2) = 9$

$$5(-1) - 7(-2) = 9$$
$$-5 + 14 = 9$$
$$9 = 9$$

The graphs of the solution sets are found in the answer section of the main text book.

13.

$$x + 5 < 9$$
$$x + 5 - 5 < 9 - 5$$
$$x < 4$$

15.

$$3x > -24$$
$$\frac{1}{3} \cdot 3x > \frac{1}{3}(-24)$$
$$x > -8$$

The graphs of the solution sets are found in the answer section of the main text book.

17.

$$2x + 1 > 10$$
$$2x + 1 - 1 > 10 - 1$$
$$2x > 9$$
$$\frac{1}{2} \cdot 2x > \frac{1}{2} \cdot 9$$
$$x > \frac{9}{2}$$

19.

$$3x + 1 < 6x - 2$$
$$3x + 1 + 2 < 6x - 2 + 2$$
$$3x + 3 < 6x$$
$$3x - 3x + 3 < 6x - 3x$$
$$3 < 3x$$
$$\frac{1}{3} \cdot 3 < \frac{1}{3} \cdot 3x$$
$$1 < x \quad \text{or} \quad x > 1$$

21.

$$-5 \leq x \leq 3$$

23.

$$-8 < 3x + 1 < 1$$
$$-8 - 1 < 3x + 1 - 1 < 1 - 1$$
$$-9 < 3x < 0$$
$$\frac{1}{3}(-9) < \frac{1}{3} \cdot 3x < \frac{1}{3} \cdot 0$$
$$-3 < x < 0$$

25.

$$1 - 6x > 4 \quad \text{and} \quad x - 6 > 4$$
$$1 - 1 - 6x > 4 - 1 \qquad x - 6 + 6 > 4 + 6$$
$$-6x > 3 \qquad\qquad x > 10$$
$$-\frac{1}{6}(-6x) < -\frac{1}{6} \cdot 3 \quad \text{Reverse sign.}$$
$$x < -\frac{1}{2}$$

$\varnothing$ since $x < -\frac{1}{2}$ and $x > 10$ is impossible

---

27.

$F = ma$ for a; $\qquad \frac{1}{m} \cdot F = \frac{1}{m} \cdot ma$

$$\frac{F}{m} = a$$

29.

$y = mx + b$ for x

$$y - b = mx + b - b$$

$$y - b = mx$$

$$\frac{1}{m}(y - b) = \frac{1}{m} \cdot mx$$

$$\frac{y - b}{m} = x$$

31.

$y - 4x = 2$ for y

$$y - 4x + 4x = 2 + 4x$$

$$y = 2 + 4x \text{ or } y = 4x + 2$$

33.

$2x + 3y - 4z = 5$ for z

$$2x - 2x + 3y - 3y - 4z = 5 - 2x - 3y$$

$$-4z = 5 - 2x - 3y$$

$$-\frac{1}{4}(-4z) = -\frac{1}{4}(5 - 2x - 3y)$$

$$z = \frac{-5 + 2x + 3y}{4}$$

$$z = \frac{2x + 3y - 5}{4}$$

35.

$$w = \frac{A}{\ell}$$

$$w = \frac{54}{9}$$

$$w = 6 \qquad \text{Answer: } 6 \text{ m}$$

37.

$$C = \frac{5}{9}(F - 32)$$

$$C = \frac{5}{9}(68 - 32)$$

$$C = \frac{5}{9}(36)$$

$$C = 20 \qquad \text{Answer: } 20°C$$

39.

Let x = the number.

Thus, $3x - 7 = 26$

$$3x - 7 + 7 = 26 + 7$$

$$3x = 33$$

#39. continued

$$\frac{1}{3} \cdot 3x = \frac{1}{3} \cdot 33$$

$$x = 11 \qquad \text{The number is } 11.$$

41.

Three consecutive integers are x, $x + 1$, $x + 2$.

Thus, $3(x) - 2(x + 1) + (x + 2) = 22$

$$3x - 2x - 2 + x + 2 = 22$$

$$2x = 22$$

$$\frac{1}{2} \cdot 2x = \frac{1}{2} \cdot 22$$

$$x = 11$$

Answer: 11, 12, 13

43.

Let x = number of quarters.

Let $2x$ = number of dimes.

Let $3x + 2$ = number of nickels.

Thus, $25x + 10(2x) + 5(3x + 2) = 430$

$$25x + 20x + 15x + 10 = 430$$

$$60x + 10 = 430$$

$$60x + 10 - 10 = 430 - 10$$

$$60x = 420$$

$$\frac{1}{60} \cdot 60x = \frac{1}{60} \cdot 420$$

$$x = 7$$

$$2x = 14$$

$$3x + 2 = 23$$

Answer: 7 quarters, 14 dimes, 23 nickels

45.

Let x = Felicia's time.

Then $x - \frac{1}{2}$ = Harry's time.

	rate	· time	= distance
Felicia	7	x	7x
Harry	10	$x - \frac{1}{2}$	$10\left(x - \frac{1}{2}\right)$

The distance is the same.

#45. continued

$$7x = 10\left(x - \frac{1}{2}\right)$$
$$7x = 10x - 5$$
$$7x + 5 = 10x - 5 + 5$$
$$7x + 5 = 10x$$
$$7x - 7x + 5 = 10x - 7x$$
$$5 = 3x$$
$$\frac{1}{3} \cdot 5 = \frac{1}{3} \cdot 3x$$
$$\frac{5}{3} = x$$
$$\frac{7}{6} = x - \frac{1}{2}$$

Answer: It took $\frac{7}{6}$ hr or 1 hr 10 min to over take Felicia, or 4:40 p.m.

47.
$$6|x| = 9$$
$$\frac{1}{6} \cdot 6|x| = \frac{1}{6} \cdot 9$$
$$|x| = \frac{3}{2}$$
$$x = \frac{3}{2} \quad or \quad x = -\frac{3}{2}$$
$$\left\{-\frac{3}{2}, \frac{3}{2}\right\}$$

49.
$$|x - 4| = 9$$
$$x - 4 = 9 \quad or \quad x - 4 = -9$$
$$x - 4 + 4 = 9 + 4 \quad x - 4 + 4 = -9 + 4$$
$$x = 13 \quad\quad x = -5$$
$$\{-5, 13\}$$

51.
$$|1 - 5x| = 16$$
$$1 - 5x = 16 \quad or \quad 1 - 5x = -16$$
$$1 - 1 - 5x = 16 - 1 \quad 1 - 1 - 5x = -16 - 1$$
$$-5x = 15 \quad\quad -5x = -17$$
$$-\frac{1}{5}(-5x) = -\frac{1}{5}(15) \quad -\frac{1}{5}(-5x) = -\frac{1}{5}(-17)$$

#53. continued

$$x = -3 \quad\quad x = \frac{17}{5}$$
$$\left\{-3, \frac{17}{5}\right\}$$

53.
$$-3|x - 2| = 12$$
$$-\frac{1}{3}(-3|x - 2|) = -\frac{1}{3} \cdot 12$$
$$|x - 2| = -4 \quad \text{Isolate absolute value.}$$
$$\varnothing \text{ since absolute value is not negative}$$

55.
$$13 - 2|x + 4| = 7$$
$$13 - 13 - 2|x + 4| = 7 - 13$$
$$-2|x + 4| = -6$$
$$-\frac{1}{2}(-2|x + 4|) = -\frac{1}{2}(-6)$$
$$|x + 4| = 3 \quad \text{Isolate absolute value.}$$
$$x + 4 = 3 \quad or \quad x + 4 = -3$$
$$x + 4 - 4 = 3 - 4 \quad x + 4 - 4 = -3 - 4$$
$$x = -1 \quad\quad x = -7$$
$$\{-7, -1\}$$

57.
$$|3x + 5| = |3x - 7|$$
$$3x + 5 = 3x - 7 \quad or \quad 3x + 5 = -(3x - 7)$$
$$3x - 3x + 5 = 3x - 3x - 7 \quad 3x + 5 = -3x + 7$$
$$5 \neq -7 \quad\quad 3x + 5 - 5 = -3x + 7 - 5$$
$$\varnothing \quad\quad 3x = -3x + 2$$
$$3x + 3x = -3x + 3x + 2$$
$$6x = 2$$
$$\frac{1}{6} \cdot 6x = \frac{1}{6} \cdot 2$$
$$x = \frac{1}{3}$$
$$\left\{\frac{1}{3}\right\}$$

59.
$$|3x - 4| \leq 1$$
$$-1 \leq 3x - 4 \leq 1 \quad \text{Theorem 6.2.1}$$
$$-1 + 4 \leq 3x - 4 + 4 \leq 1 + 4$$
$$3 \leq 3x \leq 5$$

#59. continued

$$\frac{1}{3} \cdot 3 \le \frac{1}{3} \cdot 3x \le \frac{1}{3} \cdot 5$$

$$1 \le x \le \frac{5}{3}$$

61.

$$5|x - 2| < 10$$

$$\frac{1}{5} \cdot 5|x - 2| < \frac{1}{5} \cdot 10$$

$$|x - 2| < 2$$

$$-2 < x - 2 < 2 \qquad \text{Theorem 6.2.1}$$

$$-2 + 2 < x - 2 + 2 < 2 + 2$$

$$0 < x < 4$$

63.

$$\left|2x - \frac{1}{2}\right| < -5$$

$\varnothing$ Absolute value is not negative.

65.

$$|5x + 3| > 12$$

$$5x + 3 > 12 \quad \text{or} \quad 5x + 3 < -12 \quad \text{Theorem 6.2.2}$$

$$5x + 3 - 3 > 12 - 3 \quad 5x + 3 - 3 < -12 - 3$$

$$5x > 9 \qquad\qquad 5x < -15$$

$$\frac{1}{5} \cdot 5x > \frac{1}{5} \cdot 9 \qquad \frac{1}{5} \cdot 5x < \frac{1}{5}(-15)$$

$$x > \frac{9}{5} \qquad\qquad x < -3$$

Thus, $x < -3$ or $x > \dfrac{9}{5}$

67.

$$\left|\frac{1}{2}x + 1\right| + 3 \ge 4$$

$$\left|\frac{1}{2}x + 1\right| + 3 - 3 \ge 4 - 3$$

$$\left|\frac{1}{2}x + 1\right| \ge 1 \qquad \text{Isolate absolute value.}$$

$$\frac{1}{2}x + 1 \ge 1 \quad \text{or} \quad \frac{1}{2}x + 1 \le -1$$

$$\frac{1}{2}x + 1 - 1 \ge 1 - 1 \quad \frac{1}{2}x + 1 - 1 \le -1 - 1$$

$$\frac{1}{2}x \ge 0 \qquad\qquad \frac{1}{2}x \le -2$$

#67. continued

$$2 \cdot \frac{1}{2}x \ge 2 \cdot 0 \qquad 2 \cdot \frac{1}{2}x \le 2(-2)$$

$$x \ge 0 \qquad\qquad x \le -4$$

Thus, $x \le -4$ or $x \ge 0$

69.

$$|10x - 3| \ge -6$$

$$10x - 3 \ge -6 \quad \text{or} \quad 10x - 3 \le 6$$

$$10x - 3 + 3 \ge -6 + 3 \quad 10x - 3 + 3 \ge 6 + 3$$

$$10x \ge -3 \qquad\qquad 10x \le 9$$

$$\frac{1}{10} \cdot 10x \ge \frac{1}{10}(-3) \quad \frac{1}{10} \cdot 10x \le \frac{1}{10} \cdot 9$$

$$x \ge -\frac{3}{10} \qquad\qquad x \le \frac{9}{10}$$

Thus, $x \ge -\dfrac{3}{10}$ or $x \le \dfrac{9}{10}$

All real numbers

Chapter 2 Test Solutions

1.

$$4x - 3 = 61$$

$$4x - 3 + 3 = 61 + 3$$

$$4x = 64$$

$$\frac{1}{4} \cdot 4x = \frac{1}{4} \cdot 64$$

$$x = 16$$

Check: $4 \cdot 16 - 3 = 61$

$$64 - 3 = 61$$

$$61 = 61$$

3.

$$4(2x + 6) - (7x - 5) = 2$$

$$8x + 24 - 7x + 5 = 2$$

$$x + 29 = 2$$

$$x + 29 - 29 = 2 - 29$$

$$x = -27$$

5.

$$9(2x - 3) - 4(5 + 3x) = 6(4 + x)$$

$$18x - 27 - 20 - 12x = 24 + 6x$$

$$6x - 47 = 24 + 6x$$

$$6x - 6x - 47 = 24 + 6x - 6x$$

#5. continued

$$-47 \neq 24$$

$$\varnothing$$

7.

$$|3x + 1| = |5x - 9|$$

$3x + 1 = 5x - 9$ or $3x + 1 = -(5x - 9)$

$3x + 1 + 9 = 5x - 9 + 9$ $3x + 1 = -5x + 9$

$3x + 10 = 5x$ $3x + 1 - 1 = -5x + 9 - 1$

$3x - 3x + 10 = 5x - 3x$ $3x = -5x + 8$

$10 = 2x$ $3x + 5x = -5x + 5x + 8$

$\dfrac{1}{2} \cdot 10 = \dfrac{1}{2} \cdot 2x$ $8x = 8$

$5 = x$ $\dfrac{1}{8} \cdot 8x = \dfrac{1}{8} \cdot 8$

$x = 1$

$\{1, 5\}$

9.

$$-19 < 2x - 5 < -9$$

$$-19 + 5 < 2x - 5 + 5 < -9 + 5$$

$$-14 < 2x < -4$$

$$\dfrac{1}{2}(-14) < \dfrac{1}{2} \cdot 2x < \dfrac{1}{2}(-4)$$

$$-7 < x < -2$$

The graph is in the answer section of the main text.

11.

$$|4x + 3| > 11$$

$4x + 3 > 11$ or $4x + 3 < -11$

$4x + 3 - 3 > 11 - 3$ $4x + 3 - 3 < -11 - 3$

$4x > 8$ $4x < -14$

$\dfrac{1}{4} \cdot 4x > \dfrac{1}{4} \cdot 8$ $\dfrac{1}{4} \cdot 4x < \dfrac{1}{4}(-14)$

$x > 2$ $x < -\dfrac{14}{4}$

$x < -\dfrac{7}{2}$

Thus, $x < -\dfrac{7}{2}$ or $x > 2$

The graph is in the answer section of the main text.

13.

$3x - 6 < 2$ and $2x + 1 > -1$

$3x - 6 + 6 < 2 + 6$ $2x + 1 - 1 > -1 - 1$

#13. continued

$3x < 8$ $2x > -2$

$\dfrac{1}{3} \cdot 3x < \dfrac{1}{3} \cdot 8$ $\dfrac{1}{2} \cdot 2x > \dfrac{1}{2}(-2)$

$x < \dfrac{8}{3}$ $x > -1$

Thus, $-1 < x < \dfrac{8}{3}$

The graph is in the answer section of the main text.

15.

$$A = \left(\dfrac{h}{2}\right)(a + b) \text{ for a}$$

$$2A = 2\left(\dfrac{h}{2}\right)(a + b)$$

$$2A = h(a + b)$$

$$2A = ah + bh$$

$$2A - bh = ah + bh - bh$$

$$2A - bh = ah$$

$$\dfrac{1}{h}(2A - bh) = \dfrac{1}{h} \cdot ah$$

$$\dfrac{2A - bh}{h} = a \quad \text{or} \quad a = \dfrac{2A}{h} - b$$

17.

Let x = number of tens.

Let 2x = number of fives.

Let 83 - 3x = number of twenties.

Then $10x + 5(2x) + 20(83 - 3x) = 660$

$10x + 10x + 1660 - 60x = 660$

$-40x = -1000$

$-\dfrac{1}{40}(-40x) = -\dfrac{1}{40} \cdot (-1000)$

$x = 25$

$2x = 50$

$83 - 3x = 8$

Answer: 25 tens, 50 fives, 8 twenties

Test Your Memory

1-9.
The exercises do not need worked-out solutions.

11.
$32 \div 4 \cdot 2 = 8 \cdot 2 = 16$

13.

$$\frac{\frac{-6}{25}}{4} = \frac{-6}{25} \div 4 = \frac{-6}{25} \cdot \frac{1}{4} = -\frac{-3 \cdot \cancel{2}}{25} \cdot \frac{1}{\cancel{2} \cdot 2} = -\frac{3}{50}$$

15.

$$\frac{1}{6} + \frac{3}{20} - 1 = \frac{10}{60} + \frac{9}{60} - \frac{60}{60}$$
$$= \frac{10 + 9 - 60}{60} = -\frac{41}{60}$$

17.

$$\sqrt{5 \cdot 8 + 7^2 - 8} = \sqrt{40 + 49 - 8} = \sqrt{81} = 9$$

19.

$$-\frac{3}{8} + \left(-\frac{1}{6}\right) \div \frac{2}{3} + 7\left(-\frac{5}{8}\right) = -\frac{3}{8} - \frac{1}{4} - \frac{35}{8}$$
$$= -\frac{3}{8} - \frac{2}{8} - \frac{35}{8}$$
$$= -\frac{40}{8} = -5$$

21.

$$0.48 = \frac{48}{100} = \frac{\cancel{4} \cdot 12}{\cancel{4} \cdot 25} = \frac{12}{25}$$

23.

$$\sqrt{100 - x^2} \quad \text{for } x = 8$$
$$\sqrt{100 - 8^2} = \sqrt{100 - 64} = \sqrt{36} = 6$$

25.

$$3x^2 - 5xy - 2y^2 \quad \text{for } x = -2, y = -3$$
$$3(-2)^2 - 5(-2)(-3) - 2(-3)^2 = 3 \cdot 4 - 30 - 2 \cdot 9$$
$$= 12 - 30 - 18$$
$$= -36$$

27.

$$7x - 4 = 4x - 2$$
$$7x - 4 + 4 = 4x - 2 + 4$$
$$7x = 4x + 2$$
$$7x - 4x = 4x - 4x + 2$$
$$3x = 2$$
$$\frac{1}{3} \cdot 3x = \frac{1}{3} \cdot 2$$

#27. continued

$$x = \frac{2}{3}$$
$$\left\{\frac{2}{3}\right\}$$

Check: $7 \cdot \dfrac{2}{3} - 4 = 4 \cdot \dfrac{2}{3} - 2$

$$\frac{14}{3} - 4 = \frac{8}{3} - 2$$
$$\frac{14}{3} - \frac{12}{3} = \frac{8}{3} - \frac{6}{3}$$
$$\frac{2}{3} = \frac{2}{3}$$

29.

$$4(2x - 3) - (x + 4) = 9x - 19$$
$$8x - 12 - x - 4 = 9x - 19$$
$$7x - 16 = 9x - 19$$
$$7x - 16 + 16 = 9x - 19 + 16$$
$$7x = 9x - 3$$
$$7x - 9x = 9x - 9x - 3$$
$$-2x = -3$$
$$-\frac{1}{2}(-2x) = -\frac{1}{2}(-3)$$
$$x = \frac{3}{2}$$
$$\left\{\frac{3}{2}\right\}$$

Check: $4\left(2 \cdot \dfrac{3}{2} - 3\right) - \left(\dfrac{3}{2} + 4\right) = 9\left(\dfrac{3}{2}\right) - 19$

$$4(3 - 3) - \left(\frac{3}{2} + 4\right) = \frac{27}{2} - 19$$
$$0 - \left(\frac{11}{2}\right) = \frac{27}{2} - \frac{38}{2}$$
$$-\frac{11}{2} = -\frac{11}{2}$$

31.

$$\frac{1}{4}(x + 1) - \frac{1}{3}(2x - 2) = 1$$
$$12\left[\frac{1}{4}(x + 1) - \frac{1}{3}(2x - 2)\right] = 12 \cdot 1$$
$$3(x + 1) - 4(2x - 2) = 12$$
$$3x + 3 - 8x + 8 = 12$$
$$-5x + 11 = 12$$
$$-5x + 11 - 11 = 12 - 11$$

#31. continued

$$-5x = 1$$

$$-\frac{1}{5}(-5x) = -\frac{1}{5} \cdot 1$$

$$x = -\frac{1}{5}$$

$$\left\{-\frac{1}{5}\right\}$$

Check: $\frac{1}{4}\left(-\frac{1}{5} + 1\right) - \frac{1}{3}\left(2\left(-\frac{1}{5}\right) - 2\right) = 1$

$$\frac{1}{4}\left(\frac{4}{5}\right) - \frac{1}{3}\left(-\frac{2}{5} - 2\right) = 1$$

$$\frac{1}{5} - \frac{1}{3}\left(-\frac{12}{5}\right) = 1$$

$$\frac{1}{5} + \frac{4}{5} = 1$$

$$1 = 1$$

33.

$$3(x - 5) - 4(x - 2) = 2(x + 7)$$

$$3x - 15 - 4x + 8 = 2x + 14$$

$$-x - 7 = 2x + 14$$

$$-x - 7 - 14 = 2x + 14 - 14$$

$$-x - 21 = 2x$$

$$-x + x - 21 = 2x + x$$

$$-21 = 3x$$

$$\frac{1}{3}(-21) = \frac{1}{3} \cdot 3x$$

$$-7 = x$$

$$\{-7\}$$

Check: $3(-7 - 5) - 4(-7 - 2) = 2(-7 + 7)$

$$3(-12) - 4(-9) = 2 \cdot 0$$

$$-36 + 36 = 0$$

$$0 = 0$$

35.

$$|2x + 3| = |4x - 5|$$

$$2x + 3 = 4x - 5 \quad \text{or} \quad 2x + 3 = -(4x - 5)$$

$$2x + 3 + 5 = 4x - 5 + 5 \qquad 2x + 3 = -4x + 5$$

$$2x + 8 = 4x \qquad 2x + 3 - 3 = -4x + 5 - 3$$

$$2x - 2x + 8 = 4x - 2x \qquad 2x = -4x + 2$$

$$8 = 2x \qquad 2x + 4x = -4x + 4x + 2$$

$$\frac{1}{2} \cdot 8 = \frac{1}{2} \cdot 2x \qquad 6x = 2$$

#35. continued

$$4 = x \qquad\qquad \frac{1}{6} \cdot 6x = \frac{1}{6} \cdot 2$$

$$x = \frac{1}{3}$$

$$\left\{\frac{1}{3}, 4\right\}$$

The check is left to the student.

37.

$$4(2x + 1) > 3(x - 2)$$

$$8x + 4 > 3x - 6$$

$$8x + 4 - 4 > 3x - 6 - 4$$

$$8x > 3x - 10$$

$$8x - 3x > 3x - 3x - 10$$

$$5x > -10$$

$$\frac{1}{5}x > \frac{1}{5}(-10)$$

$$x > -2$$

The graph is in the answer section of the main text.

39.

$$|2x - 1| < 7$$

$$-7 < 2x - 1 < 7$$

$$-7 + 1 < 2x - 1 + 1 < 7 + 1$$

$$-6 < 2x < 8$$

$$\frac{1}{2}(-6) < \frac{1}{2} \cdot 2x < \frac{1}{2} \cdot 8$$

$$-3 < x < 4$$

The graph is in the answer section of the main text.

41.

$$2x - 7 < 5 \text{ and } 3x - 4 \geq -7$$

$$2x - 7 < 5 \qquad\qquad 3x - 4 + 4 \geq -7 + 4$$

$$2x - 7 + 7 < 5 + 7 \qquad\qquad 3x \geq -3$$

$$2x < 12 \qquad\qquad \frac{1}{3} \cdot 3x \geq \frac{1}{3}(-3)$$

$$x < 6 \qquad\qquad x \geq -1$$

Thus $-1 \leq x < 6$

The graph is in the answer section of the main text.

43.

$$A = \frac{1}{2}bh \quad \text{for } b$$

$$2 \cdot A = 2 \cdot \frac{1}{2}bh$$

#43. continued

$$2A = bh$$

$$\frac{1}{h} \cdot 2A = \frac{1}{h} \cdot bh$$

$$\frac{2A}{h} = b$$

45.
Let x = the number.

Thus, $\quad 4x - 1 = 15$

$$4x - 1 + 1 = 15 + 1$$

$$4x = 16$$

$$\frac{1}{4} \cdot 4x = \frac{1}{4} \cdot 16$$

$$x = 4$$

The number is 4.

47.
Let x = selling price.

Then $\quad x - 0.04x = 50,000$

$$0.96x = 50,000$$

$$\frac{1}{0.96} \cdot 0.96x = \frac{1}{0.96} \cdot 50,000$$

$$x = 52,083.33$$

Answer: $52,083.33

49.
Let x = number of fives.
Then 17-x = number of twenties.

Thus, $\quad 5x + 20(17 - x) = 160$

$$5x + 340 - 20x = 160$$

$$340 - 15x = 160$$

$$340 - 340 - 15x = 160 - 340$$

$$-15x = -180$$

$$-\frac{1}{15}(-15x) = -\frac{1}{15}(-180)$$

$$x = 12$$

Answer: 12 fives, 5 twenties

Chapter 2 Study Guide

Self-Test Exercises

I. Find the solutions of the following equations and check your answers.

1. $5x-4=11$

2. $9x+3=2x-4$ 3. $9(2x+8)=20-(x+5)$ 4. $\frac{1}{7}x+\frac{2}{7}=\frac{1}{2}x-\frac{11}{14}$

5. $4(y+3)+3(2y-1)=5(y+8)$ 6. $|x-9|=-5$ 7. $12=|2x-1|+2$ 8. $|5x-4|=|6x+1|$

II. Find and graph the solutions of the following inequalities.

9. $14(y-2)\le 4y+5$ 10. $-7\le 3x-2\le 2$ 11. $|m+2|<5$

12. $|2x+3|\le 13$ 13. $x<3$ or $x\le 7$ 14. $4x-6<2$ and $5-3x>-10$

III. Solve the following equations for the indicated variables.

15. $F=ma+kx$ for x 16. $A=\dfrac{B-C}{n}$ for B

IV. Solve the following word problems.

17. The perimeter of a rectangle is 60 m. The length is 3 m less than twice the width. Find the dimensions of the rectangle.

18. Barbara and Jim leave Seattle at the same time traveling in opposite directions. If Barbara's average speed was 42 mph and Jim's was 54 mph, how long will it take to be 312 miles apart?

19. The perimeter of a triangle is 37 feet. If the second side of the triangle is one more foot than two times the length of the first side, and the third side is three times the first side, find the length of the three sides.

20. Charles has a total of $955. He has twice as many tens as fives, and one more than three times as many twenties as fives. How many of each type does he have?

The worked-out solutions begin on the next page.

Self-Test Solutions

1.

$$5x - 4 = 11$$
$$5x - 4 + 4 = 11 + 4$$
$$5x = 15$$
$$\frac{1}{5} \cdot 5x = \frac{1}{5} \cdot 15$$
$$x = 3$$

Check: $5 \cdot 3 - 4 = 11$
$$15 - 4 = 11$$
$$11 = 11$$

2.

$$9x + 3 = 2x - 4$$
$$9x + 3 - 3 = 2x - 4 - 3$$
$$9x = 2x - 7$$
$$9x - 2x = 2x - 2x - 7$$
$$7x = -7$$
$$\frac{1}{7} \cdot 7x = \frac{1}{7}(-7)$$
$$x = -1$$

Check: $9(-1) + 3 = 2(-1) - 4$
$$-9 + 3 = -2 - 4$$
$$-6 = -6$$

3.

$$9(2x + 8) = 20 - (x + 5)$$
$$18x + 72 = 20 - x - 5$$
$$18x + 72 = 15 - x$$
$$18x + 72 - 72 = 15 - 72 - x$$
$$18x = -57 - x$$
$$18x + x = -57 - x + x$$
$$19x = -57$$
$$\frac{1}{19} \cdot 19x = \frac{1}{19}(-57)$$
$$x = -3$$

Check: $9(2(-3) + 8) = 20 - (-3 + 5)$
$$9(-6 + 8) = 20 - (2)$$
$$9(2) = 20 - 2$$
$$18 = 18$$

4.

$$\frac{1}{7}x + \frac{2}{7} = \frac{1}{2}x - \frac{11}{14} \qquad LCD = 28$$
$$28\left(\frac{1}{7}x + \frac{2}{7}\right) = 28\left(\frac{1}{2}x - \frac{11}{14}\right)$$
$$4x + 8 = 14x - 22$$
$$4x + 8 + 22 = 14x - 22 + 22$$
$$4x + 30 = 14x$$
$$4x - 4x + 30 = 14x - 4x$$
$$30 = 10x$$
$$\frac{1}{10} \cdot 30 = \frac{1}{10} \cdot 10x$$
$$3 = x$$

Check: $\frac{1}{7} \cdot 3 + \frac{2}{7} = \frac{1}{2} \cdot 3 - \frac{11}{14}$
$$\frac{3}{7} + \frac{2}{7} = \frac{3}{2} - \frac{11}{14}$$
$$\frac{5}{7} = \frac{21}{14} - \frac{11}{14}$$
$$\frac{5}{7} = \frac{10}{14}$$
$$\frac{5}{7} = \frac{5}{7}$$

5.

$$4(y + 3) + 2(3y - 1) = 5(y + 8)$$
$$4y + 12 + 6y - 2 = 5y + 40$$
$$10y + 10 = 5y + 40$$
$$10y + 10 - 10 = 5y + 40 - 10$$
$$10y = 5y + 30$$
$$10y - 5y = 5y - 5y + 30$$
$$5y = 30$$
$$\frac{1}{5} \cdot 5y = \frac{1}{5} \cdot 30$$
$$y = 6$$

6.

$$|x - 9| = -5$$

∅ Absolute value is not negative.

7.

$$12 = |2x - 1| + 2$$
$$12 - 2 = |2x - 1| + 2 - 2$$
$$10 = |2x - 1| \qquad \text{Isolate absolute value.}$$

#7. continued

$$10 = 2x - 1 \quad \text{or} \quad -10 = 2x - 1$$
$$10 + 1 = 2x - 1 + 1 \qquad -10 + 1 = 2x - 1 + 1$$
$$11 = 2x \qquad\qquad -9 = 2x$$
$$\frac{1}{2} \cdot 11 = \frac{1}{2} \cdot 2x \qquad \frac{1}{2}(-9) = \frac{1}{2} \cdot 2x$$
$$\frac{11}{2} = x \qquad\qquad -\frac{9}{2} = x$$
$$\left\{ \frac{11}{2}, -\frac{9}{2} \right\}$$

8.
$$|5x - 4| = |6x + 1|$$
$$5x - 4 = 6x + 1 \quad \text{or} \quad 5x - 4 = -(6x + 1)$$
$$5x - 4 - 1 = 6x + 1 - 1 \qquad 5x - 4 = -6x - 1$$
$$5x - 5 = 6x \qquad\qquad 5x - 4 + 1 = -6x - 1 + 1$$
$$5x - 5x - x = 6x - 5x \qquad 5x - 4 + 1 = -6x$$
$$-5 = x \qquad\qquad 5x - 5x - 3 = -6x - 5x$$
$$-3 = 11x$$
$$-11(-3) = -\frac{1}{11}(-11x)$$
$$\frac{3}{11} = x$$
$$\left\{ -5, \frac{3}{11} \right\}$$

9.
$$14(y - 2) \le 4y + 5$$
$$14y - 28 \le 4y + 5$$
$$14y - 28 + 28 \le 4y + 5 + 28$$
$$14y \le 4y + 33$$
$$14y - 4y \le 4y - 4y + 33$$
$$10y \le 33$$
$$\frac{1}{10} \cdot 10y \le \frac{1}{10} \cdot 33$$
$$y \le 3.3$$

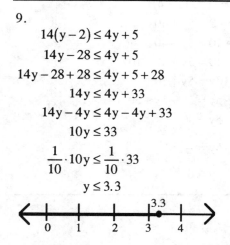

10.
$$-7 \le 3x - 2 \le 2$$
$$-7 + 2 \le 3x - 2 + 2 \le 2 + 2$$
$$-5 \le 3x \le 4$$
$$\frac{1}{3}(-5) \le \frac{1}{3} \cdot 3x \le \frac{1}{3} \cdot 4$$

#10. continued
$$-\frac{5}{3} \le x \le \frac{4}{3}$$

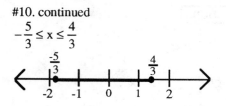

11.
$$|m + 2| < 4$$
$$-4 < m + 2 < 4$$
$$-4 - 2 < m + 2 - 2 < 4 - 2$$
$$-6 < m < 2$$

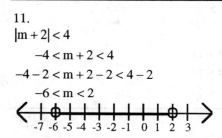

12.
$$|2x + 3| \le 13$$
$$2x + 3 \le 13 \quad \text{or} \quad 2x + 3 \ge -13$$
$$2x + 3 - 3 \le 13 - 3 \qquad 2x + 3 - 3 \ge -13 - 3$$
$$2x \le 10 \qquad\qquad 2x \ge -16$$
$$\frac{1}{2} \cdot 2x \le \frac{1}{2} \cdot 10 \qquad \frac{1}{2} \cdot 2x \ge \frac{1}{2}(-16)$$
$$x \le 5 \qquad\qquad x \ge -8$$
Thus, $-8 \le x \le 5$

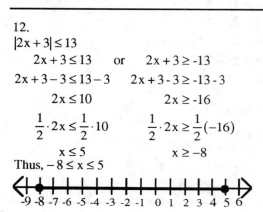

13.
Answer: $x \le 7$

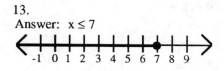

14.
$$4x - 6 < 2 \quad \text{and} \quad 5 - 3x > -10$$
$$4x - 6 + 6 < 2 + 6 \qquad 5 - 5 - 3x > -10 - 5$$
$$4x < 8 \qquad\qquad -3x > -15$$
$$\frac{1}{4} \cdot 4x < \frac{1}{4} \cdot 8 \qquad -\frac{1}{3}(-3x) < -\frac{1}{3}(-15)$$
$$x < 2 \qquad\qquad x < 5$$
Answer: $x < 2$

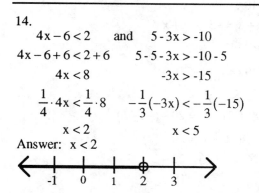

15.

$$F = ma + kx \quad \text{for x}$$
$$F - ma = ma - ma + kx$$
$$F - ma = kx$$
$$\frac{1}{k}(F - ma) = \frac{1}{k} \cdot x$$
$$\frac{F - ma}{k} = x$$

16.

$$A = \frac{B-C}{n} \quad \text{for B;} \qquad n \cdot A = n \cdot \frac{B-C}{n}$$
$$nA = B - C$$
$$nA + C = B - C + C$$
$$nA + C = B$$

17.

Let x = measurement of the width.

Let 2x - 3 = measurement of the length.

Thus, $P = 2w + 2\ell$

$$60 = 2x + 2(2x - 3)$$
$$60 = 2x + 4x - 6$$
$$60 = 6x - 6$$
$$60 + 6 = 6x - 6 + 6$$
$$66 = 6x$$
$$\frac{1}{6} \cdot 66 = \frac{1}{6} \cdot 6x$$
$$11 = x$$
$$19 = 2x - 3$$

Answer: 11 m by 19 m

18.

Let x = length of time it will take to be 312 miles apart.

	rate ·	time =	distance
Barbara	42	x	42x
Jim	54	x	54x

$$\left(\begin{matrix} \text{Barbara's} \\ \text{distance} \end{matrix} \right) + \left(\begin{matrix} \text{Jim's} \\ \text{distance} \end{matrix} \right) = 312$$

$$42x + 54x = 312$$
$$96x = 312$$
$$\frac{1}{96} \cdot 96x = \frac{1}{96} \cdot 312$$

#18. continued

$$x = 3\frac{1}{4}$$

Answer: $3\frac{1}{4}$ hr or 3 hr 15 min

19.

Let x = measurement of the first side.

Let 2x + 1 = measurement of the second side.

Let 3x = measurement of the third side.

Thus, $x + 2x + 1 + 3x = 37$

$$6x + 1 = 37$$
$$6x + 1 - 1 = 37 - 1$$
$$6x = 36$$
$$\frac{1}{6} \cdot 6x = \frac{1}{6} \cdot 36$$
$$x = 6$$
$$2x + 1 = 13$$
$$3x = 18$$

Answer: 6 ft, 13 ft, 18 ft

20.

Let x = the number of $5 bills.

Let 2x = the number of $10 bills.

Let 3x + 1 = the number of $20 bills.

Thus, $5x + 10(2x) + 20(3x + 1) = 955$

$$5x + 20x + 60x + 20 = 955$$
$$85x + 20 = 955$$
$$85x + 20 - 20 = 955 - 20$$
$$85x = 935$$
$$\frac{1}{85} \cdot 85x = \frac{1}{85} \cdot 935$$
$$x = 11$$
$$2x = 22$$
$$3x + 1 = 34$$

Answer: 11 fives, 22 tens, 34 twenties.

CHAPTER 3 POLYNOMIALS AND FACTORABLE QUADRATIC EQUATIONS

Solutions to Text Exercises

Exercises 3.1

1.-27.
The exercises do not require worked-out solutions.

29.
$$-2(9a - 2b - 10) = -2(9a) - 2(-2b) - 2(-10)$$
$$= -18a + 4b + 20$$

31.
$$-0.4(7.1x - 9.2y - z)$$
$$= -0.4(7.1x) - 0.4(-9.2y) - 0.4(-z)$$
$$= -2.84x + 3.68y + 0.4z$$

33.
$$\frac{1}{5}(25x^3 + 47y^5 - z^2)$$
$$= \frac{1}{5}(25x^3) + \frac{1}{5}(47y^5) + \frac{1}{5}(-z^2)$$
$$= 5x^3 + \frac{47}{5}y^5 - \frac{1}{5}z^2$$

35.
$$-(19q + 15r - 101s - t^2)$$
$$= -1 \cdot (19q + 15r - 101s - t^2)$$
$$= -1 \cdot (19q) + (-1)(15r) - 1 \cdot (-101s) - 1 \cdot (-t^2)$$
$$= -19q - 15r + 101s + t^2$$

37.
0, since $0 \cdot a = 0$

39.
$$-2j + 11k - 14j = -2j - 14j + 11k = -16j + 11k$$

41.
$$-12k + 13m + 42n$$

43.
$$\frac{1}{3}y + \frac{1}{4}z + \frac{2}{9}y - 3z = \frac{1}{3}y + \frac{2}{9}y + \frac{1}{4}z - 3z$$

#43. continued
$$= \frac{3}{9}y + \frac{2}{9}y + \frac{1}{4}z - \frac{12}{4}z$$
$$= \frac{5}{9}y - \frac{11}{4}z$$

45.
$$x^3 - 14x + 3 + 2x^2 - 15 = x^3 + 2x^2 - 14x + 3 - 15$$
$$= x^3 + 2x^2 - 14x - 12$$

47.
$$\frac{1}{5}x^2y - \frac{2}{3}yx^2 + \frac{1}{4}x = \frac{3}{15}x^2y - \frac{10}{15}x^2y + \frac{1}{4}x$$
$$= -\frac{7}{15}x^2y + \frac{1}{4}x$$

49.
$$(5x^3 - x + 2) + (2x^3 + 4x^2 - 2x + 9)$$
$$= 5x^3 + 2x^3 + 4x^2 - x - 2x + 2 + 9$$
$$= 7x^3 + 4x^2 - 3x + 11$$

51.
$$(5x^2 - 2xy - 3y + y^2) + (2x - 5xy + y - 7y^2)$$
$$= 5x^2 - 2xy - 5xy - 3y + y + 2x + y^2 - 7y^2$$
$$= 5x^2 - 7xy - 2y + 2x - 6y^2$$

53.
$$(x^3 + x^2 - 2) - (4x^3 - 2x^2 + 3x - 1)$$
$$= x^3 + x^2 - 2 - 4x^3 + 2x^2 - 3x + 1$$
$$= x^3 - 4x^3 + x^2 + 2x^2 - 3x - 2 + 1$$
$$= -3x^3 + 3x^2 - 3x - 1$$

55.
$$(-4a^2 - 4a + b) - (2a + 3b - b^2 + 9a^2)$$
$$= -4a^2 - 4a + b - 2a - 3b + b^2 - 9a^2$$
$$= -4a^2 - 9a^2 + b^2 - 4a - 2a + b - 3b$$
$$= -13a^2 - 6a + b^2 - 2b$$

57.

$(x^2 - 5x + 6) - (3x^2 + x - 8) + (4x^2 - 13)$

$= x^2 - 5x + 6 - 3x^2 - x + 8 + 4x^2 - 13$

$= x^2 - 3x^2 + 4x^2 - 5x - x + 6 + 8 - 13$

$= 2x^2 - 6x + 1$

59.

$$\begin{array}{r} 5x^2 + 8x + 19 \\ + (-7x^2 - 9x + 1) \\ \hline -2x^2 - x + 20 \end{array}$$

61.

$$\begin{array}{r} 8a - 3b + 4 \\ + (-9a + 5b - 4) \\ \hline -a + 2b \end{array}$$

63.

$$\begin{array}{r} -3x^2 + 2x + 11 \\ x^2 - 2x - 9 \\ \hline -2x^2 + 2 \end{array}$$

65.

$$\begin{array}{r} 7ab + 8cd - 6 \\ -7ab + 8cd - 2 \\ \hline 16cd - 8 \end{array}$$

67.

$(7x + 3y - z) - (2x - 9y + 14z)$

$= 7x + 3y - z - 2x + 9y - 14z$

$= 7x - 2x + 3y + 9y - z - 14z$

$= 5x + 12y - 15z$

69.

$-4(5x^2 + 3x - 1) + 2(5x^2 + 3x - 1)$

$= -20x^2 - 12x + 4 + 10x^2 + 6x - 2$

$= -20x^2 + 10x^2 - 12x + 6x + 4 - 2$

$= -10x^2 - 6x + 2$

71.

$3(x^2 y - 4x + 3) - 2(y^2 x + 3z - 7)$

$= 3x^2 y - 12x + 9 - 2xy^2 - 6z + 14$

#71. continued

$= 3x^2 y - 2xy^2 - 12x - 6z + 23$

73.

$0.02(7x^2 - 0.01y^2) - 0.5(10x^2 + 2.1y^2)$

$= 0.14x^2 - 0.0002y^2 - 5x^2 - 1.05y^2$

$= 4.86x^2 - 1.0502y^2$

75.

$2(3m + 7n) - (9j - 10k) = 6m + 14n - 9j + 10k$

77.

$\dfrac{5}{6}\left(2a + \dfrac{3}{10}b\right) - \dfrac{1}{9}\left(5a - \dfrac{3}{2}b\right)$

$= \dfrac{5}{3}a + \dfrac{1}{4}b - \dfrac{5}{9}a + \dfrac{1}{6}b$

$= \dfrac{5}{3}a - \dfrac{5}{9}a + \dfrac{1}{4}b + \dfrac{1}{6}b$

$= \dfrac{10}{9}a + \dfrac{10}{24}b = \dfrac{10}{9}a + \dfrac{5}{12}b$

79.

a. Area of whole rectangle before cutting $= zy$

Area of one of the cut outs $= x^2$

There are four of them.

Thus, area $= zy - 4x^2$

b. $V = \ell \cdot w \cdot h$

$\ell = y - 2x$

$w = z - 2x$

$h = x$

Thus, $V = x(y - 2x)(z - 2x)$

81.

$\begin{aligned} \text{Cost} &= 5x + 7y + 120 \\ &= 5(20) + 7(50) + 120 \\ &= 100 + 350 + 120 \\ &= 570\text{¢} \quad \text{or} \quad \$5.70 \end{aligned}$

$\begin{aligned} \text{Income} &= 3x + 2y \\ &= 3(20) + 2(50) \\ &= 60 + 100 \\ &= 160\text{¢} \quad \text{or} \quad \$1.60 \end{aligned}$

$\begin{aligned} \text{Profit / Loss} &= 3x + 2y - (5x + 7y + 120) \\ &= 3x + 2y - 5x - 7y - 120 \\ &= -2x - 5y - 120 \end{aligned}$

Exercises 3.2

1.
$$x^9 \cdot x^4 = x^{9+4} = x^{13}$$

3.
$$y^{17} \cdot y = y^{17+1} = y^{18}$$

5.
$$\left(x^2\right)^5 = x^{2 \cdot 5} = x^{10}$$

7.
$$\left(y^8\right)^7 = y^{8 \cdot 7} = y^{56}$$

9.
$$(xy)^9 = x^9 y^9$$

11.
$$\left(x^2 y^9\right)^4 = x^{2 \cdot 4} y^{9 \cdot 4} = x^8 y^{36}$$

13.
$$5x^3 y \cdot 8x^4 y^9 = 5 \cdot 8 \cdot x^3 \cdot x^4 \cdot y \cdot y^9 = 40x^7 x^{10}$$

15.
$$-3x^5 y^2 \cdot 14xy^2 = -3 \cdot 14 \cdot x^5 \cdot x \cdot y^2 \cdot y^2 = -42x^6 y^4$$

17.
$$2x^2 y^5 \cdot 5xy^{17} \cdot 9x^6 y^8 = 2 \cdot 5 \cdot 9 \cdot x^2 \cdot x \cdot x^6 \cdot y^5 \cdot y^{17} \cdot y^8$$
$$= 90x^9 y^{30}$$

19.
$$\left(3xy^3\right)^2 = 3^{1 \cdot 2} x^{1 \cdot 2} y^{3 \cdot 2} = 3^2 x^2 y^6 = 9x^2 y^6$$

21.
$$\left(-2x^3 y^{12}\right)^4 = (-2)^4 x^{3 \cdot 4} y^{12 \cdot 4} = 16x^{12} y^{48}$$

23.
$$\left(2xy^2\right)^3 \left(3x^2 y^3\right)^3 = 2^3 x^3 y^6 \cdot 3^3 x^6 y^9$$
$$= 8x^3 y^6 \cdot 27x^6 y^9$$
$$= 8 \cdot 27 \cdot x^3 \cdot x^6 \cdot y^6 \cdot y^9$$
$$= 216x^9 y^{15}$$

25.
$$\left(5x^2 y^3\right)^2 \left(-2x^4 y^2\right)^3 = 5^2 x^4 y^6 \cdot (-2)^3 x^{12} y^6$$
$$= 25x^4 y^6 \cdot (-8)x^{12} y^6$$
$$= 25(-8) \cdot x^4 \cdot x^{12} \cdot y^6 \cdot y^6$$
$$= -200x^{16} y^{12}$$

27.
$$6a^2 b(2a - 10) = 6a^2 b \cdot 2a + 6a^2 b(-10)$$
$$= 12a^3 b - 60a^2 b$$

29.
$$10x^2 \left(5x^2 - 3x + 4\right)$$
$$= 10x^2 \cdot 5x^2 + 10x^2(-3x) + 10x^2 \cdot 4$$
$$= 50x^4 - 30x^3 + 40x^2$$

31.
$$-7m^3 n^7 \left(8m - 9n^2 + 15mn^3\right)$$
$$= -7m^3 n^7 \cdot 8m - 7m^3 n^7 \left(-9n^2\right) - 7m^3 n^7 \cdot 15mn^3$$
$$= -56m^4 n^7 + 63m^3 n^9 - 105m^4 n^{10}$$

33.

F O   F   O   I   L
$$(8x + 3)(9x + 4) = 8x \cdot 9x + 8x \cdot 4 + 3 \cdot 9x + 3 \cdot 4$$
$$= 72x^2 + 32x + 27x + 12$$
$$= 72x^2 + 59x + 12$$

35.

F O   F   O   I   L
$$(5x - 11)(3x - 1) = 5x \cdot 3x + 5x(-1) - 11 \cdot 3x - 11(-1)$$
$$= 15x^2 - 5x - 33x + 11$$
$$= 15x^2 - 38x + 11$$

37.

F   O
$$(5x + 2y)(x - 3y)$$

F   O   I   L
$$= 5x \cdot x + 5x(-3y) + 2y \cdot x + 2y(-3y)$$

#37. continued
$$= 5x^2 - 15xy + 2xy - 6y^2$$
$$= 5x^2 - 13xy - 6y^2$$

39.

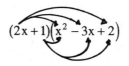

$$(6m - 3n)(5m + 6n)$$

$$\begin{matrix} F & O & I & L \end{matrix}$$
$$= 6m \cdot 5m + 6m \cdot 6n - 3n \cdot 5m - 3n \cdot 6n$$
$$= 30m^2 + 36mn - 15mn - 18n^2$$
$$= 30m^2 + 21mn - 18n^2$$

41.
$$(4a + 1)(4a - 1) = (4a)^2 - (1)^2 = 16a^2 - 1$$

43.
$$(2x + 9)(2x - 9) = (2x)^2 - (9)^2 = 4x^2 - 81$$

45.
$$\left(x^2 + y^2\right)\left(x^2 - y^2\right) = \left(x^2\right)^2 - \left(y^2\right)^2 = x^4 - y^4$$

47.
$$\left(\frac{x}{2} - 2y\right)\left(\frac{x}{2} + 2y\right) = \left(\frac{x}{2}\right)^2 - (2y)^2 = \frac{x^2}{4} - 4y^2$$

49.
$$(x + 4)^2 = x^2 + 2(4x) + 4^2 = x^2 + 8x + 16$$

51.
$$(x - 7)^2 = x^2 + 2(-7x) + (-7)^2 = x^2 - 14x + 49$$

53.
$$(3x + 4y)^2 = (3x)^2 + 2(3x \cdot 4y) + (4y)^2$$
$$= 9x^2 + 24xy + 16y^2$$

55.
$$\left(\frac{1}{2}a - 3b\right)^2 = \left(\frac{1}{2}a\right)^2 + 2\left(\frac{1}{2}a(-3b)\right) + (-3b)^2$$
$$= \frac{1}{4}a^2 - 3ab + 9b^2$$

57.
$$\left(x^2 + 1\right)^2 = \left(x^2\right)^2 + 2\left(x^2 \cdot 1\right) + (1)^2 = x^4 + 2x^2 + 1$$

59.

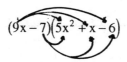

$$(x - 2)\left(x^2 + 2x + 4\right)$$

$$= x \cdot x^2 + x \cdot 2x + x \cdot 4 - 2 \cdot x^2 - 2 \cdot 2x - 2 \cdot 4$$
$$= x^3 + 2x^2 + 4x - 2x^2 - 4x - 8$$
$$= x^3 - 8$$

61.

$$(2x + 1)\left(x^2 - 3x + 2\right)$$

$$2x \cdot x^2 + 2x(-3x) + 2x \cdot 2 + 1 \cdot x^2 + 1(-3x) + 1 \cdot 2$$
$$= 2x^3 - 6x^2 + 4x + x^2 - 3x + 2$$
$$= 2x^3 - 5x^2 + x + 2$$

63.

$$(9x - 7)\left(5x^2 + x - 6\right)$$

$$= 9x \cdot 5x^2 + 9x \cdot x + 9x(-6) - 7 \cdot 5x^2 - 7 \cdot x - 7(-6)$$
$$= 45x^3 + 9x^2 - 54x - 35x^2 - 7x + 42$$
$$= 45x^3 - 26x^2 - 61x + 42$$

65.

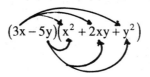

$$(3x - 5y)\left(x^2 + 2xy + y^2\right)$$

$$= 3x \cdot x^2 + 3x \cdot 2xy + 3x \cdot y^2 - 5y \cdot x^2$$
$$\quad - 5y \cdot 2xy - 5y \cdot y^2$$

#65. continued

$$= 3x^3 + 6x^2y + 3xy^2 - 5x^2y - 10xy^2 - 5y^3$$

$$= 3x^3 + x^2y - 7xy^2 - 5y^3$$

67.

$$
\begin{array}{r}
x^2 + 2x - 1 \\
x \quad \underline{5x^2 - 4x + 3} \\
5x^4 + 10x^3 - 5x^2 \\
-4x^3 - 8x^2 + 4x \\
\underline{3x^2 + 6x - 3} \\
5x^4 + 6x^3 - 10x^2 + 10x - 3
\end{array}
$$

69.

$$
\begin{array}{r}
2x^2 - 4x - 9 \\
\underline{20x^2 + 7} \\
40x^4 - 80x^3 - 180x^2 \\
14x^2 - 28x - 63 \\
\underline{} \\
40x^4 - 80x^3 - 166x^2 - 28x - 63
\end{array}
$$

71.

$$2x^2(3x - 2)(x^2 - 5x - 5)$$

$$= 2x^2(3x^3 - 15x^2 - 15x - 2x^2 + 10x + 10)$$

$$= 6x^5 - 30x^4 - 30x^3 - 4x^4 + 20x^3 + 20x^2$$

$$= 6x^5 - 34x^4 - 10x^3 + 20x^2$$

73.

$$(2y + 5)(y - 3)(2y + 1)$$

$$= (2y^2 - y - 15)(2y + 1)$$

$$= 4y^3 - 2y^2 - 30y + 2y^2 - y - 15$$

$$= 4y^3 - 31y - 15$$

75.

$$(x + 3)(x - 2)(2x^2 + x - 5) = (x^2 + x - 6)(2x^2 + x - 5)$$

$$
\begin{array}{r}
x^2 + x - 6 \\
\underline{2x^2 + x - 5} \\
2x^4 + 2x^3 - 12x^2 \\
x^3 + x^2 - 6x \\
\underline{-5x^2 - 5x + 30} \\
2x^4 + 3x^3 - 16x^2 - 11x + 30
\end{array}
$$

77.

$$x^{2n+1} \cdot x^{n+4} = x^{(2n+1)+(n+4)} = x^{3n+5}$$

79.

$$\left(x^{2n}\right)^{n+5} = x^{2n \times (n+5)} = x^{2n^2 + 10n}$$

81.

$$
\begin{array}{cccc}
 & F & O \quad I & L \\
\end{array}
$$

$$\left(x^n + 2\right)\left(x^n - 4\right) = x^n \cdot x^n - 4 \cdot x^n + 2x^n - 2 \cdot 4$$

$$= x^{2n} - 2x^n - 8$$

83.

$$\left(x^n + 3\right)^2 = \left(x^n\right)^2 + 2\left(x^n \cdot 3\right) + (3)^2 = x^{2n} + 6x^n + 9$$

Exercises 3.3

1.

$$10x^2 - 35 = 5 \cdot 2x^2 - 5 \cdot 7 \qquad \text{GCF} = 5$$

$$= 5\left(2x^2 - 7\right)$$

3.
There is no GCF other than 1.

5.

$$8x^3 + 32x^2y - 20x^2z \qquad \text{GCF} = 4x^2$$

$$= 4x^2 \cdot 2x + 4x^2 \cdot 8y + 4x^2(-5z)$$

$$= 4x^2(2x + 8y - 5z)$$

7.

$$3x^3y^3 - 15x^5y^3 + 9x^4y^5 \qquad \text{GCF} = 3x^3y^3$$

$$= 3x^3y^3 \cdot 1 + 3x^3y^3\left(-5x^2\right) + 3x^3y^3 \cdot 3xy^2$$

$$= 3x^3y^3\left(1 - 5x^2 + 3xy^2\right)$$

9.

$$-6x^5 - 18x^3 - 10x^2 \qquad \text{GCF} = -2x^2$$

$$= -2x^2 \cdot 3x^3 - 2x^2 \cdot 9x - 2x^2 \cdot 5$$

$$= -2x^2\left(3x^3 + 9x + 5\right)$$

11.

$$x(3x - 5) - 7(3x - 5) \qquad \text{GCF} = 3x - 5$$

$$= (3x - 5)(x - 7)$$

#11. continued

Note: If $w = 3x - 5$,

then $x \cdot w - 7 \cdot w = w(x - 7) = (3x - 5)(x - 7)$

13.

$2x(x - 3) + 1 \cdot (x - 3)$ GCF $= x - 3$

$= (x - 3)(2x + 1)$

Note: If $w = x - 3$,

then $2x \cdot w + 1 \cdot w = w(2x + 1)$

$= (x - 3)(2x + 1)$

15.

$h(2i - 1) + 3k(2i - 1)$ GCF $= 2i - 1$

$= (2i - 1)(h + 3k)$

Note: If $w = 2i - 1$,

then $h \cdot w + 3k \cdot w = w(h + 3k)$

$= (2i - 1)(h + 3k)$

17.

$4x(x - 2y) - 6y(x - 2y)$ GCF $= x - 2y$

$= (x - 2y) \underbrace{(4x - 6y)}_{\text{This factor can be factored further.}}$

$= (x - 2y)(2 \cdot 2x - 2 \cdot 3y)$ GCF $= 2$

$= (x - 2y) \cdot 2(2x - 3y)$ or $2(x - 2y)(2x - 3y)$

19.

$15x^3(y + 8) + 5x^2(y + 8)$ GCF $= y + 8$

$= (y + 8) \underbrace{(15x^3 + 5x^2)}_{\text{This factor can be factored further.}}$

$= (y + 8)(5x^2 \cdot 3x + 5x^2 \cdot 1)$ GCF $= 5x^2$

$= (y + 8) \cdot 5x^2(3x + 1)$ or $5x^2(y + 8)(3x + 1)$

21.

$8y^2z(1 - 5x) - 6yz^3(1 - 5x)$ GCF $= 1 - 5x$

$(1 - 5x) \underbrace{(8y^2z - 6yz^3)}_{\text{This factor can be factored further.}}$

$= (1 - 5x)(2yz \cdot 4y - 2yz \cdot 3z^2)$ GCF $= 2yz$

$= (1 - 5x) \cdot 2yz(4y - 3z^2)$ or $2yz(1 - 5x)(4y - 3z^2)$

23.

$3x^2(x + y)^2 + 12x(x + y)$ GCF $= 3x(x + y)$

$= 3x(x + y) \cdot x(x + y) + 3x(x + y) \cdot 4$

$= 3x(x + y)[x(x + y) + 4]$

25.

$2ax + 3bx + 2ay + 3by$

$= 2ax + 2ay + 3bx + 3by$ GCF $= 2a$ and $3b$

$= 2a(x + y) + 3b(x + y)$ GCF $= x + y$

$= (x + y)(2a + 3b)$

27.

$12ac - 4ad - 3bc + bd$ GCF $= 4a$ and $- b$

$= 4a \cdot 3c + 4a \cdot (-d) - b \cdot 3c - b(-d)$

$= 4a(3c - d) - b(3c - d)$

$= (3c - d)(4a - b)$

29.

$6cf - 2cg + 6df - 2dg$ GCF $= 2c$ and $2d$

$= 2c \cdot 3f + 2c(-g) + 2d \cdot 3f + 2d(-g)$

$= 2c(3f - g) + 2d(3f - g)$

$= (3f - g) \underbrace{(2c + 2d)}_{\text{This factor can be factored further.}}$ GCF $= 2$

$= (3f - g) \cdot 2(c + d)$ or $2(3f - g)(c + d)$

31.

$2x^2 - 4xy - x + 2y$ GCF $= 2x$ and -1

$= 2x \cdot x + 2x(-2y) - 1 \cdot x - 1(-2y)$

$= 2x(x - 2y) - 1 \cdot (x - 2y)$

$= (x - 2y)(2x - 1)$

33.

$4n + 15mn - 3n^2 - 20m$

$= 4n - 3n^2 - 20m + 15mn$ GCF $= n$ and $- 5m$

$= n(4 - 3n) - 5m(4 - 3n)$

$= (4 - 3n)(n - 5m)$ or $(3n - 4)(5m - n)$

35.

$6x^2 - 105y + 15xy - 42x$

$= 6x^2 - 42x + 15xy - 105y$ GCF $= 6x$ and $15y$

#35. continued

$$= 6x \cdot x - 6x \cdot 7 + 15y \cdot x - 15y \cdot 7$$

$$= 6x(x-7) + 15y(x-7) \qquad \text{GCF} = x-7$$

$$= (x-7) \qquad \underbrace{(6x+15y)}_{\text{This factor can be factored further.}} \qquad \text{GCF} = 3$$

$$= (x-7)(3 \cdot 2x + 3 \cdot 5y)$$

$$= (x-7) \cdot 3(2x+5y) \quad \text{or} \quad 3(x-7)(2x+5y)$$

37.

$$3x^2y - 5x^2y^2 - 12xy^2 + 20xy^3$$

$$= 3x^2y - 12xy^2 - 5x^2y + 20xy^3 \quad \text{GCF} = 3xy \text{ and}$$
$$\qquad\qquad\qquad\qquad\qquad -5xy^2$$

$$= 3xy \cdot x + 3xy(-4y) - 5xy^2 \cdot x - 5xy^2(-4y)$$

$$= 3xy(x-4y) - 5xy^2(x-4y) \qquad \text{GCF} = x - 4y$$

$$= (x-4y) \qquad \underbrace{\left(3xy - 5xy^2\right)}_{\text{This factor can be factored further}} \qquad \text{GCF} = xy$$

$$= (x-4y)(xy \cdot 3 + xy(-5y))$$

$$= (x-4y) \cdot xy(3-5y) \quad \text{or} \quad xy(x-4y)(3-5y)$$

39.

$$12x^3 - 9x^2 + 4x - 3 \qquad \text{GCF} = 3x^2$$

$$= 3x^2 \cdot 4x + 3x^2(-3) + 4x - 3$$

$$= 3x^2(4x-3) + 1 \cdot (4x-3)$$

$$= (4x-3)\left(3x^2+1\right)$$

41.

$$3y^3 + 7y^2 + 18y + 42 \qquad\qquad \text{GCF} = y^2 \text{ and } 6$$

$$= y^2 \cdot 3y + y^2 \cdot 7 + 6 \cdot 3y + 6 \cdot 7$$

$$= y^2(3y+7) + 6(3y+7) \qquad \text{GCF} = 3y+7$$

$$= (3y+7)(y^2+6)$$

43.

$$6x^6 - 3x^5 - 12x^4 + 6x^3 \qquad \text{GCF} = 3x^5 \text{ and} - 6x^3$$

$$= 3x^5 \cdot 2x + 3x^5(-1) - 6x^3 \cdot 2x - 6x^3(-1)$$

$$= 3x^5(2x-1) - 6x^3(2x-1) \qquad \text{GCF} = 2x-1$$

$$= (2x-1) \qquad \underbrace{\left(3x^5 - 6x^3\right)}_{\text{This factor can be factored further}} \qquad \text{GCF} = 3x^3$$

$$= (2x-1)(3x^3 \cdot x^2 + 3x^3(-2))$$

$$= (2x-1) \cdot 3x^3(x^2-2) \quad \text{or} \quad 3x^3(2x-1)(x^2-2)$$

45.

$$8x^6 + 12x^4 + 40x^3 + 60x \qquad \text{GCF} = 4x^4 \text{ and } 20x$$

$$= 4x^2 \cdot 2x^2 + 4x^4 \cdot 3 + 20x \cdot 2x^2 + 20x \cdot 3$$

$$= 4x^4\left(2x^2+3\right) + 20x\left(2x^2+3\right) \qquad \text{GCF} = 2x^2+3$$

$$= \left(2x^2+3\right) \qquad \underbrace{\left(4x^4 + 20x\right)}_{\text{This factor can be factored further}} \qquad \text{GCF} = 4x$$

$$= \left(2x^2+3\right)\left(4x \cdot x^3 + 4x \cdot 5\right)$$

$$= \left(2x^2+3\right) \cdot 4x\left(x^3+5\right) \quad \text{or} \quad 4x\left(2x^2+3\right)\left(x^3+5\right)$$

47.

$$12x^4 + 7xy - 21x^3 - 4x^2y$$

$$= 12x^4 - 4x^2y - 21x^3 + 7xy \quad \text{GCF} = 4x^2 \text{ and}$$
$$\qquad\qquad\qquad\qquad\qquad -7x$$

$$= 4x^2 \cdot 3x^2 + 4x^2(-y) - 7x \cdot 3x^2 - 7x(-y)$$

$$= 4x^2\left(3x^2-y\right) - 7x\left(3x^2-y\right) \qquad \text{GCF} = 3x^2-y$$

$$= \left(3x^2-y\right) \qquad \underbrace{\left(4x^2 - 7x\right)}_{\text{This factor can be factored further}} \qquad \text{GCF} = x$$

$$= \left(3x^2-y\right) \cdot x(4x-7) \quad \text{or} \quad x\left(3x^2-y\right)(4x-7)$$

49.

$$54b^3 + 6a^4 - 27a^3b - 12ab^2$$

$$= 54b^3 - 12ab^2 - 27a^3b + 6a^4 \qquad \text{GCF} = 6b^2 \text{ and}$$
$$\qquad\qquad\qquad\qquad\qquad -3a^3$$

$$= 6b^2 \cdot 9b + 6b^2(-a) - 3a^3 \cdot 9b - 3a^3(-2a)$$

$$= 6b^2(9b-2a) - 3a^3(9b-2a)$$

$$= (9b-2a) \qquad \underbrace{\left(6b^2 - 3a^3\right)}_{\text{This factor can be factored further}} \qquad \text{GCF} = 3$$

$$= (9b-2a)\left(3 \cdot 2b^2 - 3 \cdot a^3\right)$$

$$= (9b-2a) \cdot 3\left(2b^2-a^3\right) \quad \text{or} \quad 3(9b-2a)\left(2b-a^3\right)$$

51.

$$x^{2n}y^{2n} - x^ny^n \qquad\qquad \text{GCF} = x^ny^n$$

$$= x^ny^n \cdot x^ny^n + x^ny^n(-1)$$

$$= x^ny^n\left(x^ny^n - 1\right)$$

53.

$x^{n+4} + x^4 \qquad$ GCF $= x^4$

$x^4 \cdot x^n + x^4 \cdot 1$

$x^4 \left(x^n + 1 \right)$

Exercises 3.4

1.

$x^2 - 9 = (x+3)(x-3)$

Note: $\quad a^2 - b^2 = (a+b)(a-b)$

3.

$z^2 - 64 = (z+8)(z-8)$

Note: $\quad a^2 - b^2 = (a+b)(a-b)$

5.

$36z^2 - 49b^2 = (6z+7b)(6z-7b)$

Note: $\quad a^2 - b^2 = (a+b)(a-b)$

7.

$2x^2 - 50 = 2\left(x^2 - 25\right) = 2(x+5)(x-5)$

Note: $\quad a^2 - b^2 = (a+b)(a-b)$

9.

$27x^3 y - 12xy^3 = 3xy\left(9x^2 - 4y^2\right)$

$\qquad\qquad\qquad = 3xy(3x+2y)(3x-2y)$

Note: $\quad a^2 - b^2 = (a+b)(a-b)$

11.

$4x^2 + 9 \quad$ This binomial is prime.

13.

$3x^2 + 48 = 3\left(x^2 + 16\right)$

Both factors are prime and cannot be factored further.

15.

$81y^4 - 16z^4 = \left(9y^2 + 4z^2\right)\left(9y^2 - 4z^2\right)$

$\qquad\qquad\qquad = \left(9y^2 + 4z^2\right)(3y+2z)(3y-2z)$

Note: $\quad a^2 - b^2 = (a+b)(a-b)$

17.

$(x-3)^2 - 49 = \left[(x-3)+7\right]\left[(x-3)-7\right]$

$\qquad\qquad\qquad = (x+4)(x-10) \qquad$ Simplify

Note: $\quad a^2 - b^2 = (a+b)(a-b)$

19.

$(x+4)^2 - 4y^2 = \left[(x+4)+2y\right]\left[(x+4)-2y\right]$

$\qquad\qquad\qquad = (x+4+2y)(4+x-2y)$

Note: $\quad a^2 - b^2 = (a+b)(a-b)$

21.

$(2x-3y)^2 - 9z^2$

$= \left[(2x-3y)+3z\right]\left[(2x-3y)-3z\right]$

$= (2x-3y+3z)(2x-3y-3z)$

Note: $\quad a^2 - b^2 = (a+b)(a-b)$

23.

$y^3 - 64 = (y-4)\left(y^2 + 4\cdot y + 4^2\right)$

$\qquad\qquad = (y-4)\left(y^2 + 4y + 16\right)$

Note: $\quad x^3 - y^3 = (x-y)\left(x^2 + xy + y^2\right)$

25.

$8x^3 + 1 = (2x+1)\left((2x)^2 - 1\cdot 2x + 1^2\right)$

$\qquad\qquad = (2x+1)\left(4x^2 - 2x + 1\right)$

Note: $\quad x^3 + y^3 = (x+y)\left(x^2 - xy + y^2\right)$

27.

$8m^3 + 27n^3 = (2m+3n)\left((2m)^2 - 2m\cdot 3n + (3n)^2\right)$

$\qquad\qquad\qquad = (2m+3n)\left(4m^2 - 6mn + 9n^2\right)$

Note: $\quad x^3 + y^3 = (x+y)\left(x^2 - xy + y^2\right)$

29.

$64x^3 - 27y^3 = (4x-3y)\left((4x)^2 + 4x\cdot 3y + (3y)^2\right)$

$\qquad\qquad\qquad = (4x-3y)\left(16x^2 + 12xy + 9y^2\right)$

Note: $\quad x^3 - y^3 = (x-y)\left(x^2 + xy + y^2\right)$

31.

$$375y^3 - 3 = 3(125y^3 - 1)$$
$$= 3(5y - 1)\left((5y)^2 + 1 \cdot 5y + 1^2\right)$$
$$= 3(5y - 1)\left(25y^2 + 5y + 1\right)$$

Note: $x^3 - y^3 = (x - y)\left(x^2 + xy + y^2\right)$

33.

$$54r^3 + 128s^3 = 2\left(27r^3 + 64s^3\right)$$
$$= 2(3r + 4s)\left((3r)^2 - 3r \cdot 4s + (4s)^2\right)$$
$$= 2(3r + 4s)\left(9r^2 - 12rs + 16s^2\right)$$

Note: $x^3 + y^3 = (x + y)\left(x^2 - xy + y^2\right)$

35.

$$(y - 2)^3 - 27 = \left[(y - 2) - 3\right]\left[(y - 2)^2 + 3(y - 2) + 3^2\right]$$
$$= (y - 5)\left(y^2 - 4y + 4 + 3y - 6 + 9\right)$$
$$= (y - 5)\left(y^2 - y + 7\right)$$

Note: $x^3 - y^3 = (x - y)\left(x^2 + xy + y^2\right)$

37.

$$(2x - 1)^3 - (x + 5)^3$$
$$= \left[(2x - 1) - (x + 5)\right]$$
$$\quad \cdot \left[(2x - 1)^2 + (2x - 1)(x + 5) + (x + 5)^2\right]$$
$$= (2x - 1 - x - 5)$$
$$\quad \cdot \left(4x^2 - 4x + 1 + 2x^2 + 9x - 5 + x^2 + 10x + 25\right)$$
$$= (x - 6)\left(7x^2 + 15x + 21\right)$$

Note: $x^3 - y^3 = (x - y)\left(x^2 + xy + y^2\right)$

39.

$$(x + y)^3 + (2x - y)^3$$
$$= \left[(x + y) + (2x - y)\right]$$
$$\quad \cdot \left[(x + y)^2 - (x + y)(2x - y) + (2x - y)^2\right]$$
$$= 3x\left[x^2 + 2xy + y^2 - \left(2x^2 + xy - y^2\right)\right.$$
$$\quad \left. + 4x^2 - 4xy + y^2\right]$$
$$= 3x\left(x^2 + 2xy + y^2 - 2x^2 - xy + y^2\right.$$
$$\quad \left. + 4x^2 - 4xy + y^2\right)$$

#39. continued

$$= 3x\left(3x^2 - 3xy + 3y^2\right) \qquad \text{GCF} = 3$$
$$= 3x \cdot 3\left(x^2 - xy + y^2\right)$$
$$= 9x\left(x^2 - xy + y^2\right)$$

Note: $x^3 + y^3 = (x + y)\left(x^2 - xy + y^2\right)$

41.

$$x^6 - y^6 = \left(x^3 + y^3\right)\left(x^3 - y^3\right) \qquad \text{Factor difference}$$
$$\text{of two squares}$$

Now we have the sum and difference of two perfect cubes.

$$\left(x^3 + y^3\right)\left(x^3 - y^3\right)$$
$$= (x + y)\left(x^2 - xy + y^2\right)(x - y)\left(x^2 + xy + y^2\right)$$
$$= (x + y)(x - y)\left(x^2 - xy + y^2\right)\left(x^2 + xy + y^2\right)$$

Note: $x^3 - y^3 = (x - y)\left(x^2 + xy + y^2\right)$

43.

$$x^2(2x - 3) - 1 \cdot (2x - 3) \qquad \text{GCF} = (2x - 3)$$
$$= (2x - 3)\left(x^2 - 1\right) = (2x - 3)(x + 1)(x - 1)$$

45.

$$r^3(r - 6) - 8(r - 6) \qquad \text{GCF} = (r - 6)$$
$$= (r - 6)\left(r^3 - 8\right) = (r - 6)(r - 2)\left(r^2 + 2r + 4\right)$$

47.

$$5x^3 + 2x^2 - 45x - 18 \qquad \text{GCF} = x^2 \text{ and } -9$$
$$= x^2(5x + 2) - 9(5x + 2) \qquad \text{GCF} = 5x + 2$$
$$= (5x + 2)\left(x^2 - 9\right) = (5x + 2)(x + 3)(x - 3)$$

49.

$$3a^3 + 4b^3 - 4a^2b - 3ab^2$$
$$= 3a^3 - 3ab^2 - 4a^2b + 4b^3 \qquad \text{GCF } = 3a \text{ and } -4b$$
$$= 3a\left(a^2 - b^2\right) - 4b\left(a^2 - b^2\right) \qquad \text{GCF } = a^2 - b^2$$
$$= \left(a^2 - b^2\right)(3a - 4b) = (a + b)(a - b)(3a - 4b)$$

51.

$3x^4 - 5x^3 - 3x + 5 \qquad$ GCF $= x^3$ and -1

$= x^3(3x - 5) - 1(3x - 5) \qquad$ GCF $= 3x - 5$

$= (3x - 5)(x^3 - 1) = (3x - 5)(x - 1)(x^2 + x + 1)$

53.

$4y^4 - 3y^3 + 32y - 24 \qquad$ GCF $= y^3$ and 8

$= y^3(4y - 3) + 8(4y - 3) \qquad$ GCF $= 4y - 3$

$= (4y - 3)(y^3 + 8) = (4y - 3)(y + 2)(y^2 - 2y + 4)$

55.

$4x^5 - x^3 - 32x^2 + 8 \qquad$ GCF $= x^3$ and -8

$= x^3(4x^2 - 1) - 8(4x^2 - 1) \qquad$ GCF $= 4x^2 - 1$

$= (4x^2 - 1)(x^3 - 8)$

$= (2x + 1)(2x - 1)(x - 2)(x^2 + 2x + 4)$

57.

$7x^5 - 7x^3 - 28x^3y^2 + 28xy^2 \quad$ GCF $= 7x^3$ and

$\qquad\qquad\qquad\qquad\qquad\qquad -28xy^2$

$= 7x^3(x^2 - 1) - 28xy^2(x^2 - 1) \;$ GCF $= x^2 - 1$

$= (x^2 - 1)(7x^3 - 28xy^2) \qquad$ GCF $= 7x$

$= (x^2 - 1)[7x(x^2 - 4y^2)]$

$= (x^2 - 1) \cdot 7x(x^2 - 4y^2)$

$= 7x(x^2 - 1)(x^2 - 4y^2)$

$= 7x(x + 1)(x - 1)(x + 2y)(x - 2y)$

59.

$5x^3y^4 + 5x^3y - 40y^7 - 40y^4 \quad$ GCF $= 5x^3y$ and

$\qquad\qquad\qquad\qquad\qquad\qquad -40y^4$

$= 5x^3y(y^3 + 1) - 40y^4(y^3 + 1) \;$ GCF $= y^3 + 1$

$= (y^3 + 1)(5x^3y - 40y^4) \qquad$ GCF $= 5y$

$= (y^3 + 1)[5y(x^3 - 8y^3)]$

$= (y^3 + 1) \cdot 5y(x^3 - 8y^3)$

$= 5y(y^3 + 1)(x^3 - 8y^3)$

$= 5y(y + 1)(y^2 - y + 1)(x - 2y)(x^2 + 2xy + 4y^2)$

61.

$x^3 + 6x^2 + 12x + 8$

$= x^3 + 8 + 6x^2 + 12x$

$= (x + 2)(x^2 - 2x + 4) + 6x(x + 2) \quad$ GCF $= x + 2$

$= (x + 2)(x^2 - 2x + 4 + 6x)$

$= (x + 2)(x^2 - 4x + 4) = (x + 2)(x + 2)(x + 2)$

63.

$x^{2n} - 1 = (x^n + 1)(x^n - 1)$

65.

$x^{4n} - 81 = (x^{2n} + 9)(x^{2n} - 9)$

$\qquad\qquad = (x^{2n} + 9)(x^n + 3)(x^n - 3)$

67.

$x^{n+1} + 5x^n + 3x + 15 \qquad$ GCF $= x^n$ and 3

$= x^n(x + 5) + 3(x + 5) \qquad$ GCF $= x + 5$

$= (x + 5)(x^n + 3)$

69.

$3x^3 - 12x + x^2y - 4y \qquad$ GCF $= 3x$ and y

$= 3x(x^2 - 4) + y(x^2 - 4) \qquad$ GCF $= x^2 - 4$

$= (x^2 - 4)(3x + y) = (x + 2)(x - 2)(3x + y)$

71.

$3x^3y^2 + 6xy^2 - 15xy^3 \qquad$ GCF $= 3xy^2$

$= 3xy^2(x^2 + 2 - 5y)$

73.

$3xy^4(2x - 5) + 12y^5(2x - 5) \qquad$ GCF $= 2x - 5$

$= (2x - 5)(3xy^4 + 12y^5) \qquad$ GCF $= 3y^4$

$= (2x - 5)[3y^4(x + 4y)]$

$= (2x - 5) \cdot 3y^4(x + 4y) = 3y^4(2x - 5)(x + 4y)$

75.

$5x^3 - 5x = 5x(x^2 - 1) \qquad$ GCF $= 5x$

$\qquad\qquad = 5x(x + 1)(x - 1)$

77.

$$625 - m^4 = (25 + m^2)(25 - m^2)$$
$$= (25 + m^2)(5 + m)(5 - m)$$

Exercises 3.5

1.

$x^2 - 9x + 14$

Factors	Sum
$-2(-7) = 14$	$-2 + (-7) = -9$

Thus, $x^2 - 9x + 14 = (x - 2)(x - 7)$

3.

$x^2 - 2x - 35$

Factors	Sum
$-7 \cdot 5 = -35$	$-7 + 5 = -2$

Thus, $x^2 - 2x - 35 = (x - 7)(x + 5)$

5.

$x^2 - 9x + 8$

Factors	Sum
$-1(-8) = 8$	$-1 + (-8) = -9$

Thus, $x^2 - 9x + 8 = (x - 1)(x - 8)$

7.

$x^2 + 7x - 30$

Factors	Sum
$10(-3) = -30$	$10 + (-3) = 7$

Thus, $x^2 + 7x - 30 = (x + 10)(x - 3)$

9.

$x^2 - x + 7$ is prime since 7 will not result in two factors whose sum is -1.

11.

$x^2 - 5x - 36$

Factors	Sum
$4(-9) = -36$	$4 + (-9) = -5$

Thus, $x^2 - 5x - 36 = (x + 4)(x - 9)$

13.

$x^2 - 6x + 9$

Factors	Sum
$-3(-3) = 9$	$-3 + (-3) = -6$

Thus, $x^2 - 6x + 9 = (x - 3)(x - 3)$ or $(x - 3)^2$

15.

$x^2 + 18x + 81$

Factors	Sum
$9(9) = 81$	$9 + 9 = 18$

Thus, $x^2 + 18x + 81 = (x + 9)(x + 9) = (x + 9)^2$

17.

$3x^2 + 24x + 21 = 3(x^2 + 8x + 7)$

Next, we factor the second factor.

Factors	Sum
$7 \cdot 1$	$7 + 1 = 8$

Thus, $3(x^2 + 8x + 7) = 3(x + 7)(x + 1)$

19.

$-4x^2 + 12x + 112 = -4(x^2 - 3x - 28)$

Next, we factor the second factor.

Factors	Sum
$4(-7) = -28$	$4 + (-7) = -3$

Thus, $-4(x^2 - 3x - 28) = -4(x + 4)(x - 7)$

21.

$-2x^3 - 16x^2 - 32x = -2x(x^2 + 8x + 16)$

Next, we factor the second factor.

Factors	Sum
$4 \cdot 4 = 16$	$4 + 4 = 8$

Thus, $-2x(x^2 + 8x + 16) = -2x(x + 4)(x + 4)$
$$= -2x(x + 4)^2$$

23.

$6x^3 + 18x^2 - 6x = 6x(x^2 + 3x - 1)$

The second factor is prime since the -1 will not result in two factors whose sum is 3.

25.

$$3x^4y^2 + 9x^3y^2 + 3x^2y^2 = 3x^2y^2\left(x^2 + 3x + 1\right)$$
Prime

27.

$$x^2 - 5xy + 6y^2$$

Factors	Sum
$-2(-3) = 6$	$-2 + (-3) = -5$

Thus, $x^2 - 5x + 6y^2 = (x - 2y)(x - 3y)$

29.

$$x^2 + 6xy + 5y^2$$

Factors	Sum
$5 \cdot 1 = 5$	$5 + 1 = 6$

Thus, $x^2 + 6xy + 5y^2 = (x + 5y)(x + y)$

31.

$$x^2 + 8xy - 20y^2$$

Factors	Sum
$10(-2) = -20$	$10 + (-2) = 8$

Thus, $x^2 + 8xy - 20y^2 = (x + 10y)(x - 2y)$

33.

$$6x^2 - 13x - 5 = (6x + h)(x + k) \quad \text{or}$$
$$(3x + h)(2x + k)$$

<u>Possible Factor Combinations</u>
$$h \cdot k = -5$$

$1(-5)$	$-1 \cdot 5$	Signs are
$-5 \cdot 1$	$5(-1)$	opposite

After a little or much trial and error, the factor combination that gives the correct product is 1 and -5 with $(3x + h)(2x + h)$.

So, $6x^2 - 13x - 5 = (3x + 1)(2x - 5)$.

35.

$2x^2 + 5x - 4$ is prime since the product of 2 and -4 will not result in two factors whose sum is 5.

37.

$$12x^2 + 25x - 7 = (12x + h)(x + k) \quad \text{or}$$
$$(2x + h)(6x + k) \quad \text{or}$$
$$(3x + h)(4x + k)$$

<u>Possible Factor Combinations</u>
$$h \cdot k = -7$$

$1(-7)$	$-1 \cdot 7$	Signs are
$-7 \cdot 1$	$7(-1)$	opposite

After a little or much trial and error, the factor combination that gives the correct product is 7 and -1 with $(3x + h)(4x + k)$.

So, $12x^2 + 25x - 7 = (3x + 7)(4x - 1)$.

39.

$6x^2 - 9x - 4$ is prime since the product of 6 and -4 will not result in two factors whose sum is -9.

41.

$$9x^2 - 15x + 4 = (9x + h)(x + k) \quad \text{or}$$
$$(3x + h)(3x + k)$$

<u>Possible Factor Combinations</u>
$$h \cdot k = 4$$

$-1(-4)$	$-4(-1)$	Signs are
$-2(-2)$		negative

After a little of much trial and error, the factor combination of that gives the correct product is -1 and -4 with $(3x + h)(3x + k)$.

So, $9x^2 - 15x + 4 = (3x - 1)(3x - 4)$.

43.

$$8x^2 + 30x - 27 = (8x + h)(x + k) \quad \text{or}$$
$$(2x + h)(4x + k)$$

<u>Possible Factor Combinations</u>
$$h \cdot k = -27$$

$-27 \cdot 1$	$1(-27)$	$3(-9)$	$-9 \cdot 3$
$27(-1)$	$-27 \cdot 1$	$9(-3)$	$-3(9)$

After a little or much trial and error, the futher combination that gives the correct product is 9 and -3 with $(2x + h)(4x + k)$.

So, $8x^2 + 30x - 27 = (2x + 9)(4x - 3)$.

45.

$$9x^2 + 30x + 25 = (9x + h)(x + k) \quad \text{or}$$
$$(3x + h)(3x + k)$$

Possible Factor Combinations

$$h \cdot k = 25$$

$25 \cdot 1$	$1 \cdot 25$	$5 \cdot 5$

Both signs are positive

After a little or much trial and error, the factor combination that gives the correct product is 5 and 5 with $(3x + h)(3x + h)$.

So, $9x^2 + 30x + 25 = (3x + 5)(3x + 5) = (3x + 5)^2$.

47.

$8x^2 - 7x + 20$ is prime since the product of 8 and 20 will not result in two factors whose sum is -7.

49.

$$9x^2 + 17x - 2 = (9x + h)(x + k) \quad \text{or}$$
$$(3x + h)(3x + k)$$

Possible Factor Combinations

$$h \cdot k = -2$$

$-2 \cdot 1$	$1(-2)$	$-1 \cdot 2$	$2(-1)$

Both signs are negative

After a little or much trial and error, the factor combination that gives the correct product is -1 and 2 with $(9x + h)(x + k)$.

So, $9x^2 + 17x - 2 = (9x - 1)(x + 2)$.

51.

$$25x^2 + 10x + 1 = (25x + h)(x + k) \quad \text{or}$$
$$(5x + h)(5x + k)$$

Possible Factor Combinations

$$h \cdot k = 1$$

$1 \cdot 1$

Both signs are positive

After a little or much trial and error, the factor combination that gives the correct product is 1 and 1 with $(5x + h)(5x + k)$.

So, $25x^2 + 10x + 1 = (5x + 1)(5x + 1) = (5x + 1)^2$.

53.

$$6x^2 - 13x - 15 = (6x + h)(x + k) \quad \text{or}$$
$$(2x + h)(3x + k)$$

Possible Factor Combinations

$$h \cdot k = -15$$

$1(-15)$	$-1(15)$	$15(-1)$
$-15 \cdot 1$	$3(-5)$	$-3 \cdot 5$
$5(-3)$	$-5 \cdot 3$	

The signs are opposite

After a little or much trial and error, the factor combination that gives the correct product is 5 and -3 with $(6x + h)(x + k)$.

So, $6x^2 - 13x - 15 = (6x + 5)(x - 3)$.

55.

$$16x^2 - 26x - 12 = 2(8x^2 - 13x - 6)$$

Next, we factor the second factor.

$$8x^2 - 13x - 6 = (8x + h)(x + k) \quad \text{or}$$
$$(2x + h)(4x + k)$$

Possible Factor Combinations

$$h \cdot k = -6$$

$-6 \cdot 1$	$6(-1)$	$-1 \cdot 6$
$1(-6)$	$2(-3)$	$-2 \cdot 3$
$3(-2)$	$-3 \cdot 2$	

The signs are opposite

After a little or much trial and error, the factor combination that gives the correct product is 3 and -2 with $(8x + h)(x + k)$.

So, $16x^2 - 26x - 12 = 2(8x + 3)(x - 2)$.

57.

$$20x^2 - 60x - 35 = 5(4x^2 - 12x - 7)$$

Next, we factor the second factor.

$$4x^2 - 12x - 7 = (4x + h)(x + k) \quad \text{or}$$
$$(2x + h)(2x + k)$$

Possible Factor Combinations

$$h \cdot k = -7$$

$7(-1)$	$-7 \cdot 1$	$1(-7)$	$-1(7)$

The signs are opposite

After little or much trial and error, the factor

#57. continued

combination that gives the correct product is

-7 and 1 with $(2x + h)(2x + k)$.

So, $20x^2 - 60x - 35 = 5(2x - 7)(2x + 1)$.

59.

$10x^3 - 14x^2 + 8x = 2x(5x^2 - 7x + 4)$

The second factor is prime since the product of 5 and 4 will not result in two factors whose sum is - 7.

61.

$15x^3 + 42x^2 - 9x = 3x(5x^2 + 14x - 3)$

Next, we factor the second factor.

$5x^2 + 14x - 3 = (5x + h)(x + k)$

Possible Factor Combination

$h \cdot k = -3$

$1(-3) \quad -1 \cdot 3 \quad 3(-1) \quad -3 \cdot 1 \qquad$ The signs

are opposite

After a little or much trial and error, the factor combination that gives the correct product is

-1 and 3 with $(5x + h)(x + k)$.

So, $15x^3 + 42x^2 - 9x = 3x(5x - 1)(x + 3)$.

63.

$27x^2 + 72xy + 48y^2 = 3(9x^2 + 24xy + 16y^2)$

Next, we factor the second factor.

$9x^2 + 24xy + 16y^2 = (9x + hy)(x + ky)$ or

$\qquad\qquad\qquad (3x + hy)(3x + ky)$

Possible Factor Combinations

$h \cdot k = 16$

$1 \cdot 16 \quad 16 \cdot 1 \quad 4 \cdot 4 \qquad$ Signs are

both positive

After a little or much trial and error, the factor combination that gives the correct product is

4 and 4 with $(3x + hy)(3x + ky)$.

So, $27x^2 + 72xy + 48y^2 = 3(3x + 4y)(3x + 4y)$

$\qquad\qquad\qquad\qquad = 3(3x + 4y)^2$

65.

$9x^2 - 16xy - 4y^2 = (9x + hy)(x + ky)$ or

$\qquad\qquad\qquad (3x + hy)(3x + ky)$

Possible Factor Combinations

$h \cdot k = -4$

$1(-4) \quad -1 \cdot 4 \quad 4(-1) \qquad$ Signs are

$-4(1) \quad 2(-2) \quad -2(2) \qquad$ opposite

After little or much trial and error, the factor combination that gives the correct product is

2 and - 2 with $(9x + hy)(x + ky)$.

So, $9x^2 - 16xy - 4y^2 = (9x + 2y)(x - 2y)$.

67.

$8x^2 - 14xy + 6y^2 = 2(4x^2 - 7xy + 3y^2)$

Next, we factor the second factor.

$4x^2 - 7xy + 3y^2 = (4x + hy)(x + ky)$ or

$\qquad\qquad\qquad (2x + hy)(2x + ky)$

Possible Factor Combinations

$h \cdot k = 3$

$-1(-3) \quad -3(-1) \qquad$ Signs are

both negative

After a little or much trial and error, the factor combination that gives the correct product is

-3 and - 1 with $(4x + hy)(x + ky)$.

So, $4x^2 - 7xy + 3y^2 = 2(4x - 3y)(x - y)$.

69.

$x^4 + 12x^2 + 27$

Factors   Sum

$3 \cdot 9 = 27 \qquad 3 + 9 = 12$

So, $x^4 + 12x^2 + 27 = (x^2 + 3)(x^2 + 9)$

71.

$x^4 - x^2 - 12$

Factors   Sum

$3(-4) = -12 \qquad 3 + (-4) = -1$

So, $x^4 - x^2 - 12 = (x^2 + 3)(x^2 - 4)$

$\qquad\qquad\qquad = (x^2 + 3)(x + 2)(x - 2)$

73.
$$4x^4 + 4x^2 - 15 = \left(4x^2 + h\right)\left(x^2 + k\right) \quad \text{or}$$
$$\left(2x^2 + h\right)\left(2x^2 + k\right)$$

Possible Factor Combinations
$$h \cdot k = -15$$

$1(-15)$ $-1 \cdot 15$ $15(-1)$ Signs are

$-15 \cdot 1$ $3(-5)$ $-3 \cdot 5$ both negative

$5(-3)$ $-5(3)$

After a little or much trial and error, the factor combination that gives the correct product is

5 and -3 with $\left(2x^2 + h\right)\left(2x^2 + k\right)$.

So, $4x^4 + 4x^2 - 15 = \left(2x^2 + 5\right)\left(2x^2 - 3\right)$.

75.
$$(3x + 1)^2 - 7(3x + 1) + 10 \qquad \text{Let } u = 3x + 1$$

Then we factor $u^2 - 7u + 10$

Factors	Sum
$-2(-5) = 10$	$-2 + (-5) = -7$

Thus, $u^2 - 7u + 10 = (u - 2)(u - 5)$.

So, $(3x + 1)^2 - 7(3x + 1) + 10 = (3x + 1 - 2)(3x + 1 - 5)$

$$= (3x - 1)(3x - 4)$$

77.
$$6(2x - 1)^2 + (2x - 1) - 12 \qquad \text{Let } u = 2x - 1$$

Then we factor $6u^2 + u - 12$.

Possible Factor Combinations
$$h \cdot k = -12$$

$1(-12)$ $-1 \cdot 12$ $-12 \cdot 1$

$12(-1)$ $2(-6)$ $-2 \cdot 6$ Signs are

$6(-2)$ $-6 \cdot 2$ $3(-4)$ opposite

$-3 \cdot 4$ $4(-3)$ $-4 \cdot 3$

After a little or much trial and error, the factor combination that gives the correct product is

3 and -4 with $(2u + h)(3u + h)$ or $(2u + 3)(3u - 4)$.

So, $6(2x - 1)^2 + (2x - 1) - 12 = \left[2(2x - 1) + 3\right]$
$$\left[3(2x - 1) - 4\right]$$
$$= (4x - 2 + 3)(6x - 3 - 4)$$
$$= (4x + 1)(6x - 7)$$

79.
$$6(x + 2y)^2 - 19(x + 2y) + 8 \quad \text{Let } u = x + 2y$$

Then we factor $6u^2 - 19u + 8$

$$6u^2 - 19u + 8 = (6u + h)(u + k) \quad \text{or}$$
$$(2u + h)(3u + k)$$

Possible Factor Combinations
$$h \cdot k = 8$$

$-1(-8)$ $-8(-1)$ $-2(-4)$ Signs are

$-4(-2)$ both negative

After a little or much trial and error, the factor combination that gives the correct product is

-1 and -8 with $(2u + h)(3u + k)$ or $(2u - 1)(3u - 8)$.

So,
$$6(x + 2y)^2 - 19(x + 2y) + 8 = \left[2(x + 2y) - 1\right]$$
$$\left[3(x + 2y) - 8\right]$$
$$= (2x + 4y - 1)(3x + 6y - 8)$$

81.
$$x^2 + 6x + 9 - y^2 = \left(x^2 + 6x + 9\right) - y^2$$
$$= (x + 3)^2 - y^2$$
$$= \left[(x + 3) + y\right]\left[(x + 3) - y\right]$$
$$= (x + 3 + y)(x + 3 - y)$$

83.
$$4x^2 - y^2 - 14y - 49 = 4x^2 - \left(y^2 + 14y + 49\right)$$
$$= 4x^2 - (y + 7)^2$$
$$= \left[2x + (y + 7)\right]\left[2x - (y + 7)\right]$$
$$= (2x + y + 7)(2x - y - 7)$$

85.
$$4x^2 + 4xy + y^2 - z^2 + 2z - 1$$
$$= \left(4x^2 + 4xy + y^2\right) - \left(z^2 - 2z + 1\right)$$
$$= (2x + y)^2 - (z - 1)^2$$
$$= \left[(2x + y) + (z - 1)\right]\left[(2x + y) - (z - 1)\right]$$
$$= (2x + y + z - 1)(2x + y - z + 1)$$

87.

$$x^2 - 2x + 1 - y^2 - 2yz - z^2$$

$$= \left(x^2 - 2x + 1\right) - \left(y^2 + 2yz + z^2\right)$$

$$= (x-1)^2 - (y+z)^2$$

$$= \left[(x-1) + (y+z)\right]\left[(x-1) - (y+z)\right]$$

$$= (x-1+y+z)(x-1-y-z)$$

89.

$$9x^2 + 12x + 4 - y^2 + 4yz - 4z^2$$

$$= \left(9x^2 + 12x + 4\right) - \left(y^2 - 4yz + 4z^2\right)$$

$$= (3x+2)^2 - (y-2z)^2$$

$$= \left[(3x+2) + (y-2z)\right]\left[(3x+2) - (y-2z)\right]$$

$$= (3x+2+y-2z)(3x+2-y+2z)$$

91.

$$x^2 - x - 12 + 5xy - 20y$$

$$= \left(x^2 - x - 12\right) + 5xy - 20y$$

$$= (x+3)(x-4) + 5y(x-4) \qquad \text{GCF} = \text{x - 4}$$

$$= (x\text{ - }4)\left[(x+3) + 5y\right]$$

$$= (x-4)(x+3+5y)$$

93.

$$2x^2 - 11x + 5 - 6kx + 3k$$

$$= \left(2x^2 - 11x + 5\right) - 6kx + 3k$$

$$= (2x-1)(x-5) - 3k(2x-1) \qquad \text{GCF} = 2x\text{ - }1$$

$$= (2x\text{ - }1)\left[(x-5) - 3k\right] = (2x-1)(x-5-3k)$$

95.

$$x^2 + 4xy + 3y^2 + xz + yz$$

$$= \left(x^2 + 4xy + 3y^2\right) + xz + yz$$

$$= (x+3y)(x+y) + z(x+y) \qquad \text{GCF} = x + y$$

$$= (x+y)\left[(x+3y) + z\right] = (x+y)(x+3y+z)$$

97.

$$x^{2n} - 8x^n + 15$$

#97. continued

Factors	Sum
$-3(-5) = 15$	$-3 - 5 = -8$

Thus, $x^{2n} - 8x^n + 15 = \left(x^n - 3\right)\left(x^n - 5\right)$

99.

$$2x^{2n} - 5x^n - 3$$

Product

$$2(-3) = -6$$

Factors	Sum
$-6(1) = -6$	$-6 + 1 = -5$

Replace $-5x^n$ with $-6x^n + x^n$

Thus, $2x^{2n} - 5x^n - 3 = 2x^{2n} - 6x^n + x^n - 3$

$$= 2x^n\left(x^n - 3\right) + \left(x^n - 3\right)$$

$$= \left(x^n - 3\right)\left(2x^n + 1\right)$$

101.

$$3x^2 + 11x + 10$$

Product

$$3 \cdot 10 = 30$$

Factors	Sum
$5 \cdot 6 = 30$	$5 + 6 = 11$

Replace 11x with $5x + 6x$.

Thus, $3x^2 + 11x + 10 = 3x^2 + 5x + 6x + 10$

$$= x(3x + 5) + 2(3x + 5)$$

$$= (3x + 5)(x + 2)$$

103.

$$4x^2 + x - 14$$

Product

$$4(-14) = -56$$

Factors	Sum
$-7 \cdot 8 = -56$	$-7 + 8 = 1$

Replace x with $-7x + 8x$.

Thus, $4x^2 + x - 14 = 4x^2 - 7x + 8x - 14$

$$= x(4x - 7) + 2(4x - 7)$$

$$= (4x - 7)(x + 2)$$

105.

$$6x^2 - 7x - 10$$

#105. continued

Product

$$6(-10) = -60$$

Factors	Sum
$-12 \cdot 5 = -60$	$-12 + 5 = -7$

Replace $-7x$ with $-12x + 5x$.

Thus, $6x^2 - 7x - 10 = 6x^2 - 12x + 5x - 10$
$$= 6x(x - 2) + 5(x - 2)$$
$$= (x - 2)(6x + 5)$$

107.

$$9x^2 - 9x - 4$$

Product

$$9(-4) = -36$$

Factors	Sum
$-12 \cdot 3 = -36$	$-12 + 3 = -9$

Replace $-9x$ with $-12x + 3x$.

Thus, $9x^2 - 9x - 4 = 9x^2 - 12x + 3x - 4$
$$= 3x(3x - 4) + 1(3x - 4)$$
$$= (3x - 4)(3x + 1)$$

109.

$$4x^2 + 41x + 10$$

Product

$$4 \cdot 10 = 40$$

Factors	Sum
$40 \cdot 1 = 40$	$40 + 1 = 41$

Replace $41x$ with $40x + x$.

Thus, $4x^2 + 41x + 10 = 4x^2 + 40x + x + 10$
$$= 4x(x + 10) + 1(x + 10)$$
$$= (x + 10)(4x + 1)$$

111.

$$12x^2 - 25x - 7$$

Product

$$12(-7) = -84$$

Factors	Sum
$-28 \cdot 3 = -84$	$-28 + 3 = -25$

Replace $-25x$ with $-28x + 3x$.

Thus, $12x^2 - 25x - 7 = 12x^2 - 28x + 3x - 7$
$$= 4x(3x - 7) + 1(3x - 7)$$
$$= (3x - 7)(4x + 1)$$

113.

$$8x^2 - 21x + 10$$

Product

$$8 \cdot 10 = 80$$

Factors	Sum
$-16(-5) = 80$	$-16 + (-5) = -21$

Replace $-21x$ with $-16x - 5x$.

Thus, $8x^2 - 21x + 10 = 8x^2 - 16x - 5x + 10$
$$= 8x(x - 2) - 5(x - 2)$$
$$= (x - 2)(8x - 5)$$

115.

$$12x^2 - 28x + 15$$

Product

$$12 \cdot 15 = 180$$

Factors	Sum
$-10(-18) = 180$	$-10 + (-18) = -28$

Replace $-28x$ with $-10x - 18x$.

Thus, $12x^2 - 28x + 15 = 12x^2 - 10x - 18x + 15$
$$= 2x(6x - 5) - 3(6x - 5)$$
$$= (6x - 5)(2x - 3)$$

117.

$$12x^2 - 19x - 10$$

Product

$$12(-10) = -120$$

Factors	Sum
$-24 \cdot 5 = -120$	$-24 + 5 = -19$

Replace $-19x$ with $-24x + 5x$.

Thus, $12x^2 - 19x - 10 = 12x^2 - 24x + 5x - 10$
$$= 12x(x - 2) + 5(x - 2)$$
$$= (x - 2)(12x + 5)$$

119.

$$36 - 4y^2 = 4\left(9 - y^2\right) \qquad \text{GCF} = 4$$
$$= 4(3 + y)(3 - y)$$

121.

$12x^2 + 18x - 12 = 6\left(2x^2 + 3x - 2\right)$ GCF $= 6$

Next, we factor the second factor.

Product

$2(-2) = -4$

Factors   Sum

$-1 \cdot 4 = -4$ $-1 + 4 = 3$

Replace $3x$ with $-x + 4x$.

Thus $2x^2 + 3x - 2 = 2x^2 - x + 4x - 2$

$\qquad = x(2x-1) + 2(2x-1)$

$\qquad = (2x-1)(x+2)$

So, $12x^2 + 18x - 12 = 6(2x-1)(x+2)$.

123.

$y^3 + 64 = (y+4)\left(y^2 - 4y + 16\right)$

125.

$y^3 + 8y^2 - y - 8$ GCF $= y^2$ and -1

$= y^2(y+8) - 1(y+8)$ GCF $= y+8$

$= (y+8)\left(y^2 - 1\right) = (y+8)(y+1)(y-1)$

127.

$9x^2 - 15x + 4$

Product

$9 \cdot 4 = 36$

Factors   Sum

$-12(-3) = 36$ $-12 + (-3) = -15$

Replace $-15x$ with $-12x - 3x$.

Thus, $9x^2 - 15x + 4 = 9x^2 - 12x - 3x + 4$

$\qquad = 3x(3x-4) - 1(3x-4)$

$\qquad = (3x-4)(3x-1)$

Exercises 3.6

1.

$x^2 - 3x + 2$

Factors   Sum

$-1(-2) = 2$ $-1 + (-2) = -3$

Thus, $x^2 - 3x + 2 = (x-1)(x-2)$

3.

$2x^2 + 7x + 3$

Products

$2 \cdot 3 = 6$

Factors   Sum

$6 \cdot 1 = 6$ $6 + 1 = 7$

Replace $7x$ with $6x + x$.

Thus, $2x^2 + 7x + 3 = 2x^2 + 6x + x + 3$

$\qquad = 2x(x+3) + 1(x+3)$

$\qquad = (x+3)(2x+1)$

5.

$8x^3 + 27y^3 = (2x+3y)\left((2x)^2 - 2x \cdot 3y + (3y)^2\right)$

$\qquad = (2x+3y)\left(4x^2 - 6xy + 9y^2\right)$

7.

$2x^3 + 5x^2 + 6x + 15$ GCF $= x^2$ and 3

$= x^2(2x+5) + 3(2x+5)$ GCF $= 2x+5$

$= (2x+5)\left(x^2 + 3\right)$

9.

$3(2x-1)^2 - 11(2x-1) - 20$ Let $u = 2x-1$

Thus, we factor $3u^2 - 11u - 20$.

Product

$3(-20) = -60$

Factors   Sum

$-15 \cdot 4$ $-15 + 4 = -11$

Replace $-11u$ with $-15u + 4u$.

Thus, $3u^2 - 11u - 20 = 3u^2 - 15u + 4u - 20$

$\qquad = 3u(u-5) + 4(u-5)$

$\qquad = (u-5)(3u+4)$

So, $3(2x-1)^2 - 11(2x-1) - 20$

$\qquad = \left[(2x-1) - 5\right]\left[3(2x-1) + 4\right]$

$\qquad = (2x-1-5)(6x-3+4)$

$\qquad = (2x-6)(6x+1)$

$\qquad = 2(x-3)(6x+1)$

11.

$2x^3 - 20x^2 + 18x \qquad \text{GCF} = 2x$

$= 2x(x^2 - 10x + 9)$

$= 2x(x - 9)(x - 1) \qquad \text{Factor } x^2 - 10x + 9$

13.

$5x^2 - 3x - 2$

Product

$5(-2) = -10$

Factors   Sum

$-5 \cdot 2 = -10 \qquad -5 + 2 = -3$

Replace $-3x$ with $-5x + 2x$.

Thus, $5x^2 - 3x - 2 = 5x^2 - 5x + 2x - 2$

$\qquad\qquad\qquad = 5x(x - 1) + 2(x - 1)$

$\qquad\qquad\qquad = (x - 1)(5x + 2)$

15.

$x^2 - xy - 12y^2$

Factors   Sum

$-4 \cdot 3 = -12 \qquad -4 + 3 = -1$

Thus, $x^2 - xy - 12y^2 = (x - 4y)(x + 3y)$

17.

$5(x + 2y)^2 + 31(x + 2y) + 6 \qquad \text{Let } u = x + 2y$

Thus, we factor $5u^2 + 31u + 6$

Product

$5 \cdot 6 = 30$

Factors   Sum

$30 \cdot 1 = 30 \qquad 30 + 1 = 31$

Replace $31u$ with $30u + u$.

Thus, $5u^2 + 31u + 6 = 5u^2 + 30u + u + 6$

$\qquad\qquad\qquad\qquad = 5u(u + 6) + 1(u + 6)$

$\qquad\qquad\qquad\qquad = (u + 6)(5u + 1)$

So, $5(x + 2y)^2 + 31(x + 2y) + 6$

$= \left[(x + 2y) + 6\right]\left[5(x + 2y) + 1\right]$

$= (x + 2y + 6)(5x + 10y + 1)$

19.

$x^2 - 36 = x^2 - 6^2 = (x + 6)(x - 6)$

21.

$18x^3y^3 - 6x^2y^4 - 4xy^5 \qquad \text{GCF} = 2xy^3$

$= 2xy^3(9x^2 - 3xy - 2y^2)$

Next, we factor the second factor.

Product

$9(-2) = -18$

Factors   Sum

$3(-6) = -18 \qquad 3 + (-6) = -3$

Replace $-3xy$ with $3xy - 6xy$.

Thus, $9x^2 - 3xy - 2y^2 = 9x^2 + 3xy - 6xy - 2y^2$

$\qquad\qquad\qquad\qquad = 3x(3x + y) - 2y(3x + y)$

$\qquad\qquad\qquad\qquad = (3x + y)(3x - 2y)$

So, $18x^3y^3 - 6x^2y^4 - 4xy^5 = 2xy^3(3x + y)(3x - 2y)$

23.

$4x^2 - (2y + 3)^2 = (2x)^2 - (2y + 3)^2$

$\qquad\qquad\qquad = \left[2x + (2y + 3)\right]\left[2x - (2y + 3)\right]$

$\qquad\qquad\qquad = (2x + 2y + 3)(2x - 2y - 3)$

25.

$x^2 + 6x + 9 - 16y^2 = (x^2 + 6x + 9) - 16y^2$

$\qquad\qquad\qquad\qquad = (x + 3)^2 - 16y^2$

$\qquad\qquad\qquad\qquad = (x + 3)^2 - (4y)^2$

$\qquad\qquad\qquad\qquad = \left[(x + 3) + 4y\right]\left[(x + 3) - 4y\right]$

$\qquad\qquad\qquad\qquad = (x + 3 + 4y)(x + 3 - 4y)$

27.

$64x^3 - 1 = (4x)^3 - (1)^3 = (4x - 1)\left((4x)^2 + 4x \cdot 1 + 1^2\right)$

$\qquad\qquad\qquad\qquad = (4x - 1)(16x^2 + 4x + 1)$

29.

$y^2 + 2y - 8$

Factors   Sum

$-2 \cdot 4 = -8 \qquad -2 + 4 = 2$

Thus, $y^2 + 2y - 8 = (y - 2)(y + 4)$

31.

$4(x - y)^2 - 8(x - y) + 3 \qquad \text{Let } u = x - y$

#31. continued

Thus, we factor $4u^2 - 8u + 3$.

Product

$4 \cdot 3 = 12$

Factors	Sum
$-6(-2) = 12$	$-6 + (-2) = -8$

Replace $-8u$ with $-6u - 2u$

Thus, $4u^2 - 8u + 3 = 4u^2 - 6u - 2u + 3$

$\qquad = 2u(2u - 3) - 1(2u - 3)$

$\qquad = (2u - 3)(2u - 1)$

So, $4(x - y)^2 - 8(x - y) + 3$

$\qquad = [2(x - y) - 3][2(x - y) - 1]$

$\qquad = (2x - 2y - 3)(2x - 2y - 1)$

33.

$x^2 + 6xy + 9y^2$

Factors	Sum
$3 \cdot 3 = 9$	$3 + 3 = 6$

Thus, $x^2 + 6xy + 9y^2 = (x + 3y)(x + 3y) = (x + 3y)^2$

35.

$4x^4 - 17x^2 - 50$

Product

$4(-50) = -200$

Factors	Sum
$8(-25) = -200$	$8 - 25 = -17$

Replace $-17x^2$ with $8x^2 - 25x^2$.

Thus, $4x^4 - 17x^2 - 50 = 4x^4 + 8x^2 - 25x^2 - 50$

$\qquad = 4x^2(x^2 + 2) - 25(x^2 + 2)$

$\qquad = (x^2 + 2)(4x^2 - 25)$

$\qquad = (x^2 + 2)(2x + 5)(2x - 5)$

37.

$x^2 - 4x + 4 - 9y^2 = (x^2 - 4x + 4) - 9y^2$

$\qquad = (x - 2)^2 - 9y^2$

$\qquad = (x - 2)^2 - (3y)^2$

$\qquad = [(x - 2) + 3y][(x - 2) - 3y]$

$\qquad = (x - 2 + 3y)(x - 2 - 3y)$

39.

$25x^2 - 16y^2 = (5x)^2 - (4y)^2 = (5x + 4y)(5x - 4y)$

41.

$3x^3 + x^2y - 3x - y \qquad$ GCF $= x^2$ and -1

$= x^2(3x + y) - 1 \cdot (3x + y) \qquad$ GCF $= 3x + y$

$= (3x + y)(x^2 - 1) = (3x + y)(x + 1)(x - 1)$

43.

$(x + 3)^2 - 9y^2 = (x + 3)^2 - (3y)^2$

$\qquad = [(x + 3) + 3y][(x + 3) - 3y]$

$\qquad = (x + 3 + 3y)(x + 3 - 3y)$

45.

$3x^2 - 8xy + 4y^2$

Product

$3 \cdot 4 = 12$

Factors	Sum
$-2(-6) = 12$	$-2 + (-6) = -8$

Replace $-8xy$ with $-2xy - 6xy$.

Thus, $3x^2 - 8xy + 4y^2 = 3x^2 - 2xy - 6xy + 4y^2$

$\qquad = x(3x - 2y) - 2y(3x - 2y)$

$\qquad = (3x - 2y)(x - 2y)$

47.

$18a^3b^3 - 12ab^5 - 6ab^3 \qquad$ GCF $= 6ab^3$

$= 6ab^3(3a^2 - 2b^2 - 1)$

The second factor is prime since the product of 3 and -1 will not result in two factors whise sum is -2.

49.

$8x^2 - 6xy - 9y^2$

Product

$8(-9) = -72$

Factors	Sum
$-12 \cdot 6 = -72$	$-12 + 6 = -6$

Replace $-6xy$ with $-12xy + 6xy$.

Thus, $8x^2 - 6xy - 9y^2 = 8x^2 - 12xy + 6xy - 9y^2$

$\qquad = 4x(2x - 3y) + 3y(2x - 3y)$

$\qquad = (2x - 3y)(4x + 3y)$

51.

$$4x^2 - 24x + 36 = 4\left(x^2 - 6x + 9\right)$$

Next we factor the second factor.

Factors	Sum
$-3(-3) = 9$	$-3 + (-3) = -6$

Thus, $x^2 - 6x + 9 = (x - 3)(x - 3) = (x - 3)^2$

So, $4x^2 - 24x + 36 = 4(x - 3)^2$

53.

$$x^4 - 10x^2 + 9$$

Factors	Sum
$-1(-9) = 9$	$-1 + (-9) = -10$

Thus, $x^4 - 10x^2 + 9 = \left(x^2 - 1\right)\left(x^2 - 9\right)$

$$= (x + 1)(x - 1)(x + 3)(x - 3)$$

55.

$$x^2 + 2xy + y^2 - m^2 - 6m - 9$$
$$= \left(x + y\right)^2 - \left(m + 3\right)^2$$
$$= \left[(x + y) + (m + 3)\right]\left[(x + y) - (m + 3)\right]$$
$$= (x + y + m + 3)(x + y - m - 3)$$

57.

$$(x + 4)^3 + 125y^3$$
$$= (x + 4)^3 + (5y)^3$$
$$= \left[(x + 4) + 5y\right]\left[(x + 4)^2 - 5y(x + 4) + (5y)^2\right]$$
$$= (x + 4 + 5y)\left(x^2 + 8x + 16 - 5xy - 20y + 25y^2\right)$$

Exercises 3.7

1.-19.

The exercises do not require worked - out solutions.

21.

$$x^2 + 6x + 8 = 0$$

$(x + 2)(x + 1) = 0$	Step 2
$x + 2 = 0$ or $x + 4 = 0$	Step 3
$x = -2$ $\quad x = -4$	Step 4

A check verifies that the solution set is $\{-2, -4\}$.

23.

$$x^2 - 2x - 35 = 0$$

$(x - 7)(x + 5) = 0$	
$x - 7 = 0$ or $x + 5 = 0$	Step 2
$x = 7$ $\quad x = -5$	Step 3

A check verifies that the solution set is $\{7, -5\}$.

25.

$$2x^2 + 12x - 32 = 0$$

$x^2 + 6x - 16 = 0$	Multiply by $\frac{1}{2}$
$(x - 2)(x + 8) = 0$	Step 2
$x - 2 = 0$ or $x + 8 = 0$	Step 3
$x = 2$ $\quad x = -8$	Step 4

A check verifies that the solution set is $\{2, -8\}$.

27.

$$x^2 + 2x = 0$$

$x(x + 2) = 0$	Step 2
$x = 0$ or $x + 2 = 0$	Step 3
$x = 0$ $\quad x = -2$	Step 4

A check verifies that the solution set is $\{0, -2\}$.

29.

$$6x^2 - 2x = 0$$

$2x(3x - 1) = 0$	Step 2
$2x = 0$ or $3x - 1 = 0$	Step 3
$x = 0$ $\quad 3x = 1$	Step 4
$\quad x = \frac{1}{3}$	

A check verifies that this solutions set is $\left\{0, \frac{1}{3}\right\}$

31.

$$x^2 - 25 = 0$$

$(x + 5)(x - 5) = 0$	Step 2
$x + 5 = 0$ or $x - 5 = 0$	Step 3
$x = -5$ $\quad x = 5$	Step 4

A check verifies that the solution set is $\{-5, 5\}$

33.
$$4x^2 - 9 = 0$$
$$(2x + 3)(2x - 3) = 0 \qquad \text{Step 2}$$
$$2x + 3 = 0 \quad \text{or} \quad 2x - 3 = 0 \quad \text{Step 3}$$
$$2x = -3 \qquad\qquad 2x = 3$$
$$x = -\frac{3}{2} \qquad\qquad x = \frac{3}{2} \quad \text{Step 4}$$

A check verifies that the solution set is $\left\{-\frac{3}{2}, \frac{3}{2}\right\}$

35.
$$32x^2 - 2 = 0$$
$$16x^2 - 1 = 0 \qquad \text{Multiply by } \frac{1}{2}$$
$$(4x + 1)(4x - 1) = 0 \qquad \text{Step 2}$$
$$4x + 1 = 0 \quad \text{or} \quad 4x - 1 = 0 \quad \text{Step 3}$$
$$4x = -1 \qquad\qquad 4x = 1$$
$$x = -\frac{1}{4} \qquad\qquad x = \frac{1}{4} \quad \text{Step 4}$$

A check verifies that the solution set is $\left\{-\frac{1}{4}, \frac{1}{4}\right\}$

37.
$$2x^2 - 7x - 4 = 0$$
$$(2x + 1)(x - 4) = 0 \qquad \text{Step 2}$$
$$2x + 1 = 0 \quad \text{or} \quad x - 4 = 0 \quad \text{Step 3}$$
$$2x = -1 \qquad\qquad x = 4 \quad \text{Step 4}$$
$$x = -\frac{1}{2}$$

A check verifies that the solution set is $\left\{-\frac{1}{2}, 4\right\}$

39.
$$8x^2 + 19x + 6 = 0$$
$$(8x + 3)(x + 2) = 0 \qquad \text{Step 2}$$
$$8x + 3 = 0 \quad \text{or} \quad x + 2 = 0 \quad \text{Step 3}$$
$$8x = -3 \qquad\qquad x = -2 \quad \text{Step 4}$$
$$x = -\frac{3}{8}$$

A check verifies that the solution set is $\left\{-\frac{3}{8}, -2\right\}$

41.
$$6x^2 - 41x - 7 = 0$$
$$(6x + 1)(x - 7) = 0 \qquad \text{Step 2}$$
$$6x + 1 = 0 \quad \text{or} \quad x - 7 = 0 \quad \text{Step 3}$$
$$6x = -1 \qquad\qquad x = 7 \quad \text{Step 4}$$
$$x = -\frac{1}{6}$$

A check verifies that the solution set is $\left\{-\frac{1}{6}, 7\right\}$

43.
$$8x^2 + 2x - 3 = 0$$
$$(4x + 3)(2x - 1) = 0 \qquad \text{Step 2}$$
$$4x + 3 = 0 \quad \text{or} \quad 2x - 1 = 0 \quad \text{Step 3}$$
$$4x = -3 \qquad\qquad 2x = 1$$
$$x = -\frac{3}{4} \qquad\qquad x = \frac{1}{2} \quad \text{Step 4}$$

A check verifies that the solution set is $\left\{-\frac{3}{4}, \frac{1}{2}\right\}$

45.
$$12x^2 - 31x + 9 = 0$$
$$(4x - 9)(3x - 1) = 0 \qquad \text{Step 2}$$
$$4x - 9 = 0 \quad \text{or} \quad 3x - 1 = 0 \quad \text{Step 3}$$
$$4x = 9 \qquad\qquad 3x = 1$$
$$x = \frac{9}{4} \qquad\qquad x = \frac{1}{3} \quad \text{Step 4}$$

A check verifies that the solution set is $\left\{\frac{9}{4}, \frac{1}{3}\right\}$

47.
$$20x^2 - 28x + 8 = 0$$
$$5x^2 - 7x + 2 = 0 \qquad \text{Multiply by } \frac{1}{4}$$
$$(5x - 2)(x - 1) = 0 \qquad \text{Step 2}$$
$$5x - 2 = 0 \quad \text{or} \quad x - 1 = 0 \quad \text{Step 3}$$
$$5x = 2 \qquad\qquad x = 1 \quad \text{Step 4}$$
$$x = \frac{2}{5}$$

A check verifies that the solution set is $\left\{\frac{2}{5}, 1\right\}$

49.

$x^2 + 6x + 9 = 0$

$(x+3)^2 = 0$ Step 2

$x + 3 = 0$ Step 3

$x = -3$ Step 4

A check verifies that the solution set is {-3}

51.

$2x^2 - 16x + 32 = 0$

$x^2 - 8x + 16 = 0$ Multiply by $\dfrac{1}{2}$

$(x-4)^2 = 0$ Step 2

$x - 4 = 0$ Step 3

$x = 4$ Step 4

A check verifies that the solution set is {4}

53.

$x^2 + 5x - 4 = 10$

$x^2 + 5x - 14 = 0$ Subtract 10

$(x+7)(x-2) = 0$ Step 2

$x + 7 = 0$ or $x - 2 = 0$ Step 3

$x = -7$ $x = 2$ Step 4

A check verifies that the solution set is {-7,2}

55.

$2x^2 - x = 15$

$2x^2 - x - 15 = 0$ Subtract 15

$(2x+5)(x-3) = 0$ Step 2

$2x + 5 = 0$ or $x - 3 = 0$ Step 3

$2x = -5$ $x = 3$ Step 4

$x = -\dfrac{5}{2}$

A check verifies that the solution set is $\left\{ -\dfrac{5}{2}, 3 \right\}$

57.

$8x^2 - 13x - 9 = 5x - 4$

$8x^2 - 18x - 5 = 0$ Subtract 5x, add 4

$(4x+1)(2x-5) = 0$ Step 2

$4x + 1 = 0$ or $2x - 5 = 0$ Step 3

$4x = -1$ $2x = 5$ Step 4

$x = -\dfrac{1}{4}$ $x = \dfrac{5}{2}$

#57. continued

A check verifies that the solution set is $\left\{ -\dfrac{1}{4}, \dfrac{5}{2} \right\}$

59.

$40 - 9x = 20x - 3x^2$

$3x^2 - 29x + 40 = 0$ Add $3x^2$, subtract 20x

$(3x-5)(x-8) = 0$ Step 2

$3x - 5 = 0$ or $x - 8 = 0$ Step 3

$3x = 5$ $x = 8$

$x = \dfrac{5}{3}$

A check verifies that the solution set is $\left\{ \dfrac{5}{3}, 8 \right\}$

61.

$3x^2 + 5x + 20 = 2x^2 - 3x + 4$

$x^2 + 8x + 16 = 0$ Step 1

$(x+4)^2 = 0$ Step 2

$x + 4 = 0$ Step 3

$x = -4$ Step 4

A check verifies that the solution set is {-4}

63.

$10x^2 + 5x - 10 = x^2 + 5x - 6$

$9x^2 - 4 = 0$ Step 1

$(3x+2)(3x-2) = 0$ Step 2

$3x + 2 = 0$ or $3x - 2 = 0$ Step 3

$3x = -2$ $3x = 2$

$x = -\dfrac{2}{3}$ $x = \dfrac{2}{3}$ Step 4

A check verifies that the solution set is $\left\{ -\dfrac{2}{3}, \dfrac{2}{3} \right\}$

65.

$9x^2 + 7x - 7 = 3x(x-4)$

$9x^2 + 7x - 7 = 3x^2 - 12x$

$6x^2 + 19x - 7 = 0$ Step 1

$(3x-1)(2x+7) = 0$ Step 2

$3x - 1 = 0$ or $2x + 7 = 0$ Step 3

$3x = 1$ $2x = -7$

$x = \dfrac{1}{3}$ $x = -\dfrac{7}{2}$

#65. continued

A check verifies that the solution set is $\left\{\dfrac{1}{3}, -\dfrac{7}{2}\right\}$

67.

$3x^2 + 2x - 24 = x(2x - 3)$

$3x^2 + 2x - 24 = 2x^2 - 3x$ Step 1

$(x + 8)(x - 3) = 0$ Step 2

$x + 8 = 0$ or $x - 3 = 0$ Step 3

$x = -8$ $x = 3$ Step 4

A check verifies that the solution set is $\{-8, 3\}$

69.

$3x^2 - 8x - 31 = (2x + 3)(x - 7)$

$3x^2 - 8x - 31 = 2x^2 - 11x - 21$

$x^2 + 3x - 10 = 0$ Step 1

$(x - 2)(x + 5) = 0$ Step 2

$x - 2 = 0$ or $x + 5 = 0$ Step 3

$x = 2$ $x = -5$ Step 4

A check verifies that the solution set is $\{2, -5\}$

71.

$21x^2 + 6x - 11 = (3x - 1)(2x - 1)$

$21x^2 + 6x - 11 = 6x^2 - 5x + 1$

$15x^2 + 11x - 12 = 0$ Step 1

$(5x - 3)(3x + 4) = 0$ Step 2

$5x - 3 = 0$ or $3x + 4 = 0$ Step 3

$5x = 3$ $3x = -4$

$x = \dfrac{3}{5}$ $x = -\dfrac{4}{3}$

A check verifies that the solution set is $\left\{\dfrac{3}{5}, -\dfrac{4}{3}\right\}$

73.

$(2x + 5)(x - 4) = (x + 7)(x - 4)$

$2x^2 - 3x - 20 = x^2 + 3x - 28$

$x^2 - 6x + 8 = 0$ Step 1

$(x - 2)(x - 4) = 0$ Step 2

$x - 2 = 0$ or $x - 4 = 0$ Step 3

$x = 2$ $x = 4$ Step 4

A check verifies that the solution set is $\{2, 4\}$

75.

$(3x + 2)(3x - 2) = (3x + 2)(2x + 7)$

$9x^2 - 4 = 6x^2 + 25x + 14$

$3x^2 - 25x - 18 = 0$ Step 1

$(3x + 2)(x - 9) = 0$ Step 2

$3x + 2 = 0$ or $x - 9 = 0$ Step 3

$3x = -2$ $x = 9$ Step 4

$x = -\dfrac{2}{3}$

A check verifies that the solution set is $\left\{-\dfrac{2}{3}, 9\right\}$.

Exercises 3.8

1.

Let x = 1st positive integer.

Two consecutive integers are x and x + 1.

Thus, $x(x + 1) = 72$

$x^2 + x = 72$

$x^2 + x - 72 = 0$

$(x + 9)(x - 8) = 0$

$x + 9 = 0$ or $x - 8 = 0$

~~$x = -9$~~ $x = 8$

The consecutive positive integers are 8 and 9.

Note: -9 is not positive.

3.

Let x = 1st even integer.

Two consecutive even integers are x and x + 2.

Thus, $x(x + 2) = 48$

$x^2 + 2x = 48$

$x^2 + 2x - 48 = 0$

$(x - 6)(x + 8) = 0$

$x - 6 = 0$ or $x + 8 = 0$

$x = 6$ $x = -8$

There are two possible sets of consecutive even integers: 6 and 8 or -6 and -8.

Note: $x = 6$ $x = -8$

        $x + 2 = 8$ $x + 2 = -6$

5.

Let x = 1st positive integer.

Two consecutive positive integers are x and x + 1.

Their squares are x^2 and $(x + 1)^2$.

#5. continued

Thus, $x^2 + (x+1)^2 = 25$

$x^2 + x^2 + 2x + 1 = 25$

$2x^2 + 2x - 24 = 0$

$x^2 + x - 12 = 0$ Multiply by $\frac{1}{2}$

$(x+4)(x-3) = 0$

$x + 4 = 0$ or $x - 3 = 0$

$\cancel{x = -4}$ or $x = 3$

The positive integers are 3 and 4.

7.

Let x = 1st number.

Then $2x + 1 =$ 2nd number.

Thus, $x(2x+1) = 21$

$2x^2 + x = 21$

$2x^2 + x - 21 = 0$

$(2x+7)(x-3) = 0$

$2x + 7 = 0$ or $x - 3 = 0$

$x = -\dfrac{7}{2}$ $x = 3$

$2x + 1 = -6$ $2x + 1 = 7$

There are two possible sets of numbers:

3 and 7   or $-\dfrac{7}{2}$ and -6.

9.

Let x = the larger number.

Then x - 7 = the smaller number.

Thus, $x(x-7) = -12$

$x^2 - 7x + 12 = 0$

$(x-3)(x-4) = 0$

$x - 3 = 0$ or $x - 4 = 0$

$x = 3$ $x = 4$

$x - 7 = -4$ $x - 7 = -3$

There are two possible sets of numbers:
3 and -4 or 4 and -3.

11.

Let x = width of the rectangle.

Then $4x + 2 =$ length of the rectangle.

Thus, $A = w \cdot \ell$

$42 = x(4x+2)$

$42 = 4x^2 + 2x$

#11. continued

$0 = 4x^2 + 2x - 42$

$0 = 2x^2 + x - 21$ Multiply by $\frac{1}{2}$

$0 = (2x+7)(x-3)$

$2x + 7 = 0$ or $x - 3 = 0$

$\cancel{x = -\dfrac{7}{2}}$ $x = 3$

$4x + 2 = 14$

Answer: 3 ft x 14 ft

13.

Let x = measurement of one leg.

Then x + 2 = measurement of the other leg.

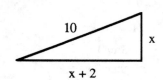

Thus, $x^2 + (x+2)^2 = 10^2$

$x^2 + x^2 + 4x + 4 = 100$

$2x^2 + 4x - 96 = 0$ Multiply by $\frac{1}{2}$

$x^2 + 2x - 48 = 0$

$(x-6)(x+8) = 0$

$x - 6 = 0$ or $x + 8 = 0$

$x = 6$ $\cancel{x = -8}$

$x + 2 = 8$

Answer: 6 in and 8 in

15.

Let x = measurement of the height.

Then $2x + 1 =$ measurement of the base.

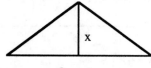

Thus, $A = \dfrac{1}{2}bh$

$14 = \dfrac{1}{2}x(2x+1)$

$28 = x(2x+1)$ Multiply by 2

$28 = 2x^2 + x$

#15. continued

$$0 = 2x^2 + x - 28$$
$$0 = (2x - 7)(x + 4)$$
$$2x - 7 = 0 \quad \text{or} \quad x + 4 = 0$$
$$x = \frac{7}{2} \qquad \cancel{x = 4}$$
$$2x + 1 = 8$$

Answer: height $= \frac{7}{2}$ in; base $= 8$ in

17.

$$h = -16t^2 + 384t \qquad \text{Let } t = 5$$
$$h = -16 \cdot 5^2 + 384 \cdot 5$$
$$h = -400 + 1920$$
$$h = 1520$$

Answer: 1520 ft after 5 sec

$$h = -16t^2 + 384t \qquad \text{Let } t = 12$$
$$h = -16 \cdot 12^2 + 384 \cdot 12$$
$$h = -2304 + 4608$$
$$h = 2304$$

Answer: 2304 ft after 12 sec

$$h = -16t^2 + 384t \qquad \text{Let } h = 0$$
$$0 = -16t^2 + 384t$$
$$0 = t^2 - 24t \qquad \text{Multiply by } -\frac{1}{16}$$
$$0 = t(t - 24)$$
$$t = 0 \quad \text{or} \quad t - 24 = 0$$
$$t = 24$$

Answer: 24 sec

19.

$$h = -16t^2 + 256 \qquad \text{Let } h = 0$$
$$0 = -16t^2 + 256$$
$$0 = t^2 - 16 \qquad \text{Multiply by } -\frac{1}{16}$$
$$0 = (t + 4)(t - 4)$$
$$t + 4 = 0 \quad \text{or} \quad t - 4 = 0$$
$$\cancel{t = 4} \qquad t = 4$$

Answer: After 4 sec

21.

$$h = -2.6t^2 + 41.6 \qquad \text{Let } h = 0$$
$$0 = -2.6t^2 + 41.6$$
$$0 = t^2 - 16 \qquad \text{Multiply by } -\frac{1}{2.6}$$
$$0 = (t + 4)(t - 4)$$
$$t + 4 = 0 \quad \text{or} \quad t - 4 = 0$$
$$\cancel{t = 4} \qquad t = 4$$

Answer: After 4 sec

23.

Let x = width of the tile border.

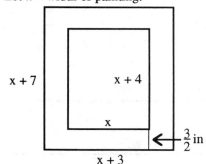

$$14x + 2x$$

The length and width of the pool and tile border
is $14 + 2x$ and $8 + 2x$.

The area of the tile border will equal the total area
of the pool and border minus the area of the pool.

Thus, $75 = (14 + 2x)(8 + 2x) - 8 \cdot 14$

$$75 = 112 + 44x + 4x^2 - 112$$
$$75 = 4x^2 + 44x$$
$$0 = 4x^2 + 44x - 75$$
$$0 = (2x - 3)(2x + 25)$$
$$2x - 3 = 0 \quad \text{or} \quad 2x + 25 = 0$$
$$x = \frac{3}{2} \qquad \cancel{x = -\frac{25}{2}}$$

Answer: $\frac{3}{2}$ ft

25.

Let x = width of painting.

#25. continued

Let $x + 2\left(\dfrac{3}{2}\right)$ or $x + 3$ = the width of the frame.

Let $x + 4 + 2\left(\dfrac{3}{2}\right)$ or $x + 7$ = the length of the frame.

Let $(x + 7)(x + 3)$ = area of the picture and frame.

Thus, $(x + 7)(x + 3) = 117$

$$x^2 + 10x + 21 = 117$$

$$x^2 + 10x - 96 = 0$$

$$(x - 6)(x + 16) = 0$$

$x - 6 = 0 \quad \text{or} \quad x + 16 = 0$

$x = 6 \qquad\qquad \cancel{x = -16}$

Answer: 6 in x 10 in

Review Exercises

1.-13.
The exercises do not require worked - out solutions.

15.

$$\left(x^2 - 3xy + 2y + 5y^2\right) - \left(4x^2 + 6xy - 2y^2 - 3x\right)$$

$$= x^2 - 3xy + 2y + 5y^2 - 4x^2 - 6xy + 2y^2 + 3x$$

$$= x^2 - 4x^2 - 3xy - 6xy + 2y + 5y^2 + 2y^2 + 3x$$

$$= -3x^2 - 9xy + 2y + 7y^2 + 3x$$

17.

$$-\left(5a + 3 - \dfrac{b}{2}\right) + \left(\dfrac{2}{3}\right)\left(9a - \dfrac{1}{4} - 3b\right)$$

$$= -5a - 3 + \dfrac{b}{2} + 6a - \dfrac{1}{6} - 2b$$

$$= -5a + 6a - 3 - \dfrac{1}{6} + \dfrac{b}{2} - 2b$$

$$= a - \dfrac{19}{6} - \dfrac{3b}{2}$$

19.

$$\begin{array}{r} 5x^2 - 8xy + 2y^2 \\ + \left(3x^2 + 8xy - 9y^2\right) \\ \hline 8x^2 \qquad\quad - 7y^2 \end{array}$$

21.

$$x^{12} \cdot x^5 = x^{12+5} = x^{17}$$

23.

$$4a^2b \cdot 7a^3b^8 = 4 \cdot 7 \cdot a^2 \cdot a^3 \cdot b \cdot b^8$$

$$= 28a^5b^9$$

25.

$$\left(3xy^4\right)^2 \cdot \left(-2x^3y^2\right)^3 = 3^2 \, x^2y^8 \cdot (-2)^3x^9y^6$$

$$= 3^2 \cdot (-2)^3 \cdot x^2 \cdot x^9 \cdot y^8y^6$$

$$= 9 \cdot (-8) \cdot x^{11} \cdot y^{14}$$

$$= -72x^{11}y^{14}$$

27.

$$-2x^2\left(5x^2 - x + 3\right) = -2x^2 \cdot 5x^2 - 2x^2(-x) - 2x^2 \cdot 3$$

$$= -10x^4 + 2x^3 - 6x^2$$

29.

$$\begin{array}{cccc} \text{F} & \text{O} & \text{I} & \text{L} \end{array}$$

$$(x + 5y)(4x + 7y) = x \cdot 4x + x \cdot 7y + 5y \cdot 4x + 5y \cdot 7y$$

$$= 4x^2 + 7xy + 20xy + 35y^2$$

$$= 4x^2 + 27xy + 35y^2$$

31.

$$(2a + 9)(2a - 9) = (2a)^2 - (9)^2 = 4a^2 - 81$$

33.

$$(x - 5)(x + 5) = x^2 - 5^2 = x^2 - 25$$

35.

$$(y - 1)^2 = y^2 - 2(y \cdot 1) + 1^2 = y^2 - 2y + 1$$

37.

$$(4x + 1)\left(16x^2 - 4x + 1\right)$$

$$\begin{array}{r} 16x^2 - 4x + 1 \\ 4x + 1 \\ \hline 64x^3 - 16x^2 + 4x \\ 16x^2 - 4x + 1 \\ \hline 64x^3 \qquad\qquad + 1 \end{array}$$

Answer: $64x^3 + 1$

39.

$y^2 + 2y + 4$

$y^2 - 2y + 4$

$\overline{}$

$y^4 + 2y^3 + 4y^2$

$\quad - 2y^3 - 4y^2 - 8y$

$\quad\quad\quad 4y^2 + 8y + 16$

$\overline{}$

$y^4 \quad\quad + 4y^2 \quad\quad + 16$

Answer: $y^4 + 4y^2 + 16$

41.

$10x^3y^2z^4 - 8x^2y^5z^3 - 14x^4yz^7 \quad\quad \text{GCF} = 2x^2yz^3$

$2x^2yz^3\left(5xyz - 4y^4 - 7x^2z^4\right)$

43.

$6a^3x - 3a^2x^2 + 3ax \quad\quad \text{GCF} = 3ax$

$3ax\left(2a^2 - ax + 1\right)$

45.

$x^2 - 9x = x(x - 9) \quad\quad \text{GCF} = x$

47.

$x^2 - 36 = x^2 - (6)^2 = (x + 6)(x - 6)$

49.

$y^2 + 16 \quad\quad \text{Prime}$

51.

$2x^3 - 18x = 2x\left(x^2 - 9\right) \quad\quad \text{GCF} = 2x$

$\quad\quad\quad\quad = 2x(x + 3)(x - 3)$

53.

$y^4 - 81 = \left(y^2\right) - (9)^2 = \left(y^2 + 9\right)\left(y^2 - 9\right)$

$\quad\quad\quad\quad\quad\quad = \left(y^2 + 9\right)(y + 3)(y - 3)$

55.

$x^2 - 3x + 4$ is prime since no two factors of 4 results in a sum of - 3.

57.

$x^2 - 3xy - 10y^2$

Note: $-5 \cdot 2 = -10$ and $-5 + 2 = -3$

Thus, $x^2 - 3xy - 10y^2 = (x - 5y)(x + 2y)$.

59.

$x^2 - 14x + 49 = (x - 7)^2$ since

$-7(-7) = 49$ and $-7 + (-7) = -14$.

61.

$3x^2 - 13x + 12 \quad\quad \text{Product: } 3 \cdot 12 = 36$

Note: $-9(-4) = 36$ and $-9 + (-4) = -13$

Thus, $3x^2 - 13x + 12 = 3x^2 - 9x - 4x + 12$

$\quad\quad\quad\quad\quad\quad = 3x(x - 3) - 4(x - 3)$

$\quad\quad\quad\quad\quad\quad = (x - 3)(3x - 4)$

63.

$18x^2 + 27x + 4 \quad\quad \text{Product: } 18 \cdot 4 = 72$

Note: $3 \cdot 24 = 72$ and $3 + 24 = 27$

Thus, $18x^2 + 27x + 4 = 18x^2 + 3x + 24x + 4$

$\quad\quad\quad\quad\quad\quad = 3x(6x + 1) + 4(6x + 1)$

$\quad\quad\quad\quad\quad\quad = (6x + 1)(3x + 4)$

65.

$18x^3 + 3x^2 - 45x = 3x\left(6x^2 + x - 15\right) \quad\quad \text{GCF} = 3x$

We factor the second factor.

$6x^2 + x - 15 \quad\quad \text{Product: } 6(-15) = -90$

Note: $10(-9) = -90$ and $10 + (-9) = 1$

Thus, $6x^2 + x - 15 = 6x^2 + 10x - 9x - 15$

$\quad\quad\quad\quad\quad\quad = 2x(3x + 5) - 3(3x + 5)$

$\quad\quad\quad\quad\quad\quad = (2x - 3)(3x + 5)$

So, $18x^3 + 3x^2 - 45x = 3x(2x - 3)(3x + 5)$.

67.

$x^4 + 4x^2 - 5$

Note: $5(-1) = -5$ and $5 + (-1) = 4$

Thus, $x^4 + 4x^2 - 5 = \left(x^2 + 5\right)\left(x^2 - 1\right)$

$\quad\quad\quad\quad\quad\quad = \left(x^2 + 5\right)(x + 1)(x - 1)$

69.

$(3x+1)^2 + 4(3x+1) - 5$ Let $u = 3x+1$

Thus, $u^2 + 4u - 5 = (u+5)(u-1)$

So, $(3x+1)^2 + 4(3x+1) - 5 = (3x+1+5)(3x+1-1)$

$\qquad\qquad = (3x+6)(3x)$

$\qquad\qquad = 3x(3x+6)$

$\qquad\qquad = 3x[3(x+2)]$

$\qquad\qquad = 9x(x+2)$

71.

$x^3 + 27 = (x+3)(x^2 - 3\cdot x + 3^2)$

$\qquad = (x+3)(x^2 - 3x + 9)$

73.

$y^3 - 1 = (y-1)(y^2 + y + 1)$

75.

$(x+2y)^3 + (3x-2y)^3$

$= [(x+2y) + (3x-2y)]$

$\quad \cdot [(x+2y)^2 - (x+2y)(3x-2y) + (3x-2y)^2]$

$= (x+2y+3x-2y)$

$\quad \cdot [x^2 + 4xy + 4y^2 - (3x^2 + 4xy - 4y^2)$

$\qquad\qquad + 9x^2 - 12xy + 4y^2]$

$= 4x(x^2 + 4xy + 4y^2 - 3x^2 - 4xy + 4y^2$

$\qquad\qquad + 9x^2 - 12xy + 4y^2)$

$= 4x(7x^2 - 12xy + 12y^2)$

77.

$5xy^2(3x+y) + 1\cdot(3x+7)$ GCF $= 3x+y$

$= (3x+y)(5xy^2 + 1)$

79.

$8a^3(a+2b) + b^3(a+2b)$ GCF $= a+2b$

$= (a+2b)(8a^3 + b^3)$

$= (a+2b)(2a+b)(4a^2 - 2ab + b^2)$

81.

$8xy + 15 + 6y + 20x$

$= 8xy + 20x + 6y + 15$ GCF $= 2x$ and 3

$= 4x(2y+5) + 3(2y+5)$

$= (2y+5)(4x+3)$

83.

$12x^2y - 27y^3 - 8x^2 + 18y^2$ GCF $= 3y$ and -2

$= 3y(4x^2 - 9y^2) - 2(4x^2 - 9y^2)$

$= (4x^2 - 9y^2)(3y-2)$

$= (2x+3y)(2x-3y)(3y-2)$

85.

$x^2 - 8xy + 16y^2 - 25z^2$

$= (x^2 - 8xy + 16y^2) - 25z^2$

$= (x-4y)^2 - 25z^2$

$= [(x-4y) + 5z][(x-4y) - 5z]$

$= (x-4y+5z)(x-4y-5z)$

87.

$x^3 + 6x^2 + 12x + 8$

$= 6x^2 + 12x + x^3 + 8$

$= 6x(x+2) + (x+2)(x^2 - 2x + 4)$ GCF $= x+2$

$= (x+2)(6x + x^2 - 2x + 4)$

$= (x+2)(x^2 + 4x + 4) = (x+2)(x+2)(x+2)$

$\qquad\qquad = (x+2)^3$

89.

$12x^2 - 8x - 15$ Product: $12(-15) = -180$

Note: $10(-18) = -180$ and $10 + (-18) = -8$

Thus, $12x^2 - 8x - 15 = 12x^2 + 10x - 18x - 15$

$\qquad\qquad = 2x(6x+5) - 3(6x+5)$

$\qquad\qquad = (6x+5)(2x-3)$

91.

$16x^2 - 10x - 9$ Product: $16(-9) = -144$

Note: $-18\cdot 8 = -144$ and $-18 + 8 = -10$

#91. continued

Thus, $16x^2 - 10x - 9 = 16x^2 - 18x + 8x - 9$

$$= 2x(8x - 9) + 1(8x - 9)$$
$$= (8x - 9)(2x + 1)$$

93.-99.

The exercises do not require worked - out solutions.

101.

$x^2 - 11x - 26 = 0$

$(x + 2)(x - 13) = 0$

$x + 2 = 0$ or $x - 13 = 0$

  $x = -2$ $x = 13$

A check verifies that the solution set is $\{-2, 13\}$.

103.

$2x^2 - 18x + 28 = 0$

  $x^2 - 9x + 14 = 0$ Multiply by $\dfrac{1}{2}$

  $(x - 2)(x - 7) = 0$

$x - 2 = 0$ or $x - 7 = 0$

  $x = 2$ $x = 7$

A check verifies that the solution set is $\{2, 7\}$.

105.

$4x^2 - 16x = 0$

  $4x(x - 4) = 0$

$4x = 0$ or $x - 4 = 0$

  $x = 0$ $x = 4$

A check verifies that the solution set is $\{0, 4\}$.

107.

  $x^2 - 36 = 0$

$(x + 6)(x - 6) = 0$

$x + 6 = 0$ or $x - 6 = 0$

  $x = -6$ $x = 6$

A check verifies that the solution set is $\{-6, 6\}$.

109.

  $3x^2 - 12 = 0$

  $x^2 - 4 = 0$ Multiply by $\dfrac{1}{3}$

$(x + 2)(x - 2) = 0$

#109. continued

$x + 2 = 0$ or $x - 2 = 0$

  $x = -2$ $x = 2$

A check verifies that the solution set is $\{-2, 2\}$.

111.

$x^2 - 8x + 16 = 0$

  $(x - 4)^2 = 0$

  $x - 4 = 0$

    $x = 4$

A check verifies that the solution set is $\{4\}$.

113.

$3x^2 - 17x + 24 = 0$

$(3x - 8)(x - 3) = 0$

$3x - 8 = 0$ or $x - 3 = 0$

  $x = \dfrac{8}{3}$ $x = 3$

A check verifies that the solution set is $\left\{\dfrac{8}{3}, 3\right\}$.

115.

  $6x^2 + 17x + 7 = 0$

$(3x + 7)(2x + 1) = 0$

$3x + 7 = 0$ or $2x + 1 = 0$

  $x = -\dfrac{7}{3}$ $x = -\dfrac{1}{2}$

A check verifies that the solution set is $\left\{-\dfrac{7}{3}, -\dfrac{1}{2}\right\}$.

117.

$$(x + 4)^2 = 25$$
$$(x + 4)^2 - 25 = 0$$
$$[(x + 4) + 5][(x + 4) - 5] = 0$$
$$(x + 4 + 5)(x + 4 - 5) = 0$$
$$(x + 9)(x - 1) = 0$$

$x + 9 = 0$ or $x - 1 = 0$

  $x = -9$ $x = 1$

A check verifies that the solution set is $\{-9, 1\}$.

119.

$$6x(x+2) = x-5$$
$$6x^2 + 12x = x - 5$$
$$6x^2 + 11x + 5 = 0$$
$$(x+1)(6x+5) = 0$$
$$x+1 = 0 \quad \text{or} \quad 6x+5 = 0$$
$$x = -1 \qquad x = -\frac{5}{6}$$

A check verifies that the solution set is $\left\{-1, -\frac{5}{6}\right\}$.

121.

$$2x^2 - 9x - 22 = 6x + 5$$
$$2x^2 - 15x - 27 = 0$$
$$(2x+3)(x-9) = 0$$
$$2x+3 = 0 \quad \text{or} \quad x-9 = 0$$
$$x = -\frac{3}{2} \qquad x = 9$$

A check verifies that the solution set is $\left\{-\frac{3}{2}, 9\right\}$.

123.

$$8x^2 - 9x - 18 = (x-3)(x+2)$$
$$8x^2 - 9x - 18 = x^2 - x - 6$$
$$7x^2 - 8x - 12 = 0$$
$$(7x+6)(x-2) = 0 \qquad \text{Multiply by } \frac{1}{2}$$
$$7x+6 = 0 \quad \text{or} \quad x-2 = 0$$
$$x = -\frac{6}{7} \qquad x = 2$$

A check verifies that the solution set is $\left\{-\frac{6}{7}, 2\right\}$.

125.

$$(x+6)(x-5) = 5(x+6)$$
$$x^2 + x - 30 = 5x + 30$$
$$x^2 - 4x - 60 = 0$$
$$(x+6)(x-10) = 0$$
$$x+6 = 0 \quad \text{or} \quad x-10 = 0$$
$$x = -6 \qquad x = 10$$

A check verifies that the solution set is $\{-6, 10\}$.

127.

$$(5x+2)(x-4) = (2x+1)(x-14)$$
$$5x^2 - 18x - 8 = 2x^2 - 27x - 14$$
$$3x^2 + 9x + 6 = 0$$
$$x^2 + 3x + 2 = 0$$
$$(x+2)(x+1) = 0$$
$$x+2 = 0 \quad \text{or} \quad x+1 = 0$$
$$x = -2 \qquad x = -1$$

A check verifies that the solution set is $\{-2, -1\}$.

129.

Let x = 1st positive integer.

Then x + 1 = 2nd positive integer.

Thus, $x(x+1) = 132$

$$x^2 + x - 132 = 0$$
$$(x+12)(x-11) = 0$$
$$x+12 = 0 \quad \text{or} \quad x-11 = 0$$
$$\cancel{x = -12} \qquad x = 11$$
$$x + 1 = 12$$

Answer: 11 and 12

131.

Let x = width of rectangle.

Then 2x - 1 = length of rectangle.

Thus, $A = w \cdot \ell$

$$36 = x(2x-1)$$
$$36 = 2x^2 - x$$
$$0 = 2x^2 - x - 36$$
$$0 = (2x-9)(x+4)$$
$$2x-9 = 0 \quad \text{or} \quad \cancel{x+4=0}$$
$$x = \frac{9}{2}$$
$$2x-1 = 8$$

Answer: $\frac{9}{2}$ ft x 8 ft

133.

Let x = base of triangle.

Then 3x + 2 = height of triangle.

Thus, $A = \frac{1}{2}bh$

$$28 = \frac{1}{2}x(3x+2)$$
$$56 = x(3x+2) \qquad \text{Multiply by 2}$$

78

#133. continued

$$56 = 3x^2 + 2x$$

$$0 = 3x^2 + 2x - 56$$

$$0 = (x - 4)(3x + 14)$$

$$x - 4 = 0 \quad \text{or} \quad 3x + 14 = 0$$

$$x = 4 \qquad \qquad \cancel{x = -\dfrac{14}{3}}$$

$$3x + 2 = 14$$

Answer: base = 4 cm, height = 14 cm

135.

$$h = -16t^2 + 16t \qquad \text{Let } t = \frac{1}{2}$$

$$h = -16\left(\frac{1}{2}\right)^2 + 16\left(\frac{1}{2}\right)$$

$$= -16\left(\frac{1}{4}\right) + 16\left(\frac{1}{2}\right)$$

$$= -4 + 8 = 4$$

Answer: 4 ft high

$$h = -16t^2 + 16t \qquad \text{Let } h = 0$$

$$0 = -16t^2 + 16t$$

$$0 = t^2 - t \qquad \text{Multiply by } -\frac{1}{16}$$

$$0 = t(t - 1)$$

$$t = 0 \quad \text{or} \quad t - 1 = 0$$

$$t = 1$$

Answer: after 1 sec

137.

Let x = width of the concrete border.

13x + 2x

The length and width of the pool and the concrete border is 13 + 2x and 8 + 2x.

The area of the concrete border will equal the total area of the pool and border minus the area of the pool.

Thus, $(13 + 2x)(8 + 2x) - 8 \cdot 13 = 100$

$$104 + 42x + 4x^2 - 104 = 100$$

$$4x^2 + 42x - 100 = 0$$

$$2x^2 + 21x - 50 = 0$$

$$(x - 2)(2x + 25) = 0$$

#137. continued

$$x - 2 = 0 \quad \text{or} \quad 2x + 25 = 0$$

$$x = 2 \qquad \qquad \cancel{x = -\dfrac{25}{2}x}$$

Answer: 2 yds

Chapter 3 Test Solutions

1.

The exercise does not require a worked-out solution.

3.

$$\left(2a^2b + 5a - 7b - a^2\right) - \left(7a + 2ab^2 + 4a^2 - 5b + 3ab\right)$$

$$= 2a^2b + 5a - 7b - a^2 - 7a - 2ab^2 - 4a^2 + 5b - 3ab$$

$$= 2a^2b + 5a - 7a - 7b + 5b - a^2 - 4a^2 - 2ab^2 - 3ab$$

$$= 2a^2b - 2a - 2b - 5a^2 - 2ab^2 - 3ab$$

5.

$$-5x^2y^4\left(2x^3y^2 - 3x^6y\right)$$

$$= -5x^2y^4 \cdot 2x^3y^2 - 5x^2y^4\left(-3x^6y\right)$$

$$= -5 \cdot 2x^2x^3y^4 \cdot y^2 - 5(-3)x^2x^6y^4 \cdot y$$

$$= -10x^5y^6 + 15x^8y^5$$

7.

$$(3x + 9)\left(2x^2 - 6x + 11\right)$$

$$\begin{array}{r} 2x^2 - 6x + 11 \\ 3x + 9 \\ \hline 6x^3 - 18x^2 + 33x \\ 18x^2 - 54x + 99 \\ \hline 6x^3 \qquad -21x + 99 \end{array}$$

Answer: $6x^3 - 21x + 99$

9.

$$3x^2 + 10x + 8 \qquad \text{Product: } 3(8) = 24$$

Note: $4 \cdot 6 = 24$ and $4 + 6 = 10$

Thus, $3x^2 + 10x + 8 = 3x^2 + 4x + 6x + 8$

$$= x(3x + 4) + 2(3x + 4)$$

$$= (3x + 4)(x + 2)$$

11.

$x^2 - 4x + 5$ is prime.

13.
$$16x^3 - 48x^2 - x + 3 = 16x^2(x-3) - 1 \cdot (x-3)$$
$$= (x-3)(16x^2 - 1)$$
$$= (x-3)(4x+1)(4x-1)$$

15.
$$9x^2 - 12xy + 4y^2 - 25 = (9x^2 - 12xy + 4y^2) - 25$$
$$= (3x - 2y)^2 - 5^2$$
$$= [(3x-2y)+5][(3x-2y)-5]$$
$$= (3x - 2y - 5)(3x - 2y + 5)$$

17.
$$10a^4x^3 - 5a^2x + 5a = 5a \cdot 2a^3x^3 - 5a \cdot ax + 5a \cdot 1$$
$$= 5a(2a^3x^3 - ax + 1)$$

19.
$$9x^2 - 4 = (3x)^2 - (2)^2 = (3x+2)(3x-2)$$

21.
$$x^2 - 8x - 20 = 0$$
$$(x+2)(x-10) = 0$$
$$x + 2 = 0 \quad \text{or} \quad x - 10 = 0$$
$$x = -2 \qquad\qquad x = 10$$

A check will verify that the solution set is $\{-2, 10\}$.

23.
$$3x(3x+5) = -4$$
$$9x^2 + 15x + 4 = 0$$
$$(3x+4)(3x+1) = 0$$
$$3x + 4 = 0 \quad \text{or} \quad 3x + 1 = 0$$
$$x = -\frac{4}{3} \qquad\qquad x = -\frac{1}{3}$$

A check will verify that the solution set is $\left\{-\frac{4}{3}, -\frac{1}{3}\right\}$.

25.
$$h = -16t^2 - 368t + 800 \qquad \text{Let } h = 0$$
$$0 = -16t^2 - 368t + 800$$
$$0 = t^2 + 23t - 50$$
$$0 = (t-2)(t+25)$$

#25. continued
$$t - 2 = 0 \quad \text{or} \quad t + 25 = 0$$
$$t = 2 \qquad\qquad \cancel{t = -25}$$

Answer: after 2 sec

Test Your Memory

1.-3.
The exercises do not require worked-out solutions.

5.
$$-6 - 4 - (-2) = -6 - 4 + 2 = -8$$

7.
$$\frac{2}{3} + \frac{3}{4} - 1 = \frac{8}{12} + \frac{9}{12} - \frac{12}{12} \qquad \text{LCD} = 12$$
$$= \frac{17}{12} - \frac{12}{12} = \frac{5}{12}$$

9.
$$\sqrt{289 - 9x^2} \quad \text{for } x = 5$$
$$\sqrt{289 - 9 \cdot 5^2} = \sqrt{289 - 9 \cdot 25} = \sqrt{289 - 225} = \sqrt{64} = 8$$

11.
$$4x^2 - |x - 5| \quad \text{for } x = -2$$
$$4(-2)^2 - |-2 - 5| = 4 \cdot 4 - |-7| = 16 - 7 = 9$$

13.
$$(5x^3 - 3x^2 + 7) - (2x^2 - 8x - 4)$$
$$= 5x^3 - 3x^2 + 7 - 2x^2 + 8x + 4$$
$$= 5x^3 - 3x^2 - 2x^2 + 8x + 7 + 4$$
$$= 5x^3 - 5x^2 + 8x + 11$$

15.
$$(7x + 2y)(x - 3y) = 7x \cdot x + 7x(-3y) + 2y \cdot x + 2y(-3y)$$
$$= 7x^2 - 21xy + 2xy - 6y^2$$
$$= 7x^2 - 19xy - 6y^2$$

17.
$$8x^2 - 14x - 15 \qquad \text{Product: } 8(-15) = -120$$
Note: $6(-20) = -120$ and $6 + (-20) = -14$

#17. continued

Thus, $8x^2 - 14x - 15 = 8x^2 + 6x - 20x - 15$

$$= 2x(4x+3) - 5(4x+3)$$
$$= (4x+3)(2x-5)$$

19.

$375a^4 + 24a = 3a(125a^3 + 8)$

$$= 3a((5a)^3 + 2^3)$$
$$= 3a(5a+2)(25a^2 - 10a + 4)$$

21.

$8x^3 + 12x^2y - 4x = 4x(2x^2 + 3xy - 1)$

The second factor is prime.

23.

$4m^2 - 12mn + 9n^2 - 4$

$$= (4m^2 - 12mn + 9n^2) - 4$$
$$= (2m - 3n)^2 - 2^2$$
$$= [(2m-3n)+2][(2m-3n)-2]$$
$$= (2m-3n+2)(2m-3n-2)$$

25.

$x^2 - 11x + 18 = 0$

$(x-2)(x-9) = 0$

$x - 2 = 0$ or $x - 9 = 0$

$x = 2$ $x = 9$

A check verifies that the solution set is $\{2, 9\}$.

27.

$3x - 7 = 6x - 1$

$-7 = 3x - 1$ Subtract $3x$

$-6 = 3x$ Add 1

$-2 = x$ Multiply by $\frac{1}{3}$

A check verifies that the solution set is $\{-2\}$.

29.

$\frac{1}{2}(2x+3) - \frac{2}{3}(x-5) = 5$ LCD $= 6$

$3(2x+3) - 4(x-5) = 30$ Multiply by 6

$6x + 9 - 4x + 20 = 30$

#29. continued

$2x + 29 = 30$ Subtract 29

$2x = 1$ Multiply by $\frac{1}{2}$

$x = \frac{1}{2}$

A check verifies that the solution set is $\left\{\frac{1}{2}.\right\}$

31.

$2(3x+1) - 4(x-3) = 7$

$6x + 2 - 4x + 12 = 7$

$2x + 14 = 7$

$2x = -7$ Subtract 14

$x = -\frac{7}{2}$ Multiply by $\frac{1}{2}$

A check verifies that the solution set is $\left\{-\frac{7}{2}\right\}$.

33.

$|2x - 4| = |x + 7|$

$2x - 4 = x + 7$ or $2x - 4 = -(x+7)$

$2x = x + 11$ $2x - 4 = -x - 7$

$x = 11$ $2x = -x - 3$

                  $3x = -3$

                  $x = -1$

A check verifies that the solution set is $\{11, -1\}$.

35.

$(2x+1)(4x-3) = -2$

$8x^2 - 2x - 3 + 2 = 0$

$(4x+1)(2x-1) = 0$

$4x + 1 = 0$ or $2x - 1 = 0$

$x = -\frac{1}{4}$ $x = \frac{1}{2}$

A check verifies that the solution set is $\left\{-\frac{1}{4}, \frac{1}{2}\right\}$.

37.

$(5x+1)(x-2) = (2x+2)(x-1)$

$5x^2 - 9x - 2 = 2x^2 - 2$

$3x^2 - 9x = 0$

$3x(x-3) = 0$

#37. continued

$3x = 0 \quad$ or $\quad x - 3 = 0$

$x = 0 \qquad\qquad x = 3$

A check verifies that the solution set is $\{0,3\}$.

39.

$5 - 3x \geq 17$

$-3x \geq 12 \qquad\qquad$ Subtract 5

$x \leq -4 \qquad\qquad$ Multiply by $-\dfrac{1}{3}$

41.

$|x - 2| > 3$

$x - 2 > 3 \quad$ or $\quad x - 2 < -3$

$x > 5 \qquad\qquad x < -1$

Answer: $x < -1$ or $x > 5$

43.

Let x = width of rectangle.

Then 3x - 2 = length of rectangle.

Thus, $A = w \cdot \ell$

$65 = x(3x - 2)$

$65 = 3x^2 - 2x$

$0 = 3x^2 - 2x - 65$

$0 = (x - 5)(3x + 13)$

$x - 5 = 0 \quad$ or $\quad 3x + 13 = 0$

$x = 5 \qquad\qquad \cancel{x = -\dfrac{13}{3}}$

Answer: 5 ft x 13 ft

45.

Let x = length of one leg.

Then x + 2 = length of other leg.

Thus, $x^2 + (x + 2)^2 = 10^2$

$x^2 + x^2 + 4x + 4 = 100$

$2x^2 + 4x - 96 = 0$

$x^2 + 2x - 48 = 0 \qquad\qquad$ Multiply by $\dfrac{1}{2}$

$(x - 6)(x + 8) = 0$

$x - 6 = 0 \quad$ or $\quad x + 8 = 0$

$x = 6 \qquad\qquad x = -8$

$x + 2 = 8$

Answer: 6 m and 8m

47.

Let x = width of the tile border.

$14x + 2x$

The area of the tile border will equal the total area of the pool and border minus the area of the pool.

Thus, $81 = (14 + 2x)(10 + 2x) - 14 \cdot 10$

$81 = 140 + 48x + 4x^2 - 140$

$81 = 4x^2 + 48x$

$0 = 4x^2 + 48x - 81$

$0 = (2x - 3)(2x + 27)$

$2x - 3 = 0 \quad$ or $\quad 2x + 27 = 0$

$x = \dfrac{3}{2} \qquad\qquad \cancel{x = -\dfrac{27}{2}}$

Answer: $\dfrac{3}{2}$ ft

49.

Let x = number of nickels.

Then 31 - x = number of dimes.

Thus, $5x + 10(31 - x) = 220$

$5x + 310 - 10x = 220$

$-5x = -90$

$x = 18$

$31 - x = 13$

Answer: 18 nickels, 13 dimes

Chapter 3 Study Guide

Self-Test Exercises

I. List the terms and coefficients of the following polynomial. State the degree of the polynomial.

1. $2x^4 + x^2 - x - 8$

II. Perform the indicated operations and simplify your answers by combining like terms.

2. $6x + 2y - 3xy - x - \dfrac{3}{4}y$ 3. $\left(3a^2b - 7a + 4b + 2a^2\right) - \left(2a - 3ab^2 - a^2 + 2b - 4ab\right)$ 4. $\left(4x^3y^2\right)^3 \left(3xy^4\right)^2$

5. $-3x^4y^2\left(4x^2y^3 - xy^6\right)$ 6. $(3x - 4y)(2x - 7y)$ 7. $(2x + 7)\left(3x^2 - 5x + 9\right)$

III. Completely factor the following polynomials.

8. $x^2 + 2xy - 3y^2$ 9. $2x^2 + 7x + 6$ 10. $27p^3 + 1$ 11. $x^2 - 8x + 7$

12. $3y - 81y^4$ 13. $25x^3 - 50x^2 - x + 2$ 14. $x^4 - 10x^2 + 9$ 15. $x^2 - 6xy + 9y^2 - 16$

16. $x^2 + 14x + 49$ 17. $12a^3x^2 - 6a^2x + 6a$ 18. $24a^2b - 6ab - 9b$ 19. $4x^2 - 25$

IV. Use factoring by grouping to factor the trinomial.

20. $5x^2 - 48x + 27$

V. Find the solution.

21. $x^2 + x - 12 = 0$ 22. $(x + 2)^2 = 9x$ 23. $x(x - 9) = 22$

VI. Use quadratic equations to find the solutions of the following problems.

24. A rectangular door has a length that is 4 ft less than 2 times the width. The area is 30 sq ft. What are the dimensions of the door?

25. A ball is cast straight up into the air at an initial velocity of 32 feet per second. Its height above the ground at a given time t is given by the formula $h = 32t - 16t^2$. Find when the ball will hit the ground.

The worked-out solutions begin on the next page

Self-Test Solutions

1.

Terms: $2x^4, x^2, -x, -8$

Coefficients: 2, 1, -1, -8

Degree: 4

2.

$$6x + 2y - 3xy - x - \frac{3}{4}y = 6x - x + 2y - \frac{3}{4}y - 3xy$$

$$= 5x + \frac{5}{4}y - 3xy$$

3.

$$\left(3a^2b - 7a + 4b + 2a^2\right) - \left(2a - 3ab^2 - a^2 + 2b - 4ab\right)$$

$$= 3a^2b - 7a + 4b + 2a^2 - 2a + 3ab^2 + a^2 - 2b + 4ab$$

$$= 3a^2b - 9a + 2b + 3a^2 + 3ab^2 + 4ab$$

4.

$$\left(4x^3y^2\right)^3\left(3xy^4\right)^2 = 4^3x^{3\cdot3}y^{2\cdot3}\cdot3^2\cdot x^2\cdot y^{4\cdot2}$$

$$= 64x^9y^6\cdot9x^2y^8$$

$$= 9\cdot64\cdot x^9\cdot x^2\cdot y^6\cdot y^8$$

$$= 576x^{11}y^{14}$$

5.

$$-3x^4y^2\left(4x^2y^3 - xy^6\right)$$

$$= -3x^4y^2\cdot4x^2y^3 - 3x^4y^2\left(-xy^6\right)$$

$$= -12x^{4+2}y^{2+3} + 3x^{4+1}y^{2+6}$$

$$= -12x^6y^5 + 3x^5y^8$$

6.

$$(3x - 4y)(2x - 7y)$$

$$\quad\quad F\quad\quad O\quad\quad I\quad\quad L$$

$$= 3x\cdot2x + 3x(-7y) - 4y\cdot2x - 4y(-7y)$$

$$= 6x^2 - 21xy - 8xy + 28y^2$$

$$= 6x^2 - 29xy + 28y^2$$

7.

$$(2x + 7)\left(3x^2 - 5x + 9\right)$$

$$\begin{array}{r} 3x^2 - 5x + 9 \\ x\quad\underline{2x + 7} \\ 6x^3 - 10x^2 + 18x \\ \underline{21x^2 - 35x + 63} \\ 6x^3 + 11x^2 - 17x + 63 \end{array}$$

8.

$$x^2 + 2xy - 3y^2$$

Two factors of -3 whose sum is 2 are -1 and 3.

Thus, $x^2 + 2xy - 3y^2 = (x - y)(x + 3y)$.

9.

$$2x^2 + 7x + 6 \quad\quad \text{Product: } 2\cdot6 = 12$$

Two factors of 12 whose sum is 7 are 3 and 4.

Thus, $2x^2 + 7x + 6 = 2x^2 + 3x + 4x + 6$

$$= x(2x + 3) + 2(2x + 3)$$

$$= (2x + 3)(x + 2)$$

10.

$$27p^3 + 1 = (3p)^3 + (1)^3 = (3p + 1)\left(9p^2 - 3p + 1\right)$$

11.

$$x^2 - 8x + 7$$

Two factors of 7 whose sum is -8 are -7 and -1.

Thus, $x^2 - 8x + 7 = (x - 7)(x - 1)$.

12.

$$3y - 81y^4 = 3y\left(1 - 27y^3\right)$$

$$= 3y\left(1 - (3y)^3\right)$$

$$= 3y(1 - 3y)\left(1 + 3y + 9y^2\right)$$

13.

$$25x^3 - 50x^2 - x + 2$$

$$= 25x^2(x - 2) - 1\cdot(x - 2) \quad\quad GCF = x - 2$$

$$= (x - 2)\left(25x^2 - 1\right) = (x - 2)(5x + 1)(5x - 1)$$

14.

$x^4 - 10x^2 + 9$

Two factors of 9 whose sum is -10 are -1 and -9.

Thus, $x^4 - 10x^2 + 9 = (x^2 - 1)(x^2 - 9)$

$$= (x+1)(x-1)(x+3)(x-3)$$

15.

$$x^2 - 6xy + 9y^2 - 16 = (x^2 - 6xy + 9y^2) - 16$$
$$= (x - 3y)^2 - (4)^2$$
$$= [(x-3y)+4][(x-3y)-4]$$
$$= (x - 3y + 4)(x - 3y - 4)$$

16.

$x^2 + 14x + 49$

Two factors of 49 whose sum is 14 are 7 and 7.

Thus, $x^2 + 14x + 49 = (x+7)(x+7) = (x+7)^2$

17.

$12a^3x^2 - 6a^2x + 6a = 6a(2a^2x^2 - ax + 1)$

18.

$24a^2b - 6ab - 9b = 3b(8a^2 - 2a - 3)$
$$= 3b(4a - 3)(2a + 1)$$

19.

$4x^2 - 25 = (2x)^2 - (5)^2 = (2x + 5)(2x - 5)$

20.

$5x^2 - 48x + 27$

Two factors of 135 whose sum is -48 are -3 and -45.

Thus, $5x^2 - 48x + 27 = 5x^2 - 3x - 45x + 27$
$$= x(5x - 3) - 9(5x - 3)$$
$$= (5x - 3)(x - 9)$$

21.

$$x^2 + x - 12 = 0$$
$$(x - 3)(x + 4) = 0$$
$$x - 3 = 0 \quad \text{or} \quad x + 4 = 0$$
$$x = 3 \qquad\qquad x = -4$$

A check verifies that the solution set is $\{3, -4\}$.

22.

$$(x + 2)^2 = 9x$$
$$x^2 + 4x + 4 = 9x$$
$$x^2 - 5x + 4 = 0 \qquad \text{Subtract } 9x$$
$$(x - 4)(x - 1) = 0$$
$$x - 4 = 0 \quad \text{or} \quad x - 1 = 0$$
$$x = 4 \qquad\qquad x = 1$$

A check verifies that the solution set is $\{4, 1\}$.

23.

$$x(x - 9) = 22$$
$$x^2 - 9x - 22 = 0$$
$$(x - 11)(x + 2) = 0$$
$$x - 11 = 0 \quad \text{or} \quad x + 2 = 0$$
$$x = 11 \qquad\qquad x = -2$$

A check verifies that the solution set is $\{11, 2\}$.

24.

Let x = width.

Then 2x - 4 = length.

Thus, $A = w \cdot \ell$

$$30 = x(2x - 4)$$
$$30 = 2x^2 - 4x$$
$$0 = 2x^2 - 4x - 30$$
$$0 = x^2 - 2x - 15 \qquad \text{Multiply by } \frac{1}{2}$$
$$0 = (x + 3)(x - 5)$$
$$x + 3 = 0 \quad \text{or} \quad x - 5 = 0$$
$$\cancel{x = -3} \qquad\qquad x = 5$$
$$2x - 4 = 6$$

Answer: 5 ft x 6ft

25.

$$h = 32t - 16t^2 \qquad \text{Let } h = 0$$
$$0 = 32t - 16t^2$$
$$0 = t^2 - 2t \qquad \text{Multiply by } -\frac{1}{16}$$
$$0 = t(t - 2)$$
$$\cancel{t = 0} \quad \text{or} \quad t - 2 = 0$$
$$t = 2$$

Answer: after 2 sec

CHAPTER 4   RATIONAL EXPRESSIONS

Solutions to Text Exercises

Exercises 4.1

1.
$$2^2 \cdot 2^3 = 2^{2+3} = 2^5 = 32$$

3.
$$\frac{5^4}{5^2} = 5^{4-2} = 5^2 = 25$$

5.
$$\frac{(-7)^4}{(-7)^5} = \frac{1}{(-7)^{5-4}} = \frac{1}{(-7)^1} = -\frac{1}{7}$$

7.
$$\left(2^2\right)^4 = 2^{2\cdot 4} = 2^8 = 256$$

9.
$$\left(4^{-1}\right)^{-2} = 4^{-1(-2)} = 4^2 = 16$$

11.
$$6^{-3} = \frac{1}{6^3} = \frac{1}{216}$$

13.
$$-8^2 = -\left(8^2\right) = -(64) = -64$$

15.
$$(-11)^2 = (-11)(-11) = 121$$

17.
$$(-3)^{-4} = \frac{1}{(-3)^4} = \frac{1}{81}$$

The product of an even number of negatives
gives a positive answer.

19.
$$-4^{-3} = \frac{-1 \cdot 4^{-3}}{1} = \frac{-1}{4^3} = -\frac{1}{64}$$

21.
$$2^{-3} + 4^{-1} = \frac{1}{2^3} + \frac{1}{4} = \frac{1}{8} + \frac{1}{4} = \frac{1}{8} + \frac{2}{8} = \frac{3}{8}$$

23.
$$4^{-1} + 2^{-2} = \frac{1}{4} + \frac{1}{2^2} = \frac{1}{4} + \frac{1}{4} = \frac{2}{4} = \frac{1}{2}$$

25.
$$\left(2^{-1} - 5^{-2}\right)^{-1} = \left(\frac{1}{2} - \frac{1}{5^2}\right)^{-1} = \left(\frac{1}{2} - \frac{1}{25}\right)^{-1}$$
$$= \left(\frac{25}{50} - \frac{2}{50}\right)^{-1}$$
$$= \left(\frac{23}{50}\right)^{-1} = \frac{50}{23}$$

27.
$$8^{-1} + 8^0 + 8 = \frac{1}{8} + 1 + 8 = \frac{1}{8} + 9 = 9\frac{1}{8} = \frac{73}{8}$$

29.
$$\left(\frac{2}{3}\right)^3 = \frac{2^3}{3^3} = \frac{8}{27}$$

31.
$$\left(\frac{9}{8}\right)^0 = 1 \quad \text{since } a^0 = 1$$

33.
$$\left(\frac{5}{8}\right)^{-1} = \frac{8}{5}$$

35.
$$\left(\frac{4}{3}\right)^{-2} = \frac{4^{-2}}{3^{-2}} = \frac{3^2}{4^2} = \frac{9}{16}$$

37.
$$\frac{5^{-2}}{3^{-3}} = \frac{3^3}{5^2} = \frac{27}{25}$$

39.

$$\frac{5^2}{3^{-2}} = \frac{5^2 \cdot 3^2}{1} = \frac{25 \cdot 9}{1} = \frac{225}{1} = 225$$

41.

$$\frac{7x^8y^2}{2x^6y} = \frac{7x^{8-6}y^{2-1}}{2} = \frac{7}{2}x^2y$$

43.

$$\left(\frac{4x^3}{2y^5}\right)^3 = \frac{4^3x^9}{2^3y^{15}} = \frac{64x^9}{8y^{15}} = \frac{8x^9}{y^{15}}$$

45.

$$\left(\frac{8x^9y^3}{4x^3y^2}\right)^2 = \left(\frac{2x^{9-3}y^{3-2}}{1}\right)^2 = \left(2x^6y\right)^2 = 2^2x^{6\cdot2}y^{1\cdot2}$$
$$= 4x^{12}y^2$$

47.

$$\left(3x^2y^{-4}\right)\left(7x^{-5}y^{-1}\right) = 3 \cdot 7x^2 \cdot x^{-5} \cdot y^{-4} \cdot y^{-1}$$
$$= 21x^{2-5} \cdot y^{-4-1}$$
$$= 21x^{-3}y^{-5}$$
$$= \frac{21x^{-3}y^{-5}}{1} = \frac{21}{x^3y^5}$$

49.

$$\left(3x^{-2}y^2\right)^{-1}\left(4xy^3\right)^{-2} = 3^{-1}x^2y^{-2} \cdot 4^{-2}x^{-2}y^{-6}$$
$$= 3^{-1} \cdot 4^{-2}x^2 \cdot x^{-2} \cdot y^{-2} \cdot y^{-6}$$
$$= 3^{-1} \cdot 4^{-2}x^0 \cdot y^{-8}$$
$$= \frac{3^{-1} \cdot 4^{-2} \cdot 1 \cdot y^{-8}}{1}$$
$$= \frac{1}{3 \cdot 4^2 y^8} = \frac{1}{48x^8}$$

51.

$$\left(8x^0y^{-2}\right)^0\left(2xy^{-2}\right)^3 = 1 \cdot \left(2xy^{-2}\right)^3 = \frac{2^3x^3y^{-6}}{1} = \frac{8x^3}{y^6}$$

53.

$$\left(\frac{5x}{2y^3}\right)^2 = \frac{5^2x^2}{2^2y^{3\cdot2}} = \frac{25x^2}{4y^6}$$

55.

$$\left(\frac{2x^4}{5y^{-2}}\right)^3 = \frac{2^3x^{4\cdot3}}{5^3y^{-2\cdot3}} = \frac{8x^{12}}{125y^{-6}} = \frac{8x^{12}y^6}{125}$$

57.

$$\left(\frac{5x^{-3}y}{3x^2y^{-2}}\right)^3 = \left(\frac{5y^{1-(-2)}}{3x^{2-(-3)}}\right)^3 = \left(\frac{5y^3}{3x^5}\right)^3 = \frac{5^3y^9}{3^3x^{5\cdot3}} = \frac{125y^9}{27x^{15}}$$

59.

$$\frac{8x^3y^{-2}}{10x^{-6}y^5} = \frac{4x^{3-(-6)}}{5y^{5-(-2)}} = \frac{4x^9}{5y^7}$$

61.

$$\frac{2^{-2}x^{-3}y^{-4}}{8^{-1}x^{-1}y^{-2}} = \frac{8}{2^2x^{-1-(-3)}y^{-2-(-4)}} = \frac{8}{4x^2y^2} = \frac{2}{x^2y^2}$$

63.

$$\frac{\left(2x^3y^{-4}\right)^4}{\left(5x^0y^7\right)^2} = \frac{2^4x^{3\cdot4}y^{-4\cdot4}}{5^2x^{0\cdot2}y^{7\cdot2}} = \frac{16x^{12}y^{-16}}{25x^0y^{14}} \quad \text{Note: } x^0 = 1$$
$$= \frac{16x^{12}}{25y^{14-(-16)}} = \frac{16x^{12}}{25y^{30}}$$

65.

$$\frac{\left(-3x^4y^7\right)^3}{\left(2x^{-4}y^3\right)^{-4}} = \frac{(-3)^3x^{4\cdot3}y^{7\cdot3}}{2^{-4}x^{-4(-4)}y^{3(-4)}} = \frac{-27x^{12}y^{21}}{2^{-4}x^{16}y^{-12}}$$
$$= \frac{-27 \cdot 2^4y^{21-(-12)}}{x^{16-12}}$$
$$= \frac{-432y^{33}}{x^4}$$

67.

$$\left(\frac{5x^4}{3y}\right)^{-2} = \frac{5^{-2}x^{4(-2)}}{3^{-2}y^{-2}} = \frac{5^{-2}x^{-8}}{3^{-2}y^{-2}} = \frac{3^2y^2}{5^2x^8} = \frac{9y^2}{25x^8}$$

69.

$$\left(\frac{3x^3}{y^{-2}}\right)^{-2}\left(\frac{4x^5}{y^2}\right)^2 = \frac{3^{-2}x^{3(-2)}}{y^{-2(-2)}} \cdot \frac{4^2x^{5\cdot2}}{y^{2\cdot2}}$$
$$= \frac{3^{-2} \cdot x^{-6}}{y^4} \cdot \frac{4^2x^{10}}{y^4}$$

#69. continued

$$= \frac{3^{-2} \cdot 4^2 x^{-6+10}}{y^{4+4}}$$

$$= \frac{3^{-2} \cdot 4^2 x^4}{y^8} = \frac{4^2 x^4}{3^2 y^8} = \frac{16 x^4}{9 y^8}$$

71.

$$\left(\frac{-4x}{y^3}\right)^{-2}\left(\frac{6x^{-2}}{y^4}\right)^2 = \frac{(-4)^{-2} x^{-2}}{y^{3(-2)}} \cdot \frac{6^2 x^{-2\cdot2}}{y^{4\cdot2}}$$

$$= \frac{(-4)^{-2} \cdot 6^2 \cdot x^{-2} \cdot x^{-4}}{y^{-6} \cdot y^8}$$

$$= \frac{(-4)^{-2} \cdot 6^2 \cdot x^{-6}}{y^2}$$

$$= \frac{6^2}{(-4)^2 x^6 y^2} = \frac{36}{16 x^6 y^2} = \frac{9}{4 x^6 y^2}$$

73.

$$\left[\frac{6x^3 y^{-2}}{9^0 x^{-5} y^{-6}}\right]^{-3} = \left[\frac{6x^{3-(-5)} y^{-2-(-6)}}{1}\right]^{-3} \quad \text{Note: } 9^0 = 1$$

$$= \left(6x^8 y^4\right)^{-3}$$

$$= 6^{-3} x^{-24} y^{-12} = \frac{1}{216 x^{24} y^{12}}$$

75.

$$\left[\left(\frac{2x^2}{3y^3}\right)^{-1}\right]^{-3} = \left(\frac{2x^2}{3y^3}\right)^{-1(-3)} = \left(\frac{2x^2}{3y^3}\right)^3 = \frac{2^3 x^{2\cdot3}}{3^3 y^{3\cdot3}}$$

$$= \frac{8x^6}{27 y^9}$$

77.

$$\frac{(2x-y)^{-3}}{(2x-y)^{-4}} = (2x-y)^{-3-(-4)} = (2x-y)^{-3+4} = (2x-y)$$

79.

$$(x+3)^{-2} = \frac{1}{(x+3)^2} = \frac{1}{x^2+6x+9}$$

81.

$$(x+y)^{-2} = \frac{1}{(x+y)^2} = \frac{1}{x^2+2xy+y^2}$$

83.

$$(x-y)(x^{-1}+y^{-1}) = (x-y)\cdot x^{-1} + (x-y)\cdot y^{-1}$$

$$= x\cdot x^{-1} - y\cdot x^{-1} + x\cdot y^{-1} - y\cdot y^{-1}$$

$$= x^0 - x^{-1}y + xy^{-1} - y^0$$

$$= 1 - \frac{y}{x} + \frac{x}{y} - 1$$

$$= \frac{x}{y} - \frac{y}{x} = \frac{x^2}{xy} - \frac{y^2}{xy} = \frac{x^2 - y^2}{xy}$$

85.

$$\left(x^{5m}\right)^2 = x^{5m\cdot2} = x^{10m}$$

87.

$$\frac{x^{-3m}}{x^{5m}} = \frac{1}{x^{5m-(-3m)}} = \frac{1}{x^{8m}}$$

89.

$$\frac{x^m \cdot x^{5m}}{x^{2m}} = \frac{x^{m+5m}}{x^{2m}} = \frac{x^{6m}}{x^{2m}} = x^{6m-2m} = x^{4m}$$

91.

$$5(2x-3)^4(x-7)^4 + 8(2x-3)^3(x-7)^5$$

$$\text{GCF} = (2x-3)^3(x-7)^4$$

$$= (2x-3)^3(x-7)^4\left(5(2x-3) + 8(x-7)\right)$$

$$= (2x-3)^3(x-7)^4(10x-15+8x-56)$$

$$= (2x-3)^3(x-7)^4(18x-71)$$

93.

$$\frac{4(x+1)^2 - 8x(x+1)}{(x+1)^4} = \frac{4(x+1)\left[(x+1) - 2x\right]}{(x+1)^4}$$

$$= \frac{4(x+1)(-x+1)}{(x+1)^4}$$

$$= \frac{4(-x+1)}{(x+1)^{4-1}} = \frac{4(1-x)}{(x+1)^3}$$

95.

$$-(x+1)(x-2)^{-2} + (x-2)^{-1} \qquad \text{GCF} = (x-2)^{-2}$$

$$= (x-2)^{-2}\left(-(x+1) + (x-2)\right)$$

$$= (x-2)^{-2}(-x-1+x-2)$$

#95. continued

$$= (x-2)^{-2}(-3)$$

$$= \frac{-3}{(x-2)^2}$$

Exercises 4.2

1.

$$\frac{8x^4y^2}{6xy^3} = \frac{2 \cdot 4x^{4-1}}{2 \cdot 3y^{3-2}} = \frac{4x^3}{3y}$$

3.

$$\frac{15x^5y^9}{3xy^3} = \frac{\cancel{3} \cdot 5x^{5-1}y^{9-3}}{\cancel{3} \cdot 1} = 5x^4y^6$$

5.

$$\frac{-9ab^2c^3}{4ab^3c} = \frac{-9\cancel{a}c^{3-1}}{4\cancel{a}b^{3-2}} = \frac{-9c^2}{4b}$$

7.

$$\frac{7x^3yz^4}{4xy^3z^8} = \frac{7x^{3-1}}{4y^{3-1}z^{8-4}} = \frac{7x^2}{4y^2z^4}$$

9.

$$\frac{-6xy^3z^2}{24x^3y^7z^8} = \frac{-1 \cdot \cancel{6}}{\cancel{6} \cdot 4x^{3-1}y^{7-3}z^{8-2}} = -\frac{1}{4x^2y^4z^6}$$

11.

$$\frac{3x^3 + x^2y}{x^2y^2 - 2xy^3} = \frac{x^2(3x+y)}{xy^2(x-2y)} = \frac{x^{2-1}(3x+y)}{y^2(x-2y)}$$

$$= \frac{x(3x+y)}{y^2(x-2y)}$$

13.

$$\frac{3x + 6y}{3x} = \frac{\cancel{3}(x+2y)}{\cancel{3} \cdot x} = \frac{x+2y}{x}$$

15.

$$\frac{4x^3y - 4x^2y}{2xy^2 - 2x^2y^2} = \frac{4x^2y(x-1)}{2xy^2(1-x)} = \frac{\cancel{2} \cdot 2x^{2-1}\cancel{(x-1)}}{\cancel{2}y^{2-1} \cdot (-1)\cancel{(x-1)}}$$

$$= \frac{2x}{-y} = -\frac{2x}{y}$$

Note: $(1-x) = -1 \cdot (-1+x) = -1 \cdot (x-1)$

17.

$$\frac{x^2 + 5x + 6}{x^2 - 4x - 21} = \frac{\cancel{(x+3)}(x+2)}{\cancel{(x+3)}(x-7)} = \frac{x+2}{x-7}$$

19.

$$\frac{6x^2 + 7x - 3}{4x^2 - 9} = \frac{\cancel{(2x+3)}(3x-1)}{\cancel{(2x+3)}(2x-3)} = \frac{3x-1}{2x-3}$$

21.

$$\frac{x^2 - 2x - 3}{x^2 + 6x + 8} \qquad \text{There are no common factors.}$$

23.

$$\frac{x^2 + 3x - 10}{4 - x^2} = \frac{(x+5)(x-2)}{-(x^2-4)} = \frac{(x+5)\cancel{(x-2)}}{-(x+2)\cancel{(x-2)}}$$

$$= -\frac{x+5}{x+2}$$

25.

$$\frac{4x^2 - 12x + 9}{2x^2 - 11x + 12} = \frac{(2x-3)\cancel{(2x-3)}}{(x-4)\cancel{(2x-3)}} = \frac{2x-3}{x-4}$$

27.

$$\frac{3x^2 + 4x}{6x^2 + 5x - 4} = \frac{x\cancel{(3x+4)}}{(2x-1)\cancel{(3x+4)}} = \frac{x}{2x-1}$$

29.

$$\frac{3x^2 + 3x - 6}{6x^2 + 30x} = \frac{3(x^2+x-2)}{6x(x+5)} = \frac{\cancel{3}(x^2+x-2)}{2 \cdot \cancel{3}x(x+5)}$$

$$= \frac{x^2 + x - 2}{2x(x+5)}$$

31.

$$\frac{2x^2 + 5x - 12}{9 - 4x^2} = \frac{(2x-3)(x+4)}{-(4x^2-9)} = \frac{\cancel{(2x-3)}(x+4)}{-(2x+3)\cancel{(2x-3)}}$$

$$= -\frac{x+4}{2x+3}$$

33.

$$\frac{x^3 + 8}{3x^2 + 2x - 8} = \frac{\cancel{(x+2)}(x^2-2x+4)}{(3x-4)\cancel{(x+2)}} = \frac{x^2 - 2x + 4}{3x-4}$$

35.

$$\frac{27x^3 - 1}{6x^2 + 4x - 2} = \frac{(3x-1)(9x^2 + 3x + 1)}{2(3x^2 + 2x - 1)}$$

$$= \frac{(3x-1)(9x^2 + 3x + 1)}{2(3x-1)(x+1)} = \frac{9x^2 + 3x + 1}{2(x+1)}$$

37.

$$\frac{2ax - ay - 2bx + by}{ax + ay - bx - by} = \frac{a(2x-y) - b(2x-y)}{a(x+y) - b(x+y)}$$

$$= \frac{(2x-y)(a-b)}{(x+y)(a-b)} = \frac{2x-y}{x+y}$$

39.

$$\frac{6ax - 8x - 3ay + 4y}{ay + 2y - 2ax - 4x} = \frac{2x(3a-4) - y(3a-4)}{y(a+2) - 2x(a+2)}$$

$$= \frac{(3a-4)(2x-y)}{(a+2)(y-2x)}$$

$$= \frac{(3a-4)(2x-y)}{(a+2)(-1)(2x-y)} = -\frac{3a-4}{a+2}$$

41.

$$\frac{x^3 + xy + 3x^2y + 3y^2}{xy^2 + x^2 + 3y^3 + 3xy} = \frac{x(x^2+y) + 3y(x^2+y)}{x(y^2+x) + 3y(y^2+x)}$$

$$= \frac{(x^2+y)(x+3y)}{(y^2+x)(x+3y)} = \frac{x^2+y}{y^2+x}$$

43.

a. $\dfrac{x^2 - x - 6}{x^2 - 4} = \dfrac{(x-3)(x+2)}{(x+2)(x-2)}$

$= \dfrac{x-3}{x-2}$

b. $x = 1; \quad \dfrac{1-3}{1-2} = \dfrac{-2}{-1} = 2$

$x = 3; \quad \dfrac{3-3}{1-2} = \dfrac{0}{-2} = 0$

$x = -2; \quad \dfrac{-2-3}{-2-2} = \dfrac{-5}{-4} = \dfrac{5}{4}$

c. $x = 1; \quad \dfrac{1^2 - 1 - 6}{1^2 - 4} = \dfrac{-6}{-3} = 2$

$x = 3; \quad \dfrac{3^2 - 3 - 6}{3^2 - 4} = \dfrac{9-3-6}{9-4} = \dfrac{0}{5} = 0$

#43. continued

$x = -2; \quad \dfrac{(-2)^2 - 1 - 6}{(-2)^2 - 4} = \dfrac{4-1-6}{4-4} = \dfrac{-3}{0}$ Undefined

---

Exercises 4.3

1.

$$\frac{5xy^2}{3a} \cdot \frac{2x^3y}{7a^4b} = \frac{5xy^2 \cdot 2x^3y}{3a \cdot 7a^4b} = \frac{10x^4y^3}{21a^5b}$$

3.

$$\frac{9xy^6}{5x^2b} \div \frac{6b^2y^8}{15b^4y} = \frac{9xy^6}{5x^2b} \cdot \frac{15b^4y}{6b^2y^8} = \frac{9xy^6 \cdot 15b^4y}{5x^2b \cdot 6b^2y^8}$$

$$= \frac{9 \cdot 15 \cdot x \cdot y^7 \cdot b^4}{5 \cdot 6 \cdot x^2 \cdot b^3 \cdot y^8}$$

$$= \frac{3 \cdot 3 \cdot 3 \cdot 5 \cdot b^{4-3}}{5 \cdot 5 \cdot 2 \cdot x^{2-1} \cdot y^{8-7}}$$

$$= \frac{9b}{2xy}$$

5.

$$\frac{4xy^3}{2xy} \div \frac{12x^2y^3}{3xy} = \frac{4xy^3}{2xy} \cdot \frac{3xy}{12x^2y^3}$$

$$= \frac{4 \cdot 3 \cdot x^2y^4}{2 \cdot 12 \cdot x^3 \cdot y^4}$$

$$= \frac{1 \cdot 12 \cdot y^4}{2 \cdot 12 \cdot x^{3-2} \cdot y^4} = \frac{1}{2x}$$

7.

$$\frac{2ab^2}{7xb} \cdot \frac{2ab^2}{4a^3b^3} = \frac{2ab^2 \cdot 2ab^2}{7xb \cdot 4a^3b^3} = \frac{4a^2b^4}{7 \cdot 4 \cdot x \cdot a^3b^4}$$

$$= \frac{1 \cdot 4b^4}{7 \cdot 4 x \cdot a^{3-2} \cdot b^4} = \frac{1}{7ax}$$

9.

$$\frac{16xy^7}{12x^3y^2} \cdot \frac{3x^2y}{4y^6} = \frac{16xy^7 \cdot 3x^2y}{12x^3y^2 \cdot 4y^6} = \frac{16 \cdot 3x^3y^8}{12 \cdot 4x^3 \cdot y^8}$$

$$= \frac{4 \cdot 4 \cdot 3x^3y^8}{3 \cdot 4 \cdot 4x^3y^8} = 1$$

11.

$$\frac{10x^2y^2}{6xy^2} \div \frac{x^2y}{3x^2y^3} = \frac{10x^2y^2}{6xy^2} \cdot \frac{3x^2y^3}{x^2y}$$

$$= \frac{10x^2y^2 \cdot 3x^2y^3}{6xy^2 \cdot x^2y}$$

$$= \frac{10 \cdot 3x^4y^5}{6x^3y^3}$$

$$= \frac{\cancel{2} \cdot 5 \cdot \cancel{3}x^{4-3}y^{5-3}}{\cancel{2} \cdot \cancel{3} \cdot 1}$$

$$= 5xy^2$$

13.

$$\frac{27ax^3}{5x^2} \div \frac{3ax^4}{5x^5y^4} = \frac{27ax^3}{5x^2} \cdot \frac{5x^5y^4}{3ax^4} = \frac{27ax^3 \cdot 5x^5y^4}{5x^2 \cdot 3ax^4}$$

$$= \frac{27 \cdot 5 \cdot a \cdot x^8y^4}{5 \cdot 3 \cdot a \cdot x^6}$$

$$= \frac{\cancel{3} \cdot 9 \cdot \cancel{5} \cdot \cancel{a} \cdot x^{8-6} \cdot y^4}{1 \cdot \cancel{5} \cdot \cancel{3} \cdot \cancel{a}}$$

$$= 9x^2y^4$$

15.

$$\frac{2x^2y}{1} \cdot \frac{5ax^3}{8y^4} = \frac{2x^2y \cdot 5ax^3}{1 \cdot 8y^4} = \frac{2 \cdot 5 \cdot a \cdot x^5 \cdot y}{8y^4}$$

$$= \frac{\cancel{2} \cdot 5 \cdot a \cdot x^5}{\cancel{2} \cdot 4y^{4-1}} = \frac{5ax^5}{4y^3}$$

17.

$$\frac{a^2bc}{1} \div \frac{abc}{2xyz^2} = \frac{a^2bc}{1} \cdot \frac{2xyz^2}{abc} = \frac{a^2bc \cdot 2xyz^2}{a \cdot b \cdot c}$$

$$= \frac{a^{2-1}\cancel{b} \cdot \cancel{c} \cdot 2xyz^2}{\cancel{b} \cdot \cancel{c}}$$

$$= 2axyz^2$$

19.

$$\frac{10x^2y^3}{7a^2b} \div \frac{5xy^5}{1} = \frac{10x^2y^3}{7a^2b} \cdot \frac{1}{5xy^5} = \frac{10x^2y^3}{7 \cdot 5a^2bxy^5}$$

$$= \frac{2 \cdot \cancel{5}x^{2-1}}{7 \cdot \cancel{5}a^2by^{5-3}}$$

$$= \frac{2x}{7a^2by^2}$$

21.

$$\frac{5xy^2}{7a^3b^4} \div \frac{3x^4y^7}{6a^2b^6} \cdot \frac{14x^5y^2}{15a^4b} = \frac{5xy^2}{7a^3b^4} \cdot \frac{6a^2b^6}{3x^4y^7} \cdot \frac{14x^5y^2}{15a^4b}$$

$$= \frac{5xy^2 \cdot 6a^2b^6 \cdot 14x^5y^2}{7a^3b^4 \cdot 3x^4y^7 \cdot 15a^4b}$$

$$= \frac{5 \cdot 6 \cdot 14 \cdot x^6 \cdot y^4a^2b^6}{7 \cdot 3 \cdot 15a^7b^5x^4y^7}$$

$$= \frac{\cancel{5} \cdot 2 \cdot \cancel{3} \cdot 2 \cdot \cancel{7}x^{6-4}b^{6-5}}{\cancel{7} \cdot 3 \cdot \cancel{3} \cdot \cancel{5}y^{7-4}a^{7-2}}$$

$$= \frac{4bx^2}{3a^5y^3}$$

23.

$$\frac{ab^2}{2x^4y^5} \cdot \frac{10xy^2}{9x^3b} \cdot \frac{6xy^3}{1} = \frac{ab^2 \cdot 10xy^2 \cdot 6xy^3}{2x^4y^5 \cdot 9a^3b}$$

$$= \frac{10 \cdot 6ab^2x^2y^5}{2 \cdot 9x^4y^5a^3b}$$

$$= \frac{\cancel{2} \cdot 5 \cdot 2 \cdot \cancel{3}b^{2-1}\cancel{y^5}}{\cancel{2} \cdot \cancel{3} \cdot 3a^{3-1}x^{4-2}\cancel{y^5}} = \frac{10b}{3a^2x^2}$$

25.

$$\frac{5x^3y^5}{8a^3b^2} \cdot \frac{14ab^7}{9xy^8} \div \frac{a^2b^2}{12xy} = \frac{5x^3y^5}{8a^3b^2} \cdot \frac{14ab^7}{9xy^8} \cdot \frac{12xy}{a^2b^2}$$

$$= \frac{5 \cdot 14 \cdot 12 \cdot x^4y^6 \cdot a \cdot b^7}{8 \cdot 9 \cdot a^5b^4 \cdot xy^8}$$

$$= \frac{5 \cdot \cancel{2} \cdot 7 \cdot \cancel{3} \cdot \cancel{4} \cdot x^{4-1} \cdot b^{7-4}}{\cancel{2} \cdot \cancel{4} \cdot 3 \cdot \cancel{3} \cdot y^{8-6} \cdot a^{5-1}}$$

$$= \frac{35b^3x^3}{3a^4y^2}$$

27.

$$\frac{48a^2b^3}{35a^4d^5} \div \frac{12b^2}{1} \cdot \frac{15c^2d}{1} = \frac{48a^2b^3}{35a^4d^5} \cdot \frac{1}{12b^2} \cdot \frac{15c^2d}{1}$$

$$= \frac{48 \cdot 15 \cdot a^2b^3c^2d}{35 \cdot 12a^4b^2d^5}$$

$$= \frac{\cancel{4} \cdot 12 \cdot \cancel{3} \cdot \cancel{5}b^{3-2}c^2}{\cancel{5} \cdot 7 \cdot \cancel{4} \cdot \cancel{3}a^{4-2}d^{5-1}} = \frac{12bc^2}{7a^2d^4}$$

29.

$$\frac{2x+2}{5x-15} \cdot \frac{x-3}{xy+y} = \frac{2\cancel{(x+1)}}{5\cancel{(x-3)}} \cdot \frac{\cancel{(x-3)}}{y\cancel{(x+1)}} = \frac{2}{5y}$$

31.

$$\frac{5m+n}{2y-2xy}\cdot\frac{6x-6}{20m+4n}=\frac{5m+n}{-2y(-1+x)}\cdot\frac{6(x-1)}{4(5m+n)}$$

$$=\frac{(5m+n)}{-2y(x-1)}\cdot\frac{2\cdot3(x-1)}{2\cdot2(5m+n)}$$

$$=-\frac{3}{4y}$$

33.

$$\frac{10x^2y^2+5xy^2}{8-4x}\div\frac{6x+3}{12x-24}$$

$$=\frac{5xy^2(2x+1)}{-4(-2+x)}\cdot\frac{12(x-2)}{3(2x+1)}$$

$$=\frac{5xy^2(2x+1)}{-1\cdot4(x-2)}\cdot\frac{3\cdot4(x-2)}{3(2x+1)}$$

$$=-5xy^2$$

35.

$$\frac{2x^2+7x+3}{2x^2+9x-5}\cdot\frac{2x^2-7x+3}{x^2-9}$$

$$=\frac{(x+3)(2x+1)}{(2x-1)(x+5)}\cdot\frac{(2x-1)(x-3)}{(x+3)(x-3)}=\frac{2x+1}{x+5}$$

37.

$$\frac{10x^2+17x+3}{2x^2+7x+6}\div\frac{5x^2+41x+8}{2x^2+x-6}$$

$$=\frac{(2x+3)(5x+1)}{(2x+3)(x+2)}\cdot\frac{(2x-3)(x+2)}{(5x+1)(x+8)}=\frac{2x-3}{x+8}$$

39.

$$\frac{10x^2-x-2}{3x^2+19x-14}\cdot\frac{x^2+11x+28}{2x^2+7x-4}$$

$$=\frac{(5x+2)(2x-1)}{(3x-2)(x+7)}\cdot\frac{(x+7)(x+4)}{(2x-1)(x+4)}=\frac{5x+2}{3x-2}$$

41.

$$\frac{x^2-1}{3x^2-x-4}\div\frac{5x-5x^2}{3x^2+2x-8}$$

$$=\frac{(x+1)(x-1)}{(3x-4)(x+1)}\cdot\frac{(3x-4)(x+2)}{-5x(-1+x)}$$

$$=\frac{(x+1)(x-1)}{(3x-4)(x+1)}\cdot\frac{(3x-4)(x+2)}{-5x(x-1)}=-\frac{x+2}{5x}$$

43.

$$\frac{2x^2-11x+15}{2x^2+5x-25}\div\frac{12x-4x^2}{x+5}$$

$$=\frac{(2x-5)(x-3)}{(2x-5)(x+5)}\cdot\frac{x+5}{-4x(x-3)}$$

$$=\frac{(2x-5)(x-3)}{(2x-5)(x+5)}\cdot\frac{(x+5)}{-4x(x-3)}=-\frac{1}{4x}$$

45.

$$\frac{6-23x-4x^2}{3x^2+10x-48}\cdot\frac{3x^2+x-24}{8x^2-6x+1}$$

$$=\frac{-(4x-1)(x+6)}{(3x-8)(x+6)}\cdot\frac{(3x-8)(x+3)}{(4x-1)(2x-1)}=-\frac{x+3}{2x-1}$$

47.

$$\frac{x^2-4x+4}{3x^2-5x-2}\div\frac{16-x^2}{3x^2-11x-4}$$

$$=\frac{(x-2)(x-2)}{(3x+1)(x-2)}\cdot\frac{(3x+1)(x-4)}{-(x+4)(x-4)}=-\frac{x-2}{x+4}$$

49.

$$\frac{3x^2-10x-8}{2x^2-3x-20}\cdot\frac{2x+5}{1}=\frac{(x-4)(3x+2)}{(2x+5)(x-4)}\cdot\frac{(2x+5)}{1}$$

$$=3x+2$$

51.

$$\frac{x^2+2x-15}{2x^2-5x-3}\div\frac{\left(x^2+x-20\right)}{1}$$

$$=\frac{(x-3)(x+5)}{(2x+1)(x-3)}\cdot\frac{1}{(x-4)(x+5)}=\frac{1}{(2x+1)(x-4)}$$

53.

$$\frac{5x^2-3xy-2y^2}{2x^2+3xy+y^2}\cdot\frac{2x^2+7xy+3y^2}{5x^2+17xy+6y^2}$$

$$=\frac{(5x+2y)(x-y)}{(2x+y)(x+y)}\cdot\frac{(2x+y)(x+3y)}{(5x+2y)(x+3y)}=\frac{x-y}{x+y}$$

55.

$$\frac{x^3-8}{x^2+2x-3}\cdot\frac{3x^2-2x-1}{3x^2-5x-2}$$

$$=\frac{(x-2)(x^2+2x+4)}{(x+3)(x-1)}\cdot\frac{(3x+1)(x-1)}{(3x+1)(x-2)}=\frac{x^2+2x+4}{x+3}$$

57.

$$\frac{x^3 + y^3}{4x^2 - 4xy + y^2} \div \frac{2x^2 - xy - 3y^2}{4x^2 - 8xy + 3y^2}$$

$$= \frac{(x+y)\left(x^2 - xy + y^2\right)}{(2x-y)(2x-y)} \cdot \frac{(2x-3y)(2x-y)}{(2x-3y)(x+y)}$$

$$= \frac{x^2 - xy + y^2}{2x - y}$$

59.

$$\frac{ax + 2bx - 3a - 6b}{2ax - 2bx + a - b} \cdot \frac{2ax - 6bx + a - 3b}{ax + 2bx + a + 2b}$$

$$= \frac{(a+2b)(x-3)}{(a-b)(2x+1)} \cdot \frac{(a-3b)(2x+1)}{(a+2b)(x+1)} = \frac{(a-3b)(x-3)}{(a-b)(x+1)}$$

61.

$$\frac{2x^3 - 10x^2 - x + 5}{x^3 - 5x^2 + 3x - 15} \div \frac{2x^3 + 6x^2 - x - 3}{x^3 + 3x^2 + 3x + 9}$$

$$= \frac{(x-5)\left(2x^2-1\right)}{(x-5)\left(x^2+3\right)} \cdot \frac{\left(x^2+3\right)(x+3)}{(x+3)\left(2x^2-1\right)} = 1$$

63.

$$\frac{2x^2 + 3x + 1}{45x^4 + 30x^3} \div \frac{x^2 - 1}{2x^2 - 3x - 9} \div \frac{2x^2 - 5x - 3}{5x - 5x^2}$$

$$= \frac{(2x+1)(x+1)}{15x^3(3x+2)} \cdot \frac{(2x+3)(x-3)}{(x+1)(x-1)} \cdot \frac{5x(1-x)}{(2x+1)(x-3)}$$

$$= \frac{(2x+1)(x+1)}{3 \cdot 5 \cdot x^{3-1}(3x+2)} \cdot \frac{(2x+3)(x-3)}{(x+1)(x-1)} \cdot \frac{-1 \cdot 5(x-1)}{(2x+1)(x-3)}$$

$$= -\frac{(2x+3)}{3x^2(3x+2)}$$

65.

$$\frac{x^3 - 1}{2x^2 - 11x + 12} \cdot \frac{2x^2 - 3x - 20}{8x^3 + 12x^2} \div \frac{2x^2 + 3x - 5}{4x^2 - 9}$$

$$= \frac{(x-1)\left(x^2 + x + 1\right)}{(2x-3)(x-4)} \cdot \frac{(2x+5)(x-4)}{4x^2(2x+3)} \cdot \frac{(2x+3)(2x-3)}{(2x+5)(x-1)}$$

$$= \frac{x^2 + x + 1}{4x^2}$$

Exercises 4.4

1.

$$\frac{3}{2x} + \frac{7}{2x} = \frac{3+7}{2x} = \frac{10}{2x} = \frac{2 \cdot 5}{2 \cdot x} = \frac{5}{x}$$

3.

$$\frac{4x}{3a} - \frac{2x}{3a} = \frac{4x - 2x}{3a} = \frac{2x}{3a}$$

5.

$$\frac{2k}{st} - \frac{9k}{st} = \frac{2k - 9k}{st} = -\frac{7k}{st}$$

7.

$$\frac{5x}{2x+3} + \frac{7x}{2x+3} = \frac{5x + 7x}{2x+3} = \frac{12x}{2x+3}$$

9.

$$\frac{x}{4x-7} - \frac{3x+5}{4x-7} = \frac{x - (3x+5)}{4x-7} = \frac{x - 3x - 5}{4x-7} = \frac{-2x - 5}{4x-7}$$

11.

$$\frac{2x}{x+1} + \frac{2}{x+1} = \frac{2x+2}{x+1} = \frac{2(x+1)}{(x+1)} = 2$$

13.

$$\frac{5x+1}{2x^2 + x - 11} - \frac{2x+9}{2x^2 + x - 11} = \frac{5x + 1 - (2x+9)}{2x^2 + x - 11}$$

$$= \frac{5x + 1 - 2x - 9}{2x^2 + x - 11}$$

$$= \frac{3x - 8}{2x^2 + x - 11}$$

15.

$$\frac{x+3}{4x^2 - 1} + \frac{x-2}{4x^2 - 1} = \frac{x + 3 + x - 2}{4x^2 - 1}$$

$$= \frac{(2x+1) \cdot 1}{(2x+1)(2x-1)} = \frac{1}{2x-1}$$

17.

$$\frac{x^2 + 2x}{x+1} + \frac{1}{x+1} = \frac{x^2 + 2x + 1}{x+1} = \frac{(x+1)(x+1)}{(x+1)} = x + 1$$

19.

$$\frac{2x^2 - 8}{x-5} + \frac{x^2 + x + 12}{5-x} = \frac{2x^2 - 8}{x-5} + \frac{x^2 + x + 12}{-(x-5)}$$

$$= \frac{2x^2 - 8}{x-5} - \frac{x^2 + x + 12}{x-5}$$

$$= \frac{2x^2 - 8 - \left(x^2 + x + 12\right)}{x-5}$$

#19. continued

$$= \frac{2x^2 - 8 - x^2 - x - 12}{x - 5}$$

$$= \frac{x^2 - x - 20}{x - 5}$$

$$= \frac{\cancel{(x - 5)}(x + 4)}{\cancel{(x - 5)}} = x + 4$$

21.

$$\frac{8}{5x} + \frac{3}{2y} \qquad LCD = 10xy$$

$$= \frac{2y}{2y} \cdot \frac{8}{5x} + \frac{3}{2y} \cdot \frac{5x}{5x} = \frac{16y}{10xy} + \frac{15x}{10xy} = \frac{16y + 15x}{10xy}$$

23.

$$\frac{11}{5x} - \frac{3c}{10y} \qquad LCD = 10xy$$

$$= \frac{2y}{2y} \cdot \frac{11}{5x} - \frac{3c}{10y} \cdot \frac{x}{x} = \frac{22y}{10xy} - \frac{3cx}{10xy} = \frac{22y - 3cx}{10xy}$$

25.

$$\frac{3}{4xy^2} + \frac{7}{3x^2y} \qquad LCD = 12x^2y^2$$

$$= \frac{3x}{3x} \cdot \frac{3}{4xy^2} + \frac{7}{3x^2y} \cdot \frac{4y}{4y}$$

$$= \frac{9x}{12x^2y^2} + \frac{28y}{12x^2y^2} = \frac{9x + 28y}{12x^2y^2}$$

27.

$$\frac{7x}{y} + \frac{2}{1} \qquad LCD = y$$

$$= \frac{7x}{y} + \frac{2}{1} \cdot \frac{y}{y} = \frac{7x}{y} + \frac{2y}{y} = \frac{7x + 2y}{y}$$

29.

$$\frac{5}{x + 2} + \frac{2}{x + 3} \qquad LCD = (x + 2)(x + 3)$$

$$= \frac{x + 3}{x + 3} \cdot \frac{5}{x + 2} + \frac{2}{x + 3} \cdot \frac{x + 2}{x + 2}$$

$$= \frac{5x + 15}{(x + 3)(x + 2)} + \frac{2x + 4}{(x + 3)(x + 2)}$$

$$= \frac{5x + 2x + 15 + 4}{(x + 3)(x + 2)} = \frac{7x + 19}{(x + 3)(x + 2)}$$

31.

$$\frac{3}{2x + 5} - \frac{5}{x - 3} \qquad LCD = (2x + 5)(x - 3)$$

$$= \frac{x - 3}{x - 3} \cdot \frac{3}{2x + 5} - \frac{5}{x - 3} \cdot \frac{2x + 5}{2x + 5}$$

$$= \frac{3x - 9}{(x - 3)(2x + 5)} - \frac{10x + 25}{(x - 3)(2x + 5)}$$

$$= \frac{3x - 9 - (10x + 25)}{(x - 3)(2x + 5)}$$

$$= \frac{3x - 9 - 10x - 25}{(x - 3)(2x + 5)} = \frac{-7x - 34}{(x - 3)(2x + 5)}$$

33.

$$\frac{2x}{x + 4} + \frac{3x}{3x - 4} \qquad LCD = (x + 4)(3x - 4)$$

$$= \frac{3x - 4}{3x - 4} \cdot \frac{2x}{x + 4} + \frac{3x}{3x - 4} \cdot \frac{x + 4}{x + 4}$$

$$= \frac{6x^2 - 8x}{(3x - 4)(x + 4)} + \frac{3x^2 + 12x}{(3x - 4)(x + 4)}$$

$$= \frac{6x^2 - 8x + 3x^2 + 12x}{(3x - 4)(x + 4)} = \frac{9x^2 + 4x}{(3x - 4)(x + 4)}$$

35.

$$\frac{2x}{5x + 3} - \frac{4x}{2x - 3} \qquad LCD = (5x + 3)(2x - 3)$$

$$= \frac{2x - 3}{2x - 3} \cdot \frac{2x}{5x + 3} - \frac{4x}{2x - 3} \cdot \frac{5x + 3}{5x + 3}$$

$$= \frac{4x^2 - 6x}{(2x - 3)(5x + 3)} - \frac{20x^2 + 12x}{(2x - 3)(5x + 3)}$$

$$= \frac{4x^2 - 6x - (20x^2 + 12x)}{(2x - 3)(5x + 3)}$$

$$= \frac{4x^2 - 6x - 20x^2 - 12x}{(2x - 3)(5x + 3)} = \frac{-16x^2 - 18x}{(2x - 3)(5x + 3)}$$

37.

$$\frac{x + 2}{2x - 3} - \frac{x - 3}{x + 6} \qquad LCD = (2x - 3)(x + 6)$$

$$= \frac{x + 6}{x + 6} \cdot \frac{x + 2}{2x - 3} - \frac{x - 3}{x + 6} \cdot \frac{2x - 3}{2x - 3}$$

$$= \frac{(x + 6)(x + 2)}{(x + 6)(2x - 3)} - \frac{(x - 3)(2x - 3)}{(x + 6)(2x - 3)}$$

$$= \frac{x^2 + 8x + 12}{(x + 6)(2x - 3)} - \frac{2x^2 - 9x + 9}{(x + 6)(2x - 3)}$$

#37. continued

$$= \frac{x^2 + 8x + 12 - \left(2x^2 - 9x + 9\right)}{(x+6)(2x-3)}$$

$$= \frac{x^2 + 8x + 12 - 2x^2 + 9x - 9}{(x+6)(2x-3)} = \frac{-x^2 + 17x + 3}{(x+6)(2x-3)}$$

39.

$$\frac{x-1}{2x+5} + \frac{3x-2}{2x-3} \qquad LCD = (2x+5)(2x-3)$$

$$= \frac{2x-3}{2x-3} \cdot \frac{x-1}{2x+5} + \frac{3x-2}{2x-3} \cdot \frac{2x+5}{2x+5}$$

$$= \frac{(2x-3)(x-1)}{(2x-3)(2x+5)} + \frac{(3x-2)(2x+5)}{(2x-3)(2x+5)}$$

$$= \frac{2x^2 - 5x + 3}{(2x-3)(2x+5)} + \frac{6x^2 + 11x - 10}{(2x-3)(2x+5)}$$

$$= \frac{8x^2 + 6x - 7}{(2x-3)(2x+5)}$$

41.

$$\frac{2}{x+1} + 3 \qquad LCD = x+1$$

$$= \frac{2}{x+1} + \frac{3}{1} \cdot \frac{x+1}{x+1} = \frac{2}{x+1} + \frac{3x+3}{x+1} = \frac{3x+5}{x+1}$$

43.

$$\frac{2x+3}{x-4} + 2 \qquad LCD = x-4$$

$$= \frac{2x+3}{x-4} + \frac{2}{1} \cdot \frac{x-4}{x-4} = \frac{2x+3}{x-4} + \frac{2x-8}{x-4} = \frac{4x-5}{x-4}$$

45.

$$\frac{3}{x^2 - 5x} + \frac{7}{25 - x^2} = \frac{3}{x(x-5)} + \frac{7}{-\left(x^2 - 25\right)}$$

$$= \frac{3}{x(x-5)} - \frac{7}{(x+5)(x-5)} \qquad LCD = x(x+5)(x-5)$$

$$= \frac{(x+5)}{(x+5)} \cdot \frac{3}{x(x-5)} - \frac{7}{(x+5)(x-5)} \cdot \frac{x}{x}$$

$$= \frac{3x+15}{x(x+5)(x-5)} - \frac{7x}{x(x+5)(x-5)}$$

$$= \frac{3x + 15 - 7x}{x(x+5)(x-5)} = \frac{-4x + 15}{x(x+5)(x-5)}$$

47.

$$\frac{x+1}{2x^2 - x} + \frac{x-11}{2x^2 + 5x - 3}$$

$$= \frac{x+1}{x(2x-1)} + \frac{x-11}{(2x-1)(x+3)} \qquad LCD = x(2x-1)(x+3)$$

$$= \frac{x+3}{x+3} \cdot \frac{x+1}{x(2x-1)} + \frac{x-11}{(2x-1)(x+3)} \cdot \frac{x}{x}$$

$$= \frac{(x+3)(x+1)}{x(x+1)(2x-1)} + \frac{x(x-11)}{x(x+3)(2x-1)}$$

$$= \frac{x^2 + 4x + 3 + x^2 - 11x}{x(x+3)(2x-1)}$$

$$= \frac{2x^2 - 7x + 3}{x(x+3)(2x-1)} = \frac{(x-3)\cancel{(2x-1)}}{x(x+3)\cancel{(2x-1)}} = \frac{x-3}{x(x+3)}$$

49.

$$\frac{2}{x^2 - 4x - 5} + \frac{5}{x^2 - 2x - 15}$$

$$= \frac{2}{(x-5)(x+1)} + \frac{5}{(x+3)(x-5)}$$

$$\qquad\qquad LCD = (x-5)(x+1)(x+3)$$

$$= \frac{x+3}{x+3} \cdot \frac{2}{(x-5)(x+1)} + \frac{5}{(x+3)(x-5)} \cdot \frac{x+1}{x+1}$$

$$= \frac{2x+6}{(x+1)(x+3)(x-5)} + \frac{5x+5}{(x+1)(x+3)(x-5)}$$

$$= \frac{7x+11}{(x+1)(x+3)(x-5)}$$

51.

$$\frac{4x}{3x^2 - 5x - 2} - \frac{1}{3x^2 + 13x + 4}$$

$$= \frac{4x}{(3x+1)(x-2)} - \frac{1}{(3x+1)(x+4)}$$

$$\qquad\qquad LCD = (3x+1)(x+4)(x-2)$$

$$= \frac{x+4}{x+4} \cdot \frac{4x}{(3x+1)(x-2)} - \frac{1}{(3x+1)(x+4)} \cdot \frac{x-2}{x-2}$$

$$= \frac{4x^2 + 16x - (x-2)}{(x+4)(x-2)(3x+1)}$$

$$= \frac{4x^2 + 16x - x + 2}{(x+4)(x-2)(3x+1)} = \frac{4x^2 + 15x + 2}{(x+4)(x-2)(3x+1)}$$

53.

$$\frac{5x}{x^2-x-6}+\frac{3}{x^2-7x+12}$$

$$=\frac{5x}{(x+2)(x-3)}+\frac{3}{(x-4)(x-3)}$$

$$\text{LCD}=(x+2)(x-3)(x-4)$$

$$=\frac{x-4}{x-4}\cdot\frac{5x}{(x+2)(x-3)}+\frac{3}{(x-4)(x-3)}\cdot\frac{x+2}{x+2}$$

$$=\frac{5x^2-20x}{(x-4)(x+2)(x-3)}+\frac{3x+6}{(x-4)(x-3)(x+2)}$$

$$=\frac{5x^2-17x+6}{(x-4)(x+2)(x-3)}$$

$$=\frac{(5x-2)(x-3)}{(x-4)(x+2)(x-3)}=\frac{5x-2}{(x-4)(x+2)}$$

55.

$$\frac{x+4}{x^2+4x+3}+\frac{x-3}{2x^2-x-3}$$

$$=\frac{x+4}{(x+3)(x+1)}+\frac{x-3}{(2x-3)(x+1)}$$

$$\text{LCD}=(x+3)(x+1)(2x-3)$$

$$=\frac{2x-3}{2x-3}\cdot\frac{x+4}{(x+3)(x+1)}+\frac{x-3}{(2x-3)(x+1)}\cdot\frac{x+3}{x+3}$$

$$=\frac{(2x-3)(x+4)}{(2x-3)(x+3)(x+1)}+\frac{(x-3)(x+3)}{(2x-3)(x+3)(x+1)}$$

$$=\frac{2x^2+5x-12+x^2-9}{(2x-3)(x+3)(x+1)}=\frac{3x^2+5x-21}{(2x-3)(x+3)(x+1)}$$

57.

$$\frac{2x+3}{x^2-1}+\frac{x-2}{x^2-6x+5}$$

$$=\frac{2x+3}{(x+1)(x-1)}+\frac{x-2}{(x-5)(x-1)}$$

$$\text{LCD}=(x+1)(x-1)(x-5)$$

$$=\frac{x-5}{x-5}\cdot\frac{2x+3}{(x+1)(x-1)}+\frac{x-2}{(x-5)(x-1)}\cdot\frac{(x+1)}{(x+1)}$$

$$=\frac{(x-5)(2x+3)}{(x-5)(x+1)(x-1)}+\frac{(x-2)(x+1)}{(x-5)(x-1)(x+1)}$$

$$=\frac{2x^2-7x-15+x^2-x-2}{(x-5)(x+1)(x-1)}=\frac{3x^2-8x-17}{(x-5)(x+1)(x-1)}$$

59.

$$\frac{4x-5}{x^2-2x-3}-\frac{3x+4}{x^2-6x+9}$$

$$=\frac{4x-5}{(x-3)(x+1)}-\frac{3x+4}{(x-3)(x-3)}$$

$$\text{LCD}=(x-3)^2(x+1)$$

$$=\frac{x-3}{x-3}\cdot\frac{4x-5}{(x-3)(x+1)}-\frac{3x+4}{(x-3)(x-3)}\cdot\frac{x+1}{x+1}$$

$$=\frac{(x-3)(4x-5)}{(x-3)^2(x+1)}-\frac{(3x+4)(x+1)}{(x-3)^2(x+1)}$$

$$=\frac{4x^2-17x+15-\left(3x^2+7x+4\right)}{(x-3)^2(x+1)}$$

$$=\frac{4x^2-17x+15-3x^2-7x-4}{(x-3)^2(x+1)}=\frac{x^2-24x+11}{(x-3)^2(x+1)}$$

61.

$$\frac{2x+9}{3x^2-7x-20}+\frac{x-12}{x^2-16}$$

$$=\frac{2x+9}{(3x+5)(x-4)}+\frac{x-12}{(x+4)(x-4)}$$

$$\text{LCD}=(3x+5)(x-4)(x+4)$$

$$=\frac{x+4}{x+4}\cdot\frac{2x+9}{(3x+5)(x-4)}+\frac{x-12}{(x+4)(x-4)}\cdot\frac{3x+5}{3x+5}$$

$$=\frac{(x+4)(2x+9)}{(x+4)(3x+5)(x-4)}+\frac{(x-12)(3x+5)}{(x+4)(x-4)(3x+5)}$$

$$=\frac{2x^2+17x+36+3x^2-31x-60}{(x+4)(x-4)(3x+5)}$$

$$=\frac{5x^2-14x-24}{(x+4)(x-4)(3x+5)}$$

$$=\frac{(5x+6)(x-4)}{(x+4)(x-4)(3x+5)}=\frac{5x+6}{(x+4)(3x+5)}$$

63.

$$\frac{3x-2}{2x^2-9x+10}-\frac{x+6}{x^2-6x+8}$$

$$=\frac{3x-2}{(2x-5)(x-2)}-\frac{x+6}{(x-2)(x-4)}$$

$$\text{LCD}=(2x-5)(x-2)(x-4)$$

$$=\frac{x-4}{x-4}\cdot\frac{3x-2}{(2x-5)(x-2)}-\frac{x+6}{(x-2)(x-4)}\cdot\frac{2x-5}{2x-5}$$

#63. continued

$$= \frac{(x-4)(3x-2)}{(x-4)(2x-5)(x-2)} - \frac{(x+6)(2x-5)}{(x-2)(x-4)(2x-5)}$$

$$= \frac{3x^2 - 14x + 8 - \left(2x^2 + 7x - 30\right)}{(x-4)(2x-5)(x-2)}$$

$$= \frac{3x^2 - 14x + 8 - 2x^2 - 7x + 30}{(x-4)(2x-5)(x-2)}$$

$$= \frac{x^2 - 21x + 38}{(x-4)(2x-5)(x-2)}$$

$$= \frac{(x-19)(x-2)}{(x-4)(2x-5)(x-2)} = \frac{x-19}{(x-4)(2x-5)}$$

65.

$$\frac{x-1}{6x^2 - 7x + 2} + \frac{x+2}{2x^2 - 7x + 3}$$

$$= \frac{x-1}{(3x-2)(2x-1)} + \frac{x+2}{(2x-1)(x-3)}$$

$$\text{LCD} = (3x-2)(2x-1)(x-3)$$

$$= \frac{x-3}{x-3} \cdot \frac{x-1}{(3x-2)(2x-1)} + \frac{x+2}{(2x-1)(x-3)} \cdot \frac{3x-2}{3x-2}$$

$$= \frac{(x-3)(x-1)}{(x-3)(3x-2)(2x-1)} + \frac{(x+2)(3x-2)}{(2x-1)(x-3)(3x-2)}$$

$$= \frac{x^2 - 4x + 3 + 3x^2 + 4x - 4}{(x-3)(3x-2)(2x-1)}$$

$$= \frac{4x^2 - 1}{(x-3)(3x-2)(2x-1)}$$

$$= \frac{(2x+1)(2x-1)}{(x-3)(3x-2)(2x-1)} = \frac{2x+1}{(x-3)(3x-2)}$$

67.

$$\frac{5x+1}{x^2+x-2} - \frac{3x+5}{x^2+2x-3}$$

$$= \frac{5x+1}{(x+2)(x-1)} - \frac{3x+5}{(x+3)(x-1)}$$

$$\text{LCD} = (x+2)(x-1)(x+3)$$

$$= \frac{x+3}{x+3} \cdot \frac{5x+1}{(x+2)(x-1)} - \frac{3x+5}{(x+3)(x-1)} \cdot \frac{x+2}{x+2}$$

$$= \frac{(x+3)(5x+1)}{(x+3)(x+2)(x-1)} - \frac{(3x+5)(x+2)}{(x+3)(x-1)(x+2)}$$

$$= \frac{5x^2 + 16x + 3 - \left(3x^2 + 11x + 10\right)}{(x+3)(x+3)(x-1)}$$

$$= \frac{5x^2 + 16x + 3 - 3x^2 - 11x - 10}{(x+3)(x+2)(x-1)}$$

#67. continued

$$= \frac{2x^2 + 5x - 7}{(x+3)(x+2)(x-1)}$$

$$= \frac{(2x+7)(x-1)}{(x+3)(x+2)(x-1)} = \frac{2x+7}{(x+3)(x+2)}$$

69.

$$\frac{x+1}{4x^2 + 4x - 15} - \frac{4x+5}{8x^2 - 10x - 3}$$

$$= \frac{x+1}{(2x-3)(2x+5)} - \frac{4x+5}{(2x-3)(4x+1)}$$

$$\text{LCD} = (2x-3)(2x+5)(4x+1)$$

$$= \frac{4x+1}{4x+1} \cdot \frac{x+1}{(2x-3)(2x+5)} - \frac{4x+5}{(2x-3)(4x+1)} \cdot \frac{2x+5}{2x+5}$$

$$= \frac{(4x+1)(x+1)}{(4x+1)(2x-3)(2x+5)} - \frac{(4x+5)(2x+5)}{(2x-3)(4x+1)(2x+5)}$$

$$= \frac{4x^2 + 5x + 1 - \left(8x^2 + 30x + 25\right)}{(4x+1)(2x-3)(2x+5)}$$

$$= \frac{4x^2 + 5x + 1 - 8x^2 - 30x - 25}{(4x+1)(2x-3)(2x+5)}$$

$$= \frac{-4x^2 - 25x - 24}{(4x+1)(2x-3)(2x+5)}$$

71.

$$\frac{3x-4}{2x^2 - 3x - 5} - \frac{4x+3}{3x^2 + 5x + 2}$$

$$= \frac{3x-4}{(2x-5)(x+1)} - \frac{4x+3}{(3x+2)(x+1)}$$

$$\text{LCD} = (2x-5)(x+1)(3x+2)$$

$$= \frac{3x+2}{3x+2} \cdot \frac{3x-4}{(2x-5)(x+1)} - \frac{4x+3}{(3x+2)(x+1)} \cdot \frac{2x-5}{2x-5}$$

$$= \frac{(3x+2)(3x-4)}{(3x+2)(2x-5)(x+1)} - \frac{(4x+3)(2x-5)}{(3x+2)(x+1)(2x-5)}$$

$$= \frac{9x^2 - 6x - 8 - \left(8x^2 - 14x - 15\right)}{(3x+2)(2x-5)(x+1)}$$

$$= \frac{9x^2 - 6x - 8 - 8x^2 + 14x + 15}{(3x+2)(2x-5)(x+1)}$$

$$= \frac{x^2 + 8x + 7}{(3x+2)(2x-5)(x+1)}$$

$$= \frac{(x+7)(x+1)}{(3x+2)(2x-5)(x+1)} = \frac{x+7}{(3x+2)(2x-5)}$$

73.

$$\frac{2x+y}{x^2-y^2}-\frac{x-y}{2x^2+3xy+y^2}$$

$$=\frac{2x+y}{(x+y)(x-y)}-\frac{x-y}{(2x+y)(x+y)}$$

$$\quad\quad\quad LCD=(x+y)(x-y)(2x+y)$$

$$=\frac{2x+y}{2x+y}\cdot\frac{2x+y}{(x+y)(x-y)}-\frac{x-y}{(2x+y)(x+y)}\cdot\frac{x-y}{x-y}$$

$$=\frac{(2x+y)^2}{(2x+y)(x+y)(x-y)}-\frac{(x-y)^2}{(2x+y)(x+y)(x-y)}$$

$$=\frac{4x^2+4xy+y^2-\left(x^2-2xy+y^2\right)}{(2x+y)(x+y)(x-y)}$$

$$=\frac{4x^2+4xy+y^2-x^2+2xy-y^2}{(2x+y)(x+y)(x-y)}$$

$$=\frac{3x^2+6xy}{(2x+y)(x+y)(x-y)}$$

75.

$$\frac{2x+y}{x^2+5xy+6y^2}-\frac{x-4y}{x^2+6xy+8y^2}$$

$$=\frac{2x+y}{(x+2y)(x+3y)}-\frac{x-4y}{(x+2y)(x+4y)}$$

$$\quad\quad\quad LCD=(x+2y)(x+3y)(x+4y)$$

$$=\frac{x+4y}{x+4y}\cdot\frac{2x+y}{(x+2y)(x+3y)}-\frac{x-4y}{(x+2y)(x+4y)}\cdot\frac{x+3y}{x+3y}$$

$$=\frac{(x+4y)(2x+y)}{(x+4y)(x+2y)(x+3y)}-\frac{(x-4y)(x+3y)}{(x+2y)(x+4y)(x+3y)}$$

$$=\frac{2x^2+9xy+4y^2-\left(x^2-xy-12y^2\right)}{(x+4y)(x+2y)(x+3y)}$$

$$=\frac{2x^2+9xy+4y^2-x^2+xy+12y^2}{(x+4y)(x+2y)(x+3y)}$$

$$=\frac{x^2+10xy+16y^2}{(x+4y)(x+2y)(x+3y)}$$

$$=\frac{(x+8y)\cancel{(x+2y)}}{(x+4y)\cancel{(x+2y)}(x+3y)}=\frac{x+8y}{(x+4y)(x+3y)}$$

77.

$$\frac{5x-y}{2x^2+xy-3y^2}-\frac{3x+2y}{x^2-2xy+y^2}$$

#77. continued

$$=\frac{5x-y}{(2x+3y)(x-y)}-\frac{3x+2y}{(x-y)(x-y)}$$

$$\quad\quad\quad LCD=(2x+3y)(x-y)^2$$

$$=\frac{x-y}{x-y}\cdot\frac{5x-y}{(2x+3y)(x-y)}-\frac{3x+2y}{(x-y)(x-y)}\cdot\frac{2x+3y}{2x+3y}$$

$$=\frac{(x-y)(5x-y)}{(2x+3y)(x-y)^2}-\frac{(3x+2y)(2x+3y)}{(2x+3y)(x-y)^2}$$

$$=\frac{5x^2-6xy+y^2-\left(6x^2+13xy+6y^2\right)}{(2x+3y)(x-y)^2}$$

$$=\frac{5x^2-6xy+y^2-6x^2-13xy-6y^2}{(2x+3y)(x-y)^2}$$

$$=\frac{-x^2-19xy-5y^2}{(2x+3y)(x-y)^2}$$

79.

$$\frac{5}{2x+2}+\frac{7}{3x+3}-\frac{1}{12}\quad\quad LCD=2^2\cdot3(x+1)$$

$$=\frac{5}{2(x+1)}+\frac{7}{3(x+1)}-\frac{1}{12}$$

$$=\frac{6}{6}\cdot\frac{5}{2(x+1)}+\frac{4}{4}\cdot\frac{7}{3(x+1)}-\frac{x+1}{x+1}\cdot\frac{1}{12}$$

$$=\frac{30}{12(x+1)}+\frac{28}{12(x+1)}-\frac{x+1}{12(x+1)}$$

$$=\frac{30+28-(x+1)}{12(x+1)}$$

$$=\frac{58-x-1}{12(x+1)}=\frac{-x+57}{12(x+1)}$$

81.

$$\frac{x}{2x^2+5x-3}+\frac{2x-1}{x^2+2x-3}-\frac{1}{2x^2-3x+1}$$

$$=\frac{x}{(2x-1)(x+3)}+\frac{2x-1}{(x+3)(x-1)}-\frac{1}{(2x-1)(x-1)}$$

$$\quad\quad\quad LCD=(2x-1)(x+3)(x-1)$$

$$=\frac{x-1}{x-1}\cdot\frac{x}{(2x-1)(x+3)}+\frac{2x-1}{2x-1}\cdot\frac{2x-1}{(x+3)(x-1)}$$

$$\quad\quad-\frac{x+3}{x+3}\cdot\frac{1}{(2x-1)(x-1)}$$

$$=\frac{x^2-x}{(x-1)(2x-1)(x+3)}+\frac{(2x-1)^2}{(2x-1)(x+3)(x-1)}$$

$$\quad\quad-\frac{x+3}{(x+3)(2x-1)(x-1)}$$

#81. continued

$$= \frac{x^2 - x + 4x^2 - 4x + 1 - x - 3}{(2x-1)(x+3)(x-1)} = \frac{5x^2 - 6x - 2}{(2x-1)(x+3)(x-1)}$$

83.

$$\frac{3x+1}{x^2+x-20} - \frac{7}{x^2+9x+20} - \frac{x-3}{x^2-16}$$

$$= \frac{3x+1}{(x-4)(x+5)} - \frac{7}{(x+4)(x+5)} - \frac{x-3}{(x+4)(x-4)}$$

$$\text{LCD} = (x+4)(x-4)(x+5)$$

$$= \frac{x+4}{x+4} \cdot \frac{3x+1}{(x-4)(x+5)} - \frac{x-4}{x-4} \cdot \frac{7}{(x+4)(x+5)}$$

$$- \frac{x+5}{x+5} \cdot \frac{x-3}{(x+4)(x-4)}$$

$$= \frac{(x+4)(3x+1)}{(x+4)(x-4)(x+5)} - \frac{7(x-4)}{(x+4)(x-4)(x+5)}$$

$$- \frac{(x+5)(x-3)}{(x+4)(x-4)(x+5)}$$

$$= \frac{3x^2+13x+4-(7x-28)-\left(x^2+2x-15\right)}{(x+4)(x-4)(x+5)}$$

$$= \frac{3x^2+13x+4-7x+28-x^2-2x+15}{(x+4)(x-4)(x+5)}$$

$$= \frac{2x^2+4x+47}{(x+4)(x-4)(x+5)}$$

85.

$$\frac{x-2}{x^2+3x+2} - \frac{x+1}{x^2-4} + \frac{2x-7}{x^2-x-2}$$

$$= \frac{x-2}{(x+2)(x+1)} - \frac{x+1}{(x+2)(x-2)} + \frac{2x-7}{(x-2)(x+1)}$$

$$\text{LCD} = (x+2)(x+1)(x-2)$$

$$= \frac{x-2}{x-2} \cdot \frac{x-2}{(x+2)(x+1)} - \frac{x+1}{x+1} \cdot \frac{x+1}{(x+2)(x-2)}$$

$$+ \frac{x+2}{x+2} \cdot \frac{2x-7}{(x-2)(x+1)}$$

$$= \frac{(x-2)^2 - (x+1)^2 + (x+2)(2x-7)}{(x-2)(x+2)(x+1)}$$

$$= \frac{x^2-4x+4-\left(x^2+2x+1\right)+2x^2-3x-14}{(x-2)(x+2)(x+1)}$$

$$= \frac{x^2-4x+4-x^2-2x-1+2x^2-3x-14}{(x-2)(x+2)(x+1)}$$

#85. continued

$$= \frac{2x^2-9x-11}{(x-2)(x+2)(x+1)}$$

$$= \frac{(2x-11)\cancel{(x+1)}}{(x-2)(x+2)\cancel{(x+1)}} = \frac{2x-11}{(x-2)(x+2)}$$

Exercises 4.5

1.

$$\frac{\dfrac{5}{7}}{\dfrac{45}{16}} = \frac{5}{7} \div \frac{45}{16} = \frac{5}{7} \cdot \frac{16}{45} = \frac{\cancel{5}}{7} \cdot \frac{16}{\cancel{5} \cdot 9} = \frac{16}{63}$$

3.

$$\frac{\dfrac{5x}{2y}}{2} = \frac{5x}{2y} \div \frac{2}{1} = \frac{5x}{2y} \cdot \frac{1}{2} = \frac{5x}{4y}$$

5.

$$\frac{\dfrac{3x}{2y^2}}{\dfrac{5x^4}{4y^3}} = \frac{3x}{2y^2} \div \frac{5x^4}{4y^3} = \frac{3x}{2y^2} \cdot \frac{4y^3}{5x^4} = \frac{3 \cdot \cancel{2} \cdot 2y^{3-2}}{\cancel{2} \cdot 5 \cdot x^{4-1}} = \frac{6y}{5x^3}$$

7.

$$\frac{\dfrac{2ab}{5xy}}{\dfrac{11x^3y}{4a^2b}} = \frac{2ab}{5xy} \div \frac{11x^3y}{4a^2b} = \frac{2ab}{5xy} \cdot \frac{4a^2b}{11x^3y} = \frac{8a^3b^2}{55x^4y^2}$$

9.

$$\frac{\dfrac{5x+3}{2x-1}}{\dfrac{x+4}{x-2}} = \frac{5x+3}{2x-1} \div \frac{x+4}{x-2} = \frac{5x+3}{2x-1} \cdot \frac{x-2}{x+4}$$

$$= \frac{(5x+3)(x-2)}{(2x-1)(x+4)}$$

11.

$$\frac{\dfrac{2x+6}{5x-7}}{\dfrac{x+3}{5x-7}} = \frac{2x+6}{5x-7} \div \frac{x+3}{5x-7} = \frac{2(x+3)}{\cancel{(5x-7)}} \cdot \frac{\cancel{(5x-7)}}{\cancel{(x+3)}} = 2$$

13.

$$\frac{\dfrac{2x^2+3x}{y^2-y}}{\dfrac{8x^4+12x^3}{3y^3-3y^2}} = \frac{2x^2+3x}{y^2-y} \div \frac{8x^4+12x^3}{3y^3-3y^2}$$

$$= \frac{x(2x+3)}{y(y-1)} \cdot \frac{3y^2(y-1)}{4x^3(2x+3)}$$

$$= \frac{\cancel{(2x+3)}}{\cancel{(y-1)}} \cdot \frac{3y^{2-1}\cancel{(y-1)}}{4 \cdot x^{3-1}\cancel{(2x+3)}} = \frac{3y}{4x^2}$$

15.

$$\frac{\dfrac{x^2+x-12}{x^2+9x+20}}{\dfrac{9-x^2}{x^2+8x+15}} = \frac{x^2+x-12}{x^2+9x+20} \div \frac{-\left(x^2-9\right)}{x^2+8x+15}$$

$$= \frac{\cancel{(x-3)}\cancel{(x+4)}}{\cancel{(x+4)}\cancel{(x+5)}} \cdot \frac{\cancel{(x+5)}\cancel{(x+3)}}{-\cancel{(x+3)}\cancel{(x-3)}} = -1$$

17.

$$\frac{\dfrac{x+7}{4x^2}}{\dfrac{x-1}{8x}} = \frac{x+7}{4x^2} \div \frac{x-1}{8x} = \frac{x+7}{4x^2} \cdot \frac{8x}{x-1}$$

$$= \frac{x+7}{\cancel{4} \cdot x^{2-1}} \cdot \frac{2 \cdot \cancel{4}}{x-1} = \frac{2(x+7)}{x(x-1)}$$

19.

$$\frac{\dfrac{2x-6}{5xy^3}}{\dfrac{12-4x}{15x^2y^5}} = \frac{2x-6}{5xy^3} \div \frac{12-4x}{15x^2y^5}$$

$$= \frac{2x-6}{5xy^3} \cdot \frac{15x^2y^5}{-4(x-3)}$$

$$= \frac{\cancel{2}\cancel{(x-3)}}{\cancel{5}} \cdot \frac{3 \cdot \cancel{5}x^{2-1}y^{5-3}}{-1 \cdot \cancel{2} \cdot 2\cancel{(x-3)}} = -\frac{3xy^2}{2}$$

21.

$$\frac{\dfrac{8xy-4y}{6xy^2}}{\dfrac{12x^2y-6xy}{9x^2y^2}} = \frac{8xy-4y}{6xy^2} \div \frac{12x^2y-6xy}{9x^2y^2}$$

$$= \frac{8xy-4y}{6xy^2} \cdot \frac{9x^2y^2}{12x^2y-6xy}$$

#21. continued

$$= \frac{4y(2x-1)}{6xy^2} \cdot \frac{9x^2y^2}{6xy(2x-1)}$$

$$= \frac{\cancel{2} \cdot \cancel{2} \cdot \cancel{(2x-1)}}{\cancel{8} \cdot \cancel{8}} \cdot \frac{\cancel{8} \cdot \cancel{3}x^2 \cdot y^3}{\cancel{2} \cdot \cancel{8} \cdot y^3 \cdot x^2 \cancel{(2x-1)}}$$

$$= 1$$

23.

$$\frac{\dfrac{x^2+5x}{x^2+x-6}}{\dfrac{2x^3}{x^2+2x-8}} = \frac{x^2+5x}{x^2+x-6} \div \frac{2x^3}{x^2+2x-8}$$

$$= \frac{x^2+5x}{x^2+x-6} \cdot \frac{x^2+2x-8}{2x^3}$$

$$= \frac{x(x+5)}{(x-2)(x+3)} \cdot \frac{(x+4)(x-2)}{2x^3}$$

$$= \frac{(x+5)}{\cancel{(x-2)}(x+3)} \cdot \frac{(x+4)\cancel{(x-2)}}{2x^{3-1}}$$

$$= \frac{(x+5)(x+4)}{2x^2(x+3)}$$

25.

$$\frac{\left(2+\dfrac{3}{x}\right)x}{\left(1-\dfrac{7}{x}\right)x} = \frac{2x+3}{x-7} \qquad \text{LCD} = x$$

27.

$$\frac{\left(\dfrac{2}{x}+\dfrac{7}{y}\right)xy}{(12)xy} = \frac{2y+7x}{12xy} \qquad \text{LCD} = xy$$

29.

$$\frac{\left(\dfrac{2}{x}+\dfrac{2}{y}\right)x^2y^2}{\left(\dfrac{4}{x^2y^2}\right)x^2y^2} \qquad \text{LCD} = x^2y^2$$

$$= \frac{2xy^2+2x^2y}{4}$$

$$= \frac{2\left(xy^2+x^2y\right)}{4}$$

$$= \frac{xy^2+x^2y}{2} = \frac{xy(y+x)}{2}$$

31.

$$\frac{\left(\dfrac{3}{x}-\dfrac{1}{2}\right)10xy}{\left(\dfrac{2}{y}-\dfrac{1}{5}\right)10xy}=\frac{30y-5xy}{20x-2xy} \qquad LCD = 10xy$$

33.

$$\frac{\left(\dfrac{1}{x}-\dfrac{1}{y}\right)x^2y^2}{\left(\dfrac{1}{x^2}-\dfrac{1}{y^2}\right)x^2y^2}=\frac{xy^2-x^2y}{y^2-x^2} \qquad LCD = x^2y^2$$

$$=\frac{xy(y-x)}{(y-x)(y+x)}=\frac{xy}{y+x}$$

35.

$$\frac{\left(\dfrac{1}{x^4}-\dfrac{1}{y^4}\right)x^4y^4}{\left(\dfrac{1}{x^2}+\dfrac{1}{y^2}\right)x^4y^4}=\frac{y^4-x^4}{x^2y^4+x^4y^2} \qquad LCD = x^4y^4$$

$$=\frac{(y^2+x^2)(y+x)(y-x)}{x^2y^2(y^2+x^2)}=\frac{y^2-x^2}{x^2y^2}$$

37.

$$\frac{\left(1-\dfrac{5}{2x}-\dfrac{3}{2x^2}\right)2x^2}{\left(2-\dfrac{15}{2x}+\dfrac{9}{2x^2}\right)2x^2} \qquad LCD = 2x^2$$

$$=\frac{2x^2-5x-3}{4x^2-15x+9}=\frac{(2x+1)(x-3)}{(4x-3)(x-3)}=\frac{2x+1}{4x-3}$$

39.

$$\frac{(1)xy}{\left(\dfrac{1}{x}+\dfrac{1}{y}\right)xy}=\frac{xy}{y+x} \qquad LCD = xy$$

41.

$$\frac{(2x-y)xy}{\left(\dfrac{2}{y}-\dfrac{1}{x}\right)xy}=\frac{xy(2x-y)}{(2x-y)} \qquad LCD = xy$$

$$= xy$$

43.

$$\frac{\left(\dfrac{5}{x}-\dfrac{2}{y}\right)xy}{(x-2)xy}=\frac{5y-2x}{xy(x-1)} \qquad LCD = xy$$

45.

$$\frac{\left(1-\dfrac{1}{9x^2}\right)9x^2}{(1-3x)9x^2}=\frac{9x^2-1}{(1-3x)9x^2} \qquad LCD = 9x^2$$

$$=\frac{-\left(1-9x^2\right)}{(1-3x)9x^2}$$

$$=\frac{-(1+3x)(1-3x)}{(1-3x)9x^2}=-\frac{3x+1}{9x^2}$$

47.

$$\frac{\left(\dfrac{1}{x}-\dfrac{1}{4}\right)4x}{(x-4)4x}=\frac{4-x}{(x-4)4x} \qquad LCD = 4x$$

$$=\frac{-(x-4)}{(x-4)4x}=-\frac{1}{4x}$$

49.

$$\frac{\left(\dfrac{3}{x^2}-\dfrac{1}{3}\right)\cdot 3x^2}{(x-3)\cdot 3x^2}=\frac{9-x^2}{3x^2(x-3)} \qquad LCD = 3x^2$$

$$=\frac{-\left(x^2-9\right)}{3x^2(x-3)}$$

$$=\frac{-(x+3)(x-3)}{3x^2(x-3)}=-\frac{x+3}{3x^2}$$

51.

$$\frac{\left(\dfrac{4}{2+h}-2\right)(2+h)}{(h)(2+h)}=\frac{4-4-2h}{h(2+h)} \qquad LCD = 2+h$$

$$=\frac{-2h}{h(2+h)}=-\frac{2}{2+h}$$

53.

$$\frac{\left(\dfrac{1}{x}-\dfrac{1}{y}\right)5xy}{\left(\dfrac{x-y}{5}\right)5xy}=\frac{5y-5x}{xy(x-y)} \qquad LCD = 5xy$$

#53. continued

$$= \frac{-5\cancel{(x-y)}}{xy\cancel{(x-y)}} = -\frac{5}{xy}$$

55.

$$\frac{\left(\dfrac{1}{x-2}+1\right)(x-2)}{\left(\dfrac{1}{x-2}-1\right)(x-2)} = \frac{1+x-2}{1-x+2} \qquad LCD = x-2$$

$$= \frac{x-1}{-x+3}$$

57.

$$\frac{4(x+3)(x-2)}{\left(\dfrac{1}{x+3}+\dfrac{5}{x-2}\right)(x+3)(x-2)} \qquad LCD = (x+3)(x-2)$$

$$= \frac{4(x+3)(x-2)}{x-2+5(x+3)} = \frac{4(x+3)(x-2)}{x-2+5x+15} = \frac{4(x+3)(x-2)}{6x+13}$$

59.

$$\frac{\left(\dfrac{1}{x-5}-\dfrac{2}{x-5}\right)(x+5)(x-5)}{\left(\dfrac{4}{x^2-25}\right)(x+5)(x-5)} \qquad LCD = x^2-25$$

$$= \frac{x+5-2(x+5)}{4} = \frac{x+5-2x-10}{4} = \frac{-x-5}{4} = -\frac{x+5}{4}$$

61.

$$\frac{\left(3-\dfrac{6}{x-2}\right)(x-2)}{\left(\dfrac{-2}{x-2}+1\right)(x-2)} = \frac{3(x-2)-6}{-2+x-2} \qquad LCD = x-2$$

$$= \frac{3x-6-6}{-4+x}$$

$$= \frac{3x-12}{x-4} = \frac{3\cancel{(x-4)}}{\cancel{(x-4)}} = 3$$

63.

$$\frac{\left(\dfrac{2}{x}+\dfrac{4}{x-1}\right)x(x-1)}{\left(\dfrac{1}{x}-\dfrac{3}{x-1}\right)x(x-1)} \qquad LCD = x(x-1)$$

$$= \frac{2(x-1)+4x}{x-1-3x} = \frac{2x-2+4x}{-2x-1} = \frac{6x-2}{-2x-1}$$

65.

$$\frac{\left(\dfrac{9}{x^2}+\dfrac{2}{x-2}\right)x^2(x-2)}{\left(\dfrac{1}{x^2}+\dfrac{1}{x-2}\right)x^2(x-2)} \qquad LCD = x^2(x-2)$$

$$= \frac{9(x-2)+2x^2}{x-2+x^2} = \frac{9x-18+2x^2}{x-2+x^2} = \frac{2x^2+9x-18}{x^2+x-2}$$

67.

$$\frac{\left(\dfrac{2}{x+1}+\dfrac{1}{x-2}\right)(x+1)(x-2)}{\left(\dfrac{3}{x+1}+\dfrac{1}{x-2}\right)(x+1)(x-2)} \qquad LCD = (x+1)(x-2)$$

$$= \frac{2(x-2)+x+1}{3(x-2)+x+1} = \frac{2x-4+x+1}{3x-6+x+1} = \frac{3x-3}{4x-5}$$

69.

$$\frac{\left(\dfrac{x}{x-1}-\dfrac{8}{x+2}\right)(x-1)(x+2)}{\left(\dfrac{-3x}{x-1}+\dfrac{5x+4}{x+2}\right)(x-1)(x+2)} \qquad LCD = (x-1)(x+2)$$

$$= \frac{x(x+2)-8(x-1)}{-3x(x+2)+(5x+4)(x-1)}$$

$$= \frac{x^2+2x-8x+8}{-3x^2-6x+5x^2-x-4}$$

$$= \frac{x^2-6x+8}{2x^2-7x-4} = \frac{\cancel{(x-4)}(x-2)}{(2x+1)\cancel{(x-4)}} = \frac{x-2}{2x+1}$$

71.

$$\frac{x^{-1}-y^{-1}}{x^{-2}-y^{-2}} = \frac{\left(\dfrac{1}{x}-\dfrac{1}{y}\right)x^2y^2}{\left(\dfrac{1}{x^2}-\dfrac{1}{y^2}\right)x^2y^2}$$

$$= \frac{xy^2-x^2y}{y^2-x^2} = \frac{xy\cancel{(y-x)}}{(y+x)\cancel{(y-x)}} = \frac{xy}{y+x}$$

73.

$$\frac{x^{-1}-(x+5)^{-1}}{5} = \frac{\left(\dfrac{1}{x}-\dfrac{1}{x+5}\right)x(x+5)}{(5)\cdot x(x+5)}$$

$$= \frac{x+5-x}{5x(x+5)} = \frac{\cancel{5}\cdot 1}{\cancel{5}x(x+5)} = \frac{1}{x(x+5)}$$

75.

$$\frac{5x^{-1}+15x^{-2}}{2+6x^{-1}} = \frac{\left(\dfrac{5}{x}+\dfrac{15}{x^2}\right)x^2}{\left(2+\dfrac{6}{x}\right)x^2}$$

$$= \frac{5x+15}{2x^2+6x} = \frac{5(x+3)}{2x(x+3)} = \frac{5}{2x}$$

77.

$$\frac{1-2x^{-1}-15x^{-2}}{1-3x^{-1}-10x^{-2}} = \frac{\left(1-\dfrac{2}{x}-\dfrac{15}{x^2}\right)x^2}{\left(1-\dfrac{3}{x}-\dfrac{10}{x^2}\right)x^2}$$

$$= \frac{x^2-2x-15}{x^2-3x-10}$$

$$= \frac{(x+3)(x-5)}{(x-5)(x+2)} = \frac{x+3}{x+2}$$

Exercises 4.6

1.

$$\frac{35x^3-49x^2+28}{7} = \frac{35x^3}{7}-\frac{49x^2}{7}+\frac{28}{7}$$

$$= 5x^3-7x^2+4$$

3.

$$\frac{5x^3+10x^2-40x}{5x} = \frac{5x^3}{5x}+\frac{10x^2}{5x}-\frac{40x}{5x} = x^2+2x-8$$

5.

$$\frac{20x^5-28x^4-4x^3}{4x^3} = \frac{20x^5}{4x^3}-\frac{28x^4}{4x^3}-\frac{4x^3}{4x^3}$$

$$= 5x^2-7x-1$$

7.

$$\frac{10x^4y^2-16x^3y^3-2x^2y^4}{2x^2y^2}$$

$$= \frac{10x^4y^2}{2x^2y^2}-\frac{16x^3y^3}{2x^2y^2}-\frac{2x^2y^4}{2x^2y^2} = 5x^2-8xy-y^2$$

9.

$$\frac{-7x^2y^3+2x^2+y}{xy^2} = \frac{-7x^2y^3}{xy^2}+\frac{2x^2}{xy^2}+\frac{y}{xy^2}$$

$$= -7xy+\frac{2x}{y^2}+\frac{1}{xy}$$

11.

$$\frac{6y^4+45x^2-7y}{10xy^2} = \frac{6y^4}{10xy^2}+\frac{45x^2}{10xy^2}-\frac{7y}{10xy^2}$$

$$= \frac{3y^2}{5x}+\frac{9x}{2y^2}-\frac{7}{10xy}$$

13.

$$\frac{3x-10x^4-72z^4}{8x^2yz} = \frac{3x}{8x^2yz}-\frac{10x^4}{8x^2yz}-\frac{72z^4}{8x^2yz}$$

$$= \frac{3}{8xyz}-\frac{5x^2}{4yz}-\frac{9z^3}{x^2y}$$

15.

$$\frac{15y^2z^5+7x^4z^3+2xy^6}{6x^2y^3z^4}$$

$$= \frac{15y^2z^5}{6x^2y^3z^4}+\frac{7x^4z^3}{6x^2y^3z^4}+\frac{2xy^6}{6x^2y^3z^4}$$

$$= \frac{5z}{2x^2y}+\frac{7x^2}{6y^3z}+\frac{y^3}{3xz^4}$$

17.

$$\begin{array}{r}
5x+4 \\
x-3{\overline{\smash{\big)}\,5x^2-11x-14}} \\
\underline{^-5x^2\pm15x} \qquad\text{Change signs} \\
4x-14 \\
\underline{^-4x\pm12} \qquad\text{Change signs} \\
-2
\end{array}$$

Quotient: $5x+4-\dfrac{2}{x-3}$

19.

$$\begin{array}{r}
6x-5 \\
4x-7{\overline{\smash{\big)}\,24x^2-62x+36}} \\
\underline{^-24x^2\pm42x} \qquad\text{Change signs} \\
-20x+36 \\
\underline{\pm20x\mp35} \qquad\text{Change signs} \\
1
\end{array}$$

Quotient: $6x-5+\dfrac{1}{4x-7}$

21.

$$5x - 3 \overline{\smash{\big)}\ 40x^2 - 44x + 17}$$
$$\phantom{5x - 3 \overline{\smash{\big)}\ }}\underline{^-40x^2 \pm 24x} \qquad \text{Change signs}$$
$$\phantom{5x - 3 \overline{\smash{\big)}\ 40x^2}}- 20x + 17$$
$$\phantom{5x - 3 \overline{\smash{\big)}\ 40x^2}}\underline{\pm 20x \mp 12} \qquad \text{Change signs}$$
$$\phantom{5x - 3 \overline{\smash{\big)}\ 40x^2 - 44x +}}5$$

Quotient: $8x - 4 + \dfrac{5}{5x - 3}$

23.

$$2x - 3 \overline{\smash{\big)}\ 18x^2 - 43x + 24}$$
$$\underline{^-18x^2 \pm 27x} \qquad \text{Change signs}$$
$$- 16x + 24$$
$$\underline{\pm 16x \mp 24} \qquad \text{Change signs}$$

Quotient: $9x - 8$

25.

$$2x + 3 \overline{\smash{\big)}\ 2x^3 - 13x^2 - 10x + 19}$$
$$\underline{^-2x^3 \mp 3x^2} \qquad \text{Change signs}$$
$$- 16x^2 - 10x$$
$$\underline{\pm 16x^2 \pm 24x} \qquad \text{Change signs}$$
$$14x + 19$$
$$\underline{^-14x \mp 21}$$
$$- 2$$

Quotient: $x^2 - 8x + 7 - \dfrac{2}{2x + 3}$

27.

$$4x + 7 \overline{\smash{\big)}\ 12x^3 + 21x^2 - 8x - 19}$$
$$\underline{^-12x^3 \mp 21x^2} \qquad \text{Change signs}$$
$$- 8x - 19$$
$$\underline{\pm 8x \pm 14} \qquad \text{Change signs}$$
$$- 5$$

Quotient: $3x^2 - 2 - \dfrac{5}{4x + 7}$

29.

$$5x + 2 \overline{\smash{\big)}\ 40x^3 + 31x^2 + 16x + 8}$$
$$\underline{^-40x^3 \mp 16x^2} \qquad \text{Change signs}$$
$$15x^2 + 16x$$
$$\underline{^-15x^2 \mp 16x} \qquad \text{Change signs}$$
$$10x + 8$$
$$\underline{^-10x \mp 4} \qquad \text{Change signs}$$
$$4$$

Quotient: $8x^2 + 3x + 2 + \dfrac{4}{5x + 2}$

31.

$$x - 2 \overline{\smash{\big)}\ 5x^2 + 0x - 13}$$
$$\underline{^-5x^2 \pm 10x} \qquad \text{Change signs}$$
$$10x - 13$$
$$\underline{^-10x \pm 20} \qquad \text{Change signs}$$
$$7$$

Quotient: $5x + 10 + \dfrac{7}{x - 2}$

33.

$$2x - 1 \overline{\smash{\big)}\ 4x^2 + 0x - 1}$$
$$\underline{^-4x^2 \pm 2x} \qquad \text{Change signs}$$
$$2x - 1$$
$$\underline{^-2x \pm 1} \qquad \text{Change signs}$$

Quotient: $2x + 1$

35.

$$3x - 2 \overline{\smash{\big)}\ 9x^2 + 0x - 5}$$
$$\underline{^-9x^2 \pm 6x} \qquad \text{Change signs}$$
$$6x - 5$$
$$\underline{^-6x \pm 4} \qquad \text{Change signs}$$
$$- 1$$

Quotient: $3x + 2 - \dfrac{1}{3x - 2}$

37.

$$5x+2\overline{)\begin{array}{c}5x-2\\25x^2+0x-4\end{array}}$$

$$\underline{{}^-25x^2\mp10x}\qquad\text{Change signs}$$
$$-10x-4$$
$$\underline{\pm10x\pm4}\qquad\text{Change signs}$$

Quotient: $5x-2$

39.

$$2x-3\overline{)\begin{array}{c}2x^2+3x+4\\4x^3+0x^2-x-7\end{array}}$$

$$\underline{{}^-4x^3\pm6x^2}\qquad\text{Change signs}$$
$$6x^2-x$$
$$\underline{{}^-6x^2\pm9x}\qquad\text{Change signs}$$
$$8x-7$$
$$\underline{{}^-8x\pm12}\qquad\text{Change signs}$$
$$5$$

Quotient: $2x^2+3x+4+\dfrac{5}{2x-3}$

41.

$$2x+6\overline{)\begin{array}{c}3x^2+x-3\\6x^3+20x^2+0x-19\end{array}}$$

$$\underline{{}^-6x^3\mp18x^2}\qquad\text{Change signs}$$
$$2x^2+0x$$
$$\underline{{}^-2x^2\mp6x}\qquad\text{Change signs}$$
$$-6x-19$$
$$\underline{\pm6x\pm18}\qquad\text{Change signs}$$
$$-1$$

Quotient: $3x^2+x-3-\dfrac{1}{2x+6}$

43.

$$x^2+x-3\overline{)\begin{array}{c}2x^2-x+4\\2x^4+x^3-3x^2+9x-13\end{array}}$$

$$\underline{{}^-2x^4\mp2x^3\pm6x^2}\qquad\text{Change signs}$$
$$-x^3+3x^2+9x$$
$$\underline{\pm x^3\pm x^2\mp3x}\qquad\text{Change signs}$$
$$4x^2+6x-13$$
$$\underline{{}^-4x^2\mp4x\pm12}\qquad\text{Change signs}$$
$$2x-1$$

Quotient: $2x^2-x+4+\dfrac{2x-1}{x^2+x-3}$

45.

$$2x^2-x-1\overline{)\begin{array}{c}x^2+2x+3\\2x^4+3x^3+3x^2+0x-4\end{array}}$$

$$\underline{{}^-2x^4\pm x^3\pm x^2}\qquad\text{Change signs}$$
$$4x^3+4x^2+0x$$
$$\underline{{}^-4x^3\pm2x^2\pm2x}\qquad\text{Change signs}$$
$$6x^2+2x-4$$
$$\underline{{}^-6x^2\pm3x\pm3}\qquad\text{Change signs}$$
$$5x-1$$

Quotient: $x^2+2x+3+\dfrac{5x-1}{2x^2-x-1}$

47.

$$x-2\overline{)\begin{array}{c}x^2+2x+4\\x^3+0x^2+0x-8\end{array}}$$

$$\underline{{}^-x^3\pm2x^2}\qquad\text{Change signs}$$
$$2x^2+0x$$
$$\underline{{}^-2x^2\pm4x}\qquad\text{Change signs}$$
$$4x-8$$
$$\underline{{}^-4x\pm8}\qquad\text{Change signs}$$

Quotient: x^2+2x+4

49.

$$3x+4 \overline{\smash{\big)}\, 27x^3 + 0x^2 + 0x + 64} \quad \overset{\displaystyle 9x^2 - 12x + 16}{}$$

$$\underline{^-27x^3 \mp 36x^2} \qquad \text{Change signs}$$
$$-36x^2 + 0x$$
$$\underline{\pm 36x^2 \pm 48x} \qquad \text{Change signs}$$
$$48x + 64$$
$$\underline{^-48x \mp 64} \qquad \text{Change signs}$$

Quotient: $9x^2 - 12x + 16$

51.

$$x+2 \overline{\smash{\big)}\, x^4 + 0x^3 + 0x^2 + 0x - 16} \quad \overset{\displaystyle x^3 - 2x^2 + 4x - 8}{}$$

$$\underline{^-x^4 \mp 2x^3} \qquad \text{Change signs}$$
$$-2x^3 + 0x^2$$
$$\underline{\pm 2x^3 \pm 4x^2} \qquad \text{Change signs}$$
$$4x^2 + 0x$$
$$\underline{^-4x^2 \mp 8x} \qquad \text{Change signs}$$
$$-8x - 16$$
$$\underline{\pm 8x \pm 16} \qquad \text{Change signs}$$

Quotient: $x^3 - 2x^2 + 4x - 8$

53.

$$x^2+2x+1 \overline{\smash{\big)}\, 3x^3 + x^2 - 7x - 3} \quad \overset{\displaystyle 3x - 5}{}$$

$$\underline{^-3x^3 \mp 6x^2 \mp 3x} \qquad \text{Change signs}$$
$$-5x^2 - 10x - 3$$
$$\underline{\pm 5x^2 \pm 10x \pm 5} \qquad \text{Change signs}$$
$$2$$

Quotient: $3x - 5 + \dfrac{2}{x^2 + 2x + 1}$

55.

$$3x^2-2x-2 \overline{\smash{\big)}\, 12x^3 - 29x^2 + 6x + 14} \quad \overset{\displaystyle 4x - 7}{}$$

$$\underline{^-12x^3 \pm 8x^2 \pm 8x} \qquad \text{Change signs}$$
$$-21x^2 + 14x + 14$$
$$\underline{\pm 21x^2 \mp 14x \mp 14} \qquad \text{Change signs}$$

Quotient: $4x - 7$

57.

$$2x^2-4 \overline{\smash{\big)}\, 2x^4 - 12x^3 + 0x^2 + 27x - 3} \quad \overset{\displaystyle x^2 - 6x + 2}{}$$

$$\underline{^-2x^4 \qquad \pm 4x^2} \qquad \text{Change signs}$$
$$-12x^3 + 4x^2 + 27x$$
$$\underline{\pm 12x^3 \qquad \mp 24x} \qquad \text{Change signs}$$
$$4x^2 + 3x - 3$$
$$\underline{^-4x^2 \qquad \pm 8} \qquad \text{Change signs}$$
$$3x + 5$$

Quotient: $x^2 - 6x + 2 + \dfrac{3x + 5}{2x^2 - 4}$

59.

$$2x^2-3 \overline{\smash{\big)}\, 2x^6 + 0x^5 + 3x^4 + 0x^3 - x^2 + 5x - 11} \quad \overset{\displaystyle x^4 + 3x^2 + 4}{}$$

$$\underline{^-2x^6 \qquad \pm 3x^4}$$
$$6x^4 + 0x^3 - x^2$$
$$\underline{^-6x^4 \qquad \pm 9x^2}$$
$$8x^2 + 5x - 11$$
$$\underline{^-8x^2 \qquad \pm 12}$$
$$5x + 1$$

Quotient: $x^4 + 3x^2 + 4 + \dfrac{5x + 1}{2x^2 - 3}$

Exercises 4.7

1.

$$\frac{x^2 - 3x + 5}{x - 2}$$

$$\underline{2} \begin{array}{rrr} 1 & -3 & 5 \\ & 2 & -2 \\ \hline 1 & -1 & 3 \end{array}$$

Quotient: $x - 1 + \dfrac{3}{x - 2}$

3.

$$\frac{2x^2 - 9}{x + 2}$$

$$\underline{-2} \begin{array}{rrr} 2 & 0 & -9 \\ & -4 & 8 \\ \hline 2 & -4 & -1 \end{array}$$

#3. continued

Quotient: $2x - 4 - \dfrac{1}{x+2}$

5.

$$\dfrac{9x^2 - 3x + 4}{x - \dfrac{1}{3}}$$

$$\dfrac{1}{3}\Big|\quad 9 \quad -3 \quad 4$$

$$\underline{\qquad\quad 3 \quad\; 0\;}$$

$$\qquad\; 9 \quad\; 0 \quad 4$$

Quotient: $9x + \dfrac{4}{x - \dfrac{1}{3}}$

7.

$$\dfrac{2x^3 - 7x^2 - x + 6}{x - 3}$$

$$3\Big|\quad 2 \quad -7 \quad -1 \quad\; 6$$

$$\underline{\qquad\quad 6 \quad -3 \quad -12}$$

$$\quad\; 2 \quad -1 \quad -4 \quad -6$$

Quotient: $2x^2 - x - 4 - \dfrac{6}{x-3}$

9.

$$\dfrac{x^3 - 7x - 4}{x + 2}$$

$$\underline{-2}\Big|\quad 1 \quad\; 0 \quad -7 \quad -4$$

$$\underline{\qquad\quad -2 \quad\; 4 \quad\; 6}$$

$$\quad\; 1 \quad -2 \quad -3 \quad\; 2$$

Quotient: $x^2 - 2x - 3 + \dfrac{2}{x+2}$

11.

$$\dfrac{3x^3 + 11x^2 - 5x}{x + 4}$$

$$\underline{-4}\Big|\quad 3 \quad 11 \quad -5 \quad\; 0$$

$$\underline{\qquad\quad -12 \quad\; 4 \quad\; 4}$$

$$\quad\; 3 \quad -1 \quad -1 \quad\; 4$$

Quotient: $3x^2 - x - 1 + \dfrac{4}{x+4}$

13.

$$\dfrac{4x^3 - 5x + 1}{x + \dfrac{1}{2}}$$

$$\underline{-\dfrac{1}{2}}\Big|\quad 4 \quad\; 0 \quad -5 \quad\; 1$$

$$\underline{\qquad\quad -2 \quad\; 1 \quad\; 2}$$

$$\quad\; 4 \quad -2 \quad -4 \quad\; 3$$

Quotient: $4x^2 - 2x - 4 + \dfrac{3}{x + \dfrac{1}{2}}$

15.

$$\dfrac{6x^3 - 4x^2 + 9x - 8}{x - \dfrac{2}{3}}$$

$$\dfrac{2}{3}\Big|\quad 6 \quad -4 \quad\; 9 \quad -8$$

$$\underline{\qquad\qquad 4 \quad\; 0 \quad\; 6}$$

$$\quad\; 6 \quad\; 0 \quad\; 9 \quad -2$$

Quotient: $6x^2 + 9 - \dfrac{2}{x - \dfrac{2}{3}}$

17.

$$\dfrac{x^4 - 2x^3 - 5x^2 + x + 5}{x + 2}$$

$$\underline{-2}\Big|\quad 1 \quad -2 \quad -5 \quad\; 1 \quad\; 5$$

$$\underline{\qquad\quad -2 \quad\; 8 \quad -6 \quad 10}$$

$$\quad\; 1 \quad -4 \quad\; 3 \quad -5 \quad 15$$

Quotient: $x^3 - 4x^2 + 3x - 5 + \dfrac{15}{x+2}$

19.

$$\dfrac{2x^4 - 3x^3 + x^2 + 7x - 5}{x - 1}$$

$$1\Big|\quad 2 \quad -3 \quad\; 1 \quad\; 7 \quad -5$$

$$\underline{\qquad\quad 2 \quad -1 \quad\; 0 \quad\; 7}$$

$$\quad\; 2 \quad -1 \quad\; 0 \quad\; 7 \quad\; 2$$

Quotient: $2x^3 - x^2 + 7 + \dfrac{2}{x-1}$

21.

$$\frac{16x^4 - 9x^2 + 4x - 2}{x + \frac{3}{4}}$$

$$-\frac{3}{4}\begin{array}{|cccccc} 16 & 0 & -9 & 4 & -2 \end{array}$$

$$\begin{array}{ccccc} & -12 & 9 & 0 & -3 \\ \hline 16 & -12 & 0 & 4 & -5 \end{array}$$

Quotient: $16x^3 - 12x^2 + 4 - \dfrac{5}{x + \frac{3}{4}}$

23.

$$\frac{x^3 + 125}{x + 5}$$

$$-5\begin{array}{|cccc} 1 & 0 & 0 & 125 \end{array}$$

$$\begin{array}{cccc} & -5 & 25 & -125 \\ \hline 1 & -5 & 25 & 0 \end{array}$$

Quotient: $x^2 - 5x + 25$

25.

$$\frac{x^6 - 64}{x - 2}$$

$$2\begin{array}{|ccccccc} 1 & 0 & 0 & 0 & 0 & 0 & -64 \end{array}$$

$$\begin{array}{ccccccc} & 2 & 4 & 8 & 16 & 32 & 64 \\ \hline 1 & 2 & 4 & 8 & 16 & 32 & 0 \end{array}$$

Quotient: $x^5 + 2x^4 + 4x^3 + 8x^2 + 16x + 32$

27.

$$\frac{x^2 - 5}{x - \sqrt{5}}$$

$$\sqrt{5}\begin{array}{|ccc} 1 & 0 & -5 \end{array}$$

$$\begin{array}{ccc} & \sqrt{5} & 5 \\ \hline 1 & \sqrt{5} & 0 \end{array}$$

Quotient: $x + \sqrt{5}$

29.

$$\frac{x^4 - 9}{x + \sqrt{3}}$$

$$-\sqrt{3}\begin{array}{|ccccc} 1 & 0 & 0 & 0 & -9 \end{array}$$

$$\begin{array}{ccccc} & -\sqrt{3} & 3 & -3\sqrt{3} & +9 \\ \hline 1 & -\sqrt{3} & 3 & -3\sqrt{3} & 0 \end{array}$$

#29. continued

Quotient: $x^3 - \sqrt{3}x^2 + 3x - 3\sqrt{3}$

Exercises 4.8

1.

$$\frac{-3}{2x} = \frac{9}{8}$$

$18x = -24$ Cross multiply

$$x = -\frac{24}{18}$$

$$x = -\frac{4}{3}$$

Solution set: $\left\{ -\dfrac{4}{3} \right\}$

3.

$$\frac{3}{2x} + \frac{1}{3x} = \frac{11}{12} \qquad LCD = 12x$$

$$\frac{12x}{1} \cdot \left(\frac{3}{2x} + \frac{1}{3x} \right) = \frac{11}{12} \cdot \frac{12x}{1}$$

$$\frac{12x}{1} \cdot \frac{3}{2x} + \frac{12x}{1} \cdot \frac{1}{3x} = \frac{11}{12} \cdot \frac{12x}{1}$$

$$18 + 4 = 11x$$

$$22 = 11x$$

$$2 = x$$

Solution set: $\{2\}$

5.

$$\frac{12x + 3}{3x + 2} = 3 - \frac{5}{3x + 2} \qquad \text{Note: } 3x + 2 \neq 0$$

$$\text{So, } x \neq -\frac{2}{3}$$

$$\frac{3x + 2}{1} \cdot \left(\frac{12x + 3}{3x + 2} \right) = \left(3 - \frac{5}{3x + 2} \right) \cdot \frac{3x + 2}{1}$$

$$\frac{\cancel{(3x + 2)}}{1} \cdot \frac{12x + 3}{\cancel{3x + 2}} = \frac{3}{1} \cdot \frac{3x + 2}{1} - \frac{5}{\cancel{(3x + 2)}} \cdot \frac{\cancel{(3x + 2)}}{1}$$

$$12x + 3 = 9x + 6 - 5$$

$$12x + 3 = 9x + 1$$

$$3x = -2$$

$$x = -\frac{2}{3}$$

Solution set: $\varnothing$ Note: $3\left(-\dfrac{2}{3} \right) + 2 = 0$

7.

$$\frac{x+1}{2} = \frac{x-5}{4}$$

$4x + 4 = 2x - 10 \qquad \text{Cross multiply}$

$2x = -14$

$x = -7$

Solution set: $\{-7\}$

9.

$$\frac{2}{2x-1} + \frac{3}{x-4} = \frac{5}{2x-1} \qquad LCD = (2x-1)(x-4)$$

$$\frac{(2x-1)(x-4)}{1} \cdot \left(\frac{2}{2x-1} + \frac{3}{x-4}\right)$$

$$= \frac{5}{2x-1} \cdot \frac{(2x-1)(x-4)}{1}$$

$$\frac{(2x-1)(x-4)}{1} \cdot \frac{2}{(2x-1)} + \frac{(2x-1)(x-4)}{1} \cdot \frac{3}{(x-4)}$$

$$= \frac{5}{2x-1} \cdot \frac{(2x-1)(x-4)}{1}$$

$2x - 8 + 6x - 3 = 5x - 20$

$8x - 11 = 5x - 20$

$3x = -9$

$x = -3$

Solution set: $\{-3\}$

11.

$$\frac{2}{x+1} - \frac{1}{3x-2} = \frac{3}{3x-2} \qquad LCD = (x+1)(3x-2)$$

$$\frac{(x+1)(3x-2)}{1} \cdot \left(\frac{2}{x+1} - \frac{1}{3x-2}\right)$$

$$= \frac{3}{3x-2} \cdot \frac{(x+1)(3x-2)}{1}$$

$$\frac{(x+1)(3x-2)}{1} \cdot \frac{2}{(x+1)} - \frac{(x+1)(3x-2)}{1} \cdot \frac{1}{(3x-2)}$$

$$= \frac{3}{3x-2} \cdot \frac{(x+1)(3x-2)}{1}$$

$6x - 4 - x - 1 = 3x + 3$

$5x - 5 = 3x + 3$

$2x = 8$

$x = 4$

Solution set: $\{4\}$

13.

$$\frac{5}{2x+1} - \frac{1}{2x-1} = \frac{6}{4x^2-1} \qquad LCD = 4x^2 - 1$$

$$\frac{(2x+1)(2x-1)}{1} \cdot \left(\frac{5}{2x+1} - \frac{1}{2x-1}\right)$$

$$= \frac{6}{4x^2-1} \cdot \frac{4x^2-1}{1}$$

$$\frac{(2x+1)(2x-1)}{1} \cdot \frac{5}{(2x+1)} - \frac{(2x+1)(2x-1)}{1} \cdot \frac{1}{(2x-1)}$$

$$= \frac{6}{4x^2-1} \cdot \frac{4x^2-1}{1}$$

$10x - 5 - 2x - 1 = 6$

$8x - 6 = 6$

$8x = 12$

$x = \frac{12}{8} \quad \text{or} \quad \frac{3}{2}$

Solution set: $\left\{\frac{3}{2}\right\}$

15.

$$1 = \frac{x+5}{x-3} - \frac{8}{x-3}$$

$$\frac{x-3}{1} \cdot 1 = \left(\frac{x+5}{x-3} - \frac{8}{x-3}\right) \cdot \frac{x-3}{1}$$

$$x - 3 = \frac{x+5}{(x-3)} \cdot \frac{(x-3)}{1} - \frac{8}{(x-3)} \cdot \frac{(x-3)}{1}$$

$x - 3 = x + 5 - 8$

$-8 = -8 \qquad \text{True}$

Solution set: All real numbers except 3.

17.

$$\frac{1}{x} - \frac{2}{4x+1} = \frac{1}{8x^2+2x} \qquad LCD = 2x(4x+1)$$

$$\frac{2x(4x+1)}{1} \cdot \left(\frac{1}{x} - \frac{2}{4x+1}\right)$$

$$= \frac{1}{8x^2+2x} \cdot \frac{8x^2+2x}{1}$$

$$\frac{2x(4x+1)}{1} \cdot \frac{1}{x} - \frac{2x(4x+1)}{1} \cdot \frac{2}{(4x+1)}$$

$$= \frac{1}{(8x^2+2x)} \cdot \frac{(8x^2+2x)}{1}$$

$8x + 2 - 4x = 1$

$4x = -1$

$x = -\frac{1}{4}$

#17. continued
Solution set: $\varnothing$ Note: $4x + 1 \neq 0$

So, $x \neq -\dfrac{1}{4}$

19.

$$\frac{2x+1}{x-2} = \frac{2x+1}{x+2} + \frac{10}{x^2-4}$$

LCD $= x^2 - 4$ or

$(x+2)(x-2)$

$$\frac{(x+2)(x-2)}{1} \cdot \frac{2x+1}{x-2}$$

$$= \frac{(x+2)(x-2)}{1}\left(\frac{2x+1}{x+2} + \frac{10}{x^2-4}\right)$$

$$\frac{(x+2)\cancel{(x-2)}}{1} \cdot \frac{2x+1}{\cancel{x-2}}$$

$$= \frac{\cancel{(x+2)}(x-2)}{1} \cdot \frac{2x+1}{\cancel{(x+2)}} + \frac{\cancel{(x^2-4)}}{1} \cdot \frac{10}{\cancel{(x^2-4)}}$$

$$(x+2)(2x+1) = (x-2)(2x+1) + 10$$

$$2x^2 + 5x + 2 = 2x^2 - 3x - 2 + 10$$

$$5x + 2 = -3x + 8$$

$$8x = 6$$

$$x = \frac{6}{8} \quad \text{or} \quad \frac{3}{4}$$

Solution set: $\left\{\dfrac{3}{4}\right\}$

21.

$$\frac{1}{x^2} - \frac{1}{6x} = \frac{1}{6}$$

LCD $= 6x^2$

$$\frac{6x^2}{1}\cdot\left(\frac{1}{x^2} - \frac{1}{6x}\right) = \frac{6x^2}{1}\cdot\frac{1}{6}$$

$$\frac{6x^2}{1}\cdot\frac{1}{\cancel{x^2}} - \frac{\cancel{6}x\cdot x}{1}\cdot\frac{1}{\cancel{6x}} = \frac{\cancel{6}x^2}{1}\cdot\frac{1}{\cancel{6}}$$

$$6 - x = x^2$$

$$0 = x^2 + x - 6$$

$$0 = (x-2)(x+3)$$

$x - 2 = 0 \quad$ or $\quad x + 3 = 0$

$\quad x = 2 \qquad\qquad x = -3$

Solution set: $\{-3, 2\}$

23.

$$\frac{5}{x^2} - \frac{1}{3x} = 2$$

LCD $= 3x^2$

$$\frac{3x^2}{1}\cdot\left(\frac{5}{x^2} - \frac{1}{3x}\right) = \frac{3x^2}{1}\cdot\frac{2}{1}$$

$$\frac{3x^2}{1}\cdot\frac{5}{\cancel{x^2}} - \frac{3x\cdot x}{1}\cdot\frac{1}{\cancel{3x}} = 6x^2$$

$$15 - x = 6x^2$$

$$0 = 6x^2 + x - 15$$

$$0 = (3x + 5)(2x - 3)$$

$3x + 5 = 0 \quad$ or $\quad 2x - 3 = 0$

$\quad x = -\dfrac{5}{3} \qquad\qquad x = \dfrac{3}{2}$

Solution set: $\left\{-\dfrac{5}{3}, \dfrac{3}{2}\right\}$

25.

$$\frac{2}{x} - \frac{3}{x^2} = \frac{1}{3}$$

LCD $= 3x^2$

$$\frac{3x^2}{1}\left(\frac{2}{x} - \frac{3}{x^2}\right) = \frac{3x^2}{1}\cdot\frac{1}{3}$$

$$\frac{3x\cdot x}{1}\cdot\frac{2}{\cancel{x}} - \frac{3x^2}{1}\cdot\frac{3}{\cancel{x^2}} = \frac{\cancel{3}x^2}{1}\cdot\frac{1}{\cancel{3}}$$

$$6x - 9 = x^2$$

$$0 = x^2 - 6x + 9$$

$$0 = (x-3)^2$$

$$x - 3 = 0$$

$$x = 3$$

Solution set: $\{3\}$

27.

$$\frac{2}{2x-1} + \frac{x}{x+4} = \frac{36}{2x^2+7x-4}$$

LCD $= 2x^2 + 7x - 4$

or $(2x-1)(x+4)$

$$\frac{(2x-1)(x+4)}{1}\cdot\left(\frac{2}{2x-1} + \frac{x}{x+4}\right)$$

$$= \frac{2x^2+7x-4}{1}\cdot\frac{36}{2x^2+7x-4}$$

$$\frac{\cancel{(2x-1)}(x+4)}{1}\cdot\frac{2}{\cancel{(2x-1)}} + \frac{(2x-1)\cancel{(x+4)}}{1}\cdot\frac{x}{\cancel{(x+4)}}$$

$$= \frac{\cancel{(2x^2+7x-4)}}{1}\cdot\frac{36}{\cancel{2x^2+7x-4}}$$

#27. continued

$$2(x+4) + x(2x-1) = 36$$

$$2x + 8 + 2x^2 - x = 36$$

$$2x^2 + x + 8 = 36$$

$$2x^2 + x - 28 = 0$$

$$(2x - 7)(x + 4) = 0$$

$$2x - 7 = 0 \quad \text{or} \quad x + 4 = 0$$

$$x = \frac{7}{2} \qquad\qquad x = -4$$

Solution set: $\left\{\frac{7}{2}\right\}$ Note: $x + 4 \neq 0$

So, $x \neq -4$

29.

$$\frac{4x}{x-1} - \frac{x}{x+3} = \frac{-12}{x^2 + 2x - 3}$$ LCD $= x^2 + 2x - 3$

or $(x-1)(x+3)$

$$\frac{(x-1)(x+3)}{1} \cdot \left(\frac{4x}{x-1} - \frac{x}{x+3}\right)$$

$$= \frac{x^2 + 2x - 3}{1} \cdot \frac{-12}{x^2 + 2x - 3}$$

$$\frac{\cancel{(x-1)}(x+3)}{1} \cdot \frac{4x}{\cancel{(x-1)}} - \frac{(x-1)\cancel{(x+3)}}{1} \cdot \frac{x}{\cancel{(x+3)}}$$

$$= \frac{\cancel{(x^2 + 2x - 3)}}{1} \cdot \frac{-12}{\cancel{(x^2 + 2x - 3)}}$$

$$4x(x+3) - x(x-1) = -12$$

$$4x^2 + 12x - x^2 + x = -12$$

$$3x^2 + 13x + 12 = 0$$

$$(3x + 4)(x + 3) = 0$$

$$3x + 4 = 0 \quad \text{or} \quad x + 3 = 0$$

$$x = -\frac{4}{3} \qquad\qquad x = -3$$

Solution set: $\left\{-\frac{4}{3}\right\}$ Note: $x + 3 \neq 0$

So, $x \neq -3$

31.

$$\frac{3x + 6}{2x + 5} = -\frac{1}{x}$$

$$3x^2 + 6x = -2x - 5 \qquad \text{Cross multiply}$$

$$3x^2 + 8x + 5 = 0$$

$$(3x + 5)(x + 1) = 0$$

#31. continued

$$3x + 5 = 0 \quad \text{or} \quad x + 1 = 0$$

$$x = -\frac{5}{3} \qquad\qquad x = -1$$

Solution set: $\left\{-\frac{5}{3}, -1\right\}$

33.

$$\frac{3x}{x-2} = \frac{x+4}{2-x}$$

$$(x-2)(x+4) = 3x(2-x) \qquad \text{Cross multiply}$$

$$x^2 + 2x - 8 = 6x - 3x^2$$

$$4x^2 - 4x - 8 = 0$$

$$x^2 - x - 2 = 0 \qquad \text{Multiply by } \frac{1}{4}$$

$$(x+1)(x-2) = 0$$

$$x + 1 = 0 \quad \text{or} \quad x - 2 = 0$$

$$x = -1 \qquad\qquad x = 2$$

Solution set: $\{-1\}$ Note: $x - 2 \neq 0$

So, $x \neq 2$

35.

$$\frac{x+3}{9x+1} = \frac{x+1}{7}$$

$$7(x+3) = (x+1)(9x+1) \qquad \text{Cross multiply}$$

$$7x + 21 = 9x^2 + 10x + 1$$

$$0 = 9x^2 + 3x - 20$$

$$0 = (3x + 5)(3x - 4)$$

$$3x + 5 = 0 \quad \text{or} \quad 3x - 4 = 0$$

$$x = -\frac{5}{3} \qquad\qquad x = \frac{4}{3}$$

Solution set: $\left\{-\frac{5}{3}, \frac{4}{3}\right\}$

37.

$$\frac{x}{x-4} = \frac{x+2}{3x-10}$$

$$x(3x - 10) = (x-4)(x+2) \qquad \text{Cross multiply}$$

$$3x^2 - 10x = x^2 - 2x - 8$$

$$2x^2 - 8x + 8 = 0$$

$$x^2 - 4x + 4 = 0 \qquad \text{Multiply by } \frac{1}{2}$$

$$(x - 2)^2 = 0$$

$$x - 2 = 0$$

$$x = 2$$

#37. continued
Solution set: $\{2\}$

39.

$$\frac{9x+3}{10x-30} = \frac{x}{x-3}$$

$(9x+3)(x-3) = x(10x-30)$ Cross multiply

$9x^2 - 24x - 9 = 10x^2 - 30x$

$0 = x^2 - 6x + 9$

$0 = (x-3)(x-3)$

$x - 3 = 0$

$x = 3$

Solution set: $\varnothing$ Note: $x - 3 \neq 0$

So, $x \neq 3$

41.

$$\frac{2}{x+3} + \frac{4}{x+1} = \frac{4}{3}$$ $\text{LCD} = 3(x+3)(x+1)$

$$\frac{3(x+3)(x+1)}{1} \cdot \left(\frac{2}{x+3} + \frac{4}{x+1} \right)$$

$$= \frac{4}{3} \cdot \frac{3(x+3)(x+1)}{1}$$

$$\frac{3\cancel{(x+3)}(x+1)}{1} \cdot \frac{2}{\cancel{(x+3)}} + \frac{3(x+3)\cancel{(x+1)}}{1} \cdot \frac{4}{\cancel{(x+1)}}$$

$$= \frac{4}{\cancel{3}} \cdot \frac{\cancel{3}(x+3)(x+1)}{1}$$

$6(x+1) + 12(x+3) = 4(x+3)(x+1)$

$6x + 6 + 12x + 36 = 4(x^2 + 4x + 3)$

$18x + 42 = 4x^2 + 16x + 12$

$0 = 4x^2 - 2x - 30$

$0 = 2x^2 - x - 15$

$0 = (2x+5)(x-3)$

$2x + 5 = 0$ or $x - 3 = 0$

$x = -\dfrac{5}{2}$ $x = 3$

Solution set: $\left\{ -\dfrac{5}{2}, 3 \right\}$

43.

$$\frac{1}{x-5} - \frac{5}{x-2} = \frac{3}{2}$$ $\text{LCD} = 2(x-5)(x-2)$

$$\frac{2(x-5)(x-2)}{1} \cdot \left(\frac{1}{x-5} - \frac{5}{x-2} \right)$$

$$= \frac{2(x-5)(x-2)}{1} \cdot \frac{3}{2}$$

#43. continued

$$\frac{2\cancel{(x-5)}(x-2)}{1} \cdot \frac{1}{\cancel{(x-5)}} - \frac{2(x-5)\cancel{(x-2)}}{1} \cdot \frac{5}{\cancel{(x-2)}}$$

$$= \frac{\cancel{2}(x-5)(x-2)}{1} \cdot \frac{3}{\cancel{2}}$$

$2(x-2) - 10(x-5) = 3(x-5)(x-2)$

$2x - 4 - 10x + 50 = 3(x^2 - 7x + 10)$

$-8x + 46 = 3x^2 - 21x + 30$

$0 = 3x^2 - 13x - 16$

$0 = (x+1)(3x-16)$

$x + 1 = 0$ or $3x - 16 = 0$

$x = -1$ $x = \dfrac{16}{3}$

Solution set: $\left\{ -1, \dfrac{16}{3} \right\}$

45.

$$\frac{x-2}{x-3} + \frac{1}{x+2} = \frac{13}{6}$$ $\text{LCD} = 6(x-3)(x+2)$

$$\frac{6(x-3)(x+2)}{1} \cdot \left(\frac{x-2}{x-3} + \frac{1}{x+2} \right)$$

$$= \frac{6(x-3)(x+2)}{1} \cdot \frac{13}{6}$$

$$\frac{6\cancel{(x-3)}(x+2)}{1} \cdot \frac{(x-2)}{\cancel{(x-3)}} + \frac{6(x-3)\cancel{(x+2)}}{1} \cdot \frac{1}{\cancel{(x+2)}}$$

$$= \frac{\cancel{6}(x-3)(x+2)}{1} \cdot \frac{13}{\cancel{6}}$$

$6(x+2)(x-2) + 6(x-3) = 13(x-3)(x+2)$

$6(x^2 - 4) + 6(x-3) = 13(x^2 - x - 6)$

$6x^2 - 24 + 6x - 18 = 13x^2 - 13x - 78$

$0 = 7x^2 - 19x - 36$

$0 = (7x+9)(x-4)$

$7x + 9 = 0$ or $x - 4 = 0$

$x = -\dfrac{9}{7}$ $x = 4$

Solution set: $\left\{ -\dfrac{9}{7}, 4 \right\}$

47.

$$\frac{x+2}{2x-1} + \frac{x+5}{x+3} = \frac{5}{3}$$ $\text{LCD} = 3(2x-1)(x+3)$

$$\frac{3(2x-1)(x+3)}{1} \cdot \left(\frac{x+2}{2x-1} + \frac{x+5}{x+3} \right)$$

$$= \frac{3(2x-1)(x+3)}{1} \cdot \frac{5}{3}$$

#47. continued

$$\frac{3(2x-1)(x+3)}{1} \cdot \frac{(x+2)}{(2x-1)} + \frac{3(2x-1)(x+3)}{1} \cdot \frac{(x+5)}{(x+3)}$$

$$= \frac{3(2x-1)(x+3)}{1} \cdot \frac{5}{3}$$

$$3(x+3)(x+2) + 3(2x-1)(x+5) = 5(2x-1)(x+3)$$

$$3(x^2+5x+6) + 3(2x^2+9x-5) = 5(2x^2+5x-3)$$

$$3x^2+15x+18+6x^2+27x-15 = 10x^2+25x-15$$

$$9x^2+42x+3 = 10x^2+25x-15$$

$$0 = x^2-17x-18$$

$$0 = (x-18)(x+1)$$

$$x-18=0 \qquad \text{or} \qquad x+1=0$$

$$x=18 \qquad\qquad x=-1$$

Solution set: $\{18,-1\}$

49.

$$\frac{x+3}{x+4} - \frac{5}{x-1} = \frac{5x-3}{x^2+3x-4} \qquad \text{LCD} = x^2+3x-4$$

$$\text{or} \quad (x+4)(x-1)$$

$$\frac{(x+4)(x-1)}{1} \cdot \left(\frac{x+3}{x+4} - \frac{5}{x-1}\right)$$

$$= \frac{x^2+3x-4}{1} \cdot \frac{5x-3}{x^2+3x-4}$$

$$\frac{(x+4)(x-1)}{1} \cdot \frac{(x+3)}{(x+4)} - \frac{(x+4)(x-1)}{1} \cdot \frac{5}{(x-1)}$$

$$= \frac{x^2+3x-4}{1} \cdot \frac{5x-3}{x^2+3x-4}$$

$$(x-1)(x+3) - 5(x+4) = 5x-3$$

$$x^2+2x-3-5x-20 = 5x-3$$

$$x^2-3x-23 = 5x-3$$

$$x^2-8x-20 = 0$$

$$(x+2)(x-10) = 0$$

$$x+2=0 \qquad \text{or} \qquad x-10=0$$

$$x=-2 \qquad\qquad x=10$$

Solution set: $\{-2,10\}$

51.

$$\frac{x+2}{2x+3} + \frac{3x+1}{x-3} = \frac{9x+5}{2x^2-3x-9} \qquad \text{LCD} = 2x^2-3x-9$$

$$\text{or } (2x+3)(x-3)$$

$$\frac{(2x+3)(x-3)}{1} \cdot \left(\frac{x+2}{2x+3} + \frac{3x+1}{x-3}\right)$$

$$= \frac{2x^2-3x-9}{1} \cdot \frac{9x+5}{2x^2-3x-9}$$

#51. continued

$$\frac{(2x+3)(x-3)}{1} \cdot \frac{(x+2)}{(2x+3)} + \frac{(2x+3)(x-3)}{1} \cdot \frac{(3x+1)}{(x-3)}$$

$$= \frac{2x^2-3x-9}{1} \cdot \frac{9x+5}{2x^2-3x-9}$$

$$(x-3)(x+2) + (2x+3)(3x+1) = 9x+5$$

$$x^2-x-6+6x^2+11x+3 = 9x+5$$

$$7x^2+10x-3 = 9x+5$$

$$7x^2+x-8 = 0$$

$$(7x+8)(x-1) = 0$$

$$7x+8=0 \qquad \text{or} \qquad x-1=0$$

$$x=-\frac{8}{7} \qquad\qquad x=1$$

Solution set: $\left\{-\frac{8}{7},1\right\}$

53.

$$\frac{x}{3x-2} - \frac{8}{2x+3} = \frac{2x^2-24x+18}{6x^2+5x-6}$$

$$\text{LCD} = 6x^2+5x-6$$

$$\text{or} \quad (3x-2)(2x+3)$$

$$\frac{(3x-2)(2x+3)}{1} \cdot \left(\frac{x}{3x-2} - \frac{8}{2x+3}\right)$$

$$= \frac{6x^2+5x-6}{1} \cdot \frac{2x^2-24x+18}{6x^2+5x-6}$$

$$\frac{(3x-2)(2x+3)}{1} \cdot \frac{x}{(3x-2)}$$

$$- \frac{(3x-2)(2x+3)}{1} \cdot \frac{8}{(2x+3)}$$

$$= \frac{6x^2+5x-6}{1} \cdot \frac{2x^2-24x+18}{6x^2+5x-6}$$

$$x(2x+3) - 8(3x-2) = 2x^2-24x+18$$

$$2x^2+3x-24x+16 = 2x^2-24x+18$$

$$2x^2-21x+16 = 2x^2-24x+18$$

$$3x = 2$$

$$x = \frac{2}{3}$$

Solution set: $\varnothing$ \qquad Note: $3x-2 \neq 0$

$$\text{So,} \quad x \neq \frac{2}{3}$$

55.

$$\frac{x-7}{2x^2+9x-5}-\frac{2x-6}{x^2-25}=\frac{x+1}{2x^2-11x+5}$$

$$\frac{x-7}{(2x-1)(x+5)}-\frac{2x-6}{(x+5)(x-5)}=\frac{x+1}{(2x-1)(x-5)}$$

$$\text{LCD} = (2x-1)(x+5)(x-5)$$

$$\frac{(2x-1)(x+5)(x-5)}{1}$$

$$\cdot\left(\frac{x-7}{(2x-1)(x+5)}-\frac{2x-6}{(x+5)(x-5)}\right)$$

$$=\frac{(2x-1)(x+5)(x-5)}{1}\cdot\frac{x+1}{(2x-1)(x-5)}$$

$$\frac{(2x-1)(x+5)(x-5)}{1}\cdot\frac{(x-7)}{(2x-1)(x+5)}$$

$$-\frac{(2x-1)(x+5)(x-5)}{1}\cdot\frac{(2x-6)}{(x+5)(x-5)}$$

$$=\frac{(2x-1)(x+5)(x-5)}{1}\cdot\frac{x+1}{(2x-1)(x-5)}$$

$$(x-5)(x-7)-(2x-1)(2x-6)=(x+5)(x+1)$$

$$x^2-12x+35-(4x^2-14x+6)=x^2+6x+5$$

$$x^2-12x+35-4x^2+14x-6=x^2+6x+5$$

$$-3x^2+2x+29=x^2+6x+5$$

$$0=4x^2+4x-24$$

$$0=x^2+x-6$$

$$0=(x+3)(x-2)$$

$$x+3=0 \quad \text{or} \quad x-2=0$$
$$x=-3 \qquad\qquad x=2$$

Solution set: $\{-3,2\}$

57.

$$\frac{3x-1}{x^2+4x+5}-\frac{3}{x^2-x-12}=\frac{2x-5}{x^2-3x-4}$$

$$\frac{3x-1}{(x+3)(x+1)}-\frac{3}{(x+3)(x-4)}=\frac{2x-5}{(x-4)(x+1)}$$

$$\text{LCD} = (x+3)(x+1)(x-4)$$

$$\frac{(x+3)(x+1)(x-4)}{1}$$

$$\cdot\left(\frac{3x-1}{(x+3)(x+1)}-\frac{3}{(x+3)(x-4)}\right)$$

$$=\frac{(x+3)(x+1)(x-4)}{1}\cdot\frac{2x-5}{(x-4)(x+1)}$$

#57. continued

$$\frac{(x+3)(x+1)(x-4)}{1}\cdot\frac{(3x-1)}{(x+3)(x+1)}$$

$$-\frac{(x+3)(x+1)(x-4)}{1}\cdot\frac{3}{(x+3)(x-4)}$$

$$=\frac{(x+3)(x+1)(x-4)}{1}\cdot\frac{(2x-5)}{(x-4)(x+1)}$$

$$(x-4)(3x-1)-3(x+1)=(x+3)(2x-5)$$

$$3x^2-13x+4-3x-3=2x+x-15$$

$$3x^2-16x+1=2x^2+x-15$$

$$x^2-17x+16=0$$

$$(x-1)(x-16)=0$$

$$x-1=0 \quad \text{or} \quad x-16=0$$
$$x=1 \qquad\qquad x=16$$

Solution set: $\{1,16\}$

59.

$$\frac{1}{2x}+\frac{3}{5x^2-10x}=\frac{1}{2x}+\frac{3}{5x(x-2)} \quad \text{LCD}=10x(x-2)$$

$$=\frac{5(x-2)}{5(x-2)}\cdot\frac{1}{2x}+\frac{3}{5x(x-2)}\cdot\frac{2}{2}$$

$$=\frac{5(x-2)}{10x(x-2)}+\frac{6}{10x(x-2)}$$

$$=\frac{5x-10+6}{10x(x-2)}=\frac{5x-4}{10x(x-2)}$$

61.

$$\frac{2x-1}{x-5}=\frac{x+2}{x+3}$$

$$(2x-1)(x+3)=(x-5)(x+2) \qquad \text{Cross multiply}$$

$$2x^2+5x-3=x^2-3x-10$$

$$x^2+8x+7=0$$

$$(x+1)(x+7)=0$$

$$x+1=0 \quad \text{or} \quad x+7=0$$
$$x=-1 \qquad\qquad x=-7$$

Solution set: $\{-1,-7\}$

63.

$$\frac{x^2+3x+2}{2x^2+7x-15}\cdot\frac{3x^2+11x-20}{x^2+4x+4}$$

$$=\frac{(x+2)(x+1)}{(2x-3)(x+5)}\cdot\frac{(3x-4)(x+5)}{(x+2)(x+2)}$$

$$=\frac{(x+1)(3x-4)}{(2x-3)(x+2)}$$

65.

$$\frac{x+1}{2x^2-3x-2}+\frac{x-5}{x^2+x-6}=0$$

$$\frac{x+1}{(2x+1)(x-2)}+\frac{x-5}{(x-2)(x+3)}=0$$

$$\text{LCD}=(2x+1)(x-2)(x+3)$$

$$\frac{(2x+1)(x-2)(x+3)}{1}$$

$$\cdot\left(\frac{x+1}{(2x+1)(x-2)}+\frac{x-5}{(x-2)(x+3)}\right)$$

$$=\frac{(2x+1)(x-2)(x+3)}{1}\cdot 0$$

$$\frac{(2x+1)(x-2)(x+3)}{1}\cdot\frac{(x+1)}{(2x+1)(x-2)}$$

$$+\frac{(2x+1)(x-2)(x+3)}{1}\cdot\frac{(x-5)}{(x-2)(x+3)}=0$$

$$(x+3)(x+1)+(2x+1)(x-5)=0$$

$$x^2+4x+3+2x^2-9x-5=0$$

$$3x^2-5x-2=0$$

$$(3x+1)(x-2)=0$$

$$3x+1=0 \quad \text{or} \quad x-2=0$$

$$x=-\frac{1}{3} \qquad x=2$$

Solution set: $\left\{-\frac{1}{3}\right\}$ Note: $x-2\neq 0$

So, $x\neq 2$

67.

$$\frac{x}{2}+\frac{x}{3}=\frac{25}{6} \qquad \text{LCD}=6$$

$$\frac{6}{1}\cdot\left(\frac{x}{2}+\frac{x}{3}\right)=\frac{25}{6}\cdot\frac{6}{1}$$

$$3x+2x=25$$

$$5x=25$$

$$x=5$$

Solution set: $\{5\}$

69.

$$\frac{8}{x+1}=\frac{9}{x+2}$$

$$8x+16=9x+9 \qquad \text{Cross multiply}$$

$$16=x+9$$

$$7=x$$

Solution set: $\{7\}$

71.

$$\frac{5}{2x+1}-\frac{3}{x-4} \qquad \text{LCD}=(2x+1)(x-4)$$

$$=\frac{x-4}{x-4}\cdot\frac{5}{2x+1}-\frac{3}{x-4}\cdot\frac{2x+1}{2x+1}$$

$$=\frac{5x-20}{(x-4)(2x+1)}-\frac{6x+3}{(x-4)(2x+1)}$$

$$=\frac{5x-20-6x-3}{(x-4)(2x+1)}=\frac{-x-23}{(x-4)(2x+1)}$$

73.

$$\frac{1}{x}-\frac{4}{5}=\frac{3}{2x} \qquad \text{LCD}=10x$$

$$\frac{10x}{1}\cdot\left(\frac{1}{x}-\frac{4}{5}\right)=\frac{10x}{1}\cdot\frac{3}{2x}$$

$$\frac{10x}{1}\cdot\frac{1}{x}-\frac{2\cdot5\cdot x}{1}\cdot\frac{4}{5}=\frac{2\cdot5\cdot x}{1}\cdot\frac{3}{2x}$$

$$10-8x=15$$

$$-8x=5$$

$$x=-\frac{5}{8}$$

Solution set: $\left\{-\frac{5}{8}\right\}$

75.

$$\frac{3}{x-2}-\frac{4}{2-x}=\frac{3}{x-2}+\frac{4}{x-2} \qquad \text{Multiply by }\frac{-1}{-1}$$

$$=\frac{3+4}{x-2}=\frac{7}{x-2}$$

77.

$$\frac{2}{x+3}+\frac{7}{x+5}=\frac{5}{3} \qquad \text{LCD}=3(x+3)(x+5)$$

$$\frac{3(x+3)(x+5)}{1}\cdot\left(\frac{2}{x+3}+\frac{7}{x+5}\right)$$

$$=\frac{3(x+3)(x+5)}{1}\cdot\frac{5}{3}$$

$$\frac{3(x+3)(x+5)}{1}\cdot\frac{2}{(x+3)}+\frac{3(x+3)(x+5)}{1}\cdot\frac{7}{(x+5)}$$

$$=\frac{3(x+3)(x+5)}{1}\cdot\frac{5}{3}$$

$$6(x+5)+21(x+3)=5(x+3)(x+5)$$

$$6x+30+21x+63=5(x^2+8x+15)$$

$$27x+93=5x^2+40x+75$$

$$0=5x^2+13x-18$$

$$0=(5x+18)(x-1)$$

#77. continued

$$5x + 18 = 0 \quad \text{or} \quad x - 1 = 0$$

$$x = -\frac{18}{5} \qquad \qquad x = 1$$

Solution set: $\left\{-\frac{18}{5}, 1\right\}$

79.

$$\frac{25x^2 - 9}{3x^2 + 12x} + \frac{10x^3 - 6x^2}{10x + 40}$$

$$= \frac{(5x+3)(5x-3)}{3x(x+4)} \cdot \frac{2 \cdot 5 \cdot (x+4)}{2x^2(5x-3)} = \frac{5(5x+3)}{3x^3}$$

81.

$$\frac{2}{x} - \frac{1}{3} + \frac{5}{2x} \qquad \qquad LCD = 6x$$

$$= \frac{6}{6} \cdot \frac{2}{x} - \frac{2x}{2x} \cdot \frac{1}{3} + \frac{3}{3} \cdot \frac{5}{2x}$$

$$= \frac{12}{6x} - \frac{2x}{6x} + \frac{15}{6x} = \frac{12 - 2x + 15}{6x} = \frac{-2x + 27}{6x}$$

83.

$$\frac{x+3}{x^2-1} - \frac{x-5}{2x^2 + 3x + 1}$$

$$= \frac{x+3}{(x+1)(x-1)} - \frac{x-5}{(2x+1)(x+1)}$$

$$LCD = (x+1)(x-1)(2x+1)$$

$$= \frac{(2x+1)}{(2x+1)} \cdot \frac{(x+3)}{(x+1)(x-1)} - \frac{(x-1)}{(x-1)} \cdot \frac{(x-5)}{(2x+1)(x+1)}$$

$$= \frac{(2x+1)(x+3)}{(2x+1)(x+1)(x-1)} - \frac{(x-1)(x-5)}{(x-1)(2x+1)(x+1)}$$

$$= \frac{2x^2 + 7x + 3 - (x^2 - 6x + 5)}{(2x+1)(x+1)(x-1)}$$

$$= \frac{2x^2 + 7x + 3 - x^2 + 6x - 5}{(2x+1)(x+1)(x-1)}$$

$$= \frac{x^2 + 13x - 2}{(2x+1)(x+1)(x-1)}$$

85.

$$\frac{x^2 - 25}{x^2 + 10x + 25} \cdot \frac{8x^2}{10 - 2x}$$

$$= \frac{(x+5)(x-5)}{(x+5)(x+5)} \cdot \frac{2 \cdot 4x^2}{-1 \cdot 2(x-5)} = -\frac{4x^2}{x+5}$$

87.

$$\frac{x}{x+3} - \frac{3}{2} = -\frac{1}{x-2} \qquad LCD = 2(x+3)(x-2)$$

$$\frac{2(x+3)(x-2)}{1} \cdot \left(\frac{x}{x+3} - \frac{3}{2}\right)$$

$$= \frac{2(x+3)(x-2)}{1} \cdot \frac{-1}{x-2}$$

$$\frac{2(x+3)(x-2)}{1} \cdot \frac{x}{x+3} - \frac{2(x+3)(x-2)}{1} \cdot \frac{3}{2}$$

$$= \frac{2(x+3)(x-2)}{1} \cdot \frac{-1}{(x-2)}$$

$$2x(x-2) - 3(x+3)(x-2) = -2(x+3)$$

$$2x^2 - 4x - 3(x^2 + x - 6) = -2x - 6$$

$$2x^2 - 4x - 3x^2 - 3x + 18 = -2x - 6$$

$$-x^2 - 7x + 18 = -2x - 6$$

$$0 = x^2 + 5x - 24$$

$$0 = (x+8)(x-3)$$

$$x + 8 = 0 \quad \text{or} \quad x - 3 = 0$$

$$x = -8 \qquad \qquad x = 3$$

Solution set: $\{-8, 3\}$

Exercises 4.9

1.
Let x = the number.

Then, $\dfrac{9-x}{11+x} = \dfrac{3}{7}$

$$63 - 7x = 33 + 3x \qquad \text{Cross multiply}$$

$$63 = 33 + 10x$$

$$30 = 10x$$

$$3 = x$$

The number is 3.

3.
Let x = the number.

Then, $\dfrac{3-x}{5-x} = \dfrac{3}{4}$

$$12 - 4x = 15 - 3x \qquad \text{Cross multiply}$$

$$12 = 15 + x$$

$$-3 = x$$

The number is -3.

5.
Let x = the numerator.

Then x + 4 = the denominator.

Thus, $\dfrac{x}{x+4}$ is the original fraction.

#5. continued

So, $\dfrac{x+3}{x+4+3} = \dfrac{1}{2}$

$\dfrac{x+3}{x+7} = \dfrac{1}{2}$

$2x+6 = x+7$ Cross multiply

$x+6 = 7$

$x = 1$

$\dfrac{x}{x+4} = \dfrac{1}{1+4} = \dfrac{1}{5}$

The original fraction is $\dfrac{1}{5}$.

7.

Let x = the number.

Let $\dfrac{1}{x}$ = the reciprocal.

Thus, $x + \dfrac{1}{x} = \dfrac{25}{12}$ LCD = $12x$

$\dfrac{12x}{1}\left(x + \dfrac{1}{x}\right) = \dfrac{12x}{1} \cdot \dfrac{25}{12}$

$12x^2 + 12 = 25x$

$12x^2 - 25x + 12 = 0$

$(4x-3)(3x-4) = 0$

$4x-3 = 0$ or $3x-4 = 0$

$x = \dfrac{3}{4}$ $x = \dfrac{4}{3}$

The numbers are $\dfrac{3}{4}$ and $\dfrac{4}{3}$.

9.

Two consecutive integers are x and $x+1$.

Their reciprocals are $\dfrac{1}{x}$ and $\dfrac{1}{x+1}$.

Thus, $\dfrac{1}{x} + \dfrac{1}{x+1} = \dfrac{7}{12}$ LCD = $12x(x+1)$

$\dfrac{12x(x+1)}{1}\left(\dfrac{1}{x} + \dfrac{1}{x+1}\right) = \dfrac{12x(x+1)}{1} \cdot \dfrac{7}{12}$

$12(x+1) + 12x = 7x(x+1)$

$12x + 12 + 12x = 7x^2 + 7x$

$24x + 12 = 7x^2 + 7x$

$0 = 7x^2 - 17x - 12$

$0 = (7x+4)(x-3)$

$7x+4 = 0$ or $x-3 = 0$

$\cancel{x = -\dfrac{4}{7}}$ $x = 3$

#9. continued

$x+1 = 3+1 = 4$

The numbers are 3 and 4.

11.

Two consecutive even integers are x and $x+2$.

Their reciprocals are $\dfrac{1}{x}$ and $\dfrac{1}{x+2}$.

Thus, $\dfrac{1}{x} + \dfrac{1}{x+2} = \dfrac{7}{24}$ LCD = $24x(x+2)$

$\dfrac{24x(x+2)}{1} \cdot \left(\dfrac{1}{x} + \dfrac{1}{x+2}\right) = \dfrac{24x(x+2)}{1} \cdot \dfrac{7}{24}$

$24(x+2) + 24x = 7x(x+2)$

$24x + 48 + 24x = 7x^2 + 14x$

$48 + 48x = 7x^2 + 14x$

$0 = 7x^2 - 34x - 48$

$0 = (7x+8)(x-6)$

$7x+8 = 0$ or $x-6 = 0$

$\cancel{x = -\dfrac{8}{7}}$ $x = 6$

$x+2 = 6+2 = 8$

Answer: The numbers are 6 and 8.

13.

$\dfrac{1}{2} + \dfrac{1}{3} = \dfrac{1}{x}$ LCD = $6x$

$\dfrac{6x}{1}\left(\dfrac{1}{2} + \dfrac{1}{3}\right) = \dfrac{6x}{1} \cdot \dfrac{1}{x}$

$3x + 2x = 6$

$5x = 6$

$x = \dfrac{6}{5}$

Answer: $\dfrac{6}{5}$ hr or $1\dfrac{1}{5}$ hr or 1 hr 12 min.

15.

$\dfrac{\frac{1}{3}}{\frac{3}{4}} + \dfrac{1}{6} = \dfrac{1}{x}$ Note: $\dfrac{1}{\frac{3}{4}} = 1 \div \dfrac{3}{4} = 1 \cdot \dfrac{4}{3}$

$\dfrac{4}{3} + \dfrac{1}{6} = \dfrac{1}{x}$ LCD = $6x$

$\dfrac{6x}{1}\left(\dfrac{4}{3} + \dfrac{1}{6}\right) = \dfrac{1}{x} \cdot \dfrac{6x}{1}$

$8x + x = 6$

$9x = 6$

$x = \dfrac{6}{9}$ or $\dfrac{2}{3}$

#15. continued

Answer: $\frac{2}{3}$ hr or 40 min.

17.

$$\frac{1}{5} + \frac{1}{x} = \frac{1}{4} \qquad LCD = 20x$$

$$\frac{20x}{1}\left(\frac{1}{5} + \frac{1}{x}\right) = \frac{1}{4} \cdot \frac{20x}{1}$$

$$4x + 20 = 5x$$

$$20 = x$$

Answer: 20 hr.

19.

Let x = Barbara's hours.
Then 2x = Gary's hours.

Thus, $\quad \frac{1}{x} + \frac{1}{2x} = \frac{1}{8} \qquad LCD = 8x$

$$\frac{8x}{1}\left(\frac{1}{x} + \frac{1}{2x}\right) = \frac{1}{8} \cdot \frac{8x}{1}$$

$$8 + 4 = x$$

$$12 = x$$

$$2x = 2 \cdot 12 = 24$$

Answer: Barbara - 12 hr; Gary - 24 hr.

21.

$$\frac{1}{4} + \frac{1}{6} - \frac{1}{12} = \frac{1}{x} \qquad LCD = 12x$$

$$\frac{12x}{1}\left(\frac{1}{4} + \frac{1}{6} - \frac{1}{12}\right) = \frac{12x}{1} \cdot \frac{1}{x}$$

$$3x + 2x - x = 12$$

$$4x = 12$$

$$x = 3$$

Answer: 3 min.

23.

Let x = 1st pipe's hours.
Then x - 1 = 2nd pipe's hours.

$$\frac{1}{x} + \frac{1}{x-1} - \frac{1}{3} = \frac{1}{2} \qquad LCD = 6x(x-1)$$

$$\frac{6x(x-1)}{1}\left(\frac{1}{x} + \frac{1}{x-1} - \frac{1}{3}\right) = \frac{6x(x-1)}{1} \cdot \frac{1}{2}$$

$$6(x-1) + 6x - 2x(x-1) = 3x(x-1)$$

$$6x - 6 + 6x - 2x^2 + 2x = 3x^2 - 3x$$

$$-2x^2 + 14x - 6 = 3x^2 - 3x$$

$$0 = 5x^2 - 17x + 6$$

$$0 = (5x - 2)(x - 3)$$

#23. continued

$$5x - 2 = 0 \quad \text{or} \quad x - 3 = 0$$

$$\cancel{x = \frac{2}{5}} \qquad\qquad x = 3$$

$$x - 1 = 3 - 1 = 2$$

Answer: 2 and 3 hr.

25.

Let x = speed of rowing in still water.

	Distance ÷	Rate	= Time
Upstream	3	x - 1	$\frac{3}{x-1}$
Downstream	7	x + 1	$\frac{7}{x+1}$

Since time is the same, then the equation is,

$$\frac{3}{x-1} = \frac{7}{x+1}$$

$$3x + 3 = 7x - 7 \qquad \text{Cross multiply}$$

$$10 = 4x$$

$$\frac{10}{4} = x$$

$$\frac{5}{2} = x$$

Answer: $2\frac{1}{2}$ mph or 2.5 mph

27.

Let x = speed of the wind.

	Distance ÷	Rate	= Time
With wind	250	90+x	$\frac{250}{90+x}$
Against wind	200	90-x	$\frac{200}{90-x}$

Since time is the same, then the equation is

$$\frac{250}{90+x} = \frac{200}{90-x}$$

$$22500 - 250x = 18000 + 200x$$

$$4500 = 450x$$

$$10 = x$$

Answer: 10 mph

29.

Let x = speed of plane in still air.

	Distance ÷	Rate	= Time
With wind	280	x + 30	$\frac{280}{x+30}$
Against wind	280	x - 30	$\frac{280}{x-30}$

#29. continued
The total flying time is 7 hr.

$$\frac{280}{x+30} + \frac{280}{x-30} = 7 \qquad LCD = (x+30)(x-30)$$

$$\frac{(x+30)(x-30)}{1}\left(\frac{280}{x+30} + \frac{280}{x-30}\right)$$
$$= 7(x+30)(x-30)$$

$$280(x-30) + 280(x+30) = 7(x^2 - 900)$$

$$280x - 8400 + 280x + 8400 = 7x^2 - 6300$$

$$560x = 7x^2 - 6300$$

$$0 = 7x^2 - 560x - 6300$$

$$0 = x^2 - 80x - 900$$

$$0 = (x-90)(x+10)$$

$$x - 90 = 0 \quad \text{or} \quad x + 10 = 0$$

$$x = 90 \qquad \cancel{x = -10}$$

Answer: 90 mph.

31.
Let x = Roachman's climbing time

	Distance ÷	Rate	= Time
Climb	$\frac{3}{4}$	x	$\frac{\frac{3}{4}}{x}$
Parachute	$\frac{3}{4}$	x+4	$\frac{\frac{3}{4}}{x+4}$

Note 1: $\dfrac{\frac{3}{4}}{x} = \dfrac{3}{4} \cdot \dfrac{1}{x} = \dfrac{3}{4x}$

Note 2: $\dfrac{\frac{3}{4}}{x+4} = \dfrac{3}{4} \cdot \dfrac{1}{x+4} = \dfrac{3}{4(x+4)}$

The total time is $\frac{5}{3}$ hr.

$$\frac{3}{4x} + \frac{3}{4(x+4)} = \frac{5}{3} \qquad LCD = 12x(x+4)$$

$$\frac{12x(x+4)}{1}\left(\frac{3}{4x} + \frac{3}{4(x+4)}\right) = \frac{5}{3} \cdot \frac{12x(x+4)}{1}$$

$$9(x+4) + 9x = 20x(x+4)$$

$$9x + 36 + 9x = 20x^2 + 80x$$

$$36 + 18x = 20x^2 + 80x$$

$$0 = 20x^2 + 62x - 36$$

$$0 = 10x^2 + 31x - 18$$

$$0 = (2x-1)(5x+18)$$

#31. continued

$$2x - 1 = 0 \qquad \text{or} \qquad 5x + 8 = 0$$

$$x = \frac{1}{2} \qquad\qquad \cancel{x = -\frac{5}{8}}$$

Answer: $\dfrac{\frac{3}{4}}{x} = \dfrac{\frac{3}{4}}{\frac{1}{2}} = \dfrac{3}{4} \div \dfrac{1}{2} = \dfrac{3}{2}$ or $1\frac{1}{2}$ hr

---

33.
Let x = Cathy's distance from home

	Distance ÷	Rate	= Time
Running	x	8	$\frac{x}{8}$
Walking	x	2	$\frac{x}{2}$

Note: $T = \dfrac{D}{R}$

The total time is $\frac{5}{4}$ hr.

$$\frac{x}{8} + \frac{x}{2} = \frac{5}{4} \qquad LCD = 8$$

$$\frac{8}{1}\left(\frac{x}{8} + \frac{x}{2}\right) = \frac{5}{4} \cdot \frac{8}{1}$$

$$x + 4x = 10$$

$$5x = 10$$

$$x = 2$$

Answer: 2 mi

35.
Let x = distance to fishing hole.

	Distance ÷	Rate	= Time
There	x	12	$\frac{x}{12}$
Back	x	16	$\frac{x}{16}$

Note: $T = \dfrac{D}{R}$

The total travel time is 7 hr.

$$\frac{x}{12} + \frac{x}{16} = 7 \qquad LCD = 48$$

$$\frac{48}{1}\left(\frac{x}{12} + \frac{x}{16}\right) = 7 \cdot 48$$

$$4x + 3x = 336$$

$$7x = 336$$

$$x = 48$$

Answer: 48 mi.

Review Exercises

1.
$$5^6 \cdot 5^{-2} = 5^{6-2} = 5^4 = 625$$

3.
$$\frac{6^5}{6^2} = 6^{5-2} = 6^3 = 216$$

5.
$$\left(2^3\right)^2 = 2^6 = 64$$

7.
$$\left(8^0\right)^{-3} = 8^0 = 1$$

9.
$$-9^2 = -(9)^2 = -81$$

11.
$$(-4)^4 = (-4)(-4)(-4)(-4) = 256$$

13.
$$-5^{-3} = -(5)^{-3} = -\frac{1}{5^3} = -\frac{1}{125}$$

15.
$$\left[\left(2^{-2}\right)^3\right]^2 = \left(2^{-2}\right)^6 = 2^{-12} = \frac{1}{2^{12}} = \frac{1}{4096}$$

17.
$$2^{-1} + 4^{-1} = \frac{1}{2} + \frac{1}{4} = \frac{2}{4} + \frac{1}{4} = \frac{3}{4}$$

19.
$$4^{-2} - 4^{-1} - 4^0 = \frac{1}{16} - \frac{1}{4} - 1 = \frac{1}{16} - \frac{4}{16} - \frac{16}{16} = -\frac{19}{16}$$

21.
$$\left(\frac{4}{5}\right)^3 = \frac{4^3}{5^3} = \frac{64}{125}$$

23.
$$\left(\frac{5}{6}\right)^{-3} = \left(\frac{6}{5}\right)^3 = \frac{6^3}{5^3} = \frac{216}{125}$$

25.
$$\left(\frac{5}{8}\right)^0 = 1$$

27.
$$\left(\frac{-3}{5}\right)^{-2} = \left(\frac{5}{-3}\right)^2 = \frac{5^2}{(-3)^2} = \frac{25}{9}$$

29.
$$\frac{4^{-2}}{4^{-6}} = 4^{-2-(-6)} = 4^4 = 256$$

31.
$$\left(5x^{-4}y^2\right)^{-2}\left(2x^3y^{-4}\right)^3 = 5^{-2}x^8y^{-4} \cdot 2^3x^9y^{-12}$$
$$= 5^{-2} \cdot 2^3 x^{17}y^{-16} = \frac{8x^{17}}{25y^{16}}$$

33.
$$\left(9x^5y^{-4}\right)^0\left(6x^{-5}y^0\right)^{-2} = 1 \cdot \left(6x^{-5} \cdot 1\right)^{-2}$$
$$= \left(6x^{-5}\right)^{-2} = 6^{-2}x^{10} = \frac{x^{10}}{36}$$

35.
$$\left(\frac{7x^2}{4y}\right)^2 = \frac{7^2x^4}{4^2y^2} = \frac{49x^4}{16y^2}$$

37.
$$\left(\frac{2x^{-5}}{5y^4}\right)^2 = \frac{2^2x^{-10}}{5^2y^8} = \frac{4}{25x^{10}y^8}$$

39.
$$\left(\frac{-6x^2}{5y^{-3}}\right)^{-3} = \frac{(-6)^{-3}x^{-6}}{5^{-3}y^9} = \frac{5^3}{(-6)^3x^6y^9} = -\frac{125}{216x^6y^9}$$

41.
$$\left(\frac{x^7}{2y^5}\right)^{-2}\left(\frac{8x^2}{3y^{-1}}\right)^{-1} = \frac{x^{-14}}{2^{-2}y^{-10}} \cdot \frac{8^{-1}x^{-2}}{3^{-1}y}$$
$$= \frac{8^{-1}x^{-16}}{2^{-2} \cdot 3^{-1}y^{-9}} = \frac{2^2 \cdot 3y^9}{8x^{16}} = \frac{3y^9}{2x^{16}}$$

43.

$$\frac{6x^{-3}y^7}{-3x^2y^2} = \frac{-2y^{7-2}}{x^{2-(-3)}} = -\frac{2y^5}{x^5}$$

45.

$$\left(\frac{6^0x^5y^{-7}}{2x^2y^2}\right)^{-2} = \frac{6^0x^{-10}y^{14}}{2^{-2}x^{-4}y^{-4}} = \frac{2^2y^{14-(-4)}}{x^{-4-(-10)}} = \frac{4y^{18}}{x^6}$$

47.

$$\left[\left(\frac{3x^4}{5y^2}\right)^{-2}\right]^{-1} = \left(\frac{3x^4}{5y^2}\right)^2 = \frac{3^2x^8}{5^2y^4} = \frac{9x^8}{25y^4}$$

49.

$$(y+4)^{-2} = \frac{1}{(y+4)^2} = \frac{1}{y^2+8x+16}$$

51.

$$\left(x^{3m}\right)^2 = x^{6m}$$

53.

$$\frac{x^{-m}\cdot x^{3m}}{x^{-5m}} = \frac{x^{2m}}{x^{-5m}} = x^{2m-(-5m)} = x^{7m}$$

55.

$$-(2x+1)(x+3)^{-2} + 2(x+3)^{-1} \qquad \text{GCF} = (x+3)^{-2}$$

$$= (x+3)^{-2}\left[-(2x+1)+2(x+3)\right]$$

$$= (x+3)^{-2}(-2x-1+2x+6)$$

$$= (x+3)^{-2}(5) = \frac{5}{(x+3)^2}$$

57.

$$\frac{12x^5y}{15xy^6} = \frac{3\cdot 4x^{5-1}}{3\cdot 5y^{6-1}} = \frac{4x^4}{5y^5}$$

59.

$$-\frac{xy^4z^7}{5xy^2z^6} = -\frac{x^{1-1}y^{4-2}z^{7-6}}{5} = -\frac{y^2z}{5} \qquad \text{Note: } x^0 = 1$$

61.

$$\frac{6x^2y - 4x^3y^5}{8xy^3 + 12x^4y^2} = \frac{2x^2y\left(3-2xy^4\right)}{4xy^2\left(2y+3x^3\right)}$$

$$= \frac{2x^{2-1}\left(3-2xy^4\right)}{2\cdot 2y^{2-1}\left(2y+3x^3\right)}$$

$$= \frac{x\left(3-2xy^4\right)}{2y\left(2y+3x^3\right)}$$

63.

$$\frac{x^2-8x+15}{5x-x^2} = \frac{(x-3)(x-5)}{-x(x-5)} = -\frac{x-3}{x}$$

65.

$$\frac{2x^2+6xy-20y^2}{x^2-25y^2} = \frac{2(x-2y)(x+5y)}{(x+5y)(x-5y)} = \frac{2(x-2y)}{x-5y}$$

67.

$$\frac{2x^2-x-15}{4x^2+4x-15} = \frac{(x-3)(2x+5)}{(2x+5)(2x-3)} = \frac{x-3}{2x-3}$$

69.

$$\frac{8a^3+b^3}{8a^2+6ab+b^2} = \frac{(2a+b)\left(4a^2-2ab+b^2\right)}{(2a+b)(4a+b)}$$

$$= \frac{4a^2-2ab+b^2}{4a+b}$$

71.

$$\frac{x^3-2x^2y-4xy^2+8y^3}{2y^2+yx-x^2} = \frac{x^2(x-2y)-4y^2(x-2y)}{-\left(x^2-yx-2y^2\right)}$$

$$= \frac{(x-2y)\left(x^2-4y^2\right)}{-(x-2y)(x+y)}$$

$$= -\frac{x^2-4y^2}{x+y}$$

73.

$$\frac{6x^4y^3}{5xy^2}\cdot\frac{10xy^5}{4x^7y} = \frac{6\cdot 10\cdot x^5y^8}{5\cdot 4x^8y^3} = \frac{2\cdot 3\cdot 2\cdot 5y^{8-3}}{5\cdot 2\cdot 2x^{8-5}} = \frac{3y^5}{x^3}$$

75.

$$10a^3bc^4 \div \frac{2ab^3}{5b^2c^5} = \frac{10a^3bc^4}{1} \cdot \frac{5b^2c^5}{2ab^3}$$

$$= \frac{10 \cdot 5a^3b^3c^9}{2ab^3}$$

$$= \frac{\cancel{2} \cdot 5 \cdot 5a^{3-1}b^{3-3}c^9}{\cancel{2}}$$

$$= 25a^2c^9 \qquad \text{Note: } b^0 = 1$$

77.

$$\frac{4a^2x^4}{5ab^3} \cdot \frac{15x^2y^3}{2xz^5} \div \frac{3ay^6}{7b^2z} = \frac{4a^2x^4}{5ab^3} \cdot \frac{15x^2y^3}{2xz^5} \cdot \frac{7b^2z}{3ay^6}$$

$$= \frac{4 \cdot 15 \cdot 7a^2b^2x^6y^3z}{5 \cdot 2 \cdot 3a^2b^3xy^6z^5}$$

$$= \frac{\cancel{2} \cdot 2 \cdot \cancel{3} \cdot \cancel{5} \cdot 7a^{2-2}x^{6-1}}{\cancel{5} \cdot \cancel{2} \cdot \cancel{3}b^{3-2}y^{6-3}z^{5-1}}$$

$$= \frac{14x^5}{by^3z^4} \qquad \text{Note: } a^0 = 1$$

79.

$$\frac{x^2+2x-15}{3x^2+15x} \cdot \frac{2x^3+6x^2}{x^2-6x+9}$$

$$= \frac{(x-3)(x+5)}{3x(x+5)} \cdot \frac{2x^2(x+3)}{(x-3)(x-3)}$$

$$= \frac{2x^2(x+3)}{3x(x-3)} = \frac{2x(x+3)}{x(x-3)}$$

81.

$$\frac{4x^2-y^2}{3x^2+10xy+3y^2} \div \frac{y^2-yx-2x^2}{x^2+4xy+3y^2}$$

$$= \frac{(2x-y)(2x+y)}{(3x+y)(x+3y)} \cdot \frac{(x+3y)(x+y)}{-(2x-y)(x+y)} = -\frac{2x+y}{3x+y}$$

83.

$$\frac{6x^2+7x-3}{8x^3+10x^2-3x} \cdot \frac{8x^2-2x}{6x^2+10x-4}$$

$$= \frac{(2x+3)(3x-1)}{x(2x+3)(4x-1)} \cdot \frac{2x(4x-1)}{2(3x-1)(x+2)} = \frac{1}{x+2}$$

85.

$$\frac{a^3-8b^3}{6b^2-ba-a^2} \div \frac{3a^3b+6a^2b^2+12ab^3}{a^2-9b^2}$$

#85. continued

$$= \frac{(a-2b)(a^2+2ab+4b^2)}{-(a-2b)(a+3b)} \cdot \frac{(a+3b)(a-3b)}{3ab(a^2+2ab+4b^2)}$$

$$= -\frac{a-3b}{3ab}$$

87.

$$\frac{2ax+xy-6a-3y}{4a^2+2ay+10a+5y} \cdot \frac{6ax-4ay+15x-10y}{3x^3-2x^2-27x+18y}$$

$$= \frac{(2a+y)(x-3)}{(2a+y)(2a+5)} \cdot \frac{(3x-2y)(2a+5)}{(x+3)(x-3)(3x-2y)} = \frac{1}{x+3}$$

89.

$$\frac{x^2+2x-35}{2x^2-3x-2} \cdot \frac{x^2-9x+14}{2x^3+10x^2} \div \frac{x^2-49}{8x^2+4x}$$

$$= \frac{(x+7)(x-5)}{(2x+1)(x-2)} \cdot \frac{(x-7)(x-2)}{2x^2(x+5)} \cdot \frac{4x(2x+1)}{(x+7)(x-7)}$$

$$= \frac{4x(x-5)}{2x^2(x+5)} = \frac{2(x-5)}{x(x+5)}$$

91.

$$\frac{5x}{4x^2y} + \frac{7x}{4x^2y} = \frac{5x+7x}{4x^2y} = \frac{12x}{4x^2y} = \frac{3}{xy}$$

93.

$$\frac{2x^2}{5x-1} + \frac{8x^2}{5x-1} = \frac{2x^2+8x^2}{5x-1} = \frac{10x^2}{5x-1}$$

95.

$$\frac{5x}{x^2-6x-7} + \frac{5}{x^2-6x-7} = \frac{5x+5}{x^2-6x-7}$$

$$= \frac{5(x+1)}{(x-7)(x+1)} = \frac{5}{x-7}$$

97.

$$\frac{2y}{y^2-y-12} - \frac{y-3}{y^2-y-12} = \frac{2y-y+3}{y^2-y-12}$$

$$= \frac{(y+3)}{(y+3)(y-4)} = \frac{1}{y-4}$$

99.

$$\frac{4}{x^2} + \frac{5}{3x} = \frac{3}{3} \cdot \frac{4}{x^2} + \frac{x}{x} \cdot \frac{5}{3x} \qquad \text{LCD} = 3x^2$$

$$= \frac{12}{3x^2} + \frac{5x}{3x^2} = \frac{5x+12}{3x^2}$$

101.

$$\frac{2x}{x-5}+\frac{3}{-(x-5)}=\frac{2x}{x-5}-\frac{3}{x-5}=\frac{2x-3}{x-5}$$

103.

$$\frac{5x}{5x-1}+\frac{2}{x+4}$$

$$=\frac{x+4}{x+4}\cdot\frac{5x}{5x-1}+\frac{5x-1}{5x-1}\cdot\frac{2}{x+4}\quad LCD=(5x-1)(x+4)$$

$$=\frac{5x^2+20x}{(x+4)(5x-1)}+\frac{10x-2}{(5x-1)(x+4)}$$

$$=\frac{5x^2+30x-2}{(x+4)(5x-1)}$$

105.

$$\frac{9}{2x-5}+2=\frac{9}{2x-5}+\frac{2x-5}{2x-5}\cdot\frac{2}{1}$$

$$=\frac{9}{2x-5}+\frac{4x-10}{2x-5}=\frac{4x-1}{2x-5}$$

107.

$$\frac{6}{2x-4}+\frac{5}{3x-6}$$

$$=\frac{6}{2(x-2)}+\frac{5}{3(x-2)}\quad LCD=6(x-2)$$

$$=\frac{3}{3}\cdot\frac{6}{2(x-2)}+\frac{2}{2}\cdot\frac{5}{3(x-2)}$$

$$=\frac{18}{6(x-2)}+\frac{10}{6(x-2)}$$

$$=\frac{28}{6(x-2)}=\frac{14}{3(x-2)}$$

109.

$$\frac{x+2}{x^2-5x-14}-\frac{x}{x^2-49}$$

$$=\frac{x+2}{(x+2)(x-7)}-\frac{x}{(x+7)(x-7)}$$

$$=\frac{x+7}{x+7}\cdot\frac{x+2}{(x+2)(x-7)}-\frac{x+2}{x+2}\cdot\frac{x}{(x+7)(x-7)}$$

$$LCD=(x+2)(x+7)(x-7)$$

$$=\frac{(x+7)(x+2)}{(x+7)(x+2)(x-7)}-\frac{x(x+2)}{(x+2)(x+7)(x-7)}$$

$$=\frac{x^2+9x+14-\left(x^2+2x\right)}{(x+2)(x+7)(x-7)}$$

#109. continued

$$=\frac{x^2+9x+14-x^2-2x}{(x+2)(x+7)(x-7)}$$

$$=\frac{7x+14}{(x+2)(x+7)(x-7)}$$

$$=\frac{7\cancel{(x+2)}}{\cancel{(x+2)}(x+7)(x-7)}=\frac{7}{(x+7)(x-7)}$$

111.

$$\frac{11x-11}{x^2-x-12}-\frac{8x-4}{x^2-5x-24}$$

$$=\frac{11x-11}{(x+3)(x-4)}-\frac{8x-4}{(x+3)(x-8)}$$

$$LCD=(x+3)(x-4)(x-8)$$

$$=\frac{x-8}{x-8}\cdot\frac{11x-11}{(x+3)(x-4)}-\frac{x-4}{x-4}\cdot\frac{8x-4}{(x+3)(x-8)}$$

$$=\frac{(x-8)(11x-11)}{(x-8)(x+3)(x-4)}-\frac{(x-4)(8x-4)}{(x-4)(x+3)(x-8)}$$

$$=\frac{11x^2-99x+88-\left(8x^2-36x+16\right)}{(x-8)(x+3)(x-4)}$$

$$=\frac{11x^2-99x+88-8x^2+36x-16}{(x-8)(x+3)(x-4)}$$

$$=\frac{3x^2-63x+72}{(x-8)(x+3)(x-4)}$$

113.

$$\frac{x+2y}{2x^2-7xy+6y^2}+\frac{3y}{4x^2-4xy-3y^2}$$

$$=\frac{x+2y}{(2x-3y)(x-2y)}+\frac{3y}{(2x+y)(2x-3y)}$$

$$LCD=(2x-3y)(x-2y)(2x+y)$$

$$=\frac{2x+y}{2x+y}\cdot\frac{x+2y}{(2x-3y)(x-2y)}$$

$$+\frac{x-2y}{x-2y}\cdot\frac{3y}{(2x+y)(2x-3y)}$$

$$=\frac{(2x+y)(x+2y)}{(2x+y)(2x-3y)(x-2y)}$$

$$+\frac{3y(x-2y)}{(x-2y)(2x+y)(2x-3y)}$$

$$=\frac{2x^2+5xy+2y^2+3xy-6y^2}{(2x+y)(2x-3y)(x-2y)}$$

#113. continued

$$= \frac{2x^2 + 8xy - 4y^2}{(2x+y)(2x-3y)(x-2y)}$$

115.

$$\frac{10x+23}{4x^2+4x-3} - \frac{4x+11}{2x^2+x-3}$$

$$= \frac{10x+23}{(2x-1)(2x+3)} - \frac{4x+11}{(2x+3)(x-1)}$$

$$\text{LCD} = (2x-1)(2x+3)(x-1)$$

$$= \frac{x-1}{x-1} \cdot \frac{10x+23}{(2x-1)(2x+3)} - \frac{2x-1}{2x-1} \cdot \frac{4x+11}{(2x+3)(x-1)}$$

$$= \frac{(x-1)(10x+23) - (2x-1)(4x+11)}{(x-1)(2x-1)(2x+3)}$$

$$= \frac{10x^2+13x-23 - \left(8x^2+18x-11\right)}{(x-1)(2x-1)(2x+3)}$$

$$= \frac{10x^2+13x-23 - 8x^2 - 18x + 11}{(x-1)(2x-1)(2x+3)}$$

$$= \frac{2x^2 - 5x - 12}{(x-1)(2x-1)(2x+3)}$$

$$= \frac{(x-4)(2x+3)}{(x-1)(2x-1)(2x+3)} = \frac{x-4}{(x-1)(2x-1)}$$

117.

$$\frac{x+2}{x^2-9} + \frac{x-4}{2x^2+7x+3} - \frac{3x}{2x^2-5x-3}$$

$$= \frac{x+2}{(x+3)(x-3)} + \frac{x-4}{(2x+1)(x+3)} - \frac{3x}{(2x+1)(x-3)}$$

$$\text{LCD} = (x+3)(x-3)(2x+1)$$

$$= \frac{2x+1}{2x+1} \cdot \frac{x+2}{(x+3)(x-3)} + \frac{x-3}{x-3} \cdot \frac{x-4}{(2x+1)(x+3)}$$

$$\qquad - \frac{x+3}{x+3} \cdot \frac{3x}{(2x+1)(x-3)}$$

$$= \frac{(2x+1)(x+2)}{(2x+1)(x+3)(x-3)} + \frac{(x-3)(x-4)}{(x-3)(2x+1)(x+3)}$$

$$\qquad - \frac{3x(x+3)}{(x+3)(2x+1)(x-3)}$$

$$= \frac{2x^2+5x+2 + x^2-7x+12 - 3x^2 - 9x}{(2x+1)(x+3)(x-3)}$$

$$= \frac{-11x+14}{(2x+1)(x+3)(x-3)}$$

119.

$$\frac{3xy^3}{7y} \div \frac{9x^5}{14xy} = \frac{3xy^3}{7y} \cdot \frac{14xy}{9x^5}$$

$$= \frac{3 \cdot 14 \cdot x^2 y^4}{7 \cdot 9 \cdot x^5 y} = \frac{3 \cdot 2 \cdot 7 y^{4-1}}{7 \cdot 3 \cdot 3 x^{5-2}} = \frac{2y^3}{3x^3}$$

121.

$$\frac{3x+7}{x-2} + \frac{x+7}{2-x} = \frac{3x+7}{x-2} \cdot \frac{-1(x-2)}{(x-2)} = \frac{3x+7}{x+7} = -\frac{3x+7}{x+7}$$

123.

$$\frac{2x^2-7x-4}{4x^2-1} + \frac{x^2-7x+12}{x^3-27}$$

$$= \frac{(2x+1)(x-4)}{(2x+1)(2x-1)} \cdot \frac{(x-3)\left(x^2+3x+9\right)}{(x-4)(x-3)}$$

$$= \frac{x^2+3x+9}{(2x-1)}$$

125.

$$\frac{6x-15}{6x-24} + \frac{10x-25}{2} = \frac{3(2x-5)}{6(x-4)} \cdot \frac{2}{5(2x-5)}$$

$$= \frac{1 \cdot 6}{6 \cdot 5(x-4)} = \frac{1}{5(x-4)}$$

127.

$$\frac{4+\dfrac{3}{x}}{\dfrac{2}{x}-3} = \frac{x\left(4+\dfrac{3}{x}\right)}{x\left(\dfrac{2}{x}-3\right)} = \frac{4x+3}{2-3x}$$

129.

$$\frac{\dfrac{3}{x^2}+\dfrac{2}{y}}{\dfrac{5}{2x^3y^2}} = \frac{\dfrac{2x^3y^2}{1}\left(\dfrac{3}{x^2}+\dfrac{2}{y}\right)}{\dfrac{2x^3y^2}{1} \cdot \dfrac{5}{2x^3y^2}} \qquad \text{LCD} = 2x^3y^2$$

$$= \frac{6xy^2 + 4x^3y}{5}$$

131.

$$\frac{2x}{\dfrac{1}{x}-\dfrac{3}{y}} = \frac{xy \cdot 2x}{\dfrac{xy}{1}\left(\dfrac{1}{x}-\dfrac{3}{y}\right)} \qquad \text{LCD} = xy$$

$$= \frac{2x^2y}{y-3x}$$

133.

$$\frac{\dfrac{1}{x}-\dfrac{1}{y}}{\dfrac{1}{x^4}-\dfrac{1}{y^4}}=\frac{\dfrac{x^4y^4}{1}\left(\dfrac{1}{x}-\dfrac{1}{y}\right)}{\dfrac{x^4y^4}{1}\left(\dfrac{1}{x^4}-\dfrac{1}{y^4}\right)} \qquad LCD=x^4y^4$$

$$=\frac{x^3y^4-x^4y^3}{y^4-x^4}$$

$$=\frac{x^3y^3\cancel{(y-x)}}{\left(y^2+x^2\right)(y+x)\cancel{(y-x)}}$$

$$=\frac{x^3y^3}{\left(y^2+x^2\right)(y+x)}$$

135.

$$\frac{\dfrac{4x}{x-1}-\dfrac{16}{3}}{x-4}=\frac{\dfrac{3(x-1)}{1}\left(\dfrac{4x}{x-1}-\dfrac{16}{3}\right)}{3(x-1)(x-4)}$$

$$=\frac{12x-16x+16}{3x^2-15x+12}$$

$$=\frac{-4x+16}{3x^2-15x+12}$$

$$=\frac{-4(x-4)}{3\left(x^2-5x+4\right)}$$

$$=\frac{-4\cancel{(x-4)}}{3\cancel{(x-4)}(x-1)}=\frac{-4}{3x-3}$$

137.

$$\frac{\dfrac{3}{5x}+\dfrac{2}{x-3}}{\dfrac{1}{5x}-\dfrac{4}{x-3}}=\frac{\dfrac{5x(x-3)}{1}\left(\dfrac{3}{5x}+\dfrac{2}{x-3}\right)}{\dfrac{5x(x-3)}{1}\left(\dfrac{1}{5x}-\dfrac{4}{x-3}\right)}$$

$$=\frac{3x-9+10x}{x-3-20x}=\frac{13x-9}{-19x-3}$$

139.

$$\frac{\dfrac{4+h}{3(2+h)}-\dfrac{2}{3}}{h}=\frac{\dfrac{3(2+h)}{1}\left(\dfrac{4+h}{3(2+h)}-\dfrac{2}{3}\right)}{3h(2+h)}$$

$$=\frac{4+h-4-2h}{3h(2+h)}$$

$$=\frac{-1\cdot h}{3h(2+h)}=\frac{-1}{3(2+h)}$$

141.

$$\frac{\dfrac{8}{2-x}+x}{2+\dfrac{2x^2}{x-2}}=\frac{\dfrac{x-2}{1}\left(\dfrac{8}{-(x-2)}+x\right)}{\dfrac{x-2}{1}\left(2+\dfrac{2x^2}{x-2}\right)}$$

$$=\frac{-8+x^2-2x}{2x-4+2x^2}$$

$$=\frac{x^2-2x-8}{2\left(x^2+x-2\right)}$$

$$=\frac{(x-4)\cancel{(x+2)}}{2\cancel{(x+2)}(x-1)}=\frac{x-4}{2(x-1)}$$

143.

$$\frac{x^{-1}-(x+3)^{-1}}{3}=\frac{\dfrac{1}{x}-\dfrac{1}{x+3}}{3}$$

$$=\frac{\dfrac{x(x+3)}{1}\left(\dfrac{1}{x}-\dfrac{1}{x+3}\right)}{3x(x+3)}$$

$$=\frac{x+3-x}{3x(x+3)}$$

$$=\frac{3}{3x(x+3)}=\frac{1}{x(x+3)}$$

145.

$$\frac{12x^2y^4}{3xy^2}-\frac{18x^3y^3}{3xy^2}-\frac{15x^4y^2}{3xy^2}=4xy^2-6x^2y-5x^3$$

147.

$$\frac{16a^4b^5}{4a^2b^3}+\frac{20a^3b^3}{4a^2b^3}+\frac{18ab^2}{4a^2b^3}=4a^2b^2+5a+\frac{9}{2ab}$$

149.

$$\begin{array}{r}
x-5 \\
2x+1\overline{\smash{\big)}\,2x^2-9x+5} \\
\underline{^-2x^2\mp x} \\
-10x+5 \\
\underline{\pm10x\pm5} \\
10
\end{array}$$

Quotient: $x-5+\dfrac{10}{2x+1}$

151.

$$
\begin{array}{r}
5x^2 + 2x + 6 \\
x-2\overline{\big)\, 5x^3 - 8x^2 + 2x + 4} \\
\underline{{}^{-}5x^3 \pm 10x^2} \\
2x^2 + 2x \\
\underline{{}^{-}2x^2 \pm 4x} \\
6x + 4 \\
\underline{{}^{-}6x \pm 12} \\
16
\end{array}
$$

Quotient: $5x^2 + 2x + 6 + \dfrac{16}{x-2}$

153.

$$
\begin{array}{r}
6x - 15 \\
2x+5\overline{\big)\, 12x^2 + 0x - 11} \\
\underline{{}^{-}12x^2 \mp 30x} \\
-30x - 11 \\
\underline{\pm 30x \pm 75} \\
64
\end{array}
$$

Quotient: $6x - 15 + \dfrac{64}{2x+5}$

155.

$$
\begin{array}{r}
6x^2 + 4x - 2 \\
3x-2\overline{\big)\, 18x^3 + 0x^2 - 14x + 6} \\
\underline{-18x^3 \pm 12x^2} \\
12x^2 - 14x \\
\underline{-12x^2 \pm 8x} \\
-6x + 6 \\
\underline{\pm 6x \mp 4} \\
2
\end{array}
$$

Quotient: $6x^2 + 4x - 2 + \dfrac{2}{3x-2}$

157.

$$
\begin{array}{r}
3x^3 - 6x^2 + 8x - 5 \\
2x+4\overline{\big)\, 6x^4 + 0x^3 - 8x^2 + 22x - 20} \\
\underline{{}^{-}6x^4 \mp 12x^3} \\
-12x^3 - 8x^2 \\
\underline{\pm 12x^3 \pm 24x^2} \\
16x^2 + 22x \\
\underline{{}^{-}16x^2 \mp 32x} \\
-10x - 20 \\
\underline{-10x - 20}
\end{array}
$$

Quotient: $3x^3 - 6x^2 + 8x - 5$

159.

$$
\begin{array}{r}
2x^2 - x - 5 \\
3x^2+x+4\overline{\big)\, 6x^4 - x^3 - 8x^2 + 0x - 10} \\
\underline{{}^{-}6x^4 \mp 2x^3 \mp 8x^2} \\
-3x^3 - 16x^2 + 0x \\
\underline{\pm 3x^3 \pm x^2 \pm 4x} \\
-15x^2 + 4x - 10 \\
\underline{\pm 15x^2 \pm 5x \pm 20} \\
9x + 10
\end{array}
$$

Quotient: $2x^2 - x - 5 + \dfrac{9x+10}{3x^2+x+4}$

161.

$$
\begin{array}{r}
x^3 + 3x^2 + 9x + 27 \\
x-3\overline{\big)\, x^4 + 0x^3 + 0x^2 + 0x - 81} \\
\underline{{}^{-}x^4 \pm 3x^3} \\
3x^3 + 0x^2 \\
\underline{{}^{-}3x^3 \pm 9x^2} \\
9x^2 + 0x \\
\underline{{}^{-}9x^2 \pm 27x} \\
27x - 81 \\
\underline{27x - 81}
\end{array}
$$

Quotient: $x^3 + 3x^2 + 9x + 27$

163.

$$3x^2 - 5 \overline{\smash{\big)}\ 12x^4 + 9x^3 - 26x^2 - 15x + 10}$$

$$\begin{array}{r} 4x^2 + 3x - 2 \\ \underline{^-12x^4 \qquad \pm 20x^2} \\ 9x^3 - 6x^2 - 15x \\ \underline{^-9x^3 \qquad \pm 15x} \\ - 6x^2 + 10 \\ \underline{- 6x^2 + 10} \end{array}$$

Quotient: $4x^2 + 3x - 2$

165.

$$4x^2 - 3 \overline{\smash{\big)}\ 4x^5 + 0x^4 + 5x^3 - 16x^2 - 6x + 12}$$

$$\begin{array}{r} x^3 + 2x - 4 \\ \underline{^-4x^5 \qquad \pm 3x^3} \\ 8x^3 - 16x^2 - 6x \\ \underline{^-8x^3 \qquad \pm 6x} \\ -16x^2 \qquad + 12 \\ \underline{-16x^2 \qquad + 12} \end{array}$$

Quotient: $x^3 + 2x - 4$

167.

$$\frac{3x^2 + 5}{x + 3}$$

$$\underline{-3}\ \begin{array}{rrr} 3 & 0 & 5 \\ & -9 & 27 \\ \hline 3 & -9 & 32 \end{array}$$

Quotient: $3x - 9 + \dfrac{32}{x + 3}$

169.

$$\frac{2x^3 - x^2 - 15x - 3}{x - 4}$$

$$\underline{4}\ \begin{array}{rrrr} 2 & -1 & -15 & -3 \\ & 8 & 28 & 52 \\ \hline 2 & 7 & 13 & 49 \end{array}$$

Quotient: $2x^2 + 7x + 13 + \dfrac{49}{x - 4}$

171.

$$\frac{x^4 - 2x^3 + 5x^2 - 4x}{x - 2}$$

#171. continued

$$\underline{2}\ \begin{array}{rrrrr} 1 & -2 & 5 & -4 & 0 \\ & 2 & 0 & 10 & 12 \\ \hline 1 & 0 & 5 & 6 & 12 \end{array}$$

Quotient: $x^3 + 5x + 6 + \dfrac{12}{x - 2}$

173.

$$\frac{6x^4 - 3x^3 - 7x - 3}{x + \dfrac{3}{2}}$$

$$\underline{-\dfrac{3}{2}}\ \begin{array}{rrrrr} 6 & -3 & 0 & -7 & -3 \\ & -9 & 18 & -27 & 51 \\ \hline 6 & -12 & 18 & -34 & 48 \end{array}$$

Quotient: $6x^3 - 12x^2 + 18x - 34 + \dfrac{48}{x + \dfrac{3}{2}}$

175.

$$\frac{x^5 - 32}{x - 2}$$

$$\underline{2}\ \begin{array}{rrrrrr} 1 & 0 & 0 & 0 & 0 & -32 \\ & 2 & 4 & 8 & 16 & 32 \\ \hline 1 & 2 & 4 & 8 & 16 & 0 \end{array}$$

Quotient: $x^4 + 2x^3 + 4x^2 + 8x + 16$

177.

$$\frac{2}{5x} - \frac{3}{4x} = \frac{7}{20} \qquad LCD = 20x$$

$$\frac{20x}{1}\left(\frac{2}{5x} - \frac{3}{4x}\right) = \frac{7}{20} \cdot \frac{20x}{1}$$

$$8 - 15 = 7x$$

$$-7 = 7x$$

$$-1 = x$$

Solution set: $\{-1\}$

179.

$$\frac{6x - 5}{3x - 2} = 1 - \frac{1}{3x - 2} \qquad LCD = 3x - 2$$

$$\frac{3x - 2}{1} \cdot \frac{6x - 5}{3x - 2} = \frac{3x - 2}{1}\left(1 - \frac{1}{3x - 2}\right)$$

$$6x - 5 = 3x - 2 - 1$$

$$6x - 5 = 3x - 3$$

$$3x = 2$$

$$x = \frac{2}{3}$$

#179. continued

Solution set: $\varnothing$ Note: $3x - 2 \neq 0$, so $x \neq \dfrac{2}{3}$

181.

$$\frac{7}{2x+3} = \frac{3}{x-4} - \frac{x+5}{(2x+3)(x-4)}$$

$$\text{LCD} = (2x+3)(x-4)$$

$$\frac{(2x+3)(x-4)}{1} \cdot \frac{7}{(2x+3)}$$

$$= \frac{(2x+3)(x-4)}{1}\left(\frac{3}{x-4} - \frac{x+5}{(2x+3)(x-4)}\right)$$

$$7(x-4) = 3(2x+3) - (x+5)$$

$$7x - 28 = 6x + 9 - x - 5$$

$$7x - 28 = 5x + 4$$

$$2x = 32$$

$$x = 16$$

Solution set: $\{16\}$

183.

$$\frac{x+1}{x-2} = 2 - \frac{x-2}{x-4}$$

$$\text{LCD} = (x-2)(x-4)$$

$$\frac{(x-2)(x-4)}{1} \cdot \frac{x+1}{x-2} = \frac{(x-2)(x-4)}{1}\left(2 - \frac{x-2}{x-4}\right)$$

$$(x-4)(x+1) = 2(x-2)(x-4) - (x-2)^2$$

$$x^2 - 3x - 4 = 2x^2 - 12x + 16 - x^2 + 4x - 4$$

$$x^2 - 3x - 4 = x^2 - 8x + 12$$

$$5x = 16$$

$$x = \frac{16}{5}$$

Solution set: $\left\{\dfrac{16}{5}\right\}$

185.

$$\frac{3}{25} + \frac{2}{5x} = \frac{1}{x^2} \qquad \text{LCD} = 25x^2$$

$$\frac{25x^2}{1}\left(\frac{3}{25} + \frac{2}{5x}\right) = \frac{25x^2}{1} \cdot \frac{1}{x^2}$$

$$3x^2 + 10x = 25$$

$$3x^2 + 10x - 25 = 0$$

$$(x+5)(3x-5) = 0$$

$$x + 5 = 0 \qquad \text{or} \qquad 3x - 5 = 0$$

$$x = -5 \qquad\qquad x = \frac{5}{3}$$

#185. continued

Solution set: $\left\{-5, \dfrac{5}{3}\right\}$

187.

$$\frac{x}{x-2} - \frac{3}{3x+2} = \frac{8}{(x-2)(3x+2)}$$

$$\frac{(x-2)(3x+2)}{1}\left(\frac{x}{x-2} - \frac{3}{3x+2}\right)$$

$$= \frac{(x-2)(3x+2)}{1} \cdot \frac{8}{(x-2)(3x+2)}$$

$$x(3x+2) - 3(x-2) = 8$$

$$3x^2 + 2x - 3x + 6 = 8$$

$$3x^2 - x - 2 = 0$$

$$(x-1)(3x+2) = 0$$

$$x - 1 = 0 \qquad \text{or} \qquad 3x + 2 = 0$$

$$x = 1 \qquad\qquad x = -\frac{3}{2}$$

Solution set: $\{1\}$ Note: $3x + 2 \neq 0$

$$\text{So,} \quad x \neq -\frac{3}{2}$$

189.

$$\frac{3x}{x-8} = \frac{-1}{x+3}$$

$$3x^2 + 9x = -x + 8 \qquad \text{Cross multiply}$$

$$3x^2 + 10x - 8 = 0$$

$$(x+4)(3x-2) = 0$$

$$x + 4 = 0 \qquad \text{or} \qquad 3x - 2 = 0$$

$$x = -4 \qquad\qquad x = \frac{2}{3}$$

Solution set: $\left\{-4, \dfrac{2}{3}\right\}$

191.

$$\frac{2x+3}{2(x+1)} = \frac{14x-41}{2(x+1)(x-3)} + \frac{3x-9}{x+1}$$

$$\frac{2(x+1)(x-3)}{1} \cdot \frac{2x+3}{2(x+1)}$$

$$= \frac{2(x+1)(x-3)}{1}\left(\frac{14x-41}{2(x+1)(x-3)} + \frac{3x-9}{x+1}\right)$$

$$(x-3)(2x+3) = 14x - 41 + 2(x-3)(3x-9)$$

$$2x^2 - 3x - 9 = 14x - 41 + 6x^2 - 36x + 54$$

$$2x^2 - 3x - 9 = 6x^2 - 22x + 13$$

$$0 = 4x^2 - 19x + 22$$

$$0 = (x-2)(4x-11)$$

#191. continued

$x - 2 = 0$ or $4x - 11 = 0$

$x = 2$ $x = \dfrac{11}{4}$

Solution set: $\left\{ 2, \dfrac{11}{4} \right\}$

193.
Let x = the denominator.
Then 2x+1 = the numerator.
So, $(2x + 1) - 21 = -2$

$$2x - 20 = -2$$
$$2x = 18$$
$$x = 9$$

Thus, $\dfrac{2x + 1}{x} = \dfrac{2 \cdot 9 + 1}{9} = \dfrac{19}{9}$

The original fraction is $\dfrac{19}{9}$.

195.
Two consecutive even integers are x and x + 2.

Their reciprocals are $\dfrac{1}{x}$ and $\dfrac{1}{x + 2}$.

So, $\dfrac{1}{x} + \dfrac{1}{x + 2} = \dfrac{13}{84}$

$$\dfrac{84x(x+2)}{1}\left(\dfrac{1}{x} + \dfrac{1}{x+2} \right) = \dfrac{84x(x+2)}{1} \cdot \dfrac{13}{84}$$

$$84(x + 2) + 84x = 13x(x + 2)$$
$$84x + 168 + 84x = 13x^2 + 26x$$
$$168x + 168 = 13x^2 + 26x$$
$$0 = 13x^2 - 142x - 168$$
$$0 = (x - 12)(13x + 14)$$

$x - 12 = 0$ or $13x + 14 = 0$

$x = 12$ $x = -\dfrac{14}{13}$

The numbers are 12 and 14.

197.
$$\dfrac{1}{3} + \dfrac{1}{2} + \dfrac{1}{x} = 1$$
$$\dfrac{6x}{1}\left(\dfrac{1}{3} + \dfrac{1}{2} + \dfrac{1}{x} \right) = 6x \cdot 1$$
$$2x + 3x + 6 = 6x$$
$$5x + 6 = 6x$$
$$6 = x$$

Answer: 6 days

199.

$$\dfrac{1}{x} + \dfrac{1}{2x} = \dfrac{1}{20}$$
$$\dfrac{20x}{1}\left(\dfrac{1}{x} + \dfrac{1}{2x} \right) = \dfrac{20x}{1} \cdot \dfrac{1}{20}$$
$$20 + 10 = x$$
$$30 = x$$

Answer: 30 min

201.
Let x = speed in still air

	Distance ÷	Rate	= Time
Trip 1	90	x - 90	$\dfrac{90}{x - 90}$
Trip 2	75	x + 90	$\dfrac{75}{x + 90}$

Note: $T = \dfrac{D}{R}$

The total trip time is $\dfrac{1}{2}$ hour.

$$\dfrac{90}{x - 90} + \dfrac{75}{x + 90} = \dfrac{1}{2}$$
$$\dfrac{2(x - 90)(x + 90)}{1}\left(\dfrac{90}{x - 90} + \dfrac{75}{x + 90} \right)$$
$$= \dfrac{2(x - 90)(x + 90)}{1} \cdot \dfrac{1}{2}$$
$$180(x + 90) + 150(x - 90) = (x - 90)(x + 90)$$
$$180x + 16,200 + 150x - 13,500 = x^2 - 8100$$
$$330x + 2700 = x^2 - 8100$$
$$0 = x^2 - 330x - 10,800$$
$$0 = (x - 360)(x + 30)$$

$x - 360 = 0$ or $x + 30 = 0$

$x = 360$ $x = -30$

Answer: 360 mph

203.
Let x = speed in still air

	Distance ÷	Rate	= Time
Down	2	x	$\dfrac{2}{x}$
Up	2	x - 18	$\dfrac{2}{x - 18}$

Note: $T = \dfrac{D}{R}$

The total time is $\dfrac{7}{30}$ of an hour.

#201. continued

$$\frac{2}{x} + \frac{2}{x-18} = \frac{7}{30}$$

$$\frac{30x(x-18)}{1}\left(\frac{2}{x} + \frac{2}{x-18}\right) = \frac{7}{30} \cdot \frac{30x(x-18)}{1}$$

$$60(x-18) + 60x = 7x(x-18)$$

$$60x - 1080 + 60x = 7x^2 - 126x$$

$$120x - 1080 = 7x^2 - 126x$$

$$0 = 7x^2 - 246x + 1080$$

$$0 = (7x - 36)(x - 30)$$

$$7x - 36 = 0 \quad \text{or} \quad x - 30 = 0$$

$$\cancel{x = \frac{36}{7}} \qquad\qquad x = 30$$

Answer: up 12 mph, down 30 mph

Chapter 4 Test Solutions

1.

$$(-3)^{-2} = \frac{1}{(-3)^2} = \frac{1}{9}$$

3.

$$4a^2b^0\left(3a^2b^3\right)^2 = 4a^2\left(3^2a^4b^6\right) \qquad \text{Note:} \quad b^0 = 1$$

$$= 4 \cdot 9a^6b^6$$

$$= 36a^6b^6$$

5.

$$\frac{-\cancel{(2b-3a)}\left(9a^2 + 6ab + 4b^2\right)}{\cancel{(2b-3a)}\left(4b^2 + 9a^2\right)} = -\frac{9a^2 + 6ab + 4b^2}{4b^2 + 9a^2}$$

7.

$$\frac{(4x+3)\cancel{(x-4)}}{\cancel{(x-4)}\cancel{(x+7)}} \cdot \frac{\cancel{(x+7)}\cancel{(x+2)}}{\cancel{(x+2)}(x-6)} = \frac{4x+3}{x-6}$$

9.

$$\frac{3x-5}{(x+3)(x-3)} - \frac{x+1}{2x(x-3)} + \frac{5}{2(x+3)}$$

$$\text{LCD} = 2x(x+3)(x-3)$$

$$= \frac{2x}{2x} \cdot \frac{3x-5}{(x+3)(x-3)} - \frac{x+3}{x+3} \cdot \frac{x+1}{2x(x-3)}$$

$$+ \frac{x(x-3)}{x(x-3)} \cdot \frac{5}{2(x+3)}$$

$$= \frac{6x^2 - 10x - (x+3)(x+1) + 5x(x-3)}{2x(x+3)(x-3)}$$

#9. continued

$$= \frac{6x^2 - 10x - x^2 - 4x - 3 + 5x^2 - 15x}{2x(x+3)(x-3)}$$

$$= \frac{10x^2 - 29x - 3}{2x(x+3)(x-3)} = \frac{(10x+1)\cancel{(x-3)}}{2x(x+3)\cancel{(x-3)}} = \frac{10x+1}{2x(x+3)}$$

11.

$$\frac{\dfrac{2x+1}{x} - \dfrac{5}{2}}{x-2} = \frac{\dfrac{2x}{1}\left(\dfrac{2x+1}{x} - \dfrac{5}{2}\right)}{2x(x-2)}$$

$$= \frac{2(2x+1) - 5x}{2x(x-2)}$$

$$= \frac{4x+2-5x}{2x(x-2)} = \frac{-x+2}{2x(x-2)} = \frac{-\cancel{(x-2)}}{2x\cancel{(x-2)}} = \frac{-1}{2x}$$

13.

$$5x-2\overline{\smash{)}\,5x^3 - 17x^2 + 26x - 8} \quad\quad \overset{\displaystyle x^2 - 3x + 4}{}$$

$$\underline{{}^-5x^3 \pm 2x^2}$$

$$-15x^2 + 26x$$

$$\underline{\pm15x^2 \mp 6x}$$

$$20x - 8$$

$$\underline{20x - 8}$$

Quotient: $x^2 - 3x + 4$

15.

$$x-3\overline{\smash{)}\,2x^5 - 7x^4 - x^3 + 12x^2 + 0x - 5} \quad\quad \overset{\displaystyle 2x^4 - x^3 - 4x^2 + 0}{}$$

$$\underline{{}^-2x^5 \pm 6x^4}$$

$$-x^4 - x^3$$

$$\underline{\pm x^4 \mp 3x^3}$$

$$-4x^3 + 12x^2$$

$$\underline{\pm 4x^3 \mp 12x^2}$$

$$+0x - 5$$

$$\underline{{}^-0x \pm 0}$$

$$-5$$

Quotient: $2x^4 - x^3 - 4x^2 - \dfrac{5}{x-3}$

17.

$$\frac{5}{3x+2} = \frac{5}{x-2}$$

$$5x - 10 = 15x + 10$$

$$-20 = 10x$$

$$-2 = x$$

#17. continued
Solution set: {-2}

19.

$$\frac{1}{18} - \frac{1}{24} = \frac{1}{x}$$

$$\frac{72x}{1}\left(\frac{1}{18} - \frac{1}{24}\right) = \frac{1}{x} \cdot \frac{72x}{1}$$

$$4x - 3x = 72$$

$$x = 72$$

Answer: 72 min

Test Your Memory

1.

$$\frac{5}{8} \cdot \frac{4}{15} + \frac{1}{2} - \frac{2}{3} \div \frac{1}{9} = \frac{1}{6} + \frac{1}{2} - 6$$

$$= \frac{1}{6} + \frac{3}{6} - \frac{36}{6} = \frac{-32}{6} = \frac{-16}{3}$$

3.

$$\frac{1}{3} + \frac{1}{(-3)^2} = \frac{1}{3} + \frac{1}{9} = \frac{3}{9} + \frac{1}{9} = \frac{4}{9}$$

5.

$$\left(\frac{6x^{-3}y^4}{x^2y}\right)^2 = \left(\frac{6y^3}{x^5}\right)^2 = \frac{6^2y^6}{x^{10}} = \frac{36y^6}{x^{10}}$$

7.

$$12x^2 + 8xy - 15y^2$$

-10 and 18 are two integers whose product is -180 and whose sum is 8.

Thus,

$$12x^2 + 8xy - 15y^2 = 12x^2 - 10xy + 18xy - 15y^2$$

$$= 2x(6x - 5y) + 3y(6x - 5y)$$

$$= (6x - 5y)(2x + 3y)$$

9.

$$4x^2 + 100y^2 = 4(x^2 + 25y^2)$$

11.

$$5x^3 - 4x^2 - 20x + 16 = x^2(5x - 4) - 4(5x - 4)$$

$$= (x^2 - 4)(5x - 4)$$

$$= (x + 2)(x - 2)(5x - 4)$$

13.

$$\left(2x^2y^4\right)^3\left(3xy^5\right)^2 = 2^3x^6y^{12} \cdot 3^2x^2y^{10}$$

$$= 8 \cdot 9 \cdot x^6x^2y^{12}y^{10} = 72x^8y^{22}$$

15.

$$(x - 4)\left(7x^2 - x - 2\right)$$

$$\begin{array}{r} 7x^2 - x - 2 \\ x - 4 \\ \hline 7x^3 - x^2 - 2x \\ -28x^2 + 4x + 8 \\ \hline 7x^3 - 29x^2 + 2x + 8 \end{array}$$

17.

$$\frac{(x-3)(x^2+3x+9)}{(x-2)(x+2)(3x-1)} \cdot \frac{(3x-1)(x+2)}{(x+6)(x-3)} = \frac{x^2+3x+9}{(x-2)(x+6)}$$

19.

$$\frac{5}{x+3} - \frac{2}{x-1} = \frac{5(x-1) - 2(x+3)}{(x+3)(x-1)}$$

$$= \frac{5x - 5 - 2x - 6}{(x+3)(x-1)} = \frac{3x - 11}{(x+3)(x-1)}$$

21.

$$\frac{x+3}{x+3} \cdot \frac{5x+2}{x+2} - \frac{4x}{(x+3)(x+2)} - \frac{x+2}{x+2} \cdot \frac{2x+4}{x+3}$$

$$\text{LCD} = (x+2)(x+3)$$

$$= \frac{(x+3)(5x+2) - 4x - (x+2)(2x+4)}{(x+3)(x+2)}$$

$$= \frac{5x^2 + 17x + 6 - 4x - 2x^2 - 8x - 8}{(x+3)(x+2)}$$

$$= \frac{3x^2 + 5x - 2}{(x+3)(x+2)} = \frac{(x+2)(3x-1)}{(x+3)(x+2)} = \frac{3x-1}{x+3}$$

23.

$$\frac{\frac{4x}{1} \cdot \left(\frac{2x+7}{x} - \frac{1}{4}\right)}{4x(x+4)} \qquad \text{LCD} = 4x$$

$$= \frac{4(2x+7) - x}{4x^2 + 16x}$$

$$= \frac{8x + 28 - x}{4x^2 + 16x} = \frac{7x + 28}{4x^2 + 16x} = \frac{7(x+4)}{4x(x+4)} = \frac{7}{4x}$$

25.

$$x^2 - 3x + 4 \overline{) 2x^4 - 5x^3 - x^2 + 25x - 26}$$

with quotient $2x^2 + x - 6$

$$\underline{{}^-2x^4 \pm 6x^3 \mp 8x^2}$$
$$x^3 - 9x^2 + 25x$$
$$\underline{{}^-x^3 \pm 3x^2 \mp 4x}$$
$$-6x^2 + 21x - 26$$
$$\underline{\pm 6x^2 \mp 18x \pm 24}$$
$$3x - 2$$

Quotient: $2x^2 + x - 6 + \dfrac{3x - 2}{x^2 - 3x + 4}$

27.

$$6x^2 + 11x + 4 = 0$$
$$(3x + 4)(2x + 1) = 0$$
$$3x + 4 = 0 \quad \text{or} \quad 2x + 1 = 0$$
$$x = -\frac{4}{3} \qquad x = -\frac{1}{2}$$

Solution set: $\left\{ -\dfrac{4}{3}, -\dfrac{1}{2} \right\}$

29.

$$(x + 2)(2x - 5) = -7$$
$$2x^2 - x - 10 = -7$$
$$2x^2 - x - 3 = 0$$
$$(x + 1)(2x - 3) = 0$$
$$x + 1 = 0 \quad \text{or} \quad 2x - 3 = 0$$
$$x = -1 \qquad x = \frac{3}{2}$$

Solution set: $\left\{ -1, \dfrac{3}{2} \right\}$

31.

$$|3x - 2| = |x + 7|$$
$$3x - 2 = x + 7 \quad \text{or} \quad 3x - 2 = -(x + 7)$$
$$2x = 9 \qquad\qquad 3x - 2 = -x - 7$$
$$x = \frac{9}{2} \qquad\qquad 4x = -5$$
$$\qquad\qquad\qquad x = -\frac{5}{4}$$

Solution set: $\left\{ \dfrac{9}{2}, -\dfrac{5}{4} \right\}$

33.

$$\frac{6}{x - 2} = \frac{3}{x + 1}$$
$$6x + 6 = 3x - 6$$
$$3x = -12$$
$$x = -4$$

Solution set: $\{-4\}$

35.

$$2(x - 4) + 5(2x - 4) = 0$$
$$2x - 8 + 10x - 20 = 0$$
$$12x - 28 = 0$$
$$12x = 28$$
$$x = \frac{28}{12} \quad \text{or} \quad \frac{7}{3}$$

Solution set: $\left\{ \dfrac{7}{3} \right\}$

37.

$$\frac{2x + 3}{x + 4} - \frac{4}{2x + 5} = \frac{-1}{(x + 4)(2x + 5)}$$

$$LCD = (x + 4)(2x + 5)$$

$$\frac{(x + 4)(2x + 5)}{1}\left(\frac{2x + 3}{x + 4} - \frac{4}{2x + 5} \right) = \frac{(x + 4)(2x + 5)}{1}$$
$$\cdot \frac{-1}{(x + 4)(2x + 5)}$$
$$(2x + 5)(2x + 3) - 4(x + 4) = -1$$
$$4x^2 + 16x + 15 - 4x - 16 = -1$$
$$4x^2 + 12x = 0$$
$$4x(x + 3) = 0$$
$$4x = 0 \quad \text{or} \quad x + 3 = 0$$
$$x = 0 \qquad\qquad x = -3$$

Solution set: $\{0, -3\}$

39.

$$\frac{6}{1}\left(\frac{1}{2}x + \frac{2}{3} \right) = \left(\frac{2}{3}x - \frac{5}{6} \right)\frac{6}{1} \qquad LCD = 6$$
$$3x + 4 = 4x - 5 \qquad\qquad \text{Multiply by 6}$$
$$9 = x$$

Solution set: $\{9\}$

41.

$5x + 7 \leq 8x + 13$

$\quad -6 \leq 3x$

$\quad -2 \leq x \quad \text{or} \quad x \geq -2$

The graph is in the answer section of the main text.

43.

Let x = measurement of the width.

Then $4x + 2$ = measurement of the length.

Thus, $\quad A = w \cdot \ell$

$\quad 72 = x(4x + 2)$

$\quad 72 = 4x^2 + 2x$

$\quad 0 = 4x^2 + 2x - 72$

$\quad 0 = 2x^2 + x - 36$

$\quad 0 = (2x + 9)(x - 4)$

$2x + 9 = 0 \quad \text{or} \quad x - 4 = 0$

$\quad \cancel{x = -\frac{9}{2}} \qquad x = 4$

Answer: 4 ft by 18 ft

45.

Let x = number of fives.

Then $5x+2$ = number of tens.

Thus, $5x + 10(5x + 2) = 350$

$\quad 5x + 50x + 20 = 350$

$\quad 55x + 20 = 350$

$\quad 55x = 330$

$\quad x = 6$

$\quad 5x + 2 = 5 \cdot 6 + 2 = 32$

Answer: 6 fives, 32 tens

47.

$\quad \dfrac{1}{2} + \dfrac{1}{6} = \dfrac{1}{x} \qquad LCD = 12x$

$\dfrac{12x}{1}\left(\dfrac{1}{2} + \dfrac{1}{6}\right) = \dfrac{12x}{1} \cdot \dfrac{1}{x}$

$\quad 6x + 2x = 12$

$\quad 8x = 12$

$\quad x = \dfrac{12}{8} \quad \text{or} \quad \dfrac{3}{2}$

Answer: $1\dfrac{1}{2}$ hours or 1 hr 30 min

49.

Let x = rowing speed in still water.

	Distance ÷	Rate	= Time
Down	12	x +2	$\dfrac{12}{x+2}$
Up	4	x - 2	$\dfrac{2}{x-18}$

The time is the same.

$\quad \dfrac{12}{x+2} = \dfrac{4}{x-2}$

$12x - 24 = 4x + 8 \qquad \text{Cross multiply}$

$\quad 8x = 32$

$\quad x = 4$

Answer: 4 mph

Chapter 4 Study Guide

Self-Test Exercises

I. Simplify each of the following. (Remember to write you answer with only positive exponents.)

1. $(-2)^{-2}$

2. $\left(\dfrac{3y^{-2}}{x^{15}}\right)^{3}$

3. $3a^{0}b^{2}\left(3a^{3}b^{0}\right)^{2}$

4. $\left(\dfrac{2x^{4}y^{-3}}{x^{3}y^{2}}\right)^{-3}$

II. Reduce the following rational expression to lowest terms.

5. $\dfrac{64a^{3}+8b^{3}}{12a^{2}-20ab+6ab-10b^{2}}$

III. Perform the indicated operations and reduce the answers to lowest terms.

6. $\dfrac{2x-3}{x^{2}+x-6}-\dfrac{x+3}{x^{2}-x-2}$

7. $\dfrac{15x^{2}-11x+2}{10x^{2}+21x-10}\cdot\dfrac{2x^{2}-7x-30}{x^{2}+3x-54}$

8. $\dfrac{2x^{2}-2xy-xy-27y^{2}}{16x^{2}-70xy-9y^{2}}\div\dfrac{8x^{2}+23xy-3y^{2}}{8x^{2}+9xy+y^{2}}$

9. $\dfrac{2}{x^{2}-16}+\dfrac{x+1}{x^{2}+8x+16}+\dfrac{3}{2x-8}$

IV. Simplify the following fractions.

10. $\dfrac{\dfrac{3x+3}{2x}}{\dfrac{x^{2}-1}{4x}}$

11. $\dfrac{\dfrac{3x-1}{x}-\dfrac{3}{5}}{x-3}$

12. $\dfrac{a^{-1}-b^{-1}}{a^{-3}+b^{-3}}$

V. Perform the indicated divisions.

13. $\dfrac{10x^{3}+6x^{2}-9x+10}{5x-2}$

14. $\dfrac{2x^{5}+x^{3}-2x-3}{x^{2}-3x+1}$

15. $\dfrac{3y^{4}-25y^{2}-18}{y-3}$

16. $\dfrac{x^{4}-2x^{3}-35x^{2}-x+1}{x+5}$

VI. Solve for x in each of the following equations.

17. $\dfrac{7}{2x+3}=\dfrac{7}{x-3}$

18. $\dfrac{1}{x-2}-\dfrac{2}{x+7}=\dfrac{4x-1}{x^{2}+5x-14}$

VII. Find the solutions to the following problems.

19. A Salmon fish hatchery tank can be filled in 18 hours using the north pipe and in 22 hours by the south pipe alone. The open drain can empty the tank in 33 hours. If both fill pipes and drain are working at once, how long would it take to fill the tank?

20. The speed of Brad's stream in front of his cabin is 4 mph. He decides to canoe 6 mi upstream to his favorite fishing hole. After fishing for two hours he canoes 12 mi down-stream to get some groceries. He notes that both trips took the same amount of time. What is the speed of the canoe in still water?

The worked-out solutions begin on the next page.

Self Test Solutions

1.

$$\frac{1}{(-2)^2} = \frac{1}{4}$$

2.

$$\left(\frac{3y^{-2}}{x^{15}}\right)^3 = \frac{3^3 y^{-6}}{x^{45}} = \frac{27}{x^{45}y^6}$$

3.

$$3a^0 b^2 \left(3a^3 b^0\right)^2 \qquad \text{Note } a^0 = 1, \ b^0 = 1$$

$$= 3b^2 \left(3a^3\right)^2 = 3b^2 \cdot 3^2 a^6 = 27a^6 b^2$$

4.

$$\left(\frac{2x^4 y^{-3}}{x^3 y^2}\right)^{-3} = \left(\frac{2x}{y^5}\right)^{-3} = \frac{2^{-3} x^{-3}}{y^{-15}} = \frac{y^{15}}{8x^3}$$

5.

$$\frac{64a^3 + 8b^3}{12a^2 - 20ab + 6ab - 10b^2}$$

$$= \frac{\cancel{(4a + 2b)}\left(16a^2 - 8ab + 4b^2\right)}{\cancel{(4a + 2b)}(3a - 5b)} = \frac{16a^2 - 8ab + 4b^2}{3a - 5b}$$

6.

$$\frac{2x - 3}{(x - 2)(x + 3)} - \frac{x + 3}{(x - 2)(x + 1)}$$

$$\text{LCD} = (x - 2)(x + 1)(x + 3)$$

$$= \frac{x + 1}{x + 1} \cdot \frac{2x - 3}{(x - 2)(x + 3)} - \frac{x + 3}{x + 3} \cdot \frac{x + 3}{(x - 2)(x + 1)}$$

$$= \frac{(x + 1)(2x - 3) - (x + 3)(x + 3)}{(x + 1)(x - 2)(x + 3)}$$

$$= \frac{2x^2 - x - 3 - x^2 - 6x - 9}{(x + 1)(x - 2)(x + 3)} = \frac{x^2 - 7x - 12}{(x + 1)(x - 2)(x + 3)}$$

7.

$$\frac{15x^2 - 11x + 2}{10x^2 + 21x - 10} \cdot \frac{2x^2 - 7x - 30}{x^2 + 3x - 54}$$

$$= \frac{\cancel{(5x - 2)}(3x - 1)}{\cancel{(2x + 5)}\cancel{(5x - 2)}} \cdot \frac{\cancel{(2x + 5)}\cancel{(x - 6)}}{\cancel{(x - 6)}(x + 9)} = \frac{3x - 1}{x + 9}$$

8.

$$\frac{2x^2 - 2xy - xy - 27y^2}{16x^2 - 70xy - 9y^2} \div \frac{8x^2 + 23xy - 3y^2}{8x^2 + 9xy + y^2}$$

$$= \frac{\cancel{(2x - 9y)}\cancel{(x + 3y)}}{\cancel{(8x + y)}\cancel{(2x - 9y)}} \cdot \frac{\cancel{(8x + y)}(x + y)}{\cancel{(x + 3y)}(8x - y)} = \frac{x + y}{8x - y}$$

9.

$$\frac{2}{(x + 4)(x - 4)} + \frac{x + 1}{(x + 4)(x + 4)} + \frac{3}{2(x - 4)}$$

$$\text{LCD} = 2(x + 4)^2(x - 4)$$

$$= \frac{2(x + 4)}{2(x + 4)} \cdot \frac{2}{(x + 4)(x - 4)} + \frac{2(x - 4)}{2(x - 4)} \cdot \frac{x + 1}{(x + 4)(x + 4)}$$

$$\qquad + \frac{2(x + 4)^2}{2(x + 4)^2} \cdot \frac{3}{2(x - 4)}$$

$$= \frac{4(x + 4) + 2(x - 4)(x + 1) + 6(x + 4)^2}{2(x + 4)^2(x - 4)}$$

$$= \frac{4x + 16 + 2x^2 - 6x - 8 + 6x^2 + 48x + 96}{2(x + 4)^2(x - 4)}$$

$$= \frac{8x^2 + 46x + 104}{2(x + 4)^2(x - 4)}$$

$$= \frac{\cancel{2}\left(4x^2 + 23x + 52\right)}{\cancel{2}(x + 4)^2(x - 4)} = \frac{4x^2 + 23x + 52}{(x + 4)^2(x - 4)}$$

10.

$$\frac{3\cancel{(x + 1)}}{2x} \cdot \frac{4x}{\cancel{(x + 1)}(x - 1)} = \frac{3 \cdot 2 \cdot \cancel{2} \cancel{x}}{\cancel{2}\cancel{x}(x - 1)} = \frac{6}{x - 1}$$

11.

$$\frac{\dfrac{5x}{1}\left(\dfrac{3x - 1}{x} - \dfrac{3}{5}\right)}{5x(x - 3)}$$

$$= \frac{5(3x - 1) - 3x}{5x(x - 3)} = \frac{15x - 5 - 3x}{5x(x - 3)} = \frac{12x - 5}{5x(x - 3)}$$

12.

$$\frac{\dfrac{a^3 b^3}{1} \cdot \left(\dfrac{1}{a} - \dfrac{1}{b}\right)}{\dfrac{a^3 b^3}{1} \cdot \left(\dfrac{1}{a^3} - \dfrac{1}{b^3}\right)} = \frac{a^2 b^3 - a^3 b^2}{b^3 - a^3}$$

#12. continued

$$= \frac{a^2 b^2 (b-a)}{(b-a)(b^2 + ab + a^2)}$$

$$= \frac{a^2 b^2}{b^2 + ab + a^2}$$

13.

$$
\begin{array}{r}
2x^2 + 2x - 1 \\
5x - 2 \overline{\smash{\big)}\ 10x^3 + 6x^2 - 9x + 10} \\
\underline{^{-}10x^3 \pm 4x^2} \\
10x^2 - 9x \\
\underline{^{-}10x^2 \pm 4x} \\
-5x + 10 \\
\underline{\pm 5x \mp 2} \\
8
\end{array}
$$

Answer: $2x^2 + 2x - 1 + \dfrac{8}{5x - 2}$

14.

$$
\begin{array}{r}
2x^3 + 6x^2 + 17x + 45 \\
x^2 - 3x + 1 \overline{\smash{\big)}\ 2x^5 + 0x^4 + x^3 + 0x^2 - 2x - 3} \\
\underline{^{-}2x^5 \pm 6x^4 \mp 2x^3} \\
6x^4 - x^3 + 0x^2 \\
\underline{^{-}6x^4 \pm 18x^3 \mp 6x^2} \\
17x^3 - 6x^2 - 2x \\
\underline{^{-}17x^3 \pm 51x^2 \mp 17x} \\
45x^2 - 19x - 3 \\
\underline{^{-}45x^2 \pm 135x \mp 45} \\
116x - 48
\end{array}
$$

Answer: $2x^3 + 6x^2 + 17x + 45 + \dfrac{116x - 48}{x^2 - 3x + 1}$

15.

$$
\begin{array}{r}
3y^3 + 9y^2 + 2y + 6 \\
y - 3 \overline{\smash{\big)}\ 3y^4 + 0y^3 - 25y^2 + 0y - 18} \\
\underline{^{-}3y^4 \pm 9y^3} \\
9y^3 - 25y^2 \\
\underline{^{-}9y^3 \pm 27y^2} \\
2y^2 + 0y \\
\underline{^{-}2y^2 \pm 6y} \\
6y - 18 \\
6y - 18
\end{array}
$$

#15. continued

Answer: $3y^3 + 9y^2 + 2y + 6$

16.

$$
\begin{array}{r}
x^3 - 7x^2 - 1 \\
x + 5 \overline{\smash{\big)}\ x^4 - 2x^3 - 35x^2 - x + 1} \\
\underline{^{-}x^4 \mp 5x^3} \\
-7x^3 - 35x^2 \\
\underline{\pm 7x^3 \pm 35x^2} \\
-x + 1 \\
\underline{\pm x \pm 5} \\
6
\end{array}
$$

Answer: $x^3 - 7x^2 - 1 + \dfrac{6}{x + 5}$

17.

$$\frac{7}{2x + 3} = \frac{7}{x - 3}$$

$$7x - 21 = 14x + 21$$

$$-42 = 7x$$

$$-6 = x$$

Solution set: $\{-6\}$

18.

$$\frac{1}{x - 2} - \frac{2}{x + 7} = \frac{4x - 1}{(x - 2)(x + 7)}$$

$$\frac{(x - 2)(x + 7)}{1}\left(\frac{1}{x - 2} - \frac{2}{x + 7}\right)$$

$$= \frac{(x - 2)(x + 7)}{1} \cdot \frac{4x - 1}{(x - 2)(x + 7)}$$

$$x + 7 - 2(x - 2) = 4x - 1$$

$$x + 7 - 2x + 4 = 4x - 1$$

$$-x + 11 = 4x - 1$$

$$12 = 5x$$

$$\frac{12}{5} = x$$

Solution set: $\left\{\dfrac{12}{5}\right\}$

19.

$$\frac{1}{18}+\frac{1}{22}-\frac{1}{33}=\frac{1}{x}$$

$$\frac{198x}{1}\left(\frac{1}{18}+\frac{1}{22}-\frac{1}{33}\right)=\frac{198x}{1}\cdot\frac{1}{x}$$

$$11x+9x-6x=198$$

$$14x=198$$

$$x=\frac{198}{14}\quad or\quad \frac{99}{7}$$

Answer: $14\frac{1}{7}$ hr

20.

Let x = speed of canoe in still water.

	Distance ÷	Rate	= Time
Down	6	x - 4	$\frac{6}{x-4}$
Up	12	x + 4	$\frac{12}{x+4}$

The two trips took the same amount of time.

$$\frac{6}{x-4}=\frac{12}{x+4}$$

$$6x+24=12x-48$$

$$72=6x$$

$$12=x$$

Answer: 12 mph

CHAPTER 5   EXPONENTIAL AND RADICAL EXPRESSIONS

Solutions to Text Exercises

Exercises 5.1

1.-36.
The exercises do not require worked - out solutions.

37.
$$4^{5/2} = \left(4^{1/2}\right)^5 = 2^5 = 32$$

39.
Not a real number.

41.
$$-8^{2/3} = -1 \cdot \left(8^{1/3}\right)^2 = -1 \cdot 2^2 = -4$$

43.
$$32^{3/5} = \left(32^{1/5}\right)^3 = (2)^3 = 8$$

45.
$$-\left(\frac{1}{16}\right)^{3/2} = -\left[\left(\frac{1}{16}\right)^{1/2}\right]^3 = -\left(\frac{1}{4}\right)^3 = -\frac{1}{64}$$

47.
$$\left(\frac{16}{81}\right)^{3/4} = \left[\left(\frac{16}{81}\right)^{1/4}\right]^3 = \left(\frac{2}{3}\right)^3 = \frac{8}{27}$$

49.
$$49^{-1/2} = \frac{1}{49^{1/2}} = \frac{1}{7}$$

51.
$$(-64)^{-2/3} = \left[(-64)^{1/3}\right]^{-2} = (-4)^{-2} = \frac{1}{(-4)^2} = \frac{1}{16}$$

53.
$$25^{-3/2} = \left(25^{1/2}\right)^{-3} = (5)^{-3} = \frac{1}{5^3} = \frac{1}{125}$$

55.
$$\left(\frac{16}{25}\right)^{-1/2} = \left(\frac{25}{16}\right)^{1/2} = \frac{5}{4}$$

57.
$$\left(\frac{1}{27}\right)^{-2/3} = \left[\left(\frac{1}{27}\right)^{1/3}\right]^{-2} = \left(\frac{1}{3}\right)^{-2} = \left(\frac{3}{1}\right)^2 = 9$$

59.
$$\left(\frac{16}{81}\right)^{-3/4} = \left[\left(\frac{16}{81}\right)^{1/4}\right]^{-3} = \left(\frac{2}{3}\right)^{-3} = \left(\frac{3}{2}\right)^3 = \frac{27}{8}$$

61.
$$25^{1/3} \cdot 25^{1/6} = 25^{2/6} \cdot 25^{1/6} = 25^{3/6} = 25^{1/2} = 5$$

63.
$$\left(7^{-1/2}\right)^4 = 7^{-4/2} = 7^{-2} = \frac{1}{7^2} = \frac{1}{49}$$

65.
$$\left(9^{2/5} \cdot 9^{11/10}\right)^2 = \left(9^{4/10} \cdot 9^{11/10}\right)^2$$
$$= \left(9^{15/10}\right)^2 = \left(9^{3/2}\right)^2 = (27)^2 = 729$$

67.
$$\frac{27^{1/2}}{27^{1/6}} = 27^{1/2 - 1/6} = 27^{1/3} = 3$$

69.
$$5^{1/3} \cdot 5^{1/5} = 5^{1/3 + 1/5} = 5^{8/15}$$

71.
$$\frac{9^{1/6} \cdot 9^{-2}}{9^{2/3}} = 9^{1/6 - 2 - 2/3}$$
$$= 9^{1/6 - 12/6 - 4/6}$$
$$= 9^{-5/2} = \left(9^{1/2}\right)^{-5} = (3)^{-5} = \frac{1}{3^5} = \frac{1}{243}$$

73.
$$\left(6^{-\frac{3}{4}}\right)^{-\frac{2}{3}} = 6^{\frac{6}{12}} = 6^{\frac{1}{2}}$$

75.
$$\left(9^{\frac{1}{20}} \cdot 32^{\frac{1}{50}}\right)^{10} = 9^{\frac{1}{2}} \cdot 32^{\frac{1}{5}} = 3 \cdot 2 = 6$$

77.
$$x^{\frac{3}{4}} \cdot x^{\frac{1}{2}} = x^{\frac{3}{4}+\frac{1}{2}} = x^{\frac{5}{4}}$$

79.
$$\left(27x^4\right)^{\frac{2}{3}} = 27^{\frac{2}{3}} \cdot x^{\frac{8}{3}} = \left(27^{\frac{1}{3}}\right)^2 x^{\frac{8}{3}} = 3^2 x^{\frac{8}{3}} = 9x^{\frac{8}{3}}$$

81.
$$\left(4x^3\right)^{\frac{1}{2}}\left(8x^{\frac{1}{3}}\right) = 4^{\frac{1}{2}} \cdot x^{\frac{3}{2}} \cdot 8 \cdot x^{\frac{1}{3}}$$
$$= 2 \cdot 8 \cdot x^{\frac{3}{2}+\frac{1}{3}} = 16x^{\frac{11}{6}}$$

83.
$$\frac{x^{\frac{5}{3}}}{x^{\frac{1}{2}}} = x^{\frac{5}{3}-\frac{1}{2}} = x^{\frac{7}{6}}$$

85.
$$\frac{2x^{-\frac{2}{3}}}{3x^{\frac{1}{4}}} = \frac{2}{3x^{\frac{1}{4}+\frac{2}{3}}} = \frac{2}{3x^{\frac{11}{12}}}$$

87.
$$\frac{\left(x^2 \cdot x^{\frac{1}{2}}\right)^3}{x^5} = \frac{\left(x^{\frac{5}{2}}\right)^3}{x^5} = \frac{x^{\frac{15}{2}}}{x^5} = x^{\frac{15}{2}-5} = x^{\frac{5}{2}}$$

89.
$$\frac{-6x^3 y^{-\frac{1}{2}}}{4x^{\frac{1}{3}} y^2} = \frac{-3x^{3-\frac{1}{3}}}{2y^{2+\frac{1}{2}}} = -\frac{3x^{\frac{8}{3}}}{2y^{\frac{5}{2}}}$$

91.
$$\frac{16^{\frac{3}{8}} x^{-\frac{1}{2}} y^{\frac{5}{3}}}{16^{\frac{1}{8}} x^{\frac{1}{4}} y^{\frac{1}{2}}} = \frac{\left(16^{\frac{1}{8}}\right)^3 y^{\frac{5}{3}-\frac{1}{2}}}{\left(16^{\frac{1}{8}}\right)^1 x^{\frac{1}{4}+\frac{1}{2}}}$$

#91. continued
$$= \frac{\left(16^{\frac{1}{8}}\right)^2 y^{\frac{7}{6}}}{x^{\frac{3}{4}}}$$
$$= \frac{16^{\frac{1}{4}} y^{\frac{7}{6}}}{x^{\frac{3}{4}}} = \frac{2y^{\frac{7}{6}}}{x^{\frac{3}{4}}}$$

93.
$$\left(\frac{4x^4}{9y^{\frac{1}{3}}}\right)^{\frac{1}{2}} = \frac{4^{\frac{1}{2}} x^2}{9^{\frac{1}{2}} y^{\frac{1}{6}}} = \frac{2x^2}{3y^{\frac{1}{6}}}$$

95.
$$\left(\frac{8x^{-\frac{1}{2}}}{125y^{\frac{1}{4}}}\right)^{-\frac{1}{3}} = \frac{8^{-\frac{1}{3}} x^{\frac{1}{6}}}{125^{-\frac{1}{3}} y^{-\frac{1}{12}}}$$
$$= \frac{125^{\frac{1}{3}} x^{\frac{1}{6}} y^{\frac{1}{12}}}{8^{\frac{1}{3}}} = \frac{5x^{\frac{1}{6}} y^{\frac{1}{12}}}{2}$$

97.
$$\frac{\left(-8x^3 y^7\right)^{\frac{2}{3}}}{6x^4 y^{-\frac{1}{3}}} = \frac{(-8)^{\frac{2}{3}} x^2 y^{\frac{14}{3}}}{6x^4 y^{-\frac{1}{3}}}$$
$$= \frac{4y^{\frac{14}{3}+\frac{1}{3}}}{6x^{4-2}}$$
$$= \frac{2y^5}{3x^2}$$

99.
$$\left(\frac{4x}{y^3}\right)^{-\frac{3}{2}}\left(\frac{8x^4}{y^{-2}}\right)^{\frac{1}{3}} = \left(\frac{y^3}{4x}\right)^{\frac{3}{2}}\left(\frac{8x^4}{y^{-2}}\right)^{\frac{1}{3}}$$
$$= \frac{y^{\frac{9}{2}}}{4^{\frac{3}{2}} x^{\frac{3}{2}}} \cdot \frac{8^{\frac{1}{3}} x^{\frac{4}{3}}}{y^{-\frac{2}{3}}}$$
$$= \frac{2y^{\frac{9}{2}+\frac{2}{3}}}{8x^{\frac{3}{2}-\frac{4}{3}}} = \frac{y^{\frac{31}{6}}}{4x^{\frac{1}{6}}}$$

101.
$$x^{\frac{1}{6}}\left(3x^{\frac{1}{3}} - 7\right) = 3x^{\frac{1}{3}+\frac{1}{6}} - 7x^{\frac{1}{6}} = 3x^{\frac{1}{2}} - 7x^{\frac{1}{6}}$$

103.

$$5x^{-3/2}\left(2x + 5x^{1/2}\right) = 10x^{1-3/2} + 25x^{1/2-3/2}$$

$$= 10x^{-1/2} + 25x^{-1}$$

$$= \frac{10}{x^{1/2}} + \frac{25}{x}$$

105.

$$7x^{-2/3}\left(3x^{-1/6} + 4xy^{3/4}\right) = 21x^{-2/3-1/6} + 28x^{1-2/3}y^{3/4}$$

$$= 21x^{-5/6} + 28x^{1/3}y^{3/4}$$

$$= \frac{21}{x^{5/6}} + 28x^{1/3}y^{3/4}$$

107.

$$x^{1/4}\left(x^4 - 3x^{3/4} + x^{-1/4}\right) = x^{4+1/4} - 3x^{3/4+1/4} + x^{-1/4+1/4}$$

$$= x^{17/4} - 3x + 1$$

109.

$$\left(x^{1/4} + 2\right)\left(x^{1/4} - 2\right) = \left(x^{1/4}\right)^2 - (2)^2 = x^{1/2} - 4$$

111.

$$\left(x^{1/2} - 5\right)^2 = \left(x^{1/2}\right)^2 - 2\left(5 \cdot x^{1/2}\right) + (5)^2$$

$$= x - 10x^{1/2} + 25$$

113.

$$\left(x^{1/3} - 2\right)\left(x^{2/3} + 2x^{1/3} + 4\right)$$

$$= x^{1/3+2/3} + 2x^{1/3+1/3} + 4x^{1/3} - 2x^{2/3} - 4x^{1/3} - 8$$

$$= x + 2x^{2/3} + 4x^{1/3} - 2x^{2/3} - 4x^{1/3} - 8$$

$$= x - 8$$

Note: the exercise is the factored form of the difference of two perfect cubes.

115.

$$x^{m/2} \cdot x^{m/2} = x^{m/2+m/2} = x^m$$

117.

$$\frac{x^{m/3}}{x^{m/4}} = x^{m/3-m/4} = x^{m/12}$$

119.

$$\left(\frac{x^{m/3} \cdot x^{-m/3}}{x^{3/m}}\right)^3 = \left(\frac{x^0}{x^{3/m}}\right)^3 = \frac{1}{x^{9/m}}$$

121.

$$\left(2^{1/2}x^{m/4}\right)^{1/2}\left(2^{1/4}x^{m/6}\right)^3 = 2^{1/4}x^{m/8} \cdot 2^{3/4}x^{m/2}$$

$$= 2^{1/4+3/4} \cdot x^{m/8+m/2}$$

$$= 2x^{5m/8}$$

123.

$$\frac{1}{2} \cdot \left(\frac{x+1}{x}\right)^{-1/2} \cdot \frac{-1}{x^2} = \frac{1}{2} \cdot \left(\frac{x}{x+1}\right)^{1/2} \cdot \frac{-1}{x^2}$$

$$= \frac{1}{2} \cdot \frac{x^{1/2}}{(x+1)^{1/2}} \cdot \frac{-1}{x^2}$$

$$= \frac{-1}{2x^{2-1/2}(x+1)^{1/2}}$$

$$= \frac{-1}{2x^{3/2}(x+1)^{1/2}}$$

125.

$$4(2x+1)^{3/2}(3x-1)^{1/3} + 3(2x+1)^{1/2}(3x-1)^{4/3}$$

$$\text{GCF} = (2x+1)^{1/2}(3x-1)^{1/3}$$

$$= (2x+1)^{1/2}(3x-1)^{1/3}[4(2x+1) + 3(3x-1)]$$

$$= (2x+1)^{1/2}(3x-1)^{1/3}(17x+1)$$

127.

$$(2x+1)^{1/2}(3x-2)^{-2/3} + (2x+1)^{-1/2}(3x-2)^{1/3}$$

$$\text{GCF} = (2x+1)^{-1/2}(3x-2)^{-2/3}$$

$$= (2x+1)^{-1/2}(3x-2)^{-2/3}[(2x+1) + (3x-2)]$$

$$= \frac{5x-1}{(2x+1)^{1/2}(3x-2)^{2/3}}$$

Exercises 5.2

1.-35.
The exercises do not require worked - out solutions.

37.
$$\sqrt{\sqrt{81}} = \left(81^{1/2}\right)^{1/2} = 81^{1/4} = 3$$

39.
$$7^{2/3} = \left(7^2\right)^{1/3} = 49^{1/3} = \sqrt[3]{49}$$

41.-45.
The exercises do not require worked - out solutions.

47.
$$(4xy)^{3/4} = \left[(4xy)^3\right]^{1/4} = \left(64x^3y^3\right)^{1/4} = \sqrt[4]{64x^3y^3}$$

49.-79.
The exercises do not require worked - out solutions.

81.
$$x^{3/2} \cdot x^{1/6} = x^{3/2 + 1/6} = x^{5/3}$$

83.
$$\left(5y^3\right)^{1/2} \cdot \left(5y^7\right)^{1/2} = 5^{1/2} y^{3/2} \cdot 5^{1/2} \cdot y^{7/2}$$
$$= 5^{1/2 + 1/2} y^{3/2 + 7/2} = 5y^5$$

85.
$$\left(-3x^4\right)^{1/3} \cdot \left(9x^8\right)^{1/3} = (-3)^{1/3} x^{4/3} \cdot 9^{1/3} x^{8/3}$$
$$= (-1)^{1/3} (3)^{1/3} \cdot \left(3^2\right)^{1/3} x^{4/3} \cdot x^{8/3}$$
$$= -1 \cdot 3^{1/3} \cdot 3^{2/3} x^{12/3}$$
$$= -1 \cdot 3 \cdot x^4 = -3x^4$$

87.
$$\frac{x^{1/4}}{x^{1/3}} = \frac{1}{x^{1/3 - 1/4}} = \frac{1}{x^{1/12}}$$

89.
$$\frac{x^{5/2}}{x^{1/2}} = x^{5/2 - 1/2} = x^2$$

91.
$$\frac{x^{7/3}}{x^{4/3}} = x^{7/3 - 4/3} = x$$

93.
$$\left(x^{1/5}\right)^{1/4} = x^{1/20}$$

95.
$$\left(x^{2/3}\right)^{1/5} = x^{2/15}$$

97.
$$\left[\left(x^{1/4}\right)^{1/3}\right]^{1/2} = \left(x^{1/4}\right)^{1/6} = x^{1/24}$$

99.-107.
The exercises do not require worked - out solutions.

109.
$$\sqrt{4x^2 + 4x + 1} = \sqrt{(2x + 1)^2} = 2x + 1$$

111.
$$\sqrt{x^2 + 6xy + 9y^2} = \sqrt{(x + 3y)^2} = x + 3y$$

Exercises 5.3

1.
$$\sqrt{5} \cdot \sqrt{2} = \sqrt{5 \cdot 2} = \sqrt{10}$$

3.
$$\sqrt{15}\sqrt{10} = \sqrt{150} = \sqrt{25 \cdot 6} = \sqrt{25} \cdot \sqrt{6} = 5\sqrt{6}$$

5.
$$\sqrt{13}\sqrt{13} = \sqrt{13^2} = 13$$

7.
$$\sqrt[3]{4} \cdot \sqrt[3]{6} = \sqrt[3]{24} = \sqrt[3]{8 \cdot 3} = \sqrt[3]{8}\sqrt[3]{3} = 2\sqrt[3]{3}$$

9.

$$\sqrt[3]{15}\sqrt[3]{50} = \sqrt[3]{750} = \sqrt[3]{125 \cdot 6} = \sqrt[3]{125}\sqrt[3]{6} = 5\sqrt[3]{6}$$

11.

$$\sqrt[4]{7}\sqrt[4]{2} = \sqrt[4]{14}$$

13.

$$\sqrt{48} = \sqrt{16 \cdot 3} = \sqrt{16}\sqrt{3} = 4\sqrt{3}$$

15.

$$\sqrt{150} = \sqrt{25 \cdot 6} = \sqrt{25}\sqrt{6} = 5\sqrt{6}$$

17.

$$\sqrt{80} = \sqrt{16 \cdot 5} = \sqrt{16}\sqrt{5} = 4\sqrt{5}$$

19.

$$\sqrt{108} = \sqrt{36 \cdot 3} = \sqrt{36}\sqrt{3} = 6\sqrt{3}$$

21.

$$\sqrt[3]{16} = \sqrt[3]{8 \cdot 2} = \sqrt[3]{8}\sqrt[3]{2} = 2\sqrt[3]{2}$$

23.

$$\sqrt[3]{-40} = \sqrt[3]{-8 \cdot 5} = \sqrt[3]{-8}\sqrt[3]{5} = -2\sqrt[3]{5}$$

25.

$$\sqrt[4]{80} = \sqrt[4]{16 \cdot 5} = \sqrt[4]{16}\sqrt[4]{5} = 2\sqrt[4]{5}$$

27.

$$\sqrt{32x^3} = \sqrt{16x^2 \cdot 2x} = \sqrt{16x^2}\sqrt{2x} = 4x\sqrt{2x}$$

29.

$$\sqrt[3]{128x^7} = \sqrt[3]{64x^6 \cdot 2x} = \sqrt[3]{64x^6}\sqrt[3]{2x} = 4x^2\sqrt[3]{2x}$$

31.

$$\sqrt[4]{64x^7} = \sqrt[4]{16x^4 \cdot 4x^3} = \sqrt[4]{16x^4}\sqrt[4]{4x^3} = 2x\sqrt[4]{4x^3}$$

33.

$$\sqrt{x^2y^2} = xy$$

35.

$$\sqrt{45x^3y^9} = \sqrt{9x^2y^8 \cdot 5xy}$$
$$= \sqrt{9x^2y^8}\sqrt{5xy} = 3xy^4\sqrt{5xy}$$

37.

$$\sqrt{32x^7y^4} = \sqrt{16x^6y^4 \cdot 2x}$$
$$= \sqrt{16x^6y^4}\sqrt{2x} = 4x^3y^2\sqrt{2x}$$

39.

$$\sqrt{75x^4y^5z^6} = \sqrt{25x^4y^4z^6 \cdot 3y}$$
$$= \sqrt{25x^4y^4z^6}\sqrt{3y} = 5x^2y^2z^3\sqrt{3y}$$

41.

$$\sqrt{144x^5y^7z^9} = \sqrt{144x^4y^6z^8 \cdot xyz}$$
$$= \sqrt{144x^4y^6z^8}\sqrt{xyz} = 12x^2y^3z^4\sqrt{xyz}$$

43.

$$\sqrt[3]{64x^3y^{10}} = \sqrt[3]{64x^3y^9 \cdot y} = \sqrt[3]{64x^3y^9}\sqrt[3]{y} = 4xy^3\sqrt[3]{y}$$

45.

$$\sqrt[3]{81x^9y^2z^4} = \sqrt[3]{27x^9z^3 \cdot 3y^2z}$$
$$= \sqrt[3]{27x^9z^3}\sqrt[3]{3y^2z} = 3x^3z\sqrt[3]{3y^2z}$$

47.

$$\sqrt[4]{32x^5y^7} = \sqrt[4]{16x^4y^4 \cdot 2xy^3}$$
$$= \sqrt[4]{16x^4y^4}\sqrt[4]{2xy^3} = 2xy\sqrt[4]{2xy^3}$$

49.

$$\frac{1}{\sqrt{3}} \cdot \frac{\sqrt{3}}{\sqrt{3}} = \frac{\sqrt{3}}{\sqrt{9}} = \frac{\sqrt{3}}{3}$$

51.

$$\sqrt{\frac{1}{2} \cdot \frac{2}{2}} = \sqrt{\frac{2}{4}} = \frac{\sqrt{2}}{\sqrt{4}} = \frac{\sqrt{2}}{2}$$

53.

$$\frac{\sqrt{3}}{\sqrt{12}} = \sqrt{\frac{3}{12}} = \sqrt{\frac{1}{4}} = \frac{\sqrt{1}}{\sqrt{4}} = \frac{1}{2}$$

55.

$$\frac{3}{\sqrt{6}} \cdot \frac{\sqrt{6}}{\sqrt{6}} = \frac{3\sqrt{6}}{6} = \frac{\sqrt{6}}{2} \qquad \text{Reduce}$$

57.

$$\frac{5}{\sqrt{x}} \cdot \frac{\sqrt{x}}{\sqrt{x}} = \frac{5\sqrt{x}}{x}$$

59.

$$\frac{4}{\sqrt{8x}} = \frac{4}{2\sqrt{2x}} = \frac{2}{\sqrt{2x}} = \frac{2}{\sqrt{2x}} \cdot \frac{\sqrt{2x}}{\sqrt{2x}} = \frac{2\sqrt{2x}}{2x} = \frac{\sqrt{2x}}{x}$$

61.

$$\sqrt[3]{\frac{1}{4}} = \sqrt[3]{\frac{1}{2^2} \cdot \frac{2}{2}} = \sqrt[3]{\frac{2}{2^3}} = \frac{\sqrt[3]{2}}{\sqrt[3]{2^3}} = \frac{\sqrt[3]{2}}{2}$$

63.

$$\sqrt{\frac{9x^2}{16y^4}} = \frac{\sqrt{9x^2}}{\sqrt{16y^4}} = \frac{3x}{4y^2}$$

65.

$$\frac{\sqrt{6x^3}}{\sqrt{24x}} = \sqrt{\frac{6x^3}{24x}} = \sqrt{\frac{x^2}{4}} = \frac{\sqrt{x^2}}{\sqrt{4}} = \frac{x}{2}$$

67.

$$\sqrt{\frac{2}{9x}} = \sqrt{\frac{2}{3^2 x} \cdot \frac{x}{x}} = \sqrt{\frac{2x}{9x^2}} = \frac{\sqrt{2x}}{\sqrt{9x^2}} = \frac{\sqrt{2x}}{3x}$$

69.

$$\sqrt{\frac{7}{2x} \cdot \frac{2x}{2x}} = \sqrt{\frac{14x}{4x^2}} = \frac{\sqrt{14x}}{\sqrt{4x^2}} = \frac{\sqrt{14x}}{2x}$$

71.

$$\sqrt{\frac{7x}{2y} \cdot \frac{2y}{2y}} = \sqrt{\frac{14xy}{4y^2}} = \frac{\sqrt{14xy}}{\sqrt{4y^2}} = \frac{\sqrt{14xy}}{2y}$$

73.

$$\sqrt{\frac{2x}{3y^3}} = \sqrt{\frac{2x}{3y^3} \cdot \frac{3y}{3y}} = \sqrt{\frac{6xy}{9y^4}} = \frac{\sqrt{6xy}}{\sqrt{9y^4}} = \frac{\sqrt{6xy}}{3y^2}$$

75.

$$\sqrt{\frac{25x^4}{8y^7} \cdot \frac{2y}{2y}} = \sqrt{\frac{50x^4 y}{16y^8}} = \frac{\sqrt{25x^4 \cdot 2y}}{\sqrt{16y^8}} = \frac{5x^2\sqrt{2y}}{4y^4}$$

77.

$$\sqrt{\frac{8x^4}{5y^2}} = \sqrt{\frac{8x^4}{5y^2} \cdot \frac{5}{5}} = \sqrt{\frac{40x^4}{25y^2}} = \frac{\sqrt{4x^4 \cdot 10}}{\sqrt{25y^2}} = \frac{2x^2\sqrt{10}}{5y}$$

79.

$$\sqrt[3]{\frac{8}{27x^3}} = \frac{\sqrt[3]{8}}{\sqrt[3]{27x^3}} = \frac{2}{3x}$$

81.

$$\sqrt[3]{\frac{7}{8x} \cdot \frac{x^2}{x^2}} = \sqrt[3]{\frac{7x^2}{8x^3}} = \frac{\sqrt[3]{7x^2}}{\sqrt[3]{8x^3}} = \frac{\sqrt[3]{7x^2}}{2x}$$

83.

$$\sqrt[3]{\frac{5}{4x^2}} = \sqrt[3]{\frac{5}{2^2 x^2} \cdot \frac{2x}{2x}} = \sqrt[3]{\frac{10x}{2^3 x^3}} = \frac{\sqrt[3]{10x}}{\sqrt[3]{2^3 x^3}} = \frac{\sqrt[3]{10x}}{2x}$$

85.

$$\sqrt[3]{\frac{16}{3x^2} \cdot \frac{3^2 x}{3^2 x}} = \sqrt[3]{\frac{16 \cdot 9x}{3^3 x^3}} = \frac{\sqrt[3]{8 \cdot 18x}}{\sqrt[3]{3^3 x^3}} = \frac{2\sqrt[3]{18x}}{3x}$$

87.

$$\sqrt[3]{\frac{3y^2}{5x^4} \cdot \frac{5^2 x^2}{5^2 x^2}} = \sqrt[3]{\frac{75x^2 y^2}{5^3 x^6}} = \frac{\sqrt[3]{75x^2 y^2}}{\sqrt[3]{5^3 x^6}} = \frac{\sqrt[3]{75x^2 y^2}}{5x^2}$$

89.

$$\sqrt[4]{\frac{16}{x^8}} = \frac{\sqrt[4]{16}}{\sqrt[4]{x^8}} = \frac{2}{x^2}$$

91.

$$\sqrt[4]{\frac{3}{5} \cdot \frac{5^3}{5^3}} = \sqrt[4]{\frac{375}{5^4}} = \frac{\sqrt[4]{375}}{\sqrt[4]{5^4}} = \frac{\sqrt[4]{375}}{5}$$

93.

$$\sqrt[4]{\frac{7x}{2y} \cdot \frac{2^3 y^3}{2^3 y^3}} = \sqrt[4]{\frac{56xy^3}{2^4 y^4}} = \frac{\sqrt[4]{56xy^3}}{\sqrt[4]{2^4 y^4}} = \frac{\sqrt[4]{56xy^3}}{2y}$$

95.

$$\sqrt[4]{\frac{5}{4y^3}} = \sqrt[4]{\frac{5}{2^2 y^3} \cdot \frac{2^2 y}{2^2 y}} = \sqrt[4]{\frac{20y}{2^4 y^4}} = \frac{\sqrt[4]{20y}}{\sqrt[4]{2^4 y^4}} = \frac{\sqrt[4]{20y}}{2y}$$

97.
$$\sqrt[6]{x^3} = x^{3/6} = x^{1/2} = \sqrt{x}$$

99.
$$\sqrt[4]{16x^2} = \left(16x^2\right)^{1/4} = 16^{1/4}x^{2/4} = 2x^{1/2} = 2\sqrt{x}$$

101.
$$\sqrt[n]{x^{4n}y^{2n}} = \left(x^{4n}y^{2n}\right)^{1/n} = x^{4n/n}y^{2n/n} = x^4y^2$$

Exercises 5.4

1.-7.
The exercises do not require worked - out solutions.

9.
$$5\sqrt{12} - 2\sqrt{27} + 3\sqrt{3} = 5\sqrt{4 \cdot 3} - 2\sqrt{9 \cdot 3} + 3\sqrt{3}$$
$$= 5 \cdot 2\sqrt{3} - 2 \cdot 3\sqrt{3} + 3\sqrt{3}$$
$$= 10\sqrt{3} - 6\sqrt{3} + 3\sqrt{3} = 7\sqrt{3}$$

11.
$$5\sqrt{8} - \sqrt{18} + 2\sqrt{32} = 5\sqrt{4 \cdot 2} - \sqrt{9 \cdot 2} + 2\sqrt{16 \cdot 2}$$
$$= 5 \cdot 2\sqrt{2} - 3\sqrt{2} + 2 \cdot 4\sqrt{2}$$
$$= 10\sqrt{2} - 3\sqrt{2} + 8\sqrt{2} = 15\sqrt{2}$$

13.
$$\sqrt[3]{32} + \sqrt[3]{108} = \sqrt[3]{8 \cdot 4} + \sqrt[3]{27 \cdot 4} = 2\sqrt[3]{4} + 3\sqrt[3]{4} = 5\sqrt[3]{4}$$

15.
$$\sqrt[4]{32} - \sqrt[4]{2} = \sqrt[4]{16 \cdot 2} - \sqrt[4]{2} = 2\sqrt[4]{2} - \sqrt[4]{2} = \sqrt[4]{2}$$

17.
$$\sqrt{27x} - 2\sqrt{12x} + 2\sqrt{48x}$$
$$= \sqrt{9 \cdot 3x} - 2\sqrt{4 \cdot 3x} + 2\sqrt{16 \cdot 3x}$$
$$= 3\sqrt{3x} - 2 \cdot 2\sqrt{3x} + 2 \cdot 4\sqrt{3x}$$
$$= 3\sqrt{3x} - 4\sqrt{3x} + 8\sqrt{3x} = 7\sqrt{3x}$$

19.
$$x\sqrt{18x} + 5\sqrt{2x^3} + 2x\sqrt{8x}$$
$$= x\sqrt{9 \cdot 2x} + 5\sqrt{x^2 \cdot 2x} + 2x\sqrt{4 \cdot 2x}$$
$$= 3x\sqrt{2x} + 5x\sqrt{2x} + 2 \cdot 2x\sqrt{2x}$$
$$= 3x\sqrt{2x} + 5x\sqrt{2x} + 4x\sqrt{2x} = 12x\sqrt{2x}$$

21.
$$\sqrt[3]{108x} - \sqrt[3]{32x} = \sqrt[3]{27 \cdot 4x} - \sqrt[3]{8 \cdot 4x}$$
$$= 3\sqrt[3]{4x} - 2\sqrt[3]{4x} = \sqrt[3]{4x}$$

23.
$$\sqrt[4]{162x} + \sqrt[4]{32x} = \sqrt[4]{81 \cdot 2x} + \sqrt[4]{16 \cdot 2x}$$
$$= 3\sqrt[4]{2x} + 2\sqrt[4]{2x} = 5\sqrt[4]{2x}$$

25.
$$\sqrt{27x} + 2\sqrt{12x} - \sqrt{150y} - 4\sqrt{24y}$$
$$= \sqrt{9 \cdot 3x} + 2\sqrt{4 \cdot 3x} - \sqrt{25 \cdot 6y} - 4\sqrt{4 \cdot 6y}$$
$$= 3\sqrt{3x} + 2 \cdot 2\sqrt{3x} - 5\sqrt{6y} - 4 \cdot 2\sqrt{6y}$$
$$= 3\sqrt{3x} + 4\sqrt{3x} - 5\sqrt{6y} - 8\sqrt{6y}$$
$$= 7\sqrt{3x} - 13\sqrt{6y}$$

27.
$$5x\sqrt{3x^2} - x^2\sqrt{300} + 2y\sqrt{8y^2} + \sqrt{8y^4}$$
$$= 5x\sqrt{x^2 \cdot 3} - x^2\sqrt{100 \cdot 3} + 2y\sqrt{4y^2 \cdot 2} + \sqrt{4y^4 \cdot 2}$$
$$= 5x \cdot x\sqrt{3} - 10x^2\sqrt{3} + 2y \cdot 2y\sqrt{2} + 2y^2\sqrt{2}$$
$$= 5x^2\sqrt{3} - 10x^2\sqrt{3} + 4y^2\sqrt{2} + 2y^2\sqrt{2}$$
$$= -5x^2\sqrt{3} + 6y^2\sqrt{2}$$

29.
$$\sqrt{112} + 2\sqrt{180} = \sqrt{16 \cdot 7} + 2\sqrt{36 \cdot 5}$$
$$= 4\sqrt{7} + 2 \cdot 6\sqrt{5} = 4\sqrt{7} + 12\sqrt{5}$$

31.
$$\sqrt[3]{54} - \sqrt[3]{40} = \sqrt[3]{27 \cdot 2} - \sqrt[3]{8 \cdot 5} = 3\sqrt[3]{2} - 2\sqrt[3]{5}$$

33.
$$\sqrt{90x} - 2\sqrt{40x} + 5\sqrt{10x^3}$$
$$= \sqrt{9 \cdot 10x} - 2\sqrt{4 \cdot 10x} + 5\sqrt{x^2 \cdot 10x}$$
$$= 3\sqrt{10x} - 4\sqrt{10x} + 5x\sqrt{10x}$$
$$= (3 - 4 + 5x)\sqrt{10x}$$
$$= (-1 + 5x)\sqrt{10x}$$

35.

$$y\sqrt{32y} + 5\sqrt{2y^3} - \sqrt{98y}$$
$$= y\sqrt{16\cdot 2y} + 5\sqrt{y^2\cdot 2y} - \sqrt{49\cdot 2y}$$
$$= 4y\sqrt{2y} + 5y\sqrt{2y} - 7\sqrt{2y}$$
$$= 9y\sqrt{2y} - 7\sqrt{2y} = (9y-7)\sqrt{2y}$$

37.

$$x\sqrt{20x^2} - x\sqrt{45} + 2\sqrt{500}$$
$$= x\sqrt{4x^2\cdot 5} - x\sqrt{9\cdot 5} + 2\sqrt{100\cdot 5}$$
$$= 2x\cdot x\sqrt{5} - 3x\sqrt{5} + 2\cdot 10\sqrt{5}$$
$$= 2x^2\sqrt{5} - 3x\sqrt{5} + 20\sqrt{5} = \left(2x^2 - 3x + 20\right)\sqrt{5}$$

39.

$$\sqrt[3]{54x^2y} - \sqrt[3]{2x^2y} = \sqrt[3]{27\cdot 2x^2y} - \sqrt[3]{2x^2y}$$
$$= 3\sqrt[3]{2x^2y} - \sqrt[3]{2x^2y}$$
$$= 2\sqrt[3]{2x^2y}$$

41.

$$4\sqrt[3]{3x^4y} - x\sqrt[3]{24xy} = 4\sqrt[3]{x^3\cdot 3xy} - x\sqrt[3]{8\cdot 3xy}$$
$$= 4x\sqrt[3]{3xy} - 2x\sqrt[3]{3xy}$$
$$= 2x\sqrt[3]{3xy}$$

43.

$$\sqrt[4]{80xy} - 3\sqrt[4]{5xy} = \sqrt[4]{16\cdot 5xy} - 3\sqrt[4]{5xy}$$
$$= 2\sqrt[4]{5xy} - 3\sqrt[4]{5xy}$$
$$= -\sqrt[4]{5xy}$$

45.

$$y^2\sqrt[4]{16xy} - y\sqrt[4]{xy^5} = y^2\sqrt[4]{16\cdot xy} - y\sqrt[4]{y^4\cdot xy}$$
$$= 2y^2\sqrt[4]{xy} - y\cdot y\sqrt[4]{xy}$$
$$= 2y^2\sqrt[4]{xy} - y^2\sqrt[4]{xy} = y^2\sqrt[4]{xy}$$

47.

$$\frac{1}{\sqrt{2}} + 5\sqrt{2} = \frac{\sqrt{2}}{2} + 5\sqrt{2} = \frac{\sqrt{2}}{2} + \frac{10\sqrt{2}}{2} = \frac{11\sqrt{2}}{2}$$

Note: $\dfrac{1}{\sqrt{2}}\cdot\dfrac{\sqrt{2}}{\sqrt{2}} = \dfrac{\sqrt{2}}{\sqrt{4}} = \dfrac{\sqrt{2}}{2}$

49.

$$\frac{3}{\sqrt{5}} + \sqrt{80} = \frac{3\sqrt{5}}{5} + 4\sqrt{5} = \frac{3\sqrt{5}}{5} + \frac{20\sqrt{5}}{5} = \frac{23\sqrt{5}}{5}$$

51.

$$\frac{1}{\sqrt[3]{2}} - \sqrt[3]{108} = \frac{\sqrt[3]{4}}{2} - 3\sqrt[3]{4} = \frac{\sqrt[3]{4}}{2} - \frac{6\sqrt[3]{4}}{2} = -\frac{5\sqrt[3]{4}}{2}$$

53.

$$\sqrt{3} + \frac{5}{\sqrt{3}} + \frac{2\sqrt{27}}{3} = \sqrt{3} + \frac{5\sqrt{3}}{3} + \frac{6\sqrt{3}}{3}$$
$$= \frac{3\sqrt{3}}{3} + \frac{5\sqrt{3}}{3} + \frac{6\sqrt{3}}{3} = \frac{14\sqrt{3}}{3}$$

55.

$$\sqrt{\frac{2}{3}} + 5\sqrt{6} - \sqrt{\frac{3}{2}} = \frac{\sqrt{6}}{3} + 5\sqrt{6} - \frac{\sqrt{6}}{2}$$
$$= \frac{2\sqrt{6}}{6} + \frac{30\sqrt{6}}{6} - \frac{3\sqrt{6}}{6} = \frac{29\sqrt{6}}{6}$$

57.

$$\sqrt{\frac{x}{2}} + 5\sqrt{2x} = \frac{\sqrt{2x}}{2} + 5\sqrt{2x}$$
$$= \frac{\sqrt{2x}}{2} + \frac{10\sqrt{2x}}{2} = \frac{11\sqrt{2x}}{2}$$

59.

$$\sqrt[3]{\frac{x}{4}} - \sqrt[3]{16x} = \frac{\sqrt[3]{2x}}{2} - 2\sqrt[3]{2x}$$
$$= \frac{\sqrt[3]{2x}}{2} - \frac{4\sqrt[3]{2x}}{2} = -\frac{3\sqrt[3]{2x}}{2}$$

61.

$$\sqrt{\frac{2}{x}} + \sqrt{\frac{x}{2}} - \sqrt{2x} = \frac{\sqrt{2x}}{x} + \frac{\sqrt{2x}}{2} - \sqrt{2x}$$
$$= \frac{2\sqrt{2x}}{2x} + \frac{x\sqrt{2x}}{2x} - \frac{2x\sqrt{2x}}{2x}$$
$$= \frac{2\sqrt{2x} - x\sqrt{2x}}{2x} = \left(\frac{2-x}{2x}\right)\sqrt{2x}$$

63.

$$\sqrt{\frac{2}{x}} + \sqrt{32x} - \sqrt{\frac{1}{2x}} = \frac{\sqrt{2x}}{x} + 4\sqrt{2x} - \frac{\sqrt{2x}}{2x}$$

$$= \frac{2\sqrt{2x}}{2x} + \frac{8x\sqrt{2x}}{2x} - \frac{\sqrt{2x}}{2x}$$

$$= \frac{\sqrt{2x} + 8x\sqrt{2x}}{2x}$$

$$= \left(\frac{1+8x}{2x}\right)\sqrt{2x} = \left(\frac{8x+1}{2x}\right)\sqrt{2x}$$

Exercises 5.5

1.

$$\sqrt{3}\left(2 + \sqrt{5}\right) = 2\sqrt{3} + \sqrt{3}\cdot\sqrt{5} = 2\sqrt{3} + \sqrt{15}$$

3.

$$2\sqrt{6}\left(7 - \sqrt{5} + \sqrt{2}\right) = 2\sqrt{6}\cdot 7 - 2\sqrt{6}\cdot\sqrt{5} + 2\sqrt{6}\sqrt{2}$$

$$= 14\sqrt{6} - 2\sqrt{30} + 2\sqrt{12}$$

$$= 14\sqrt{6} - 2\sqrt{30} + 4\sqrt{3}$$

5.

$$\sqrt{x}\left(3 + \sqrt{x}\right) = 3\sqrt{x} + \sqrt{x}\sqrt{x} = 3\sqrt{x} + x$$

7.

$$2\sqrt{3y}\left(4 - y + \sqrt{6}\right) = 2\sqrt{3y}\cdot 4 + 2\sqrt{3y}(-y) + 2\sqrt{3y}\sqrt{6}$$

$$= 8\sqrt{3y} - 2y\sqrt{3y} + 2\sqrt{18y}$$

$$= 8\sqrt{3y} - 2y\sqrt{3y} + 6\sqrt{2y}$$

9.

$$\begin{array}{cccc} \text{F} & \text{O} & \text{I} & \text{L} \end{array}$$

$$\left(2 + \sqrt{7}\right)\left(\sqrt{8} - 3\right) = 2\sqrt{8} - 6 + \sqrt{56} - 3\sqrt{7}$$

$$= 4\sqrt{2} - 6 + 2\sqrt{14} - 3\sqrt{7}$$

11.

$$\begin{array}{cccc} \text{F} & \text{O} & \text{I} & \text{L} \end{array}$$

$$\left(\sqrt{5} + \sqrt{7}\right)\left(\sqrt{10} + \sqrt{2}\right) = \sqrt{50} + \sqrt{10} + \sqrt{70} + \sqrt{14}$$

$$= 5\sqrt{2} + \sqrt{10} + \sqrt{70} + \sqrt{14}$$

13.

$$\left(2\sqrt{3} + 4\sqrt{5}\right)\left(3\sqrt{2} - 6\sqrt{7}\right)$$

#13. continued

$$\begin{array}{cccc} \text{F} & \text{O} & \text{I} & \text{L} \end{array}$$

$$= 6\sqrt{6} - 12\sqrt{21} + 12\sqrt{10} - 24\sqrt{35}$$

15.

$$\left(2\sqrt{x} - 1\right)\left(3\sqrt{x} + 4\right)$$

$$\begin{array}{cccc} \text{F} & \text{O} & \text{I} & \text{L} \end{array}$$

$$= 6\sqrt{x}\sqrt{x} + 8\sqrt{x} - 3\sqrt{x} - 4$$

$$= 6x + 5\sqrt{x} - 4$$

17.

$$\left(2\sqrt{x} + 3\sqrt{y}\right)\left(4\sqrt{x} + 5\sqrt{y}\right)$$

$$\begin{array}{cccc} \text{F} & \text{O} & \text{I} & \text{L} \end{array}$$

$$= 8\sqrt{x}\sqrt{x} + 10\sqrt{x}\sqrt{y} + 12\sqrt{x}\sqrt{y} + 15\sqrt{y}\sqrt{y}$$

$$= 8x + 10\sqrt{xy} + 12\sqrt{xy} + 15y$$

$$= 8x + 22\sqrt{xy} + 15y$$

19.

$$\left(\sqrt{5} - \sqrt{3}\right)^2 = \left(\sqrt{5}\right)^2 - 2\left(\sqrt{5}\sqrt{3}\right) + \left(\sqrt{3}\right)^2$$

$$= 5 - 2\sqrt{15} + 3 = 8 - 2\sqrt{15}$$

21.

$$\left(2\sqrt{3} + 1\right)^2 = \left(2\sqrt{3}\right)^2 + 2\left(2\sqrt{3}\cdot 1\right) + (1)^2$$

$$= 4\cdot 3 + 4\sqrt{3} + 1$$

$$= 12 + 4\sqrt{3} + 1 = 13 + 4\sqrt{3}$$

23.

$$\left(3\sqrt{2} - 6\sqrt{3}\right)^2 = \left(3\sqrt{2}\right)^2 - 2\left(3\sqrt{2}\cdot 6\sqrt{3}\right) + \left(6\sqrt{3}\right)^2$$

$$= 9\cdot 2 - 36\sqrt{6} + 36\cdot 3$$

$$= 18 - 36\sqrt{6} + 108$$

$$= 126 - 36\sqrt{6}$$

25.

$$\left(\sqrt{x} - \sqrt{2y}\right)^2 = \left(\sqrt{x}\right)^2 - 2\left(\sqrt{x}\sqrt{2y}\right) + \left(\sqrt{2y}\right)^2$$

$$= x - 2\sqrt{2xy} + 2y$$

27.

$$\left(2\sqrt{x} + 7\right)^2 = \left(2\sqrt{x}\right)^2 + 2\left(2\sqrt{x}\cdot 7\right) + (7)^2$$

$$= 4\cdot x + 28\sqrt{x} + 49$$

$$= 4x + 28\sqrt{x} + 49$$

29.

$$\left(1+\sqrt{x-1}\right)^2 = 1^2 + 2\left(1\cdot\sqrt{x-1}\right)+\left(\sqrt{x-1}\right)^2$$
$$= 1+2\sqrt{x-1}+x-1$$
$$= x+2\sqrt{x-1}$$

31.

$$\left(3-\sqrt{x+2}\right)^2 = 3^2 - 2\left(3\cdot\sqrt{x+2}\right)+\left(\sqrt{x+2}\right)^2$$
$$= 9-6\sqrt{x+2}+x+2$$
$$= x-6\sqrt{x+2}+11$$

33.

$$\left(\sqrt{x}-\sqrt{3x-2}\right)^2$$
$$= \left(\sqrt{x}\right)^2 - 2\left(\sqrt{x}\cdot\sqrt{3x-2}\right)+\left(\sqrt{3x-2}\right)^2$$
$$= x-2\sqrt{3x^2-2x}+3x-2$$
$$= 4x-2\sqrt{3x^2-2x}-2$$

35.

$$\left(\sqrt{5}-\sqrt{3}\right)\left(\sqrt{5}+\sqrt{3}\right)=\left(\sqrt{5}\right)^2-\left(\sqrt{3}\right)^2=5-3=2$$

37.

$$\left(5\sqrt{3}+\sqrt{6}\right)\left(5\sqrt{3}-\sqrt{6}\right)=\left(5\sqrt{3}\right)^2-\left(\sqrt{6}\right)^2$$
$$= 25\cdot3-6=75-6=69$$

39.

$$\left(\sqrt{x}+7\right)\left(\sqrt{x}-7\right)=\left(\sqrt{x}\right)^2-(7)^2=x-49$$

41.

$$\left(\sqrt{2x}+\sqrt{3y}\right)\left(\sqrt{2x}-\sqrt{3y}\right)=\left(\sqrt{2x}\right)^2-\left(\sqrt{3y}\right)^2$$
$$= 2x-3y$$

43.

$$\qquad\qquad\quad \text{F}\qquad \text{O}\qquad \text{I}\qquad \text{L}$$
$$\left(\sqrt[3]{3}+2\right)\left(\sqrt[3]{4}-5\right)=\sqrt[3]{12}-5\sqrt[3]{3}+2\sqrt[3]{4}-10$$

45.

$$\left(\sqrt[3]{2}+\sqrt[3]{3}\right)^2=\left(\sqrt[3]{2}\right)^2+2\left(\sqrt[3]{2}\sqrt[3]{3}\right)+\left(\sqrt[3]{3}\right)^2$$
$$= \sqrt[3]{2}\sqrt[3]{2}+2\sqrt[3]{6}+\sqrt[3]{3}\sqrt[3]{3}$$
$$= \sqrt[3]{4}+2\sqrt[3]{6}+\sqrt[3]{9}$$

47.

$$\left(\sqrt[3]{2x}-1\right)\left(\sqrt[3]{2x}+1\right)=\left(\sqrt[3]{2x}\right)^2-(1)^2$$
$$= \sqrt[3]{2x}\sqrt[3]{2x}-1$$
$$= \sqrt[3]{4x^2}-1$$

49.

$$\left(\sqrt[3]{3}-1\right)\left(\sqrt[3]{9}+\sqrt[3]{3}+1\right)=\left(\sqrt[3]{3}\right)^3-(1)^3$$
$$= 3-1=2$$

This is the factorization of the difference of two perfect cubes.

51.

$$\left(\sqrt[3]{x}-\sqrt[3]{y}\right)\left(\sqrt[3]{x^2}+\sqrt[3]{xy}+\sqrt[3]{y^2}\right)=\left(\sqrt[3]{x}\right)^3-\left(\sqrt[3]{y}\right)^3$$
$$= x-y$$

This is the factorization of the difference of two perfect cubes.

53.

$$\frac{1}{\sqrt{3}+2}\cdot\frac{\sqrt{3}-2}{\sqrt{3}-2}=\frac{\sqrt{3}-2}{3-4}=\frac{\sqrt{3}-2}{-1}=2-\sqrt{3}$$

55.

$$\frac{2}{\sqrt{3}+\sqrt{5}}\cdot\frac{\sqrt{3}-\sqrt{5}}{\sqrt{3}-\sqrt{5}}=\frac{2\sqrt{3}-2\sqrt{5}}{3-5}$$
$$= \frac{2\sqrt{3}-2\sqrt{5}}{-2}$$
$$= \frac{2\left(\sqrt{3}-\sqrt{5}\right)}{-1\cdot2}$$
$$= \sqrt{5}-\sqrt{3}$$

57.

$$\frac{x}{\sqrt{x}+\sqrt{y}}\cdot\frac{\sqrt{x}-\sqrt{y}}{\sqrt{x}-\sqrt{y}}=\frac{x\sqrt{x}-x\sqrt{y}}{\left(\sqrt{x}\right)^2-\left(\sqrt{y}\right)^2}$$
$$= \frac{x\sqrt{x}-x\sqrt{y}}{x-y}$$

59.

$$\frac{2+\sqrt{3}}{2-\sqrt{3}}\cdot\frac{2+\sqrt{3}}{2+\sqrt{3}}=\frac{\left(2+\sqrt{3}\right)^2}{4-\left(\sqrt{3}\right)^2}$$

#59. continued

$$= \frac{2^2 + 2\left(2 \cdot \sqrt{3}\right) + \left(\sqrt{3}\right)^2}{4-3}$$

$$= \frac{4 + 4\sqrt{3} + 3}{1} = 7 + 4\sqrt{3}$$

61.

$$\frac{\sqrt{x}+5}{\sqrt{x}-4} \cdot \frac{\sqrt{x}+4}{\sqrt{x}+4} = \frac{\left(\sqrt{x}+5\right)\left(\sqrt{x}+4\right)}{\left(\sqrt{x}\right)^2 - 4^2}$$

$$= \frac{x + 9\sqrt{x} + 20}{x - 16}$$

63.

$$\frac{\sqrt{3}+\sqrt{2}}{\sqrt{6}-\sqrt{3}} \cdot \frac{\sqrt{6}+\sqrt{3}}{\sqrt{6}+\sqrt{3}} = \frac{\left(\sqrt{3}+\sqrt{2}\right)\left(\sqrt{6}+\sqrt{3}\right)}{\left(\sqrt{6}\right)^2 - \left(\sqrt{3}\right)^2}$$

$$= \frac{\sqrt{18}+\sqrt{9}+\sqrt{12}+\sqrt{6}}{6-3}$$

$$= \frac{3\sqrt{2}+3+2\sqrt{3}+\sqrt{6}}{3}$$

65.

$$\frac{\sqrt{x}+\sqrt{y}}{\sqrt{x}-\sqrt{y}} \cdot \frac{\sqrt{x}+\sqrt{y}}{\sqrt{x}+\sqrt{y}} = \frac{\left(\sqrt{x}+\sqrt{y}\right)^2}{\left(\sqrt{x}\right)^2 - \left(\sqrt{y}\right)^2}$$

$$= \frac{\left(\sqrt{x}\right)^2 + 2\left(\sqrt{x}\sqrt{y}\right) + \left(\sqrt{y}\right)^2}{x-y}$$

$$= \frac{x + 2\sqrt{xy} + y}{x-y}$$

67.

$$\frac{5\sqrt{3}+7}{2\sqrt{3}-4} \cdot \frac{2\sqrt{3}+4}{2\sqrt{3}+4} = \frac{\left(5\sqrt{3}+7\right)\left(2\sqrt{3}+4\right)}{\left(2\sqrt{3}\right)^2 - 4^2}$$

$$= \frac{10 \cdot 3 + 20\sqrt{3} + 14\sqrt{3} + 28}{12-16}$$

$$= \frac{30 + 34\sqrt{3} + 28}{-4}$$

$$= \frac{58 + 34\sqrt{3}}{-4}$$

$$= \frac{2\left(29 + 17\sqrt{3}\right)}{-2 \cdot 2}$$

$$= -\frac{29 + 17\sqrt{3}}{2}$$

69.

$$\frac{2\sqrt{3}+3\sqrt{2}}{4\sqrt{2}+2\sqrt{3}} \cdot \frac{4\sqrt{2}-2\sqrt{3}}{4\sqrt{2}-2\sqrt{3}}$$

$$= \frac{\left(2\sqrt{3}+3\sqrt{2}\right)\left(4\sqrt{2}-2\sqrt{3}\right)}{\left(4\sqrt{2}\right)^2 - \left(2\sqrt{3}\right)^2}$$

$$= \frac{8\sqrt{6}-4 \cdot 3 + 12 \cdot 2 - 6\sqrt{6}}{32-12}$$

$$= \frac{8\sqrt{6}-12+24-6\sqrt{6}}{20}$$

$$= \frac{2\sqrt{6}+12}{20} = \frac{2\left(\sqrt{6}+6\right)}{2 \cdot 10} = \frac{\sqrt{6}+6}{10}$$

71.

$$\frac{2\sqrt{x}-3\sqrt{y}}{3\sqrt{x}+2\sqrt{y}} \cdot \frac{3\sqrt{x}-2\sqrt{y}}{3\sqrt{x}-2\sqrt{y}}$$

$$= \frac{\left(2\sqrt{x}-3\sqrt{y}\right)\left(3\sqrt{x}-2\sqrt{y}\right)}{\left(3\sqrt{x}\right)^2 - \left(2\sqrt{y}\right)^2}$$

$$= \frac{6x - 9\sqrt{xy} - 4\sqrt{xy} + 6y}{9x - 4y}$$

$$= \frac{6x - 13\sqrt{xy} + 6y}{9x - 4y}$$

73.

$$\frac{2\sqrt{2x}-5\sqrt{3y}}{\sqrt{3x}+\sqrt{6y}} \cdot \frac{\sqrt{3x}-\sqrt{6y}}{\sqrt{3x}-\sqrt{6y}}$$

$$= \frac{\left(2\sqrt{2x}-5\sqrt{3y}\right)\left(\sqrt{3x}-\sqrt{6y}\right)}{\left(\sqrt{3x}\right)^2 - \left(\sqrt{6y}\right)^2}$$

$$= \frac{2\sqrt{6x^2}-2\sqrt{12xy}-5\sqrt{9xy}+5\sqrt{18y^2}}{3x-6y}$$

$$= \frac{2x\sqrt{6}-2 \cdot 2\sqrt{3xy}-5 \cdot 3\sqrt{xy}+5 \cdot 3y\sqrt{2}}{3x-6y}$$

$$= \frac{2x\sqrt{6}-4\sqrt{3xy}-15\sqrt{xy}+15y\sqrt{2}}{3x-6y}$$

Exercises 5.6

1.

$$\left(\sqrt{2x+1}\right)^2 = (3)^2$$

$$2x+1 = 9 \qquad \text{Square both sides}$$

$$2x = 8$$

$$x = 4$$

Solution set: $\{4\}$

3.

$$\sqrt{x^2+7} - 4 = 0$$

$$\sqrt{x^2+7} = 4$$

$$\left(\sqrt{x^2+7}\right)^2 = 4^2$$

$$x^2 + 7 = 16$$

$$x^2 - 9 = 0$$

$$(x+3)(x-3) = 0$$

$$x+3 = 0 \qquad x-3 = 0$$

$$x = -3 \qquad x = 3$$

Solution set: $\{-3, 3\}$

5.

$$\sqrt{3x-7} + 2 = 0$$

$$\sqrt{3x-7} = -2$$

$\varnothing$ By definition, square root is not negative over the real numbers.

7.

$$\left(\sqrt[3]{3x+1}\right)^3 = (4)^3$$

$$3x+1 = 64$$

$$3x = 63$$

$$x = 21$$

Solution set: $\{21\}$

9.

$$\sqrt[3]{3x^2-10x} + 2 = 0$$

$$\left(\sqrt[3]{3x^2-10x}\right)^3 = (-2)^3$$

$$3x^2 - 10x = -8$$

$$3x^2 - 10x + 8 = 0$$

$$(3x-4)(x-2) = 0$$

#9. continued

$$3x - 4 = 0 \qquad x - 2 = 0$$

$$x = \frac{4}{3} \qquad x = 2$$

Solution set: $\left\{\frac{4}{3}, 2\right\}$

11.

$$\sqrt[4]{3x+5} = -3$$

$\varnothing$ An even root is not negative.

13.

$$\sqrt[4]{2x-3} - 1 = 0$$

$$\left(\sqrt[4]{2x-3}\right)^4 = (1)^4$$

$$2x - 3 = 1$$

$$2x = 4$$

$$x = 2$$

Solution set: $\{2\}$

15.

$$\left(\sqrt{2x+3}\right)^2 = \left(\sqrt{7-x}\right)^2$$

$$2x + 3 = 7 - x$$

$$3x = 4$$

$$x = \frac{4}{3}$$

Solution set: $\left\{\frac{4}{3}\right\}$

17.

$$\left(\sqrt{3x^2-x-6}\right)^2 = \left(\sqrt{2x^2+x+9}\right)^2$$

$$3x^2 - x - 6 = 2x^2 + x + 9$$

$$x^2 - 2x - 15 = 0$$

$$(x+3)(x-5) = 0$$

$$x+3 = 0 \qquad x-5 = 0$$

$$x = -3 \qquad x = 5$$

Solution set: $\{-3, 5\}$

19.

$$\left(\sqrt[3]{3x-2}\right)^3 = \left(\sqrt[3]{5x+7}\right)^3$$

$$3x - 2 = 5x + 7$$

$$-9 = 2x$$

$$-\frac{9}{2} = x$$

#19. continued

Solution set: $\left\{-\dfrac{9}{2}\right\}$

21.

$$\left(\sqrt[4]{4x-3}\right)^4 = \left(\sqrt[4]{x+8}\right)^4$$

$$4x - 3 = x + 8$$

$$3x = 11$$

$$x = \dfrac{11}{3}$$

Solution set: $\left\{\dfrac{11}{3}\right\}$

23.

$$\left(x^2\right) = \left(\sqrt{3x+40}\right)^2$$

$$x^2 = 3x + 40$$

$$x^2 - 3x - 40 = 0$$

$$(x+5)(x-8) = 0$$

$$x + 5 = 0 \quad \text{or} \quad x - 8 = 0$$

$$x = -5 \qquad\qquad x = 8$$

Solution set: $\{8\}$ -5 does not check.

25.

$$\left(\sqrt{2x^2+2x-3}\right)^2 = (x)^2$$

$$2x^2 + 2x - 3 = x^2$$

$$x^2 + 2x - 3 = 0$$

$$(x+3)(x-1) = 0$$

$$x + 3 = 0 \quad \text{or} \quad x - 1 = 0$$

$$x = -3 \qquad\qquad x = 1$$

Solution set: $\{1\}$ -3 does not check.

27.

$$\sqrt{4x^2-x-1} + 1 = 2x$$

$$\left(\sqrt{4x^2-x-1}\right)^2 = (2x-1)^2$$

$$4x^2 - x - 1 = 4x^2 - 4x + 1$$

$$3x = 2$$

$$x = \dfrac{2}{3}$$

Solution set: $\left\{\dfrac{2}{3}\right\}$

29.

$$\sqrt{x^2-8x+26} + 5 = x$$

$$\left(\sqrt{x^2-8x+26}\right)^2 = (x-5)^2$$

$$x^2 - 8x + 26 = x^2 - 10x + 25$$

$$2x = -1$$

$$x = -\dfrac{1}{2}$$

$\varnothing$ $-\dfrac{1}{2}$ does not check.

31.

$$\sqrt{2x+5} - x = 3$$

$$\left(\sqrt{2x+5}\right)^2 = (3+x)^2$$

$$2x + 5 = 9 + 6x + x^2$$

$$0 = x^2 + 4x + 4$$

$$0 = (x+2)^2$$

$$x + 2 = 0$$

$$x = -2$$

Solution set: $\{-2\}$

33.

$$\sqrt{x+8} - x = -4$$

$$\left(\sqrt{x+8}\right)^2 = (x-4)^2$$

$$x + 8 = x^2 - 8x + 16$$

$$0 = x^2 - 9x + 8$$

$$0 = (x-8)(x-1)$$

$$x - 8 = 0 \quad \text{or} \quad x - 1 = 0$$

$$x = 8 \qquad\qquad x = 1$$

Solution set: $\{8\}$ 1 does not check.

35.

$$\sqrt{3x+5} + 1 = 3x$$

$$\left(\sqrt{3x+5}\right)^2 = (3x-1)^2$$

$$3x + 5 = 9x^2 - 6x + 1$$

$$0 = 9x^2 - 9x - 4$$

$$0 = (3x-4)(3x+1)$$

Solution set: $\left\{\dfrac{4}{3}\right\}$ $-\dfrac{1}{3}$ does not check.

37.
$$\sqrt{3-x} - x = 3$$
$$\left(\sqrt{3-x}\right)^2 = (x+3)^2$$
$$3 - x = x^2 + 6x + 9$$
$$0 = x^2 + 7x + 6$$
$$0 = (x+1)(x+6)$$
$$x + 1 = 0 \quad \text{or} \quad x + 6 = 0$$
$$x = -1 \qquad\quad x = -6$$
Solution set: $\{-1\}$ -6 does not check.

39.
$$\sqrt{x-4} + x = 6$$
$$\left(\sqrt{x-4}\right)^2 = (6-x)^2$$
$$x - 4 = 36 - 12x + x^2$$
$$0 = x^2 - 13x + 40$$
$$0 = (x-5)(x-8)$$
$$x - 5 = 0 \quad \text{or} \quad x - 8 = 0$$
$$x = 5 \qquad\quad x = 8$$
Solution set: $\{5\}$ 8 does not check.

41.
$$\sqrt{x-16} - \sqrt{x} = -2$$
$$\left(\sqrt{x-16}\right)^2 = \left(\sqrt{x} - 2\right)^2$$
$$x - 16 = x - 4\sqrt{x} + 4$$
$$-20 = -4\sqrt{x}$$
$$5 = \sqrt{x} \qquad\qquad \text{Divide by} -4$$
$$(5)^2 = \left(\sqrt{x}\right)^2$$
$$25 = x$$
Solution set: $\{25\}$

43.
$$\sqrt{2x+7} - \sqrt{x} = 2$$
$$\left(\sqrt{2x+7}\right)^2 = \left(\sqrt{x} + 2\right)^2$$
$$2x + 7 = x + 4\sqrt{x} + 4$$
$$(x+3)^2 = \left(4\sqrt{x}\right)^2$$
$$x^2 + 6x + 9 = 16x$$
$$x^2 - 10x + 9 = 0$$
$$(x-9)(x-1) = 0$$

#43. continued
$$x - 9 = 0 \quad \text{or} \quad x - 1 = 0$$
$$x = 9 \qquad\qquad x = 1$$
Solution set: $\{1, 9\}$

45.
$$\sqrt{x-1} - \sqrt{x+4} = -1$$
$$\left(\sqrt{x-1}\right)^2 = \left(\sqrt{x+4} - 1\right)^2$$
$$x - 1 = x + 4 - 2\sqrt{x+4} + 1$$
$$x - 1 = x - 2\sqrt{x+4} + 5$$
$$-6 = -2\sqrt{x+4}$$
$$3^2 = \left(\sqrt{x+4}\right)^2 \qquad \text{Divide by} -2,$$
$$9 = x + 4 \qquad\qquad \text{square}$$
$$5 = x$$
Solution set: $\{5\}$

47.
$$\sqrt{2x+3} + \sqrt{x+2} = 2$$
$$\left(\sqrt{2x+3}\right)^2 = \left(2 - \sqrt{x+2}\right)^2$$
$$2x + 3 = 4 - 4\sqrt{x+2} + x + 2$$
$$2x + 3 = 6 - 4\sqrt{x+2} + x$$
$$(x-3)^2 = \left(-4\sqrt{x+2}\right)^2$$
$$x^2 - 6x + 9 = 16(x+2)$$
$$x^2 - 6x + 9 = 16x + 32$$
$$x^2 - 22x - 23 = 0$$
$$(x+1)(x-23) = 0$$
$$x + 1 = 0 \quad \text{or} \quad x - 23 = 0$$
$$x = -1 \qquad\qquad x = 23$$
Solution set: $\{-1\}$ 23 does not check.

49.
$$\sqrt{5-2x} - \sqrt{x+6} = 1$$
$$\left(\sqrt{5-2x}\right)^2 = \left(\sqrt{x+6} + 1\right)^2$$
$$5 - 2x = x + 6 + 2\sqrt{x+6} + 1$$
$$5 - 2x = x + 7 + 2\sqrt{x+6}$$
$$-2 - 3x = 2\sqrt{x+6}$$
$$2 + 3x = -2\sqrt{x+6} \qquad \text{Multiply by} -1$$
$$(3x+2)^2 = \left(-2\sqrt{x+6}\right)^2$$

#49. continued

$$9x^2 + 12x + 4 = 4(x + 6)$$

$$9x^2 + 12x + 4 = 4x + 24$$

$$9x^2 + 8x - 20 = 0$$

$$(x + 2)(9x - 10) = 0$$

$$x + 2 = 0 \quad \text{or} \quad 9x - 10 = 0$$

$$x = -2 \qquad\qquad x = \frac{10}{9}$$

Solution set: $\{-2\}$ $\quad \dfrac{10}{9}$ does not check.

51.

$$\left(\sqrt{x+2} + \sqrt{x-1}\right)^2 = \left(\sqrt{4x+1}\right)^2$$

$$(x+2) + 2\sqrt{(x+2)(x-1)} + (x-1) = 4x + 1$$

$$1 + 2x + 2\sqrt{x^2 + x - 2} = 4x + 1$$

$$2\sqrt{x^2 + x - 2} = 2x$$

$$\left(\sqrt{x^2 + x - 2}\right)^2 = (x)^2 \quad \text{Divide by 2}$$

$$x^2 + x - 2 = x^2$$

$$x - 2 = 0$$

$$x = 2$$

Solution set: $\{2\}$

53.

$$\left(\sqrt{2x+2} - \sqrt{5x+1}\right)^2 = \left(\sqrt{-4x}\right)^2$$

$$(2x+2) - 2\sqrt{(2x+2)(5x+1)}$$
$$+ (5x+1) = -4x$$

$$7x + 3 - 2\sqrt{10x^2 + 12x + 2} = -4x$$

$$-2\sqrt{10x^2 + 12x + 2} = -11x - 3$$

$$\left(2\sqrt{10x^2 + 12x + 2}\right)^2 = (11x + 3)^2$$

$$\text{Multiply by} -1$$

$$4\left(10x^2 + 12x + 2\right) = 121x^2 + 66x + 9$$

$$40x^2 + 48x + 8 = 121x^2 + 66x + 9$$

$$0 = 81x^2 + 18x + 1$$

$$0 = (9x + 1)(9x + 1)$$

$$9x + 1 = 0$$

$$x = -\frac{1}{9}$$

Solution set: $\left\{ -\dfrac{1}{9} \right\}$

55.

$$\left(\sqrt{x + \sqrt{x+11}}\right)^2 = (1)^2$$

$$x + \sqrt{x+11} = 1$$

$$\left(\sqrt{x+11}\right)^2 = (1-x)^2$$

$$x + 11 = 1 - 2x + x^2$$

$$0 = x^2 - 3x - 10$$

$$0 = (x-5)(x+2)$$

$$x - 5 = 0 \quad \text{or} \quad x + 2 = 0$$

$$x = 5 \qquad\qquad x = -2$$

Solution set: $\{-2\}$ $\quad 5$ does not check.

57.

$$\left(\sqrt{\sqrt{2x+3} - 3x}\right)^2 = (2)^2$$

$$\sqrt{2x+3} - 3x = 4$$

$$\left(\sqrt{2x+3}\right)^2 = (3x + 4)^2$$

$$2x + 3 = 9x^2 + 24x + 16$$

$$0 = 9x^2 + 22x + 13$$

$$0 = (x+1)(9x+13)$$

$$x + 1 = 0 \quad \text{or} \quad 9x + 13 = 0$$

$$x = -1 \qquad\qquad x = -\frac{13}{9}$$

Solution set: $\{-1\}$ $\quad -\dfrac{13}{9}$ does not check.

59.

$$\left(\sqrt[3]{2x + \sqrt{x+1}}\right)^3 = (2)^3$$

$$2x + \sqrt{x+1} = 8$$

$$\left(\sqrt{x+1}\right)^2 = (8 - 2x)^2$$

$$x + 1 = 64 - 32x + 4x^2$$

$$0 = 4x^2 - 33x + 63$$

$$0 = (x-3)(4x-21)$$

$$x - 3 = 0 \quad \text{or} \quad 4x - 21 = 0$$

$$x = 3 \qquad\qquad x = \frac{21}{4}$$

Solution set: $\{3\}$ $\quad \dfrac{21}{4}$ does not check.

61.

$$\frac{x}{\sqrt{2x-5}} = 3$$

$$\frac{\sqrt{2x-5}}{1} \cdot \frac{x}{\sqrt{2x-5}} = 3 \cdot \sqrt{2x-5}$$

$$(x)^2 = \left(3\sqrt{2x-5}\right)^2$$

$$x^2 = 9(2x-5)$$

$$x^2 = 18x - 45$$

$$x^2 - 18x + 45 = 0$$

$$(x-15)(x-3) = 0$$

$$x - 15 = 0 \quad \text{or} \quad x - 3 = 0$$

$$x = 15 \qquad x = 3$$

Solution set: $\{3, 15\}$

63.

$$\frac{1}{\sqrt{x+2}} - \sqrt{x+1} = \sqrt{x+2}$$

$$\frac{\sqrt{x+2}}{1}\left(\frac{1}{\sqrt{x+2}} - \sqrt{x+1}\right) = \sqrt{x+2}\sqrt{x+2}$$

$$1 - \sqrt{(x+2)(x+1)} = x + 2$$

$$\left(-\sqrt{x^2+3x+2}\right)^2 = (x+1)^2$$

$$x^2 + 3x + 2 = x^2 + 2x + 1$$

$$x = -1$$

Solution set: $\{-1\}$

Exercises 5.7

1.

$$\sqrt{-81} = \sqrt{81}\sqrt{-1} = 9i$$

3.

$$-\sqrt{-20} = -\sqrt{20} \cdot \sqrt{-1} = -2\sqrt{5}i$$

5.

$$3\sqrt{-48} = 3 \cdot \sqrt{48} \cdot \sqrt{-1} = 3 \cdot 4\sqrt{3}i = 12\sqrt{3}i$$

7.

$$\sqrt{-9} + \sqrt{-16} = \sqrt{9}\sqrt{-1} + \sqrt{16}\sqrt{-1} = 3i + 4i = 7i$$

9.

$$2\sqrt{-25} - 3\sqrt{-49} = 2\sqrt{25}\sqrt{-1} - 3\sqrt{49}\sqrt{-1}$$

$$= 2 \cdot 5i - 3 \cdot 7i = 10i - 21i = -11i$$

11.

$$\sqrt{-12} + \sqrt{-27} = \sqrt{12}\sqrt{-1} + \sqrt{27}\sqrt{-1}$$

$$= 2\sqrt{3}i + 3\sqrt{3}i = 5\sqrt{3}i$$

13.

$$\sqrt{81} + \sqrt{-9} = \sqrt{81} + \sqrt{9}\sqrt{-1} = 9 + 3i$$

15.

$$\sqrt{48} - \sqrt{-98} = \sqrt{48} - \sqrt{98}\sqrt{-1} = 4\sqrt{3} - 7\sqrt{2}i$$

17.

$$3\sqrt{18} - 2\sqrt{-24} = 3\sqrt{18} - 2\sqrt{24}\sqrt{-1}$$

$$= 3 \cdot 3\sqrt{2} - 2 \cdot 2\sqrt{6}i = 9\sqrt{2} - 4\sqrt{6}i$$

19.

$$\sqrt{-16} \cdot \sqrt{-9} = \sqrt{16}\sqrt{-1} \cdot \sqrt{9}\sqrt{-1}$$

$$= 4i \cdot 3i = 12i^2 = -12 \quad \text{Note: } i^2 = -1$$

21.

$$\sqrt{-13} \cdot \sqrt{-13} = \sqrt{13}\sqrt{-1} \cdot \sqrt{13}\sqrt{-1}$$

$$= \sqrt{13} \cdot \sqrt{13}i \cdot i = 13i^2 = -13$$

Note: $i^2 = -1$

23.

$$-\sqrt{-3} \cdot \sqrt{-3} = -\sqrt{3}\sqrt{-1} \cdot \sqrt{3} \cdot \sqrt{-1}$$

$$= -\sqrt{3} \cdot \sqrt{3} \cdot i \cdot i = -3i^2 = 3$$

Note: $i^2 = -1$

25.

$$-\sqrt{5} \cdot \sqrt{-5} = -\sqrt{5} \cdot \sqrt{5}\sqrt{-1} = -5i$$

27.

$$3\sqrt{-4}\sqrt{8} = 3\sqrt{4} \cdot \sqrt{-1} \cdot \sqrt{8} = 3 \cdot 2 \cdot i \cdot 2\sqrt{2} = 12\sqrt{2}i$$

29.

$$\frac{\sqrt{-81}}{\sqrt{-4}} = \frac{\sqrt{81}\sqrt{-1}}{\sqrt{4}\sqrt{-1}} = \frac{9i}{2i} = \frac{9}{2}$$

31.

$$\frac{-\sqrt{-25}}{\sqrt{-1}} = \frac{-\sqrt{25}\sqrt{-1}}{\sqrt{-1}} = \frac{-5i}{i} = -5$$

33.

$$\frac{\sqrt{-4}\sqrt{-9}}{\sqrt{-16}} = \frac{\sqrt{4}\sqrt{-1}\sqrt{9}\sqrt{-1}}{\sqrt{16}\sqrt{-1}} = \frac{2\cdot i \cdot 3 \cdot i}{4\cdot i} = \frac{3}{2}i$$

35.

$$(2+3i)+(5-6i) = 7-3i$$

37.

$$\left(\frac{1}{4}+\frac{1}{8}i\right)+\left(-\frac{1}{3}+\frac{1}{6}i\right) = \frac{1}{4}-\frac{1}{3}+\frac{1}{8}i+\frac{1}{6}i$$

$$= \frac{3}{12}-\frac{4}{12}+\frac{3}{24}i+\frac{4}{24}i$$

$$= -\frac{1}{12}+\frac{7}{24}i$$

39.

$$(-5-2i)-(3+i) = -5-2i-3-i = -8-3i$$

41.

$$\left(-\frac{2}{3}+\frac{1}{9}i\right)-\left(\frac{3}{4}-\frac{1}{4}i\right) = -\frac{2}{3}-\frac{3}{4}+\frac{1}{9}i+\frac{1}{4}i$$

$$= -\frac{8}{12}-\frac{9}{12}+\frac{4}{36}i+\frac{9}{36}i$$

$$= -\frac{17}{12}+\frac{13}{36}i$$

43.

$$7(3+9i) = 21+63i$$

45.

$$2i(4-7i) = 8i-14i^2 = 8i+14 = 14+8i$$

47.

$$(6+i)(3-4i) = 18-21i-4i^2 = 18-21i+4 = 22-21i$$

49.

$$(5-3i)(2-4i) = 10-26i+12i^2$$

$$= 10-26i-12 = -2-26i$$

51.

$$(2-5i)(2+5i) = 4-25i^2 = 4+25 = 29$$

53.

$$\left(\sqrt{3}+i\right)\left(\sqrt{3}-i\right) = \left(\sqrt{3}\right)^2 - i^2 = 3+1 = 4$$

55.

$$(2+3i)^2 = 4+12i+9i^2 = 4+12i-9 = -5+12i$$

57.

$$\left(\sqrt{3}+i\right)^2 = \left(\sqrt{3}\right)^2 + 2\sqrt{3}i + i^2$$

$$= 3+2\sqrt{3}i-1 = 2+2\sqrt{3}i$$

59.

$$\frac{5}{3i}\cdot\frac{i}{i} = \frac{5i}{3i^2} = \frac{5i}{-3} = -\frac{5}{3}i$$

61.

$$\frac{5+3i}{i}\cdot\frac{i}{i} = \frac{5i+3i^2}{i^2} = \frac{5i-3}{-1} = 3-5i$$

63.

$$\frac{4+i}{2i}\cdot\frac{i}{i} = \frac{4i+i^2}{2i^2} = \frac{4i-1}{-2} = \frac{4i}{-2}+\frac{-1}{-2} = \frac{1}{2}-2i$$

65.

$$\frac{1}{4+2i}\cdot\frac{4-2i}{4-2i} = \frac{4-2i}{16-4i^2}$$

$$= \frac{4-2i}{16+4} = \frac{4-2i}{20} = \frac{4}{20}-\frac{2}{20}i = \frac{1}{5}-\frac{1}{10}i$$

67.

$$\frac{3}{2-3i}\cdot\frac{2+3i}{2+3i} = \frac{6+9i}{4-9i^2} = \frac{6+9i}{4+9} = \frac{6+9i}{13} = \frac{6}{13}+\frac{9}{13}i$$

69.

$$\frac{2i}{1+3i}\cdot\frac{1-3i}{1-3i} = \frac{2i-6i^2}{1-9i^2}$$

$$= \frac{2i+6}{1+9} = \frac{2i+6}{10} = \frac{6}{10}+\frac{2}{10}i = \frac{3}{5}+\frac{1}{5}i$$

71.

$$\frac{-4i}{-3-5i}\cdot\frac{-3+5i}{-3+5i} = \frac{12i-20i^2}{9-25i^2}$$

$$= \frac{12i+20}{9+25}$$

$$= \frac{12i+20}{34} = \frac{20}{34}+\frac{12}{34}i = \frac{10}{17}+\frac{6}{17}i$$

73.

$$\frac{3-4i}{3+i} \cdot \frac{3-i}{3-i} = \frac{9-15i+4i^2}{9-i^2}$$

$$= \frac{9-15i-4}{9+1}$$

$$= \frac{5-15i}{10} = \frac{5}{10} - \frac{15}{10}i = \frac{1}{2} - \frac{3}{2}i$$

75.

$$\frac{9-17i}{2-i} \cdot \frac{2+i}{2+i} = \frac{18-25i-17i^2}{4-i^2}$$

$$= \frac{18-25i+17}{4+1}$$

$$= \frac{35-25i}{5} = \frac{35}{5} - \frac{25}{5}i = 7-5i$$

77.

$$\frac{-2-5i}{1-4i} \cdot \frac{1+4i}{1+4i} = \frac{-2-13i-20i^2}{1-16i^2}$$

$$= \frac{-2-13i+20}{1+16} = \frac{18-13i}{17} = \frac{18}{17} - \frac{13}{17}i$$

79.

$$\frac{-3+2i}{-5-4i} \cdot \frac{-5+4i}{-5+4i} = \frac{15-22i+8i^2}{25-16i^2}$$

$$= \frac{15-22i-8}{25+16} = \frac{7-22i}{41} = \frac{7}{41} - \frac{22}{41}i$$

81.

$$i^{91} = i \cdot i^{90} = i\left(i^2\right)^{45} = i(-1) = -i$$

83.

$$-i^{27} = -i \cdot i^{26} = -i\left(i^2\right)^{13} = -i(-1) = i$$

85.

$$-6i^{65} = -6i \cdot i^{64} = -6i \cdot \left(i^2\right)^{32} = -6i(1) = -6i$$

87.

$$2i^{22} = 2\left(i^2\right)^{11} = 2(-1) = -2$$

89.

$$i^{-15} = \frac{1}{i^{15}}$$

$$= \frac{1}{i \cdot i^{14}}$$

$$= \frac{1}{i \cdot \left(i^2\right)^7} = \frac{1}{i(-1)} = -\frac{1}{i} \cdot \frac{i}{i} = -\frac{i}{i^2} = -\frac{i}{-1} = i$$

91.

$$i^{-37} = \frac{1}{i^{37}} = \frac{1}{i \cdot i^{36}} = \frac{1}{i\left(i^2\right)^{18}} = \frac{1}{i} \cdot \frac{i}{i} = \frac{i}{i^2} = \frac{i}{-1} = -i$$

93.

$$-i^{-56} = -\frac{1}{i^{56}} = -\frac{1}{\left(i^2\right)^{28}} = -\frac{1}{1} = -1$$

95.

$$2i^{-3} = 2 \cdot \frac{1}{i^3}$$

$$= 2 \cdot \frac{1}{i \cdot i^2} = 2 \cdot \frac{1}{-i} = -\frac{2}{i} \cdot \frac{i}{i} = -\frac{2i}{i^2} = -\frac{2i}{-1} = 2i$$

97.

$$3i^{-10} = 3 \cdot \frac{1}{i^{10}} = 3 \cdot \frac{1}{\left(i^2\right)^5} = 3 \cdot \frac{1}{-1} = -3$$

Review Exercises

1.-15.
The exercises do not require worked - out solutions.

17.

$$-27^{2/3} = -\left(27^{1/3}\right)^2 = -(3)^2 = -9$$

19.

$$27^{4/3} = \left(27^{1/3}\right)^4 = (3)^4 = 81$$

21.

$$32^{2/5} = \left(32^{1/5}\right)^2 = (2)^2 = 4$$

23.
$$-4^{3/2} = -\left[(4)^{1/2}\right]^3 = -(2)^3 = -8$$

25.
$$8^{-4/3} = \left(8^{1/3}\right)^{-4} = (2)^{-4} = \frac{1}{2^4} = \frac{1}{16}$$

27.
$$-64^{-2/3} = -\left(64^{1/3}\right)^{-2} = -(4)^{-2} = -\frac{1}{4^2} = -\frac{1}{16}$$

29.
$$\left(\frac{1}{4}\right)^{-1/2} = \left(\frac{4}{1}\right)^{1/2} = 2$$

31.
$$\left(-\frac{8}{27}\right)^{-2/3} = \left(-\frac{27}{8}\right)^{2/3} = \left[\left(-\frac{27}{8}\right)^{1/3}\right]^2 = \left(-\frac{3}{2}\right)^2 = \frac{9}{4}$$

33.
$$\left(-\frac{1}{32}\right)^{-3/5} = \left[\left(-\frac{32}{1}\right)^{1/5}\right]^3 = (-2)^3 = -8$$

35.
$$8^{7/12} \cdot 8^{-1/4} = 8^{7/12 - 1/4} = 8^{4/12} = 8^{1/3} = 2$$

37.
$$\left(25^{-1/4} \cdot 25^{2/3}\right)^6 = \left(25^{-1/4 + 2/3}\right)^6$$
$$= \left(25^{5/12}\right)^6 = 25^{5/2} = 5^5 = 3125$$

39.
$$\frac{64^{-1} \cdot 64^{7/15}}{64^{-1/5}} = 64^{-15/15 + 7/15 - (-3/15)} = 64^{-5/15} = 64^{-1/3} = \frac{1}{4}$$

41.
$$\left(8^{1/12} \cdot 25^{-1/8}\right)^4 = 8^{4/12} \cdot 25^{-4/8} = 8^{1/3} \cdot 25^{-1/2} = \frac{8^{1/3}}{25^{1/2}} = \frac{2}{5}$$

43.
$$\left(9x^{2/7}\right)^{1/2} = 9^{1/2} x^{1/7} = 3x^{1/7}$$

45.
$$\frac{x^{3/8}}{x^{2/3}} = x^{3/8 - 2/3} = x^{-7/24} = \frac{1}{x^{7/24}}$$

47.
$$\frac{\left(x^{-1} \cdot x^{3/4}\right)^{-2/3}}{x^{-1/12}} = \frac{x^{2/3} \cdot x^{-1/2}}{x^{-1/12}} = x^{2/3 - 1/2 + 1/12} = x^{1/4}$$

49.
$$\frac{27^{4/9} x^{-1/3} y^{5/6}}{27^{1/9} x^{-2/7} y^{-3/4}} = 27^{4/9 - 1/9} x^{-1/3 - (-2/7)} y^{5/6 - (-3/4)}$$
$$= 27^{1/3} x^{-1/21} y^{19/12} = \frac{3y^{19/12}}{x^{1/21}}$$

51.
$$\left(\frac{2x^{-3/7}}{8y^{-1/2}}\right)^{3/2} = \left(\frac{x^{-3/7}}{4y^{-1/2}}\right)^{3/2} = \frac{x^{-9/14}}{4^{3/2} y^{-3/4}} = \frac{y^{3/4}}{8x^{9/14}}$$

53.
$$\left(\frac{8x^2}{y^{-3}}\right)^{1/3}\left(\frac{9x^4}{y^{-1}}\right)^{-3/2} = \frac{8^{1/3} x^{2/3}}{y^{-1}} \cdot \frac{9^{-3/2} x^{-6}}{y^{3/2}}$$
$$= \frac{8^{1/3} \cdot 9^{-3/2} x^{-16/3}}{y^{1/2}}$$
$$= \frac{8^{1/3}}{9^{3/2} x^{16/3} y^{1/2}} = \frac{2}{27x^{16/3} y^{1/2}}$$

55.
$$3x^{-1/6}\left(2x^{1/2} + x^{1/6}\right) = 6x^{2/6} + 3x^0 = 6x^{1/3} + 3$$

57.
$$\left(x^{1/3} + 3\right)\left(x^{1/3} - 3\right) = \left(x^{1/3}\right)^2 - (3)^2 = x^{2/3} - 9$$

59.

$$\left(x^{\frac{1}{3}} + 3\right)\left(x^{\frac{2}{3}} - 3x^{\frac{1}{3}} + 9\right)$$

$$= x^{\frac{1}{3} + \frac{2}{3}} - 3x^{\frac{1}{3} + \frac{1}{3}} + 9x^{\frac{1}{3}} + 3x^{\frac{2}{3}} - 9x^{\frac{1}{3}} + 27$$

$$= x - 3x^{\frac{2}{3}} + 9x^{\frac{1}{3}} + 3x^{\frac{2}{3}} - 9x^{\frac{1}{3}} + 27$$

$$= x + 27$$

61.

$$\left(x^{2m}\right)^{\frac{3}{2}} = x^{3m}$$

63.

$$\frac{\left(x^{\frac{m}{4}} \cdot x^{-m}\right)^3}{x^{\frac{m}{2}}} = \frac{\left(x^{-\frac{3m}{4}}\right)^3}{x^{\frac{m}{2}}}$$

$$= \frac{x^{-\frac{9m}{4}}}{x^{\frac{m}{2}}} = \frac{1}{x^{\frac{m}{2} - \left(-\frac{9m}{4}\right)}} = \frac{1}{x^{\frac{11m}{4}}}$$

65.

$$\left(5^{\frac{1}{3}} x^{\frac{m}{2}}\right)^{\frac{3}{2}} \left(5^{\frac{1}{2}} x^{\frac{m}{4}}\right)^3 = 5^{\frac{1}{2}} x^{\frac{3m}{4}} \cdot 5^{\frac{3}{2}} x^{\frac{3m}{4}}$$

$$= 5^{\frac{1}{2}} \cdot 5^{\frac{3}{2}} \cdot x^{\frac{3m}{4}} \cdot x^{\frac{3m}{4}}$$

$$= 5^2 \cdot x^{\frac{6m}{4}}$$

$$= 25 x^{\frac{3m}{2}}$$

67.

$$15(6x-1)^{\frac{3}{2}}(x+3)^{\frac{5}{3}} + \frac{5}{3}(6x-1)^{\frac{5}{2}}(x+3)^{\frac{2}{3}}$$

$$\text{GCF} = \frac{5}{3}(6x-1)^{\frac{3}{2}}(x+3)^{\frac{2}{3}}$$

$$= \frac{5}{3}(6x-1)^{\frac{3}{2}}(x+3)^{\frac{2}{3}}[9(x+3) + (6x-1)]$$

$$= \frac{5}{3}(6x-1)^{\frac{3}{2}}(x+3)^{\frac{2}{3}}(9x + 27 + 6x - 1)$$

$$= \frac{5}{3}(6x-1)^{\frac{3}{2}}(x+3)^{\frac{2}{3}}(15x + 26)$$

69.-103.
The exercises do not require worked - out solutions.

105.

$$\sqrt[3]{y} \cdot \sqrt[4]{y^3} = y^{\frac{1}{3}} \cdot y^{\frac{3}{4}} = y^{\frac{4}{12} + \frac{9}{12}} = y^{\frac{13}{12}}$$

107.

$$\frac{\sqrt[3]{x}}{\sqrt[6]{x}} = \frac{x^{\frac{1}{3}}}{x^{\frac{1}{6}}} = x^{\frac{1}{3} - \frac{1}{6}} = x^{\frac{1}{6}}$$

109.

$$\sqrt[3]{\sqrt{x}} = \left(x^{\frac{1}{2}}\right)^{\frac{1}{3}} = x^{\frac{1}{6}}$$

111.-113.
The exercises do not require worked - out solutions.

115.

$$\sqrt{x^2 + 2x + 1} = \sqrt{(x+1)^2} = x + 1$$

117.

$$\sqrt{6} \cdot \sqrt{5} = \sqrt{30}$$

119.

$$\sqrt{7} \cdot \sqrt{7} = \sqrt{49} = 7$$

121.

$$\sqrt[3]{6} \cdot \sqrt[3]{18} = \sqrt[3]{108} = \sqrt[3]{27 \cdot 4} = \sqrt[3]{27} \cdot \sqrt[3]{4} = 3\sqrt[3]{4}$$

123.

$$\sqrt{75} = \sqrt{25 \cdot 3} = \sqrt{25} \cdot \sqrt{3} = 5\sqrt{3}$$

125.

$$\sqrt{99} = \sqrt{9 \cdot 11} = \sqrt{9}\sqrt{11} = 3\sqrt{11}$$

127.

$$\sqrt[3]{81} = \sqrt[3]{27 \cdot 3} = \sqrt[3]{27}\sqrt[3]{3} = 3\sqrt[3]{3}$$

129.

$$\sqrt[4]{48} = \sqrt[4]{16 \cdot 3} = \sqrt[4]{16}\sqrt[4]{3} = 2\sqrt[4]{3}$$

131.

$$\sqrt[4]{2x^4 y^9} = \sqrt[4]{x^4 y^8 \cdot 2y} = \sqrt[4]{x^4 y^8}\sqrt[4]{2y} = xy^2\sqrt[4]{2y}$$

133.

$$\sqrt{24x^4 y^{12}} = \sqrt{4x^4 y^{12} \cdot 6} = \sqrt{4x^4 y^{12}}\sqrt{6} = 2x^2 y^6 \sqrt{6}$$

135.

$$\sqrt[3]{24x^4y^{12}} = \sqrt[3]{8x^3y^{12} \cdot 3x}$$
$$= \sqrt[3]{8x^3y^{12}}\,\sqrt[3]{3x} = 2xy^4\sqrt[3]{3x}$$

137.

$$\frac{2}{\sqrt{x}} \cdot \frac{\sqrt{x}}{\sqrt{x}} = \frac{2\sqrt{x}}{x}$$

139.

$$\frac{5}{\sqrt{3y}} \cdot \frac{\sqrt{3y}}{\sqrt{3y}} = \frac{5\sqrt{3y}}{3y}$$

141.

$$\sqrt{\frac{6x^5}{y^3} \cdot \frac{y}{y}} = \sqrt{\frac{6x^5y}{y^4}} = \frac{\sqrt{x^4 \cdot 6xy}}{\sqrt{y^4}} = \frac{x^2\sqrt{6xy}}{y^2}$$

143.

$$\sqrt{\frac{5x^2}{28y^3} \cdot \frac{7y}{7y}} = \sqrt{\frac{35x^2y}{196y^4}} = \frac{\sqrt{35x^2y}}{\sqrt{196y^4}} = \frac{x\sqrt{35y}}{14y^2}$$

145.

$$\frac{x}{\sqrt[3]{2}} \cdot \frac{\sqrt[3]{4}}{\sqrt[3]{4}} = \frac{x\sqrt[3]{4}}{\sqrt[3]{8}} = \frac{x\sqrt[3]{4}}{2}$$

147.

$$\sqrt[3]{\frac{2}{5} \cdot \frac{5^2}{5^2}} = \sqrt[3]{\frac{50}{5^3}} = \frac{\sqrt[3]{50}}{\sqrt[3]{5^3}} = \frac{\sqrt[3]{50}}{5}$$

149.

$$\sqrt[3]{\frac{7x^6}{40} \cdot \frac{25}{25}} = \sqrt[3]{\frac{175x^6}{1000}} = \frac{\sqrt[3]{x^6 \cdot 175}}{\sqrt[3]{1000}} = \frac{x^2\sqrt[3]{175}}{10}$$

151.

$$\frac{3ax}{\sqrt[4]{y}} \cdot \frac{\sqrt[4]{y^3}}{\sqrt[4]{y^3}} = \frac{3ax\sqrt[4]{y^3}}{\sqrt[4]{y^4}} = \frac{3ax\sqrt[4]{y^3}}{y}$$

153.

$$\sqrt[4]{x^6} = \sqrt[4]{x^4 \cdot x^2}$$
$$= \sqrt[4]{x^4}\,\sqrt[4]{x^2} = x\sqrt[4]{x^2} = x \cdot x^{2/4} = x \cdot x^{1/2} = x\sqrt{x}$$

155.

$$\sqrt[6]{x^4y^9} = \sqrt[6]{y^6 \cdot x^4y^3}$$
$$= \sqrt[6]{y^6}\,\sqrt[6]{x^4y^3}$$
$$= y\sqrt[6]{x^4y^3}$$
$$= y \cdot x^{4/6}y^{3/6} = y \cdot x^{2/3} \cdot y^{1/2} = y\sqrt[3]{x^2}\,\sqrt{y}$$

157.

$$15\sqrt[3]{6} + \sqrt[3]{6} - 20\sqrt[3]{6} = (15 + 1 - 20)\sqrt[3]{6} = -4\sqrt[3]{6}$$

159.

$$9x^2\sqrt{2y} + x^2\sqrt{2y} = \left(9x^2 + x^2\right)\sqrt{2y} = 10x^2\sqrt{2y}$$

161.

$$\sqrt{18} - \sqrt{98} + 5\sqrt{8} = 3\sqrt{2} - 7\sqrt{2} + 5 \cdot 2\sqrt{2}$$
$$= 3\sqrt{2} - 7\sqrt{2} + 10\sqrt{2}$$
$$= (3 - 7 + 10)\sqrt{2}$$
$$= 6\sqrt{2}$$

163.

$$\sqrt[3]{40} + 3\sqrt[3]{72} = 2\sqrt[3]{5} + 3 \cdot 2\sqrt[3]{9} = 2\sqrt[3]{5} + 6\sqrt[3]{9}$$

165.

$$\sqrt{16x^3} - x\sqrt{8x} - 2x\sqrt{9x}$$
$$= 4x\sqrt{x} - 2x\sqrt{2x} - 2 \cdot 3x\sqrt{x}$$
$$= 4x\sqrt{x} - 2x\sqrt{2x} - 6x\sqrt{x}$$
$$= -2x\sqrt{x} - 2x\sqrt{2x}$$

167.

$$4\sqrt[3]{2p^4} + p\sqrt[3]{54p} - \sqrt[3]{16p}$$
$$= 4\sqrt[3]{p^3 \cdot 2p} + p\sqrt[3]{27 \cdot 2p} - \sqrt[3]{8 \cdot 2p}$$
$$= 4p\sqrt[3]{2p} + 3p\sqrt[3]{2p} - 2\sqrt[3]{2p}$$
$$= 7p\sqrt[3]{2p} - 2\sqrt[3]{2p}$$

169.

$$\sqrt{20x^5y^6} - 7xy\sqrt{45x^3y^4} + y^2\sqrt{40x^4y}$$
$$= \sqrt{4x^4y^6 \cdot 5x} - 7xy\sqrt{9x^2y^4 \cdot 5x} + y^2\sqrt{4x^4 \cdot 10y}$$
$$= 2x^2y^3\sqrt{5x} - 7xy \cdot 3xy^2\sqrt{5x} + y^2 \cdot 2x^2\sqrt{10y}$$

#169. continued

$$= 2x^2y^3\sqrt{5x} - 21x^2y^3\sqrt{5x} + 2x^2y^2\sqrt{10y}$$

$$= -19x^2y^3\sqrt{5x} + 2x^2y^2\sqrt{10y}$$

171.

$$7x\sqrt[4]{16x^9y^4} - 3y\sqrt[4]{x^{13}} = 7x\sqrt[4]{16x^8y^4 \cdot x} - 3y\sqrt[4]{x^{12} \cdot x}$$

$$= 7x \cdot 2x^2y\sqrt[4]{x} - 3y \cdot x^3\sqrt[4]{x}$$

$$= 14x^3y\sqrt[4]{x} - 3x^3y\sqrt[4]{x}$$

$$= 11x^3y\sqrt[4]{x}$$

173.

$$\frac{2}{\sqrt[3]{5}} - \sqrt[3]{200} = \frac{2\sqrt[3]{25}}{5} - 2\sqrt[3]{25}$$

$$= \frac{2\sqrt[3]{25}}{5} - \frac{10\sqrt[3]{25}}{5} = -\frac{8\sqrt[3]{25}}{5}$$

175.

$$\sqrt{\frac{2}{5}} + \sqrt{40} - \sqrt{\frac{5}{2}} = \frac{\sqrt{10}}{5} + 2\sqrt{10} - \frac{\sqrt{10}}{2}$$

$$= \frac{2\sqrt{10}}{10} + \frac{20\sqrt{10}}{10} - \frac{5\sqrt{10}}{10}$$

$$= \frac{17\sqrt{10}}{10}$$

177.

$$\sqrt[3]{\frac{x}{2}} - \sqrt[3]{108x} = \frac{\sqrt[3]{4x}}{2} - 3\sqrt[3]{4x}$$

$$= \frac{\sqrt[3]{4x}}{2} - \frac{6\sqrt[3]{4x}}{2} = -\frac{5\sqrt[3]{4x}}{2}$$

179.

$$3\sqrt{2}\left(5 + \sqrt{10} - 3\sqrt{2}\right) = 15\sqrt{2} + 3\sqrt{20} - 9\sqrt{4}$$

$$= 15\sqrt{2} + 3 \cdot 2\sqrt{5} - 9 \cdot 2$$

$$= 15\sqrt{2} + 6\sqrt{5} - 18$$

181.

$$\overset{\text{F} \qquad \text{O} \qquad \text{I} \qquad \text{L}}{\left(\sqrt{3} + 5\right)\left(\sqrt{7} - 1\right) = \sqrt{21} - \sqrt{3} + 5\sqrt{7} - 5}$$

183.

$$\overset{\text{F} \qquad \text{O} \qquad \text{I} \qquad \text{L}}{\left(3\sqrt{5} - 2\right)\left(5\sqrt{5} + 6\right) = 15 \cdot 5 + 18\sqrt{5} - 10\sqrt{5} - 12}$$

$$= 75 + 8\sqrt{5} - 12$$

$$= 63 + 8\sqrt{5}$$

185.

$$\overset{\text{F} \qquad \text{O} \qquad \text{I} \qquad \text{L}}{\left(3\sqrt{x} - \sqrt{y}\right)\left(3\sqrt{x} + 4\sqrt{y}\right) = 9x + 12\sqrt{xy} - 3\sqrt{xy} - 4y}$$

$$= 9x + 9\sqrt{xy} - 4y$$

187.

$$\left(2\sqrt{y} + 1\right)^2 = \left(2\sqrt{y}\right)^2 + 2\left(2\sqrt{y}\right) + (1)^2 = 4y + 4\sqrt{y} + 1$$

189.

$$\left(\sqrt{x} - 2\right)\left(\sqrt{x} + 2\right) = \left(\sqrt{x}\right)^2 - (2)^2 = x - 4$$

191.

$$\left(\sqrt[3]{5} + 1\right)\left(\sqrt[3]{25} - \sqrt[3]{5} + 1\right)$$

$$= \sqrt[3]{125} - \sqrt[3]{25} + \sqrt[3]{5} + \sqrt[3]{25} - \sqrt[3]{5} + 1$$

$$= 5 - \sqrt[3]{25} + \sqrt[3]{25} + \sqrt[3]{5} - \sqrt[3]{5} + 1 = 6$$

193.

$$\frac{\sqrt{3}}{\sqrt{7} - 1} \cdot \frac{\sqrt{7} + 1}{\sqrt{7} + 1} = \frac{\sqrt{21} + \sqrt{3}}{7 - 1} = \frac{\sqrt{21} + \sqrt{3}}{6}$$

195.

$$\frac{\sqrt{6}}{\sqrt{2} - \sqrt{3}} \cdot \frac{\sqrt{2} + \sqrt{3}}{\sqrt{2} + \sqrt{3}} = \frac{\sqrt{12} + \sqrt{18}}{2 - 3}$$

$$= \frac{2\sqrt{3} + 3\sqrt{2}}{-1}$$

$$= -\left(2\sqrt{3} + 3\sqrt{2}\right) = -2\sqrt{3} - 3\sqrt{2}$$

197.

$$\frac{\sqrt{3x} + \sqrt{y}}{\sqrt{5x} + \sqrt{2y}} \cdot \frac{\sqrt{5x} - \sqrt{2y}}{\sqrt{5x} - \sqrt{2y}}$$

$$= \frac{\sqrt{15x^2} - \sqrt{6xy} + \sqrt{5xy} - \sqrt{2y^2}}{\left(\sqrt{5x}\right)^2 - \left(\sqrt{2y}\right)^2}$$

#197. continued

$$= \frac{x\sqrt{15} - \sqrt{6xy} + \sqrt{5xy} - y\sqrt{2}}{5x - 2y}$$

199.

$$\frac{4\sqrt{3x} - 5\sqrt{2y}}{3\sqrt{6x} - 6\sqrt{8y}} \cdot \frac{3\sqrt{6x} + 6\sqrt{8y}}{3\sqrt{6x} + 6\sqrt{8y}}$$

$$= \frac{12\sqrt{18x^2} + 24\sqrt{24xy} - 15\sqrt{12xy} - 30\sqrt{16y^2}}{\left(3\sqrt{6x}\right)^2 - \left(6\sqrt{8y}\right)^2}$$

$$= \frac{36x\sqrt{2} + 48\sqrt{6xy} - 30\sqrt{3xy} - 120y}{9 \cdot 6x - 36 \cdot 8y}$$

$$= \frac{36x\sqrt{2} + 48\sqrt{6xy} - 30\sqrt{3xy} - 120y}{54x - 288y}$$

$$= \frac{6\left(6x\sqrt{2} + 8\sqrt{6xy} - 5\sqrt{3xy} - 20y\right)}{6(9x - 48y)}$$

$$= \frac{6x\sqrt{2} + 8\sqrt{6xy} - 5\sqrt{3xy} - 20y}{9x - 48y}$$

201.

$$\frac{4\sqrt{3x} - 7\sqrt{2y}}{\sqrt{5} - 2} \cdot \frac{\sqrt{5} + 2}{\sqrt{5} + 2}$$

$$= \frac{4\sqrt{15x} + 8\sqrt{3x} - 7\sqrt{10y} - 14\sqrt{2y}}{5 - 4}$$

$$= 4\sqrt{15x} + 8\sqrt{3x} - 7\sqrt{10y} - 14\sqrt{2y}$$

203.

$$\sqrt{4x + 5} - 3 = 0$$

$$\left(\sqrt{4x + 5}\right)^2 = (3)^2$$

$$4x + 5 = 9$$

$$4x = 4$$

$$x = 1$$

Solution set: $\{1\}$

205.

$$\left(\sqrt[4]{x^2 - 6x}\right)^4 = (2)^4$$

$$x^2 - 6x = 16$$

$$x^2 - 6x - 16 = 0$$

$$(x + 2)(x - 8) = 0$$

$$x + 2 = 0 \quad \text{or} \quad x - 8 = 0$$

$$x = -2 \qquad x = 8$$

#205. continued
Solution set: $\{-2, 8\}$

207.

$$\left(\sqrt[3]{2 - x^2}\right)^3 = \left(\sqrt[3]{3x - 2}\right)^3$$

$$2 - x^2 = 3x - 2$$

$$0 = x^2 + 3x - 4$$

$$0 = (x + 4)(x - 1)$$

$$x + 4 = 0 \quad \text{or} \quad x - 1 = 0$$

$$x = -4 \qquad x = 1$$

Solution set: $\{-4, 1\}$

209.

$$\sqrt{x^2 - 3x + 1} + 2 = x$$

$$\left(\sqrt{x^2 - 3x + 1}\right)^2 = (x - 2)^2$$

$$x^2 - 3x + 1 = x^2 - 4x + 4$$

$$x = 3$$

Solution set: $\{3\}$

211.

$$\sqrt{13 - 3x} - 3 = 2x$$

$$\left(\sqrt{13 - 3x}\right)^2 = (2x + 3)^2$$

$$13 - 3x = 4x^2 + 12x + 9$$

$$0 = 4x^2 + 15x - 4$$

$$0 = (4x - 1)(x + 4)$$

$$4x - 1 = 0 \quad \text{or} \quad x + 4 = 0$$

$$x = \frac{1}{4} \qquad x = -4$$

Solution set: $\left\{\frac{1}{4}\right\}$ -4 does not check.

213.

$$\sqrt{x + 11} - \sqrt{2x - 1} = 1$$

$$\left(\sqrt{x + 11}\right)^2 = \left(\sqrt{2x - 1} + 1\right)^2$$

$$x + 11 = (2x - 1) + 2\sqrt{2x - 1} + 1$$

$$x + 11 = 2x + 2\sqrt{2x - 1}$$

$$(-x + 11)^2 = \left(2\sqrt{2x - 1}\right)^2$$

$$x^2 - 22x + 121 = 4(2x - 1)$$

$$x^2 - 22x + 121 = 8x - 4$$

$$x^2 - 30x + 125 = 0$$

$$(x - 5)(x - 25) = 0$$

#213. continued

$x - 5 = 0$ or $x - 25 = 0$

$x = 5$ $\qquad$ $x = 25$

Solution set: $\{5\}$ $\qquad$ 25 does not check.

215.

$$\left(\sqrt{5x-4} - \sqrt{x+5}\right)^2 = \left(\sqrt{x-3}\right)^2$$

$$(5x-4) - 2\sqrt{(5x-4)(x+5)}$$
$$+ (x+5) = x - 3$$

$$6x + 1 - 2\sqrt{5x^2 + 21x - 20} = x - 3$$

$$-2\sqrt{5x^2 + 21x - 20} = -5x - 4$$

$$\left(2\sqrt{5x^2 + 21x - 20}\right)^2 = (5x+4)^2$$

Multiply by -1

$$4\left(5x^2 + 21x - 20\right) = 25x^2 + 40x + 16$$

$$20x^2 + 84x - 80 = 25x^2 + 40x + 16$$

$$0 = 5x^2 - 44x + 96$$

$$0 = (x-4)(5x-24)$$

$x - 4 = 0$ or $5x - 24 = 0$

$x = 4$ $\qquad$ $x = \dfrac{24}{5}$

Solution set: $\left\{4, \dfrac{24}{5}\right\}$

217.

$$\frac{x}{\sqrt{2x+5}} = \frac{2}{1}$$

$$\left(2\sqrt{2x+5}\right)^2 = (x)^2 \qquad \text{Cross multiply}$$

$$4(2x+5) = x^2$$

$$8x + 20 = x^2$$

$$0 = x^2 - 8x - 20$$

$$0 = (x-10)(x+2)$$

$x - 10 = 0$ or $x + 2 = 0$

$x = 10$ $\qquad$ $x = -2$

Solution set: $\{10\}$ $\qquad$ -2 does not check.

219.

$$\sqrt{-25} + \sqrt{-81} = \sqrt{25}\sqrt{-1} + \sqrt{81}\sqrt{-1} = 5i + 9i = 14i$$

221.

$$\sqrt{-49} \cdot \sqrt{-36} = \sqrt{49}\sqrt{-1} \cdot \sqrt{36}\sqrt{-1}$$

$$= 7i \cdot 6i = 42i^2 = -42$$

223.

$$\frac{\sqrt{-36}}{\sqrt{-9}} = \frac{\sqrt{36}\sqrt{-1}}{\sqrt{9}\sqrt{-1}} = \frac{6i}{3i} = 2$$

225.

$$\sqrt{-18} + \sqrt{-8} = \sqrt{18}\sqrt{-1} + \sqrt{8}\sqrt{-1}$$

$$= 3\sqrt{2}i + 2\sqrt{2}i = 5\sqrt{2}i$$

227.

$$\sqrt{-7} \cdot \sqrt{7} = \sqrt{7}\sqrt{-1} \cdot \sqrt{7} = \sqrt{7}\sqrt{7}\sqrt{-1} = 7i$$

229.

$$(5 + 7i) + (8 - 2i) = 13 + 5i$$

231.

$$(4 - 5i) - (1 + 7i) = 4 - 5i - 1 - 7i = 3 - 12i$$

233.

$$6(5 + 8i) = 30 + 48i$$

235.

$$\begin{array}{cccc} \text{F} & \text{O} & \text{I} & \text{L} \end{array}$$

$$(5 + 2i)(7 + i) = 35 + 5i + 14i + 2i^2$$

$$= 35 + 5i + 14i - 2$$

$$= 33 + 19i$$

237.

$$(6 - 5i)(6 + 5i) = 6^2 - (5i)^2$$

$$= 36 - 25i^2 = 36 + 25 = 61$$

239.

$$\frac{5 + 3i}{2i} \cdot \frac{i}{i} = \frac{5i + 3i^2}{2i^2} = \frac{5i - 3}{-2} = \frac{3}{2} - \frac{5}{2}i$$

241.

$$\frac{4}{1 + 3i} \cdot \frac{1 - 3i}{1 - 3i} = \frac{4 - 12i}{1 - 9i^2}$$

$$= \frac{4 - 12i}{1 + 9}$$

$$= \frac{4 - 12i}{10} = \frac{4}{10} - \frac{12}{10}i = \frac{2}{5} - \frac{6}{5}i$$

243.

$$\frac{3-i}{4-i} \cdot \frac{4+i}{4+i} = \frac{12-i-i^2}{16-i^2} = \frac{12-i+1}{16+1} = \frac{13-i}{17} = \frac{13}{17} - \frac{1}{17}i$$

245.

$$i^{20} = \left(i^2\right)^{10} = 1$$

247.

$$i^{35} = i \cdot i^{34} = i \cdot \left(i^2\right)^{17} = i \cdot (-1) = -i$$

249.

$$i^{-2} = \frac{1}{i^2} = \frac{1}{-1} = -1$$

Chapter 5 Test Solutions

1.

$$9^{-\frac{1}{2}} = \frac{1}{9^{\frac{1}{2}}} = \frac{1}{3}$$

3.

$$\left(4x^6y^3\right)^{\frac{3}{2}} \cdot 3xy^2 = 4^{\frac{3}{2}} x^9 y^{\frac{9}{2}} \cdot 3xy^2$$

$$= 8 \cdot 3 \cdot x^{10} \cdot y^{\frac{13}{2}} = 24x^{10}y^{\frac{13}{2}}$$

5.

$$\sqrt{25^{-3}} = \sqrt{\frac{1}{25^3}} = \frac{\sqrt{1}}{\sqrt{25^3}} = \frac{1}{\sqrt{5^6}} = \frac{1}{5^3} = \frac{1}{125}$$

7.

$$\sqrt{72x^2y} + 5x\sqrt{8y} = \sqrt{36x^2 \cdot 2y} + 5x\sqrt{8y}$$

$$= 6x\sqrt{2y} + 5x \cdot 2\sqrt{2y}$$

$$= 6x\sqrt{2y} + 10x\sqrt{2y} = 16x\sqrt{2y}$$

9.

$$\frac{2}{\sqrt[3]{4x}} \cdot \frac{\sqrt[3]{2x^2}}{\sqrt[3]{2x^2}} = \frac{2\sqrt[3]{2x^2}}{\sqrt[3]{8x^3}} = \frac{2\sqrt[3]{2x^2}}{2x} = \frac{\sqrt[3]{2x^2}}{x}$$

11.

$$\frac{\sqrt{2}}{\sqrt{6}+2} \cdot \frac{\sqrt{6}-2}{\sqrt{6}-2} = \frac{\sqrt{12}-2\sqrt{2}}{6-4} = \frac{2\sqrt{3}-2\sqrt{2}}{2}$$

$$= \frac{2\left(\sqrt{3}-\sqrt{2}\right)}{2} = \sqrt{3} - \sqrt{2}$$

13.

$$\overset{\text{F} \quad \text{O} \quad \text{I} \quad \text{L}}{\left(2\sqrt{x}-3\right)\left(\sqrt{x}+4\right) = 2\sqrt{x^2} + 8\sqrt{x} - 3\sqrt{x} - 12}$$

$$= 2x + 5\sqrt{x} - 12$$

15.

$$\sqrt{y^2 - 3y - 9} - 3 = y$$

$$\left(\sqrt{y^2 - 3y - 9}\right)^2 = (y+3)^2$$

$$y^2 - 3y - 9 = y^2 + 6y + 9$$

$$-18 = 9y$$

$$-2 = y$$

Solution set: {-2}

17.

$$(3 + 7i) + (2 - i) = 5 + 6i$$

19.

$$\left(\frac{1}{2} - 3i\right) - (2 + 5i) = \frac{1}{2} - 3i - 2 - 5i$$

$$= \frac{1}{2} - 2 - 3i - 5i = -\frac{3}{2} - 8i$$

21.

$$i^{-5} = \frac{1}{i^5} = \frac{1}{i \cdot i^4} = \frac{1}{i\left(i^2\right)^2} = \frac{1}{i} = \frac{1}{i} \cdot \frac{i}{i} = \frac{i}{i^2} = \frac{i}{-1} = -i$$

23.

$$\overset{\text{F} \quad \text{O} \quad \text{I} \quad \text{L}}{(2 + 9i)(1 - 2i) = 2 - 4i + 9i - 18i^2}$$

$$= 2 - 4i + 9i + 18 = 20 + 5i$$

25.

$$\frac{4+12i}{2i} \cdot \frac{i}{i} = \frac{4i + 12i^2}{2i^2}$$

$$= \frac{-12 + 4i}{-2}$$

$$= 6 - 2i$$

1.

$$-\left(\frac{2}{3}\right)^2 + \frac{5}{8} + \frac{1}{4} - \frac{5}{6} = -\frac{4}{9} + \frac{5}{8} + \frac{1}{4} - \frac{5}{6}$$

$$= -\frac{4}{9} + \frac{5}{2} - \frac{5}{6}$$

$$= -\frac{8}{18} + \frac{45}{18} - \frac{15}{18}$$

$$= \frac{22}{18} = \frac{11}{9}$$

3.

$$25^{\frac{1}{2}} - 27^{-\frac{2}{3}} = 25^{\frac{1}{2}} - \frac{1}{27^{\frac{2}{3}}}$$

$$= 5 - \frac{1}{9} = \frac{45}{9} - \frac{1}{9} = \frac{44}{9}$$

5.

$$\left(\frac{2x^{-3}y^5}{x^{-6}y^{-1}}\right)^2 = \left(2x^{-3+6}y^{5+1}\right)^2$$

$$= \left(2x^3y^6\right)^2 = 2^2 x^6 y^{12} = 4x^6 y^{12}$$

7.

$$\frac{\left(x^{-\frac{2}{3}}\right)^2 x^{\frac{1}{2}}}{x^{-1}} = \frac{x^{-\frac{4}{3}} x^{\frac{1}{2}}}{x^{-1}} = x^{-\frac{4}{3} + \frac{1}{2} + 1} = x^{\frac{1}{6}}$$

9.

$$1 - 125t^3 = (1 - 5t)\left(1 + 5t + 25t^2\right)$$

11.

$$(2x + y)^2 + 7(2x + y) + 12 \qquad \text{Let } u = 2x + y$$

$$u^2 + 7u + 12 = (u + 3)(u + 4)$$

$$= [(2x + y) + 3][(2x + y) + 4]$$

$$= (2x + y + 3)(2x + y + 4)$$

13.

$$\frac{(x+1)(x-1)}{8x^3(x-1)} \cdot \frac{2x(x-5)(x+1)}{(x+1)(x+1)} = \frac{2x(x-5)}{8x^3}$$

$$= \frac{2x(x-5)}{2x \cdot 4x^2}$$

$$= \frac{x-5}{4x^2}$$

15.

$$\frac{2x+3}{(x-2)(x-2)} - \frac{x+1}{(x+3)(x-2)}$$

$$\text{LCD} = (x-2)^2(x+3)$$

$$= \frac{x+3}{x+3} \cdot \frac{2x+3}{(x-2)(x-2)} - \frac{x-2}{x-2} \cdot \frac{x+1}{(x+3)(x-2)}$$

$$= \frac{(x+3)(2x+3) - (x-2)(x+1)}{(x+3)(x-2)^2}$$

$$= \frac{2x^2 + 9x + 9 - \left(x^2 - x - 2\right)}{(x+3)(x-2)^2}$$

$$= \frac{2x^2 + 9x + 9 - x^2 + x + 2}{(x+3)(x-2)^2}$$

$$= \frac{x^2 + 10x + 11}{(x+3)(x-2)^2}$$

17.

$$\frac{x^4 y^4 \left(\frac{1}{x} - \frac{1}{y}\right)}{x^4 y^4 \left(\frac{1}{x^4} - \frac{1}{y^4}\right)} = \frac{x^3 y^4 - x^4 y^3}{y^4 - x^4} \qquad \text{LCD} = x^4 y^4$$

$$= \frac{x^3 y^3 (y - x)}{\left(y^2 + x^2\right)(y + x)(y - x)}$$

$$= \frac{x^3 y^3}{\left(y^2 + x^2\right)(y + x)}$$

19.

$$\sqrt{\frac{1}{36^2}} = \frac{\sqrt{1}}{\sqrt{36^2}} = \frac{1}{36}$$

21.

$$x\sqrt{72y} + 2x\sqrt{32y} - 10\sqrt{2x^2 y}$$

$$= x\sqrt{36 \cdot 2y} + 2x\sqrt{16 \cdot 2y} - 10\sqrt{x^2 \cdot 2y}$$

$$= 6x\sqrt{2y} + 2x \cdot 4\sqrt{2y} - 10x\sqrt{2y}$$

$$= 6x\sqrt{2y} + 8x\sqrt{2y} - 10x\sqrt{2y}$$

$$= 4x\sqrt{2y}$$

23.

$$\sqrt{\frac{3x}{5y} \cdot \frac{5y}{5y}} = \sqrt{\frac{15xy}{25y^2}} = \frac{\sqrt{15xy}}{\sqrt{25y^2}} = \frac{\sqrt{15xy}}{5y}$$

25.

$$\left(2\sqrt{x} - 7\right)^2 = \left(2\sqrt{x}\right)^2 - 2\left(2\sqrt{x} \cdot 7\right) + 7^2$$
$$= 4x - 28\sqrt{x} + 49$$

27.

$$\sqrt{-20} + 2\sqrt{-45} = \sqrt{20}\sqrt{-1} + 2\sqrt{45}\sqrt{-1}$$
$$= 2\sqrt{5}i + 6\sqrt{5}i = 8\sqrt{5}i$$

29.

$$\text{F} \quad \text{O} \quad \text{I} \quad \text{L}$$
$$(5 + i)(2 - 4i) = 10 - 20i + 2i - 4i^2$$
$$= 10 - 20i + 2i + 4 = 14 - 18i$$

31.

$$\frac{4}{x-1} = \frac{5}{x+2}$$
$$5x - 5 = 4x + 8$$
$$x = 13$$

Solution set: $\{13\}$

33.

$$2x^2 - 5x - 12 = 0$$
$$(2x + 3)(x - 4) = 0$$
$$2x + 3 = 0 \quad \text{or} \quad x - 4 = 0$$
$$x = -\frac{3}{2} \qquad x = 4$$

Solution set: $\left\{-\frac{3}{2}, 4\right\}$

35.

$$\left(\sqrt{3x + 7}\right)^2 = (4)^2$$
$$3x + 7 = 16$$
$$3x = 9$$
$$x = 3$$

Solution set: $\{3\}$

37.

$$|3x - 4| = |2x - 1|$$
$$3x - 4 = 2x - 1 \quad \text{or} \quad 3x - 4 = -(2x - 1)$$
$$x = 3 \qquad\qquad 3x - 4 = -2x + 1$$
$$5x = 5$$
$$x = 1$$

Solution set: $\{1, 3\}$

39.

$$\frac{2x}{x+3} - \frac{1}{x-2} = \frac{-5}{(x+3)(x-2)} \quad \text{LCD} = (x+3)(x-2)$$
$$\frac{(x+3)(x-2)}{1}\left(\frac{2x}{x+3} - \frac{1}{x-2}\right)$$
$$= \frac{(x+3)(x-2)}{1} \cdot \frac{-5}{(x+3)(x-2)}$$
$$2x(x-2) - (x+3) = -5$$
$$2x^2 - 4x - x - 3 = -5$$
$$2x^2 - 5x + 2 = 0$$
$$(2x - 1)(x - 2) = 0$$
$$2x - 1 = 0 \quad \text{or} \quad x - 2 = 0$$
$$x = \frac{1}{2} \qquad\qquad x = 2$$

Solution set: $\left\{\frac{1}{2}\right\}$ Note: $x - 2 \neq 0$

So, $x \neq 2$

41.

$$\left(\sqrt{2x+3} - 1\right)^2 = \left(\sqrt{x+5}\right)^2$$
$$(2x+3) - 2\sqrt{2x+3} + 1 = x + 5$$
$$2x + 4 - 2\sqrt{2x+3} = x + 5$$
$$\left(-2\sqrt{2x+3}\right)^2 = (-x+1)^2$$
$$4(2x+3) = x^2 - 2x + 1$$
$$8x + 12 = x^2 - 2x + 1$$
$$0 = x^2 - 10x - 11$$
$$0 = (x - 11)(x + 1)$$
$$x - 11 = 0 \quad \text{or} \quad x + 1 = 0$$
$$x = 11 \qquad\qquad x = -1$$

Solution set: $\{11\}$ -1 does not check.

43.

$$(x+8)(x+5) = (2x+13)(x+4)$$
$$x^2 + 13x + 40 = 2x^2 + 21x + 52$$
$$0 = x^2 + 8x + 12$$
$$0 = (x+2)(x+6)$$
$$x + 2 = 0 \quad \text{or} \quad x + 6 = 0$$
$$x = -2 \qquad\qquad x = -6$$

Solution set: $\{-2, -6\}$

45.

Let x = width.
Then, $3x-1$ = length.

#45. continued

Thus, $p = 2w + 2\ell$

$94 = 2x + 2(3x - 1)$

$94 = 2x + 6x - 2$

$94 = 8x - 2$

$96 = 8x$

$12 = x$

$3x - 1 = 3 \cdot 12 - 1 = 35$

Answer: 12 ft by 35 ft

47.

Let x = number of dimes.

Then, 22-x = number of quarters.

Thus, $10x + 25(22 - x) = 430$

$10x + 550 - 25x = 430$

$-15x = -120$

$x = 8$

$22 - x = 22 - 8 = 14$

Answer: 8 dimes, 14 quarters

49.

$$\frac{1}{4} + \frac{1}{x} = \frac{1}{3} \qquad LCD = 12x$$

$$\frac{12x}{1}\left(\frac{1}{4} + \frac{1}{x}\right) = \left(\frac{1}{3}\right)\frac{12x}{1}$$

$$3x + 12 = 4x$$

$$12 = x$$

Answer: 12 hr

Chapter 5 Study Guide

Self-Test Exercises

I. Simplify each of the following. (Remember to write your answer with only positive exponents.)

1. $25^{-\frac{1}{2}}$

2. $-27^{\frac{2}{3}}$

3. $\left(9x^3y^6\right)^{\frac{3}{2}} \cdot 2xy^5$

4. $\dfrac{\left(p^{\frac{1}{5}}\right)^{-3} \cdot p^{\frac{3}{5}}}{p^{-\frac{2}{5}}}$

II. Simplify each of the following. Assume that all variables represent positive numbers.

5. $\sqrt{49^{-3}}$

6. $\sqrt{72y^3}$

7. $\sqrt{50xy^2} + 3y\sqrt{18x}$

8. $\sqrt[3]{54x^7y^8}$

9. $\dfrac{3}{\sqrt[3]{9m}}$

10. $\sqrt{\dfrac{7}{x^5y^2}}$

11. $\dfrac{\sqrt{3}}{\sqrt{10}-3}$

12. $\dfrac{\sqrt{56}}{\sqrt{14}}$

13. $\left(3\sqrt{x}-2\right)\left(\sqrt{x}-7\right)$

14. $\left(\sqrt{5}+3\right)^2$

III. Find the solution(s) of the following radical equations.

15. $\sqrt{x^2+8}-2=x$

16. $\sqrt{2x-3}-\sqrt{x+7}=-2$

IV. Perform the indicated operations. Express all answers in the form $a+bi$.

17. $(2-5i)(3+i)$

18. $\sqrt{-25}+\sqrt{-121}$

19. $\left(\dfrac{1}{4}+5i\right)-(4+3i)$

20. $\sqrt{-36}\cdot\sqrt{-25}$

21. i^{-7}

22. $\dfrac{\sqrt{-81}}{\sqrt{-36}}$

23. $(-3-2i)(-3+2i)$

24. $(3-2i)^2$

25. $\dfrac{17i}{2+5i}$

The worked-out solutions begin on the next page.

Self-Test Solutions

1.
$$25^{-\frac{1}{2}} = \frac{1}{25^{\frac{1}{2}}} = \frac{1}{5}$$

2.
$$-27^{\frac{2}{3}} = -\left[(27)^{\frac{1}{3}}\right]^2 = -(3)^2 = -9$$

3.
$$9^{\frac{3}{2}} x^{\frac{9}{2}} y^9 \cdot 2xy^5 = 27 \cdot 2 \cdot x^{\frac{9}{2}+1} y^{9+5} = 54 x^{11\frac{1}{2}} y^{14}$$

4.
$$\frac{\left(p^{\frac{1}{5}}\right)^{-3} \cdot p^{\frac{3}{5}}}{p^{-\frac{2}{5}}} = \frac{p^{-\frac{3}{5}} \cdot p^{\frac{3}{5}}}{p^{-\frac{2}{5}}} = p^{-\frac{3}{5}+\frac{3}{5}+\frac{2}{5}} = p^{\frac{2}{5}}$$

5.
$$\sqrt{49^{-3}} = \sqrt{\left(7^2\right)^{-3}} = \sqrt{7^{-6}} = \sqrt{\frac{1}{7^6}} = \frac{\sqrt{1}}{\sqrt{7^6}} = \frac{1}{7^3} = \frac{1}{343}$$

6.
$$\sqrt{72y^3} = \sqrt{36y^2 \cdot 2y} = \sqrt{36y^2} \cdot \sqrt{2y} = 6y\sqrt{2y}$$

7.
$$\sqrt{25y^2 \cdot 2x} + 3y\sqrt{9 \cdot 2x}$$
$$= 5y\sqrt{2x} + 3 \cdot 3y\sqrt{2x}$$
$$= 5y\sqrt{2x} + 9y\sqrt{2x} = 14y\sqrt{2x}$$

8.
$$\sqrt[3]{54x^7y^8} = \sqrt[3]{27x^6y^6 \cdot 2xy^2}$$
$$= \sqrt[3]{27x^6y^6}\sqrt[3]{2xy^2} = 3x^2y^2\sqrt[3]{2xy^2}$$

9.
$$\frac{3}{\sqrt[3]{9m}} \cdot \frac{\sqrt[3]{3m^2}}{\sqrt[3]{3m^2}} = \frac{3\sqrt[3]{3m^2}}{\sqrt[3]{27m^3}} = \frac{3\sqrt[3]{3m^2}}{3m} = \frac{\sqrt[3]{3m^2}}{m}$$

10.
$$\sqrt{\frac{7}{x^5y^2} \cdot \frac{x}{x}} = \sqrt{\frac{7x}{x^6y^2}} = \frac{\sqrt{7x}}{\sqrt{x^6y^2}} = \frac{\sqrt{7x}}{x^3y}$$

11.
$$\frac{\sqrt{3}}{\sqrt{10}-3} \cdot \frac{\sqrt{10}+3}{\sqrt{10}+3} = \frac{\sqrt{30}+3\sqrt{3}}{10-9}$$
$$= \frac{\sqrt{30}+3\sqrt{3}}{1} = \sqrt{30}+3\sqrt{3}$$

12.
$$\frac{\sqrt{56}}{\sqrt{14}} = \frac{\sqrt{4 \cdot 14}}{\sqrt{14}} = \frac{2\sqrt{14}}{\sqrt{14}} = 2$$

13.
$$\begin{array}{cccc} \text{F} & \text{O} & \text{I} & \text{L} \end{array}$$
$$\left(3\sqrt{x}-2\right)\left(\sqrt{x}-7\right) = 3\sqrt{x^2} - 21\sqrt{x} - 2\sqrt{x} + 14$$
$$= 3x - 23\sqrt{x} + 14$$

14.
$$\left(\sqrt{5}+3\right)^2 = \left(\sqrt{5}\right)^2 + 2\left(3\sqrt{5}\right) + (3)^2$$
$$= 5 + 6\sqrt{5} + 9 = 14 + 6\sqrt{5}$$

15.
$$\left(\sqrt{x^2+8}\right)^2 = (x+2)^2$$
$$x^2 + 8 = x^2 + 4x + 4$$
$$4 = 4x$$
$$1 = x$$
Solution set: $\{1\}$

16.
$$\left(\sqrt{2x-3}\right)^2 = \left(\sqrt{x+7}-2\right)^2$$
$$2x - 3 = (x+7) - 4\sqrt{x+7} + 4$$
$$2x - 3 = x + 11 - 4\sqrt{x+7}$$
$$x - 14 = -4\sqrt{x+7}$$
$$(x-14)^2 = \left(-4\sqrt{x+7}\right)^2$$
$$x^2 - 28x + 196 = 16(x+7)$$
$$x^2 - 28x + 196 = 16x + 112$$
$$x^2 - 44x + 84 = 0$$
$$(x-2)(x-42) = 0$$
$$x - 2 = 0 \quad \text{or} \quad x - 42 = 0$$
$$x = 2 \qquad\qquad x = 42$$
Solution set: $\{2\}$ 42 does not check.

17.

$$F \quad O \quad I \quad L$$
$$(2 - 5i)(3 + i) = 6 + 2i - 15i - 5i^2$$
$$= 6 + 2i - 15i + 5 = 11 - 13i$$

18.

$$\sqrt{-25} + \sqrt{-121} = \sqrt{25}\sqrt{-1} + \sqrt{121}\sqrt{-1} = 5i + 11i = 16i$$

19.

$$\left(\frac{1}{4} + 5i\right) - (4 + 3i) = \frac{1}{4} + 5i - 4 - 3i = -\frac{15}{4} + 2i$$

20.

$$\sqrt{-36} \cdot \sqrt{-25} = \sqrt{36}\sqrt{-1} \cdot \sqrt{25}\sqrt{-1}$$
$$= 6i \cdot 5i = 30i^2 = -30$$

21.

$$i^{-7} = \frac{1}{i^7}$$
$$= \frac{1}{i \cdot i^6}$$
$$= \frac{1}{i \cdot \left(i^2\right)^3} = \frac{1}{i(-1)} = -\frac{1}{i} = -\frac{1}{i} \cdot \frac{i}{i} = -\frac{i}{i^2} = -\frac{i}{-1} = i$$

22.

$$\frac{\sqrt{-81}}{\sqrt{-36}} = \frac{\sqrt{81}\sqrt{-1}}{\sqrt{36}\sqrt{-1}} = \frac{9i}{6i} = \frac{3}{2}$$

23.

$$(-3 - 2i)(-3 + 2i) = (-3)^2 - (2i)^2$$
$$= 9 - 4i^2 = 9 + 4 = 13$$

24.

$$(3 - 2i)^2 = 3^2 - 2(6i) + (2i)^2$$
$$= 9 - 12i + 4i^2 = 9 - 12i - 4 = 5 - 12i$$

25.

$$\frac{17i}{2 + 5i} \cdot \frac{2 - 5i}{2 - 5i} = \frac{34i - 85i^2}{(2)^2 - (5i)^2}$$
$$= \frac{34i + 85}{4 - 25i^2}$$
$$= \frac{34i + 85}{4 + 25} = \frac{34i + 85}{29} = \frac{85}{29} + \frac{34}{29}i$$

CHAPTER 6   RELATIONS AND FUNCTIONS

Solutions to Text Exercises

Exercises 6.1

1.-21.
The answers are in the answer section of the main text.

Exercises 6.2

1.
The exercise does not require a worked-out solution.

3.

$x = \dfrac{1}{2};$ $y = \sqrt{2 \cdot \dfrac{1}{2} - 1} = \sqrt{1-1} = 0;$ $\left(\dfrac{1}{2}, 0\right)$

$x = 1;$ $y = \sqrt{2 \cdot 1 - 1} = \sqrt{2-1} = \sqrt{1} = 1;$ $(1,1)$

$x = 3;$ $y = \sqrt{2 \cdot 3 - 1} = \sqrt{6-1} = \sqrt{5};$ $\left(3, \sqrt{5}\right)$

$x = 5;$ $y = \sqrt{2 \cdot 5 - 1} = \sqrt{10-1} = \sqrt{9} = 3;$ $(5,3)$

5.

$x = -2;$ $y = |-2| + 3 = 2 + 3 = 5;$ $(-2,5)$

$x = 0;$ $y = |0| + 3 = 3;$ $(0,3)$

$x = 2;$ $y = |2| + 3 = 2 + 3 = 5;$ $(2,5)$

7.-39.
The exercises do not require worked-out solutions.

Exercises 6.3

1.
$f(x) = 2x - 3$

$f(-2) = 2(-2) - 3 = -7$

$f\left(-\dfrac{1}{2}\right) = 2\left(-\dfrac{1}{2}\right) - 3 = -1 - 3 = -4$

$f(0) = 2(0) - 3 = -3$

$f\left(\dfrac{1}{3}\right) = 2\left(\dfrac{1}{3}\right) - 3 = \dfrac{2}{3} - 3 = -\dfrac{7}{3}$

$f(1) = 2(1) - 3 = 2 - 3 = -1$

$f(a) = 2 \cdot a - 3 = 2a - 3$

$f(a+h) = 2(a+h) - 3 = 2a + 2h - 3$

3.
$f(x) = x^2 - 2$

$f(-2) = (-2)^2 - 2 = 4 - 2 = 2$

$f\left(-\dfrac{1}{2}\right) = \left(-\dfrac{1}{2}\right)^2 - 2 = \dfrac{1}{4} - 2 = -\dfrac{7}{4}$

$f(0) = (0)^2 - 2 = -2$

$f\left(\dfrac{1}{3}\right) = \left(\dfrac{1}{3}\right)^2 - 2 = \dfrac{1}{9} - 2 = -\dfrac{17}{9}$

$f(1) = (1)^2 - 2 = -1$

$f(a) = a^2 - 2$

$f(a+h) = (a+h)^2 - 2 = a^2 + 2ah + h^2 - 2$

5.
$f(x) = 2x^3$

$f(-2) = 2(-2)^3 = 2(-8) = -16$

$f\left(-\dfrac{1}{2}\right) = 2\left(-\dfrac{1}{2}\right)^3 = 2\left(-\dfrac{1}{8}\right) = -\dfrac{1}{4}$

$f(0) = 2(0)^3 = 0$

$f\left(\dfrac{1}{3}\right) = 2\left(\dfrac{1}{3}\right)^3 = 2\left(\dfrac{1}{27}\right) = \dfrac{2}{27}$

$f(1) = 2(1)^3 = 2$

$f(a) = 2a^3$

$f(a+h) = 2(a+h)^3 = 2a^3 + 6a^2h + 6ah^2 + 2h^3$

7.
$f(x) = -2$ for all values of x.

9.
$(f+g)(x) = \left(2x^2 + 5x - 1\right) + (x - 2) = 2x^2 + 6x - 3$

$(f-g)(x) = \left(2x^2 + 5x - 1\right) - (x - 2)$

$\qquad = 2x^2 + 5x - 1 - x + 2 = 2x^2 + 4x + 1$

$(f \cdot g)(x) = \left(2x^2 + 5x - 1\right)(x - 2) = 2x^3 + x^2 - 11x + 2$

$\left(\dfrac{f}{g}\right)(x) = \dfrac{2x^2 + 5x - 1}{x - 2}$

11.

$(f + g)(x) = (x^2 + 5) + (x^2 - 9) = 2x^2 - 4$

$(f - g)(x) = (x^2 + 5) - (x^2 - 9) = x^2 + 5 - x^2 + 9 = 14$

$(f \cdot g)(x) = (x^2 + 5)(x^2 - 9) = x^4 - 4x^2 - 45$

$\left(\dfrac{f}{g}\right)(x) = \dfrac{x^2 + 5}{x^2 - 9}$

13.

$(f + g)(x) = (2x + 3) + (x - 11) = 3x - 8$

$(f - g)(x) = (2x + 3) - (x - 11)$

$\qquad = 2x + 3 - x + 11 = x + 14$

$(f \cdot g)(x) = (2x + 3)(x - 11) = 2x^2 - 19x - 33$

$\left(\dfrac{f}{g}\right)(x) = \dfrac{2x + 3}{x - 11}$

15.

$(f + g)(x) = (x^3 + 3x - 5) + (2x + 1) = x^3 + 5x - 4$

$(f - g)(x) = (x^3 + 3x - 5) - (2x + 1)$

$\qquad = x^3 + 3x - 5 - 2x - 1 = x^3 + x - 6$

$(f \cdot g)(x) = (x^3 + 3x - 5)(2x + 1)$

$\qquad = 2x^4 + x^3 + 6x^2 - 7x - 5$

$\left(\dfrac{f}{g}\right)(x) = \dfrac{x^3 + 3x - 5}{2x + 1}$

17.

$f(-2) = 3(-2) - 1 = -6 - 1 = -7$

19.

$m(-3) = (-3)^2 - 4 = 9 - 4 = 5$

21.

$5f(-2) - 4g(3) = 5[3(-2) - 1] - 4[3 \cdot 3^2 + 5 \cdot 3 - 1]$

$\qquad = 5(-6 - 1) - 4(27 + 15 - 1)$

$\qquad = 5(-7) - 4(41)$

$\qquad = -35 - 164$

$\qquad = -199$

23.

$[f(-2)]^3 = [3(-2) - 1]^3 = (-7)^3 = -343$

25.

$\left(\dfrac{m(-3)}{g(3)}\right)^2 = \left(\dfrac{(-3)^2 - 4}{3(3)^2 + 5 \cdot 3 - 1}\right)^2$

$\qquad = \left(\dfrac{9 - 4}{27 + 15 - 1}\right)^2 = \left(\dfrac{5}{41}\right)^2 = \dfrac{25}{1681}$

27.

$(m \cdot n)(-1) = (x^2 - 4)(2x + 1)$

$\qquad = [(-1)^2 - 4][2(-1) + 1]$

$\qquad = (1 - 4)(-2 + 1)$

$\qquad = -3(-1) = 3$

29.

$(g - f)(0) = (3x^2 + 5x - 1) - (3x - 1)$

$\qquad = (3 \cdot 0^2 + 5 \cdot 0 - 1) - (3 \cdot 0 - 1)$

$\qquad = -1 + 1 = 0$

31.

$g(x) - m(x) = (3x^2 + 5x - 1) - (x^2 - 4)$

$\qquad = 3x^2 + 5x - 1 - x^2 + 4$

$\qquad = 2x^2 + 5x + 3$

33.

$\dfrac{m(x)}{n(x)} = \dfrac{x^2 - 4}{2x + 1}$

35.

$g(2x) = 3(2x)^2 + 5(2x) - 1$

$\qquad = 12x^2 + 10x - 1$

37.

$n(3x) = 2(3x) + 1 = 6x + 1$

39.

$f(a) + f(h) = 3a - 1 + 3h - 1 = 3a + 3h - 2$

41.

$\dfrac{f(a + h) - f(a)}{h} = \dfrac{3(a + h) - 1 - (3a - 1)}{h}$

$\qquad = \dfrac{3a + 3h - 1 - 3a + 1}{h}$

#41. continued

$$= \frac{3\cancel{h}}{\cancel{h}} = 3$$

43.

$$n(a) + n(h) = 2a + 1 + 2h + 1 = 2a + 2h + 2$$

45.

$$\frac{n(a+h) - n(a)}{h} = \frac{2(a+h) + 1 - (2a+1)}{h}$$

$$= \frac{2a + 2h + 1 - 2a - 1}{h}$$

$$= \frac{2\cancel{h}}{\cancel{h}} = 2$$

47.

$$g(a) + g(h) = 3a^2 + 5a - 1 + 3h^2 + 5h - 1$$

$$= 3a^2 + 5a + 3h^2 + 5h - 2$$

49.

$$\frac{g(a+h) - g(a)}{h}$$

$$= \frac{3(a+h)^2 + 5(a+h) - 1 - \left(3a^2 + 5a - 1\right)}{h}$$

$$= \frac{3\left(a^2 + 2ah + h^2\right) + 5a + 5h - 1 - 3a^2 - 5a + 1}{h}$$

$$= \frac{3a^2 + 6ah + 3h^2 + 5a + 5h - 1 - 3a^2 - 5a + 1}{h}$$

$$= \frac{6ah + 3h^2 + 5h}{h} = \frac{\cancel{h}(6a + 3h + 5)}{\cancel{h}} = 6a + 3h + 5$$

51.

$$f\big(g(3)\big) \quad g(3) = 3 \cdot 3^2 + 5 \cdot 3 - 1 = 41$$

$$f(41) = 3 \cdot 41 - 1 = 122$$

53.

$$f\big(m(-3)\big) \quad m(-3) = (-3)^2 - 4 = 9 - 4 = 5$$

$$f(5) = 3 \cdot 5 - 1 = 15 - 1 = 14$$

Exercises 6.4

1.-15.
The graphs are in the answer section of the main text.

17.

$$m = \frac{y_2 - y_1}{x_2 - x_1} \qquad m = \frac{-3 - 0}{0 - 6} = \frac{-3}{-6} = \frac{1}{2}$$

19.

$$m = \frac{y_2 - y_1}{x_2 - x_1} \qquad m = \frac{-5 - 1}{3 + 4} = \frac{-6}{7} = -\frac{6}{7}$$

21.

$$m = \frac{y_2 - y_1}{x_2 - x_1} \qquad m = \frac{2 + 7}{6 - 3} = \frac{9}{3} = 3$$

23.

$$m = \frac{y_2 - y_1}{x_2 - x_1} \qquad m = \frac{-6 - 0}{\frac{1}{3} - 0} = \frac{-6}{\frac{1}{3}} = -18$$

25.

$$m = \frac{y_2 - y_1}{x_2 - x_1} \qquad m = \frac{-\frac{1}{4} - \frac{3}{8}}{-\frac{7}{15} - \frac{1}{5}} = \frac{-\frac{5}{8}}{-\frac{10}{15}} = \frac{15}{16}$$

27.

$$m = \frac{y_2 - y_1}{x_2 - x_1} \qquad m = \frac{2 - 2}{9 - 5} = \frac{0}{4} = 0$$

29.-35.
The graphs are in the answer section of the main text.

37.
$$2x - y = 6$$
$$-y = -2x + 6$$
$$y = 2x - 6$$

x	y
0	-6
1	-4

$$m = \frac{-6 + 4}{0 - 1} = \frac{-2}{-1} = 2$$

39.
$$y = -2x$$

x	y
0	0
1	-2

$$m = \frac{0 + 2}{0 - 1} = -2$$

41.

$x + 3y = 0$

$3y = -x$

$y = -\dfrac{1}{3}x$

x	y
0	0
3	-1

$m = \dfrac{0+1}{0-3} = -\dfrac{1}{3}$

43.

$5x - 10y = -1$

$-10y = -5x - 1$

$y = \dfrac{1}{2}x + \dfrac{1}{10}$

x	y
0	$\dfrac{1}{10}$
2	$\dfrac{11}{10}$

$m = \dfrac{\dfrac{1}{10} - \dfrac{11}{10}}{0 - 2} = \dfrac{-1}{-2} = \dfrac{1}{2}$

45.

$\dfrac{2}{3}x - \dfrac{1}{5}y = -3$

$-\dfrac{1}{5}y = -\dfrac{2}{3}x - 3$

$y = \dfrac{10}{3}x + 15$

x	y
0	15
-3	5

$m = \dfrac{15-5}{0+3} = \dfrac{10}{3}$

47.

$y = 2x - \dfrac{8}{5}$

x	y
0	$-\dfrac{8}{5}$
1	$\dfrac{2}{5}$

#47. continued

$m = \dfrac{-\dfrac{8}{5} - \dfrac{2}{5}}{0 - 1} = \dfrac{-2}{-1} = 2$

49.

$y = -5$

x	y
0	-5
1	-5

$m = \dfrac{-5+5}{0-1} = 0$

51.

$x = \dfrac{7}{4}$ Slope is undefined.

53.-63.

The graphs are in the answer section of the main text.

Exercises 6.5

1.-7.

The exercises do not require worked - out solutions.

9.

$y = 3x - 4$ $m = 3, b = -4$

11.

$3x - 5y = 7$

$-5y = -3x + 7$

$y = \dfrac{3}{5}x - \dfrac{7}{5}$ $m = \dfrac{3}{5}, b = -\dfrac{7}{5}$

13.

$5y - 6x = 8$

$5y = 6x + 8$

$y = \dfrac{6}{5}x + \dfrac{8}{5}$ $m = \dfrac{6}{5}, b = \dfrac{8}{5}$

15.

$y = \dfrac{2}{7}x - \dfrac{4}{7}$ $m = \dfrac{2}{7}, b = -\dfrac{4}{7}$

17.

$y = 0x + 3$ $m = 0, b = 3$

172

19.

$$y + 2 = 3(x+1)$$
$$y + 2 = 3x + 3$$
$$y = 3x + 1$$

21.

$$y - 8 = \frac{9}{2}(x-3)$$
$$y - 8 = \frac{9}{2}x - \frac{27}{2}$$
$$y = \frac{9}{2}x - \frac{27}{2} + 8$$
$$y = \frac{9}{2}x - \frac{11}{2}$$

23.

$$y + \frac{1}{3} = \frac{4}{3}\left(x - \frac{5}{8}\right)$$
$$y + \frac{1}{3} = \frac{4}{3}x - \frac{5}{6}$$
$$y = \frac{4}{3}x - \frac{5}{6} - \frac{1}{3}$$
$$y = \frac{4}{3}x - \frac{7}{6}$$

25.

$$y - 5 = 0(x+6)$$
$$y - 5 = 0$$
$$y = 5$$

27.

$$(-2,4) \quad \text{and} \quad (-5,7)$$
$$m = \frac{7-4}{-5+2} = \frac{3}{-3} = -1$$
$$y - 4 = -1 \cdot (x+2)$$
$$y - 4 = -x - 2$$
$$y = -x + 2$$

29.

$$(-8,6) \quad \text{and} \quad (4,-3)$$
$$m = \frac{-3-6}{4+8} = \frac{-9}{12} = -\frac{3}{4}$$
$$y + 3 = -\frac{3}{4}(x-4)$$
$$y + 3 = -\frac{3}{4}x + 3$$

#29. comtinued

$$y = -\frac{3}{4}x$$

31.

$$(0,0) \quad \text{and} \quad (-2,3)$$
$$m = \frac{3-0}{-2-0} = -\frac{3}{2}$$
$$y - 0 = -\frac{3}{2}(x-0)$$
$$y = -\frac{3}{2}x$$

33.

$$\left(2, -\frac{1}{8}\right) \quad \text{and} \quad \left(\frac{1}{2}, \frac{5}{4}\right)$$
$$m = \frac{\frac{5}{4} + \frac{1}{8}}{\frac{1}{2} - 2} = \frac{\frac{11}{8}}{-\frac{3}{2}} = -\frac{11}{12}$$
$$y + \frac{1}{8} = -\frac{11}{12}(x-2)$$
$$y + \frac{1}{8} = -\frac{11}{12}x + \frac{11}{6}$$
$$y = -\frac{11}{12}x + \frac{11}{6} - \frac{1}{8}$$
$$y = -\frac{11}{12}x + \frac{41}{24}$$

35.

$$(8,-4) \quad \text{and} \quad (-3,-4)$$
$$m = \frac{-4+4}{-3-8} = \frac{0}{-11} = 0$$
$$y + 4 = 0(x+3)$$
$$y + 4 = 0$$
$$y = -4$$

37.

$$(3,0) \quad \text{and} \quad (0,-2)$$
$$m = \frac{-2-0}{0-3} = \frac{2}{3}$$
$$y - 0 = \frac{2}{3}(x-3)$$
$$y = \frac{2}{3}x - 2$$

39.

$(1,-2),\quad m=3$

$y+2=3(x-1)$

$y+2=3x-3$

$y=3x-5$

41.

$(3,11),\quad m=-\dfrac{5}{4}$

Note: $5x+4y=-8$

$4y=-5x-8$

$y=-\dfrac{5}{4}x-2$

$y-11=-\dfrac{5}{4}(x-3)$

$y-11=-\dfrac{5}{4}x+\dfrac{15}{4}$

$y=-\dfrac{5}{4}x+\dfrac{59}{4}$

43.

$\left(\dfrac{1}{4},-\dfrac{3}{5}\right),\quad m=\dfrac{3}{4}$

Note: $y=\dfrac{6x-13}{8}$

$y=\dfrac{6}{8}x-\dfrac{13}{8}$

$y=\dfrac{3}{4}x-\dfrac{13}{8}$

$y+\dfrac{3}{5}=\dfrac{3}{4}\left(x-\dfrac{1}{4}\right)$

$y+\dfrac{3}{5}=\dfrac{3}{4}x-\dfrac{3}{16}$

$y=\dfrac{3}{4}x-\dfrac{3}{16}-\dfrac{3}{5}$

$y=\dfrac{3}{4}x-\dfrac{63}{80}$

45.

$(-8,-9)$ This is a vertical line, so $x=-8$.

47.

$(3,-7)$ This is a horizontal line, so $y=-7$.

49.

$(1,-2),\quad m=-\dfrac{1}{3}$

#49. continued

$y+2=-\dfrac{1}{3}(x-1)$

$y+2=-\dfrac{1}{3}x+\dfrac{1}{3}$

$y=-\dfrac{1}{3}x+\dfrac{1}{3}-2$

$y=-\dfrac{1}{3}x-\dfrac{5}{3}$

51.

$(3,11),\quad m=\dfrac{4}{5}$ See #41 above.

$y-11=\dfrac{4}{5}(x-3)$

$y-11=\dfrac{4}{5}x-\dfrac{12}{5}$

$y=\dfrac{4}{5}x-\dfrac{12}{5}+11$

$y=\dfrac{4}{5}x+\dfrac{43}{5}$

53.

$\left(\dfrac{1}{4},-\dfrac{3}{5}\right),\quad m=-\dfrac{4}{3}$ See #43 above.

$y+\dfrac{3}{5}=-\dfrac{4}{3}\left(x-\dfrac{1}{4}\right)$

$y+\dfrac{3}{5}=-\dfrac{4}{3}x+\dfrac{1}{3}$

$y=-\dfrac{4}{3}x+\dfrac{1}{3}-\dfrac{3}{5}$

$y=-\dfrac{4}{3}x-\dfrac{4}{15}$

55.

$(-8,-9),$ The line is a horizontal line, so $y=-9$.

57.

$(3,-7)$ The line is a vertical line, so $x=3$.

59.-69.

The answers and graphs are in the answer section of the main text.

Exercises 6.6

1.
$(6,3)$ and $(6,-5)$

$d = \sqrt{(6-6)^2 + (3+5)^2} = \sqrt{0^2 + 8^2} = \sqrt{64} = 8$

$M = \left(\dfrac{6+6}{2}, \dfrac{3-5}{2}\right) = \left(\dfrac{12}{2}, \dfrac{-2}{2}\right) = (6,-1)$

3.
$(-1,3)$ and $(5,11)$

$d = \sqrt{(-1-5)^2 + (3-11)^2} = \sqrt{(-6)^2 + (-8)^2}$
$\qquad = \sqrt{36+64} = \sqrt{100} = 10$

$M = \left(\dfrac{-1+5}{2}, \dfrac{3+11}{2}\right) = \left(\dfrac{4}{2}, \dfrac{14}{2}\right) = (2,7)$

5.
$(1,-5)$ and $(-2,5)$

$d = \sqrt{(1+2)^2 + (-5-5)^2} = \sqrt{3^2 + (-10)^2} = \sqrt{9+100}$
$\qquad = \sqrt{109}$

$M = \left(\dfrac{1-2}{2}, \dfrac{-5+5}{2}\right) = \left(-\dfrac{1}{2}, 0\right)$

7.
$(-5,9)$ and $(3,-1)$

$d = \sqrt{(-5-3)^2 + (9+1)^2} = \sqrt{(-8)^2 + 10^2}$
$\qquad = \sqrt{64+100} = \sqrt{164} = 2\sqrt{41}$

$M = \left(\dfrac{-5+3}{2}, \dfrac{9-1}{2}\right) = \left(\dfrac{-2}{2}, \dfrac{8}{2}\right) = (-1,4)$

9.
$(0,3)$ and $(-2,-6)$

$d = \sqrt{(0+2)^2 + (3+6)^2} = \sqrt{2^2 + 9^2} = \sqrt{4+81} = \sqrt{85}$

$M = \left(\dfrac{0-2}{2} + \dfrac{3-6}{2}\right) = \left(-1, -\dfrac{3}{2}\right)$

11.
$(-2,1)$, $(4,7)$

$d = \sqrt{(-2-4)^2 + (1-7)^2} = \sqrt{(-6)^2 + (-6)^2}$
$\qquad = \sqrt{36+36} = \sqrt{72} = 6\sqrt{2}$

$M = \left(\dfrac{-2+4}{2}, \dfrac{1+7}{2}\right) = \left(\dfrac{2}{2}, \dfrac{8}{2}\right) = (1,4)$

13.
$\left(2, \dfrac{1}{3}\right)$, $\left(-\dfrac{1}{2}, -3\right)$

$d = \sqrt{\left(2+\dfrac{1}{2}\right)^2 + \left(\dfrac{1}{3}+3\right)^2} = \sqrt{\left(\dfrac{5}{2}\right)^2 + \left(\dfrac{10}{3}\right)^2}$

$\qquad = \sqrt{\dfrac{25}{4} + \dfrac{100}{9}}$

$\qquad = \sqrt{\dfrac{225}{36} + \dfrac{400}{36}}$

$\qquad = \sqrt{\dfrac{625}{36}} = \dfrac{25}{6}$

$M = \left(\dfrac{2-\dfrac{1}{2}}{2}, \dfrac{\dfrac{1}{3}-3}{2}\right) = \left(\dfrac{\dfrac{3}{2}}{2}, \dfrac{-\dfrac{8}{3}}{2}\right) = \left(\dfrac{3}{4}, -\dfrac{4}{3}\right)$

15.
$(-2.1, 4.3)$, $(-8.5, -6)$

$d = \sqrt{(-2.1+8.5)^2 + (4.3+6)^2} = \sqrt{(6.4)^2 + (10.3)^2}$
$\qquad = \sqrt{40.96 + 106.09}$
$\qquad = \sqrt{147.05} = 12.126$

$M = \left(\dfrac{-2.1-8.5}{2}, \dfrac{4.3-6}{2}\right) = \left(\dfrac{-10.6}{2}, \dfrac{-1.7}{2}\right)$
$\qquad = (-5.3, -0.85)$

17.
$\left(\sqrt{12}, \sqrt{20}\right)$, $\left(\sqrt{3}, \sqrt{5}\right)$

$d = \sqrt{\left(\sqrt{12}-\sqrt{3}\right)^2 + \left(\sqrt{20}-\sqrt{5}\right)^2} = \sqrt{\left(\sqrt{3}\right)^2 + \left(\sqrt{5}\right)^2}$
$\qquad = \sqrt{3+5}$
$\qquad = \sqrt{8} = 2\sqrt{2}$

$M = \left(\dfrac{2\sqrt{3}+\sqrt{3}}{2}, \dfrac{2\sqrt{5}+\sqrt{5}}{2}\right) = \left(\dfrac{3\sqrt{3}}{2}, \dfrac{3\sqrt{5}}{2}\right)$

19.
Let $B = (x,y)$, $A = (9,4)$

Then $M = \left(\dfrac{x+9}{2}, \dfrac{y+4}{2}\right)$

So, $\dfrac{x+9}{2} = 4$ and $\dfrac{y+4}{2} = -1$

$\qquad x+9 = 8 \qquad\quad y+4 = -2$

$\qquad\quad x = -1 \qquad\qquad y = -6$

#19. continued

Thus, $B = (-1, -6)$

21.

Let C be that point. $(-2, -6)$, $(8, 10)$

$$C = \left(\frac{x_1 + 3x_2}{4}, \frac{y_1 + 3y_2}{4} \right)$$

$$C = \left(\frac{-2 + 24}{4}, \frac{-6 + 30}{4} \right)$$

$$C = \left(\frac{11}{2}, 6 \right)$$

23.

$A = (-1, 6)$, $B = (1, 2)$, $C = (-5, -1)$

Step 1: $d(A, B) = \sqrt{(-1-1)^2 + (6-2)^2}$

$\qquad\qquad = \sqrt{4 + 16} = \sqrt{20}$

$\qquad d(B, C) = \sqrt{(1+5)^2 + (2+1)^2}$

$\qquad\qquad = \sqrt{36 + 9} = \sqrt{45}$

$\qquad d(A, C) = \sqrt{(-1+5)^2 + (6+1)^2}$

$\qquad\qquad = \sqrt{16 + 49} = \sqrt{65}$

If ABC is a right triangle, $\overline{AC}$ would be the hypotenuse since it is the longest side.

Step 2: $\qquad a^2 + b^2 = c^2$

$$\left(\sqrt{20} \right)^2 + \left(\sqrt{45} \right)^2 = \left(\sqrt{65} \right)^2$$

$$20 + 45 = 65$$

$$65 = 65$$

Therefore, ABC is a right triangle.

25.

$$d\left(\overline{AB} \right) = \sqrt{(2+4)^2 + (5+1)^2}$$

$$= \sqrt{36 + 36} = 6\sqrt{2}$$

$$d\left(\overline{BC} \right) = \sqrt{(-4+6)^2 + (-1-7)^2}$$

$$= \sqrt{4 + 64} = 2\sqrt{17}$$

$$d\left(\overline{AC} \right) = \sqrt{(2+6)^2 + (5-7)^2}$$

$$= \sqrt{64 + 4} = 2\sqrt{17}$$

Since $d\left(\overline{BC} \right) = d\left(\overline{AC} \right)$, ABC is an isosceles triangle.

27.

Step 1: Show opposite sides are equal.

$$d(A, B) = \sqrt{(2-4)^2 + (3-2)^2}$$

$$= \sqrt{4+1} = \sqrt{5}$$

$$d(C, D) = \sqrt{(8-6)^2 + (10-11)^2}$$

$$= \sqrt{4+1} = \sqrt{5}$$

$$d(B, C) = \sqrt{(4-8)^2 + (2-10)^2}$$

$$= \sqrt{16+64} = \sqrt{80}$$

$$d(A, D) = \sqrt{(2-6)^2 + (3-11)^2}$$

$$= \sqrt{16+64} = \sqrt{80}$$

Yes, $d(A, B) = d(C, D)$ and $d(B, C) = d(A, D)$

Step 2: Show that adjacent sides are perpendicular.

$$m(A, B) = \frac{2-3}{4-2} = -\frac{1}{2}$$

$$m(B, C) = \frac{10-2}{8-4} = 2$$

$$m(C, D) = \frac{10-11}{8-6} = -\frac{1}{2}$$

$$m(A, D) = \frac{3-11}{2-6} = 2$$

Since adjacent sides are perpendicular, ADCD is a rectangle.

29.

$$10 = \sqrt{(x-2)^2 + (-3-5)^2}$$

$$10 = \sqrt{x^2 - 4x + 4 + 64}$$

$$10 = \sqrt{x^2 - 4x + 68}$$

$$100 = x^2 - 4x + 68$$

$$0 = x^2 - 4x - 32$$

$$0 = (x+4)(x-8)$$

$$x + 4 = 0 \quad \text{or} \quad x - 8 = 0$$

$$x = -4 \qquad\qquad x = 8$$

Answer: $(-4, -3)$ or $(8, -3)$

31.

Step 1: Show that the distance from each point to the center are equal.

$$d(P, C) = \sqrt{(-1-2)^2 + (3+1)^2}$$

$$= \sqrt{9 + 16} = \sqrt{25} = 5$$

#31. continued

$$d(Q,C) = \sqrt{(7-2)^2 + (-1+1)^2}$$
$$= \sqrt{25} = 5$$

$$d(R,C) = \sqrt{(6-2)^2 + (2+1)^2}$$
$$= \sqrt{16+9} = 5$$

Since the distances are equal, all the points lie on a circle whose center is C (2,-1).

Step 2: $5 = \sqrt{(x-2)^2 + (y+1)}$

$25 = (x-2)^2 + (y+1)^2$

33.

A(0,0), B(b,c), $M(x_1, y_1)$, C(a,0) and $N(x_2, y_2)$

Step 1: $M(x_1, y_1) = \left(\dfrac{b+0}{2}, \dfrac{c+0}{2}\right)$

So, $x_1 = \dfrac{b}{2}$, $y_1 = \dfrac{c}{2}$

Thus, $M(x_1, y_1) = M\left(\dfrac{b}{2}, \dfrac{c}{2}\right)$

Step 2: $N(x_2, y_2) = N\left(\dfrac{a+b}{2}, \dfrac{c-0}{2}\right)$

So, $x_2 = \dfrac{a+b}{2}$, $y_2 = \dfrac{c}{2}$

Thus, $N(x_2, y_2) = N\left(\dfrac{a+b}{2}, \dfrac{c}{2}\right)$

Step 3: $d(M,N) = \sqrt{\left(\dfrac{b}{2} - \dfrac{a+b}{2}\right)^2 + \left(\dfrac{c}{2} - \dfrac{c}{2}\right)^2}$

$= \sqrt{\left(\dfrac{-a}{2}\right)^2} = \dfrac{a}{2}$

$d(A,C) = \sqrt{(0-a)^2 + (0-0)^2}$

$= a$

Thus, $d(M,N) = \dfrac{1}{2} d(A,C)$

Exercises 6.7

1.-27.
The graphs are in the answer section of the main text.

Exercises 6.8

1.
$y = kx$; $32 = 4k$
$8 = k$

3.
$y = k\sqrt[3]{x}$; $2 = k\sqrt[3]{64}$
$2 = 4k$
$\dfrac{1}{2} = k$

5.
$y = \dfrac{k}{x^2}$; $\dfrac{1}{4} = \dfrac{k}{6^2}$
$\dfrac{1}{4} = \dfrac{k}{36}$
$4k = 36$
$k = 9$

7.
$y = k\sqrt{x}\sqrt[3]{w}$; $-30 = k\sqrt{25}\sqrt[3]{27}$
$-30 = k \cdot 5 \cdot 3$
$-30 = 15k$
$-2 = k$

9.
$y = \dfrac{x^3}{\sqrt{w}}k$; $3 = \dfrac{2^3}{\sqrt{16}}k$
$3 = \dfrac{8}{4}k$
$3 = 2k$
$\dfrac{3}{2} = k$

11.
$C = kr$; $k = \dfrac{C}{r}$
$k = \dfrac{10\pi}{5} = 2\pi$

13.
$A = kE^2$; $k = \dfrac{A}{E^2}$
$k = \dfrac{96}{4^2} = 6$

15.

$$A = kH(B_1 + B_2); \quad k = \frac{A}{H(B_1 + B_2)}$$

$$k = \frac{18}{3(5+7)} = \frac{18}{36} = \frac{1}{2}$$

17.

$$H = \frac{kV}{R^2}; \quad k = \frac{HR^2}{V}$$

$$k = \frac{7 \cdot 6^2}{84\pi} = \frac{252}{84\pi} = \frac{3}{\pi}$$

19.

$$y = kx^3$$

Find k | Find y
$32 = 8k$ | $y = 4 \cdot 3^3$
$4 = k$ | $y = 108$

21.

$$y = \frac{k}{x}$$

Find k | Find y
$2 = \frac{k}{5}$ | $y = \frac{10}{4}$
$10 = k$ | $y = \frac{5}{2}$

23.

$$y = xw^3 k$$

Find k | Find y
$162 = 1 \cdot 3^3 k$ | $y = 6 \cdot 2^2 \cdot 6$
$6 = k$ | $y = 144$

25.

$$y = \frac{x}{\sqrt{w}} k$$

Find k | Find y
$6 = \frac{3}{\sqrt{16}} k$ | $y = \frac{6}{\sqrt{25}} \cdot 8$
$6 = \frac{3}{4} k$ | $y = \frac{6}{5} \cdot 8$
$8 = k$ | $y = \frac{48}{5}$

27.

$$y = \frac{xw^2}{z} \cdot k$$

Find k | Find y
$4 = \frac{3 \cdot 2^2}{6} k$ | $y = \frac{4 \cdot 3^2}{12} \cdot 2$
$4 = 2k$ | $y = \frac{36}{12} \cdot 2$
$2 = k$ | $y = 3 \cdot 2 = 6$

29.

$$d = kt^2$$

Find k | Find d
$4 = \left(\frac{1}{2}\right)^2 k$ | $d = 16 \cdot 3^2$
$4 = \frac{1}{4} k$ | $d = 144$
$16 = k$

Answer: 144 ft

31.

$$F = kc$$

Find k | | Find F
$12 = 2k$ Note: $32 - 30 = 2$ | | $F = 6 \cdot 7$
$6 = k$ | | $F = 42$
| | Answer: 42 lb

33.

$$I = RTk$$

Find k | | Find I
$180 = 0.05 \cdot 3k$ | | $I = 0.06 \cdot \frac{5}{4} \cdot 1200$
$180 = 0.15k$ | | $I = 90$
$1200 = k$ | | Answer: $90

35.

$$G = \frac{k}{h}$$

Find k | Find G
$0.25 = \frac{k}{8}$ | $G = \frac{2}{2}$
$2 = k$ | $G = 1$
| Answer: 100%

37.

$$V = \frac{t}{p}k$$

Find k

$$80 = \frac{400}{12}k$$

$$960 = 400k$$

$$\frac{12}{5} = k$$

Find p

$$54 = \frac{450}{p} \cdot \frac{12}{5}$$

$$54 = \frac{1080}{p}$$

$$54p = 1080$$

$$p = 20$$

Answer: 20 lb per
sq in

39.

$$F = \frac{m_1 m_2}{d^2}k$$

Find k

$$5 \times 10^{-18} = \frac{1.5 \times 10^{-6} \times 1.8 \times 10^{-6}}{\left(6.0 \times 10^{-6}\right)^2}k$$

$$5 \times 10^{-18} = \frac{2.7 \times 10^{-12}}{36 \times 10^{-12}} \cdot k$$

$$5 \times 10^{-18} = .075k$$

$$\frac{5 \times 10^{-18}}{7.5 \times 10^{-2}} = k$$

$$6.67 \times 10^{-17} = k$$

Find F

$$F = \frac{5.0 \times 10^1 \times 9.05 \times 10^1}{\left(5.79 \times 10^{-2}\right)^2} \cdot 6.67 \times 10^{-17}$$

$$F = \frac{45.25 \times 10^2}{33.5241 \times 10^{-4}} \cdot 6.67 \times 10^{-7}$$

$$F = \frac{45.25 \times 10^2 \times 6.67 \times 10^{-17}}{33.5241 \times 10^{-4}}$$

$$F = \frac{45.25 \times 6.67}{33.5241} \times 10^{-11}$$

$$F = 9 \times 10^{-11}$$

Exercises 6.9

1.-27.
The graphs are in the answer section of the main text.

Review Exercises

1.-13.
The graphs and answer are in the answer section of
the main text.

15.
$$g(2) = 5 \cdot 2 - 2 = 8$$

17.
$$[f(-3)]^2 = \left[4 \cdot (-3)^2 - 3(-3) + 7\right]^2$$
$$= (36 + 9 + 7)^2 = (52)^2 = 2704$$

19.
$$f(x) + g(x) = 4x^2 - 3x + 7 + 5x - 2 = 4x^2 + 2x + 5$$

21.
$$g(a) + g(h) = 5a - 2 + 5h - 2 = 5a + 5h - 4$$

23.
$$\frac{g(a+h) - g(a)}{h} = \frac{5(a+h) - 2 - (5a - 2)}{h}$$
$$= \frac{5a + 5h - 2 - 5a + 2}{h} = \frac{5h}{h} = 5$$

25.-27.
The graphs are in the answer section of the main text.

31.
$$m = \frac{0 - 2}{-3 - 0} = \frac{-2}{-3} = \frac{2}{3}$$

33.
$$m = \frac{7 - 0}{3 - 0} = \frac{7}{3}$$

35.
$$m = \frac{7 + 4}{2 - 2} = \frac{11}{0} \qquad \text{undefined}$$

37.
$$y = 2x - 1$$

x	y
0	-1
1	1

$$m = \frac{-1 - 1}{0 - 1} = \frac{-2}{-1} = 2$$

39.

$x + 2y = 5$

$2y = -x + 5$

$y = -\dfrac{1}{2}x + \dfrac{5}{2}$

x	y
0	$\dfrac{5}{2}$
1	2

$m = \dfrac{\dfrac{5}{2} - 2}{0 - 1} = \dfrac{\dfrac{1}{2}}{-1} = -\dfrac{1}{2}$

41.

$x = -2$ Slope is undefined.

43.

$y + 3 = 4(x - 5)$

$y + 3 = 4x - 20$

$y = 4x - 23$

45.

$y - 1 = 0(x + 2)$

$y - 1 = 0$

$y = 1$

47.

$(0, 2), \quad (-3, 0)$

$m = \dfrac{0 - 2}{-3 - 0} = \dfrac{2}{3}$

$y - 2 = \dfrac{2}{3}(x - 0)$

$y - 2 = \dfrac{2}{3}x$

$y = \dfrac{2}{3}x + 2$

49.

$(0, 0), \quad (3, 7)$

$m = \dfrac{7 - 0}{3 - 0} = \dfrac{7}{3}$

$y - 0 = \dfrac{7}{3}(x - 0)$

$y = \dfrac{7}{3}x$

51.

$(2, -4), \quad (2, 7)$

$m = \dfrac{7 + 4}{2 - 2} = \dfrac{11}{0}$ Undefined

This is a vertical line through $x = 2$.

53.

$(3, -4); \quad m = 3$

$y + 4 = 3(x - 3)$

$y + 4 = 3x - 9$

$y = 3x - 13$

55.

$(7, -1); \quad m = 0$ Note: x axis has slope of 0.

$y + 1 = 0(x - 7)$

$y + 1 = 0$

$y = -1$

57.

$(-5, 2); \quad m = -\dfrac{4}{5}$

$y - 2 = -\dfrac{4}{5}(x + 5)$

$y - 2 = -\dfrac{4}{5}x - 4$

$y = -\dfrac{4}{5}x - 2$

59.

$(4, -6);$ slope is undefined.

The line is vertical passing through $x = 4$.

61.

$(-1, 2), \quad (5, 10)$

$d = \sqrt{(-1 - 5)^2 + (2 - 10)^2} = \sqrt{36 + 64} = \sqrt{100} = 10$

$M = \left(\dfrac{-1 + 5}{2}, \dfrac{2 + 10}{2} \right) = (2, 6)$

63.

$(-2, -5), \quad (4, 3)$

$d = \sqrt{(-2 - 4)^2 + (-5 - 3)^2} = \sqrt{36 + 64} = \sqrt{100} = 10$

$M = \left(\dfrac{-2 + 4}{2}, \dfrac{-5 + 3}{2} \right) = (1, -1)$

65.

$(1,-3), \quad (9,-3)$

$d = \sqrt{(1-9)^2 + (-3+3)^2} = \sqrt{64} = 8$

$M = \left(\dfrac{1+9}{2}, \dfrac{-3-3}{2} \right) = (5,-3)$

67.-71.

The graphs are in the answer section of the main text.

73.

a. $A = ks^2$

b. $9\sqrt{3} = k \cdot 6^2$

$\dfrac{9\sqrt{3}}{36} = k$

$\dfrac{\sqrt{3}}{4} = k$

75.

$I = kP$

Find k	Find I

$2000 = 100,000k \qquad I = 165,000 \cdot \dfrac{1}{50}$

$\dfrac{1}{50} = k \qquad\qquad I = 3,300$

Answer: $3,300$ persons

77.-83.

The graphs and answer are in the answer section of the main text.

Chapter 6 Test Solution

1.

Domain = {all real numbers}

Range = $\{y | y \geq -3\}$

3.

$f(-3) = \dfrac{3}{-3-2} = -\dfrac{3}{5}$

5.

$[f(3)]^2 = \left(\dfrac{3}{3-2} \right)^2 = (3)^2 = 9$

7.

Domain = $\{x | x \neq 2\}$ since $x - 2 \neq 0$ and $x \neq 2$

9.

$$\dfrac{g(a+h) - g(a)}{h} = \dfrac{(a+h)^2 + 1 - (a^2 + 1)}{h}$$

$$= \dfrac{a^2 + 2ah + h^2 + 1 - a^2 - 1}{h}$$

$$= \dfrac{\cancel{h}(2a+h)}{\cancel{h}} = 2a + h$$

11.

The graph is in the answer section of the main text.

13.

$(-4,1); \quad m = \dfrac{1}{4}$

$y - 1 = \dfrac{1}{4}(x+4)$

$y - 1 = \dfrac{1}{4}x + 1$

$y = \dfrac{1}{4}x + 2$

15.

$(1,-2), \quad \left(\dfrac{7}{2}, 4 \right)$

$d = \sqrt{\left(1 - \dfrac{7}{2}\right)^2 + (-2-4)^2} = \sqrt{\left(\dfrac{-5}{2}\right)^2 + (-6)^2}$

$$= \sqrt{\dfrac{25}{4} + 36}$$

$$= \sqrt{\dfrac{25}{4} + \dfrac{144}{4}}$$

$$= \sqrt{\dfrac{169}{4}} = \dfrac{13}{2}$$

$M = \left(\dfrac{1 + \dfrac{7}{2}}{2}, \dfrac{-2+4}{2} \right) = \left(\dfrac{\dfrac{9}{2}}{2}, 1 \right) = \left(\dfrac{9}{4}, 1 \right)$

17.

The graph is in the answer section of the main text.

Test Your Memory

1.

$$3^0 + 2^{-1} - \left(\frac{2}{3}\right)^{-2} = 1 + \frac{1}{2} - \frac{9}{4} = \frac{3}{2} - \frac{9}{4} = -\frac{3}{4}$$

3.

$$\frac{\left(3x^{-2}y^3\right)^2}{\left(2x^4y^{-3}\right)^{-3}} = \frac{3^2 x^{-4} y^6}{2^{-3} x^{-12} y^9} = \frac{3^2 \cdot 2^3 x^{-4+12}}{y^{9-6}} = \frac{72x^8}{y^3}$$

5.

$$\frac{3x\left(x^2 - 3x + 9\right)}{9x^2(3x-7)(x-1)} \cdot \frac{(3x-7)(x+3)(x-3)}{(x+3)\left(x^2 - 3x + 9\right)}$$

$$= \frac{3x(x-3)}{3 \cdot 3x \cdot x(x-1)} = \frac{x-3}{3x(x-1)}$$

7.

$$\frac{\left(\dfrac{3x}{x+1} - 4\right) \cdot \dfrac{x+1}{1}}{\left(\dfrac{x}{x+1} - 1\right) \cdot \dfrac{x+1}{1}} = \frac{3x - 4(x+1)}{x - (x+1)}$$

$$= \frac{3x - 4x - 4}{x - x - 1} = \frac{-x - 4}{-1} = x + 4$$

9.

$$2\sqrt{45x} - \sqrt{500x} + \sqrt{20x}$$
$$= 2\sqrt{9 \cdot 5x} - \sqrt{100 \cdot 5x} + \sqrt{4 \cdot 5x}$$
$$= 6\sqrt{5x} - 10\sqrt{5x} + 2\sqrt{5x}$$
$$= -2\sqrt{5x}$$

11.

$$\sqrt{\frac{7x}{4y} \cdot \frac{y}{y}} = \sqrt{\frac{7xy}{4y^2}} = \frac{\sqrt{7xy}}{\sqrt{4y^2}} = \frac{\sqrt{7xy}}{2y}$$

13.

$$\left(4\sqrt{x} + 3\right)^2 = \left(4\sqrt{x}\right)^2 + 2\left(3 \cdot 4\sqrt{x}\right) + (3)^2$$
$$= 16x + 24\sqrt{x} + 9$$

15.

$$\sqrt{-4} \cdot \sqrt{-36} = \sqrt{4}\sqrt{-1} \cdot \sqrt{36}\sqrt{-1} = 2i \cdot 6i = 12i^2 = -12$$

17.
The answer is in the answer section of the main text.

19.
$$f(-2) = 5(-2) - 4 = -10 - 4 = -14$$

21.
$$[f(3)]^2 = (5 \cdot 3 - 4)^2 = (15 - 4)^2 = 121$$

23.
$$g(2a) = (2a)^2 + 7 = 4a^2 + 7$$

25.
The graph is in the answer section of the main text.

27.
$(3, 9)$, $(-5, -1)$

$$m = \frac{-1 - 9}{-5 - 3} = \frac{-10}{-8} = \frac{5}{4}$$

$$y - 9 = \frac{5}{4}(x - 3)$$

$$y - 9 = \frac{5}{4}x - \frac{15}{4}$$

$$y = \frac{5}{4}x - \frac{15}{4} + 9$$

$$y = \frac{5}{4}x + \frac{21}{4}$$

29.
$(-4, -7)$; $m = -\dfrac{4}{3}$

$$y + 7 = -\frac{4}{3}(x + 4)$$

$$y + 7 = -\frac{4}{3}x - \frac{16}{3}$$

$$y = -\frac{4}{3}x - \frac{16}{3} - 7$$

$$y = -\frac{4}{3}x - \frac{37}{3}$$

31.
$$4x - 1 < 9$$
$$4x < 10$$
$$x < \frac{10}{4}$$
$$x < \frac{5}{2}$$

The graph is in the answer section of the main text.

33.
The graph is in the answer section of the main text.

35.
$$\left(\sqrt{2x-3}\right)^2 = (7)^2$$
$$2x - 3 = 49$$
$$2x = 52$$
$$x = 26$$
Solution set: $\{26\}$

37.
$$|5x - 2| = 6$$
$$5x - 2 = 6 \quad \text{or} \quad 5x - 2 = -6$$
$$5x = 8 \qquad\qquad 5x = -4$$
$$x = \frac{8}{5} \qquad\qquad x = -\frac{4}{5}$$
Solution set: $\left\{\frac{8}{5}, -\frac{4}{5}\right\}$

39.
$$\frac{(x+2)(x-1)}{1}\left(\frac{3x-1}{x+2} + \frac{3}{x-1}\right)$$
$$= \left(\frac{x+15}{(x+2)(x-1)}\right)\frac{(x+2)(x-1)}{1}$$
$$(x-1)(3x-1) + 3(x+2) = x+15$$
$$3x^2 - 4x + 1 + 3x + 6 = x + 15$$
$$3x^2 - x + 7 = x + 15$$
$$3x^2 - 2x - 8 = 0$$
$$(3x+4)(x-2) = 0$$
$$3x + 4 = 0 \qquad x - 2 = 0$$
$$x = -\frac{4}{3} \qquad x = 2$$
Solution set: $\left\{-\frac{4}{3}, 2\right\}$

41.
$$\sqrt{2x} + 1 = 4x$$
$$\left(\sqrt{2x}\right)^2 = (4x - 1)^2$$
$$2x = 16x^2 - 8x + 1$$
$$0 = 16x^2 - 10x + 1$$
$$0 = (2x - 1)(8x - 1)$$

#41. continued
$$2x - 1 = 0 \quad \text{or} \quad 8x - 1 = 0$$
$$x = \frac{1}{2} \qquad\qquad x = \frac{1}{8}$$
Solution set: $\left\{\frac{1}{2}\right\}$ $\qquad$ $\frac{1}{8}$ does not check.

43.
$$\frac{x+2}{1}\left(1 + \frac{6}{x+2}\right) = \left(\frac{2x+10}{x+2}\right)\frac{x+2}{1}$$
$$x + 2 + 6 = 2x + 10$$
$$x + 8 = 2x + 10$$
$$-2 = x$$
Solution set: $\varnothing$ $\qquad$ Note: $x + 2 \neq 0$
$\qquad\qquad\qquad\qquad\qquad$ So, $x \neq -2$

45.
Let x = height
Then $2x + 4$ = base
So, $A = \frac{1}{2}bh$
$$63 = \frac{1}{2}x(2x + 4)$$
$$63 = x^2 + 2x$$
$$0 = x^2 + 2x - 63$$
$$0 = (x - 7)(x + 9)$$
$$x - 7 = 0 \quad \text{or} \quad x + 9 = 0$$
$$x = 7 \qquad\qquad \cancel{x = -9}$$
$$2x + 4 = 2 \cdot 7 + 4 = 18$$
Answer: Height: 7 in $\quad$ Base: 18 in

47.
$$\frac{1}{8} - \frac{1}{10} = \frac{1}{x}$$
$$\frac{40x}{1}\left(\frac{1}{8} - \frac{1}{10}\right) = \frac{1}{x} \cdot \frac{40x}{1}$$
$$5x - 4x = 40$$
$$x = 40$$
Answer: 40 hours

49.
$$d = kt^2$$

Find k $\qquad\qquad$ Find d
$144 = 3^2 k \qquad\qquad d = 16 \cdot 5^2$
$16 = x \qquad\qquad\quad d = 400$
$\qquad\qquad\qquad\qquad$ Answer: 400 ft

Chapter 6 Study Guide

Self-Test Exercises

I. For the relations defined by the following equations, find the domains and ranges and graph them.

1. $y = x^2 + 1$

2. $y = \sqrt{x - 2}$

II. Let $f(x) = 3/(x-2)$ and $g(x) = x^2 + 1$. Find the following.

3. $f(-2)$

4. $g(0)$

5. $[g(2)]^{-1}$

6. $g(2x)$

7. Domain of g

8. $\left(\dfrac{f}{g}\right)(x)$

9. $f(-x)$

10. $g(a-b)$

III. Graph the following lines.

11. $5x - 2y = 8$; also find both intercepts and slope.

12. Passing through $(-4, 5)$ with slope -2.

IV. Find the equations of the following lines. Write the equations in slope-intercept form, if possible.

13. Through $(10,3)$ and $(0,-2)$

14. Through $(-2, 4)$ and perpendicular to $4x - 2y = 8$

15. Through $(-4,-5)$ and parallel to $2x + y = -3$

16. Find the length and midpoint of the line segment with endpoints $(1,10)$ and $(7,2)$.

V. Graph the solution sets of the following inequalities.

17. $-2 \leq 2x + y$

18. $\begin{cases} 2y - x \leq 4 \\ x > 5 \end{cases}$

19. P varies inversely as the square square of R. If P=10 when R=6, find P when R=20.

20. The volume of a liquid varies inversely as its pressure. If the pressure 200 mm of mercury when the volume is 800 cc, find the volume when the pressure is 25 mm of mercury.

The worked-out solutions begin on the next page.

Self-Test Solutions

1.

$y = x^2 + 1$

Domain = {all real numbers}

Range = $\{y \mid y \geq 1\}$

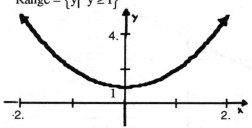

2.

$y = \sqrt{x - 2}$

Domain = $\{x \mid x \geq 2\}$

Range = $\{y \mid y \geq 0\}$

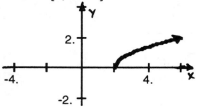

3.

$f(-2) = \dfrac{3}{(-2-2)} = \dfrac{3}{-4} = -\dfrac{3}{4}$

4.

$g(0) = 0^2 + 1 = 1$

5.

$[g(2)]^{-1} = (2^2 + 1)^{-1} = (5)^{-1} = \dfrac{1}{5}$

6.

$g(2x) = (2x)^2 + 1 = 4x^2 + 1$

7.

{all real numbers}

8.

$\left(\dfrac{f}{g}\right)(x) = \dfrac{\frac{3}{x-2}}{x^2+1} = \dfrac{3}{(x-2)(x^2+1)}$

9.

$f(-x) = \dfrac{3}{-x-2} = -\dfrac{3}{x+2}$

10.

$g(a-b) = (a-b)^2 + 1$

$\qquad\quad = a^2 - 2ab + b^2 + 1$

11.

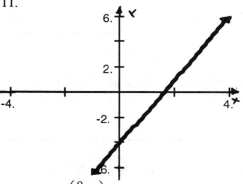

x intercept: $\left(\dfrac{8}{5}, 0\right)$ y intercept: $(0, -4)$

$m = \dfrac{5}{2}$

12.

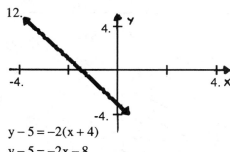

$y - 5 = -2(x + 4)$

$y - 5 = -2x - 8$

$\quad y = -2x - 3$

13.

$(10, 3), (0, -2)$

$m = \dfrac{-2-3}{0-10} = \dfrac{-5}{-10} = \dfrac{1}{2}$

#13. continued

$$y + 2 = \frac{1}{2}(x - 0)$$

$$y + 2 = \frac{1}{2}x$$

$$y = \frac{1}{2}x - 2$$

14.

$$(-2, 4); \ m = -\frac{1}{2}$$

$$y - 4 = -\frac{1}{2}(x + 2)$$

$$y - 4 = -\frac{1}{2}x - 1$$

$$y = -\frac{1}{2}x + 3$$

15.

$$(-4, -5); \ m = -2$$

$$y + 5 = -2(x + 4)$$

$$y + 5 = -2x - 8$$

$$y = -2x - 13$$

16.

$$(1, 10) \text{ and } (7, 2)$$

$$d = \sqrt{(1 - 7)^2 + (10 - 2)^2}$$

$$= \sqrt{(-6)^2 + (8)^2} = \sqrt{36 + 64} = \sqrt{100} = 10$$

$$M = \left(\frac{1 + 7}{2}, \frac{10 + 2}{2}\right) = (4, 6)$$

17.

$$-2 \le 2x + y$$

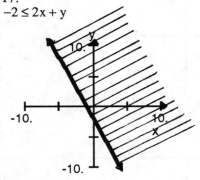

18.

$$\begin{cases} 2y - x \le 4 \\ x > 5 \end{cases}$$

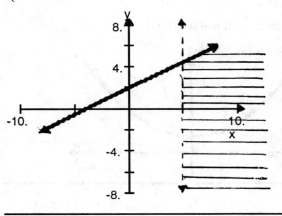

19.

$$P = \frac{k}{R^2}$$

Find k	Find P
$10 = \dfrac{k}{6^2}$	$P = \dfrac{360}{(20)^2}$
$10 = \dfrac{k}{36}$	$P = \dfrac{360}{400}$
$360 = k$	$P = \dfrac{9}{10}$

20.

$$V = \frac{k}{P}$$

Find k	Find V
$800 = \dfrac{k}{200}$	$V = \dfrac{160,000}{25}$
$160,000 = k$	$V = 6,400$
	Answer: $6,400$ cc

CHAPTER 7 QUADRATIC AND HIGHER-DEGREE EQUATIONS AND INEQUALITIES

Solutions to Text Exercises

1.-41.
The graphs are in the answer section of the main text.

The graphs are in the answer section of the main text.

1.

$x^2 + 6x + \underline{9}$ Note: $\left(\dfrac{6}{2}\right)^2 = 9$

3.

$x^2 - 22x + \underline{121}$ Note: $\left(\dfrac{22}{2}\right)^2 = 121$

5.

$x^2 + 3x + \dfrac{9}{\underline{4}}$ Note: $\left(\dfrac{3}{2}\right)^2 = \dfrac{9}{4}$

7.

$x^2 - 7x + \dfrac{49}{\underline{4}}$ Note: $\left(\dfrac{7}{2}\right)^2 = \dfrac{49}{4}$

9.
$y = x^2 - 2x + 5$

$y = x^2 - 2x + 1 + 5 - 1$ Note: $\left(\dfrac{-2}{2}\right)^2 = 1$

$y = (x - 1)^2 + 4$ Vertex $(1, 4)$

11.

$y = x^2 + 2x + 1 + 1 - 1$ Note: $\left(\dfrac{2}{2}\right)^2 = 1$

$y = (x + 1)^2 + 0$ Vertex $(-1, 0)$

13.

$y = x^2 + 10x + 25 + 22 - 25$ Note: $\left(\dfrac{10}{2}\right)^2 = 25$

$y = (x + 5)^2 - 3$ Vertex $(-5, -3)$

The graphs are in the answer section of the main text.

15.

$y = x^2 - 4x + 4 + 2 - 4$ Note: $\left(\dfrac{-4}{2}\right)^2 = 4$

$y = (x - 2)^2 - 2$ Vertex $(2, -2)$

17.
$y = -x^2 + 6x \qquad - 7$

$y = -\left(x^2 - 6x + 9\right) - 7 + 9$ Note: $\left(\dfrac{-6}{2}\right)^2 = 9$

$y = -(x - 3)^2 + 2$ Vertex $(3, 2)$

19.
$y = -x^2 - 6x \qquad + 1$

$y = -\left(x^2 + 6x + 9\right) + 1 + 9$ Note: $\left(\dfrac{6}{2}\right)^2 = 9$

$y = -(x + 3)^2 + 10$ Vertex $(-3, 10)$

21.
$y = -x^2 + 6x \qquad - 9$

$y = -\left(x^2 - 6x + 9\right) - 9 + 9$ Note: $\left(\dfrac{-6}{2}\right)^2 = 9$

$y = -(x - 3)^2 + 0$ Vertex $(3, 0)$

23.
$y = -x^2 - 4x \qquad - 3$

$y = -\left(x^2 + 4x + 4\right) - 3 + 4$ Note: $\left(\dfrac{4}{2}\right)^2 = 4$

$y = -(x + 2)^2 + 1$ Vertex $(-2, 1)$

25.
$y = 2x^2 - 12x \qquad + 17$

$y = 2\left(x^2 - 6x + 9\right) + 17 - 18$ Note: $\left(\dfrac{-6}{2}\right) = 9$

$y = 2(x - 3)^2 - 1$ Vertex: $(3, -1)$

CHAPTER 7 QUADRATIC AND HIGHER-DEGREE EQUATIONS AND INEQUALITIES

The graphs are in the answer section of the main text.

27.

$y = 2x^2 - 16x \quad + 32$

$y = 2(x^2 - 8x + 16) + 32 - 32 \quad$ Note: $\left(\dfrac{-8}{2}\right)^4 = 16$

$y = 2(x - 4)^2 + 0 \quad\quad\quad$ Vertex $(4, 0)$

29.

$y = 3x^2 - 12x \quad + 17$

$y = 3(x^2 - 4x + 4) + 17 - 12 \quad$ Note: $\left(\dfrac{-4}{2}\right)^2 = 4$

$y = 3(x - 2)^2 + 5 \quad\quad\quad$ Vertex $(2, 5)$

31.

$y = 3x^2 + 12x \quad + 5$

$y = 3(x^2 + 4x + 4) + 5 - 12 \quad$ Note: $\left(\dfrac{-4}{2}\right)^2 = 4$

$y = 3(x + 2)^2 - 7 \quad\quad\quad$ Vertex $(-2, -7)$

33.

$y = -2x^2 + 12x \quad - 23$

$y = -2(x^2 - 6x + 9) - 23 + 18 \quad$ Note: $\left(\dfrac{-6}{2}\right)^2 = 9$

$y = -2(x - 3)^2 - 5 \quad\quad\quad$ Vertex $(3, -5)$

35.

$y = -4x^2 - 40x \quad - 92$

$y = -4(x^2 + 10x + 25) - 92 + 100 \quad$ Note: $\left(\dfrac{10}{2}\right)^2 = 25$

$y = -4(x + 5)^2 + 8 \quad\quad\quad$ Vertex $(-5, 8)$

37.

$y = -3x^2 - 6x \quad - 3$

$y = -3(x^2 + 2x + 1) - 3 + 3 \quad$ Note: $\left(\dfrac{2}{2}\right)^2 = 1$

$y = -3(x + 1)^2 + 0 \quad\quad\quad$ Vertex $(-1, 0)$

The graphs are in the answer section of the main text.

39.

$y = x^2 - x + \dfrac{1}{4} + 1 - \dfrac{1}{4} \quad$ Note: $\left(\dfrac{-1}{2}\right)^2 = \dfrac{1}{4}$

$y = \left(x - \dfrac{1}{2}\right)^2 + \dfrac{3}{4} \quad\quad$ Vertex $\left(\dfrac{1}{2}, \dfrac{3}{4}\right)$

41.

$y = x^2 + x + \dfrac{1}{4} + 2 - \dfrac{1}{4} \quad$ Note: $\left(\dfrac{1}{2}\right)^2 = \dfrac{1}{4}$

$y = \left(x + \dfrac{1}{2}\right)^2 + \dfrac{7}{4} \quad\quad$ Vertex $\left(-\dfrac{1}{2}, \dfrac{7}{4}\right)$

43.

$y = x^2 - \dfrac{4x}{3} + \dfrac{4}{9} + 1 - \dfrac{4}{9}$

$\quad\quad$ Note: $\left(\dfrac{-4}{3} \cdot \dfrac{1}{2}\right)^2 = \left(-\dfrac{2}{3}\right)^2 = \dfrac{4}{9}$

$y = \left(x - \dfrac{2}{3}\right)^2 + \dfrac{5}{9} \quad\quad$ Vertex $\left(\dfrac{2}{3}, \dfrac{5}{9}\right)$

45.

$y = x^2 + \dfrac{5}{2}x + \dfrac{25}{16} + 2 - \dfrac{25}{16} \quad$ Note: $\left(\dfrac{5}{2} \cdot \dfrac{1}{2}\right)^2 = \dfrac{25}{16}$

$y = \left(x + \dfrac{5}{4}\right)^2 + \dfrac{7}{16} \quad\quad$ Vertex $\left(-\dfrac{5}{4}, \dfrac{7}{16}\right)$

47.

$y = 2x^2 + 2x \quad - 1$

$y = 2\left(x^2 + x + \dfrac{1}{4}\right) - 1 - \dfrac{1}{2} \quad$ Note: $\left(\dfrac{1}{2}\right)^2 = \dfrac{1}{4}$

$y = 2\left(x + \dfrac{1}{2}\right)^2 - \dfrac{3}{2} \quad\quad$ Vertex $\left(-\dfrac{1}{2}, -\dfrac{3}{2}\right)$

49.

$y = 3x^2 - 3x \quad - 1$

$y = 3\left(x^2 - x + \dfrac{1}{4}\right) - 1 - \dfrac{3}{4} \quad$ Note: $\left(\dfrac{-1}{2}\right)^2 = \dfrac{1}{4}$

$y = 3\left(x - \dfrac{1}{2}\right)^2 - \dfrac{7}{4} \quad\quad$ Vertex $\left(\dfrac{1}{2}, -\dfrac{7}{4}\right)$

The graphs are in the answer section of the main text.

51.

$y = -2x^2 - x \qquad - 1$

$y = -2\left(x^2 + \dfrac{1}{2}x + \dfrac{1}{16}\right) - 1 + \dfrac{1}{8}$ Note: $\left(-\dfrac{1}{2} \cdot \dfrac{1}{2}\right)^2 = \dfrac{1}{16}$

$y = -2\left(x + \dfrac{1}{4}\right)^2 - \dfrac{7}{8} \qquad$ Vertex $\left(-\dfrac{1}{4}, -\dfrac{7}{8}\right)$

53.

$y = -2x^2 + \dfrac{8}{3}x \qquad + \dfrac{1}{9}$

$y = -2\left(x^2 - \dfrac{4}{3}x + \dfrac{4}{9}\right) + \dfrac{1}{9} + \dfrac{8}{9}$ Note: $\left(\dfrac{-4}{3} \cdot \dfrac{1}{2}\right)^2 = \dfrac{4}{9}$

$y = -2\left(x - \dfrac{2}{3}\right)^2 + 1 \qquad$ Vertex $\left(\dfrac{2}{3}, 1\right)$

55.-63.
The graphs are in the answer section of the main text.

65.

$y = -3x^2 + 720x - 17,800$

$y = -3\left(x^2 - 240x + 14,400\right) - 17,800 + 43,200$

$y = -3(x - 120)^2 + 25,400$

Answer: 120 dolls, $25,400

67.
Let x = width.
Then $120 - 2x$ = length.
Let y = area.

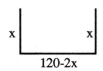

120-2x

Thus $y = x(120 - 2x)$

$\qquad y = 120x - 2x^2$

$\qquad y = -2x^2 + 120x$

$\qquad y = -2\left(x^2 - 60x + 900\right) + 1800$

$\qquad y = -2(x - 30)^2 + 1800$

Vertex $(30, 1800)$
Maximum at $x = 30$
Note: $120 - 2x = 120 - 2 \cdot 30 = 60$
Answer: 30 ft by 60 ft

69.
$y = 2(x - 25)^2 + 1250$
25 ft by 50 ft

71.
$h = -16t^2 + 384t$

$h = -16\left(t^2 - 24t + 144\right) + 2304$

$h = -16(t - 12)^2 + 2304 \qquad$ Maximum at $t = 12$.

Answer: Maximum height at 12 sec.
$\qquad\qquad$ Maximum height = 2304 ft.

$0 = -16t^2 + 384t$

$0 = -16t(t - 24)$

$-16t = 0 \quad$ or $\quad t - 24 = 0$

$\qquad t = 0 \qquad\qquad t = 24$

Answer: Hits ground after 24 sec.

Exercises 7.3

1.-7.
Exercises do not require worked - out solutions.

9.
$x^2 = -196$

$x = \pm\sqrt{-196}$

$x = \pm 14i \qquad$ Note: $\sqrt{-196} = \sqrt{196}\sqrt{-1} = 14i$

$\{\pm 14i\}$

11.
$x^2 = -121$

$x = \pm\sqrt{-121}$

$x = \pm 11i \qquad$ Note: $\sqrt{-121} = \sqrt{121}\sqrt{-1} = 11i$

13.
$x^2 - 64 = 0$

$\qquad x^2 = 64$

$\qquad x = \pm\sqrt{64}$

$\qquad x = \pm 8$

15.
$x^2 + 9 = 0$

$\qquad x^2 = -9$

$\qquad x = \pm\sqrt{-9}$

$\qquad x = \pm 3i$

17.
$$4x^2 + 81 = 0$$
$$4x^2 = -81$$
$$x^2 = -\frac{81}{4}$$
$$x = \pm\sqrt{-\frac{81}{4}}$$
$$x = \pm\frac{9}{2}i \qquad \text{Note: } \sqrt{\frac{-81}{4}} = \frac{\sqrt{-81}}{\sqrt{4}} = \frac{9i}{2}$$

19.
$$36x^2 - 121 = 0$$
$$36x^2 = 121$$
$$x^2 = \frac{121}{36}$$
$$x = \pm\sqrt{\frac{121}{36}}$$
$$x = \pm\frac{11}{6}$$

21.
$$5x^2 - 8 = 0$$
$$5x^2 = 8$$
$$x^2 = \frac{8}{5}$$
$$x = \pm\sqrt{\frac{8}{5}}$$
$$x = \pm\frac{2\sqrt{10}}{5} \qquad \text{Note: } \sqrt{\frac{8}{5}\cdot\frac{5}{5}} = \frac{\sqrt{40}}{5} = \frac{2\sqrt{10}}{5}$$

23.
$$7x^2 - 2 = 0$$
$$7x^2 = 2$$
$$x^2 = \frac{2}{7}$$
$$x = \pm\sqrt{\frac{2}{7}}$$
$$x = \pm\frac{\sqrt{14}}{7} \qquad \text{Note: } \sqrt{\frac{2}{7}\cdot\frac{7}{7}} = \frac{\sqrt{14}}{7}$$

25.
$$9x^2 + 1 = 5$$
$$9x^2 = 4$$
$$x^2 = \frac{4}{9}$$
$$x = \pm\sqrt{\frac{4}{9}}$$
$$x = \pm\frac{2}{3}$$

27.
$$4x^2 + 8 = 4$$
$$4x^2 = -4$$
$$x^2 = -1$$
$$x = \pm\sqrt{-1}$$
$$x = \pm i$$

29.
$$2x^2 - 3 = 8$$
$$2x^2 = 11$$
$$x^2 = \frac{11}{2}$$
$$x = \pm\sqrt{\frac{11}{2}}$$
$$x = \pm\frac{\sqrt{22}}{2} \qquad \text{Note: } \sqrt{\frac{11}{2}\cdot\frac{2}{2}} = \frac{\sqrt{22}}{2}$$

31.
$$(x - 5)^2 = 9$$
$$x - 5 = \pm 3$$
$$x = 5 \pm 3; \qquad x = 5 + 3 = 8; \quad x = 5 - 3 = 2$$

33.
$$(2x - 7)^2 = 1$$
$$2x - 7 = \pm 1$$
$$2x = 7 \pm 1$$
$$x = \frac{7 \pm 1}{2}; \qquad x = \frac{7 + 1}{2} = 4; \quad x = \frac{7 - 1}{2} = 3$$

35.

$$(2x-1)^2 = 0$$
$$2x - 1 = 0$$
$$2x = 1$$
$$x = \frac{1}{2}$$

37.

$$(4x-5)^2 = 6$$
$$4x - 5 = \pm\sqrt{6}$$
$$4x = 5 \pm \sqrt{6}$$
$$x = \frac{5 \pm \sqrt{6}}{4}$$

39.

$$(2x+5)^2 = -49$$
$$2x + 5 = \pm\sqrt{-49}$$
$$2x = -5 \pm 7i$$
$$x = \frac{-5 \pm 7i}{2}$$

41.

$$(x-1)^2 = -75$$
$$x - 1 = \pm\sqrt{-75}$$
$$x = 1 \pm 5\sqrt{3}i$$

43.

$$\left(5x + \frac{2}{3}\right)^2 = \frac{5}{9}$$
$$5x + \frac{2}{3} = \pm\sqrt{\frac{5}{9}}$$
$$5x = -\frac{2}{3} \pm \frac{\sqrt{5}}{3}$$
$$x = \frac{1}{5}\left(-\frac{2}{3} \pm \frac{\sqrt{5}}{3}\right)$$
$$x = -\frac{2}{15} \pm \frac{\sqrt{5}}{15}$$
$$x = \frac{-2 \pm \sqrt{5}}{15}$$

45.

$$(4x-1)^2 + 6 = 10$$
$$(4x-1)^2 = 4$$
$$4x - 1 = \pm 2$$
$$4x = 1 \pm 2$$
$$x = \frac{1 \pm 2}{4}; \quad x = \frac{1+2}{4} = \frac{3}{4}; \quad x = \frac{1-2}{4} = -\frac{1}{4}$$

47.

$$(3x+1)^2 + 1 = 8$$
$$(3x+1)^2 = 7$$
$$3x + 1 = \pm\sqrt{7}$$
$$3x = -1 \pm \sqrt{7}$$
$$3x = -1 \pm \sqrt{7}$$
$$x = \frac{-1 \pm \sqrt{7}}{3}$$

49.

$$(2x+3)^2 - 1 = -8$$
$$(2x+3)^2 = -7$$
$$2x + 3 = \pm\sqrt{-7}$$
$$2x = -3 \pm \sqrt{7}i$$
$$x = \frac{-3 \pm \sqrt{7}i}{2}$$

51.

$$x^2 - 6x - 16 = 0$$
$$x^2 - 6x + 9 = 16 + 9 \qquad \text{Note: } \left(\frac{-6}{2}\right)^2 = 9$$
$$(x-3)^2 = 25$$
$$x - 3 = \pm\sqrt{25}$$
$$x = 3 \pm 5; \qquad x = 3 + 5 = 8; \quad x = 3 - 5 = -2$$

53.

$$x^2 + 10x + 21 = 0$$
$$x^2 + 10x + 25 = -21 + 25 \qquad \text{Note: } \left(\frac{10}{2}\right)^2 = 25$$
$$(x+5)^2 = 4$$
$$x + 5 = \pm 2$$
$$x = -5 \pm 2; \qquad x = -5 + 2 = -3;$$
$$x = -5 - 2 = -7$$

55.

$$4x^2 + 12x = 0$$

$$x^2 + 3x = 0 \qquad \text{Multiply by } \frac{1}{4}$$

$$x^2 + 3x + \frac{9}{4} = \frac{9}{4} \qquad \text{Note: } \left(\frac{3}{2}\right)^2 = \frac{9}{4}$$

$$\left(x + \frac{3}{2}\right)^2 = \frac{9}{4}$$

$$x + \frac{3}{2} = \pm\frac{3}{2}$$

$$x = -\frac{3}{2} \pm \frac{3}{2}; \qquad x = -\frac{3}{2} + \frac{3}{2} = 0$$

$$x = -\frac{3}{2} - \frac{3}{2} = -3$$

57.

$$9x^2 - 42x - 32 = 0$$

$$x^2 - \frac{42}{9}x - \frac{32}{9} = 0 \qquad \text{Multiply by } \frac{1}{9}$$

$$x^2 - \frac{14}{3}x + \frac{49}{9} = \frac{32}{9} + \frac{49}{9} \qquad \text{Note: } \left(\frac{14}{3} \cdot \frac{1}{2}\right)^2 = \frac{49}{9}$$

$$\left(x - \frac{7}{3}\right)^2 = \frac{81}{9}$$

$$x = \frac{7}{3} \pm \frac{9}{3}; \qquad x = \frac{7}{3} + \frac{9}{3} = \frac{16}{3}$$

$$x = \frac{7}{3} - \frac{9}{3} = -\frac{2}{3}$$

59.

$$x^2 + 4x - 1 = 0$$

$$x^2 + 4x + 4 = 1 + 4$$

$$(x + 2)^2 = 5$$

$$x + 2 = \pm\sqrt{5}$$

$$x = -2 \pm \sqrt{5}$$

61.

$$x^2 - 2x - 7 = 0$$

$$x^2 - 2x + 1 = 7 + 1$$

$$(x - 1)^2 = 8$$

$$x - 1 = \pm 2\sqrt{2}$$

$$x = 1 \pm 2\sqrt{2}$$

63.

$$4x^2 + 4x - 2 = 0$$

$$x^2 + x - \frac{1}{2} = 0 \qquad \text{Multiply by } \frac{1}{4}$$

$$x^2 + x + \frac{1}{4} = \frac{1}{2} + \frac{1}{4}$$

$$\left(x + \frac{1}{2}\right)^2 = \frac{3}{4}$$

$$x + \frac{1}{2} = \pm\frac{\sqrt{3}}{2}$$

$$x = -\frac{1}{2} \pm \frac{\sqrt{3}}{2}$$

65.

$$9x^2 + 24x - 16 = 0$$

$$x^2 + \frac{8}{3}x - \frac{16}{9} = 0 \qquad \text{Multiply by } \frac{1}{9}$$

$$x^2 + \frac{8}{3}x + \frac{16}{9} = \frac{16}{9} + \frac{16}{9} \qquad \text{Note: } \left(\frac{8}{3} \cdot \frac{1}{2}\right)^2 = \frac{16}{9}$$

$$\left(x + \frac{4}{3}\right)^2 = \frac{32}{9}$$

$$x + \frac{4}{3} = \pm\frac{4\sqrt{2}}{3}$$

$$x = -\frac{4}{3} \pm \frac{4\sqrt{2}}{3}$$

$$x = \frac{-4 \pm 4\sqrt{2}}{3}$$

67.

$$x^2 + 4x + 29 = 0$$

$$x^2 + 4x + 4 = -29 + 4$$

$$(x + 2)^2 = -25$$

$$x + 2 = \pm\sqrt{-25}$$

$$x = -2 \pm 5i$$

69.

$$x^2 - 2x + 37 = 0$$

$$x^2 - 2x + 1 = -37 + 1$$

$$(x - 1)^2 = -36$$

$$x - 1 = \pm\sqrt{-36}$$

$$x = 1 \pm 6i$$

71.

$$4x^2 - 4x + 17 = 0$$

$$x^2 - x + \frac{17}{4} = 0 \qquad \text{Multiply by } \frac{1}{4}$$

$$x^2 - x + \frac{1}{4} = -\frac{17}{4} + \frac{1}{4}$$

$$\left(x - \frac{1}{2}\right)^2 = -\frac{16}{4}$$

$$x - \frac{1}{2} = \pm\sqrt{-4}$$

$$x = \frac{1}{2} \pm 2i$$

$$x = \frac{1 \pm 4i}{2}$$

73.

$$9x^2 + 12x + 5 = 0$$

$$x^2 + \frac{4}{3}x + \frac{5}{9} = 0 \qquad \text{Multiply by } \frac{1}{9}$$

$$x^2 + \frac{4}{3}x + \frac{4}{9} = -\frac{5}{9} + \frac{4}{9}$$

$$\left(x + \frac{2}{3}\right)^2 = -\frac{1}{9}$$

$$x + \frac{2}{3} = \pm\sqrt{-\frac{1}{9}}$$

$$x = -\frac{2}{3} \pm \frac{1}{3}i$$

$$x = \frac{-2 \pm i}{3}$$

75.

$$x^2 - x - 2 = 0$$

$$x^2 - x + \frac{1}{4} = 2 + \frac{1}{4}$$

$$\left(x - \frac{1}{2}\right)^2 = \frac{9}{4}$$

$$x - \frac{1}{2} = \pm\sqrt{\frac{9}{4}}$$

$$x = \frac{1}{2} \pm \frac{3}{2}; \quad x = \frac{1}{2} + \frac{3}{2} = 2; \quad x = \frac{1}{2} - \frac{3}{2} = -1$$

77.

$$x^2 + 5x + 6 = 0$$

$$x^2 + 5x + \frac{25}{4} = -6 + \frac{25}{4}$$

$$\left(x + \frac{5}{2}\right)^2 = \frac{1}{4}$$

$$x + \frac{5}{2} = \pm\sqrt{\frac{1}{4}}$$

$$x = -\frac{5}{2} \pm \frac{1}{2}; \quad x = -\frac{5}{2} + \frac{1}{2} = -3;$$

$$x = -\frac{5}{2} - \frac{1}{2} = -2$$

79.

$$2x^2 + x - 6 = 0$$

$$x^2 + \frac{1}{2}x - 3 = 0 \qquad \text{Multiply by } \frac{1}{2}$$

$$x^2 + \frac{1}{2}x + \frac{1}{16} = 3 + \frac{1}{16}$$

$$\left(x + \frac{1}{4}\right)^2 = \frac{49}{16}$$

$$x + \frac{1}{4} = \pm\sqrt{\frac{49}{16}}$$

$$x = -\frac{1}{4} \pm \frac{7}{4}; \quad x = -\frac{1}{4} + \frac{7}{4} = \frac{3}{2};$$

$$x = -\frac{1}{4} - \frac{7}{4} = -2$$

81.

$$2x^2 - 3x - 2 = 0$$

$$x^2 - \frac{3}{2}x - 1 = 0 \qquad \text{Multiply by } \frac{1}{2}$$

$$x^2 - \frac{3}{2}x + \frac{9}{16} = 1 + \frac{9}{16}$$

$$\left(x - \frac{3}{4}\right)^2 = \frac{26}{16}$$

$$x - \frac{3}{4} = \pm\sqrt{\frac{25}{16}}$$

$$x = \frac{3}{4} \pm \frac{5}{4}; \quad x = \frac{3}{4} + \frac{5}{4} = 2$$

$$x = \frac{3}{4} - \frac{5}{4} = -\frac{1}{2}$$

83.
$$4x^2 + 12x + 7 = 0$$
$$x^2 + 3x + \frac{7}{4} = 0 \qquad \text{Multiply by } \frac{1}{4}$$
$$x^2 + 3x + \frac{9}{4} = -\frac{7}{4} + \frac{9}{4}$$
$$\left(x + \frac{3}{2}\right)^2 = \frac{1}{2}$$
$$x + \frac{3}{2} = \pm\sqrt{\frac{1}{2}}$$
$$x = -\frac{3}{2} \pm \frac{\sqrt{2}}{2}$$
$$x = \frac{-3 \pm \sqrt{2}}{2}$$

85.
$$9x^2 - 12x + 1 = 0$$
$$x^2 - \frac{4}{3}x + \frac{1}{9} = 0 \qquad \text{Multiply by } \frac{1}{9}$$
$$x^2 - \frac{4}{3}x + \frac{4}{9} = -\frac{1}{9} + \frac{4}{9}$$
$$\left(x - \frac{2}{3}\right)^2 = \frac{1}{3}$$
$$x - \frac{2}{3} = \pm\sqrt{\frac{1}{3}}$$
$$x = \frac{2}{3} \pm \frac{\sqrt{3}}{3}$$
$$x = \frac{2 \pm \sqrt{3}}{3}$$

87.
$$16x^2 - 8x - 7 = 0$$
$$x^2 - \frac{1}{2}x - \frac{7}{16} = 0 \qquad \text{Multiply by } \frac{1}{16}$$
$$x^2 - \frac{1}{2}x + \frac{1}{16} = \frac{7}{16} + \frac{1}{16}$$
$$\left(x - \frac{1}{4}\right)^2 = \frac{1}{2}$$
$$x - \frac{1}{4} = \pm\sqrt{\frac{1}{2}}$$
$$x = \frac{1}{4} \pm \frac{\sqrt{2}}{2}$$
$$x = \frac{1 \pm 2\sqrt{2}}{4}$$

89.
$$16x^2 + 24x - 63 = 0$$
$$x^2 + \frac{3}{2}x - \frac{63}{16} = 0 \qquad \text{Multiply by } \frac{1}{16}$$
$$x^2 + \frac{3}{2}x + \frac{9}{16} = \frac{63}{16} + \frac{9}{16}$$
$$\left(x + \frac{3}{4}\right)^2 = \frac{72}{16}$$
$$x + \frac{3}{4} = \pm\sqrt{\frac{72}{16}}$$
$$x = -\frac{3}{4} \pm \frac{6\sqrt{2}}{4}$$
$$x = \frac{-3 \pm 6\sqrt{2}}{4}$$

91.
$$2x^2 - 10x + 37 = 0$$
$$x^2 - 5x + \frac{37}{2} = 0 \qquad \text{Multiply by } \frac{1}{2}$$
$$x^2 - 5x + \frac{25}{4} = -\frac{37}{2} + \frac{25}{4}$$
$$\left(x - \frac{5}{2}\right)^2 = -\frac{49}{4}$$
$$x - \frac{5}{2} = \pm\sqrt{\frac{-49}{4}}$$
$$x = \frac{5}{2} \pm \frac{7}{2}i$$
$$x = \frac{5 \pm 7i}{2}$$

93.
$$9x^2 + 6x + 5 = 0$$
$$x^2 + \frac{2}{3}x + \frac{5}{9} = 0 \qquad \text{Multiply by } \frac{1}{9}$$
$$x^2 + \frac{2}{3}x + \frac{1}{9} = -\frac{5}{9} + \frac{1}{9}$$
$$\left(x + \frac{1}{3}\right)^2 = \frac{-4}{9}$$
$$x + \frac{1}{3} = \pm\sqrt{-\frac{4}{9}}$$
$$x = -\frac{1}{3} \pm \frac{2}{3}i$$
$$x = \frac{-1 \pm 2i}{3}$$

95.

$12x^2 - 12x + 7 = 0$

$x^2 - x + \dfrac{7}{12} = 0$ Multiply by $\dfrac{1}{12}$

$x^2 - x + \dfrac{1}{4} = -\dfrac{7}{12} + \dfrac{1}{4}$

$\left(x - \dfrac{1}{2}\right)^2 = -\dfrac{1}{3}$

$x - \dfrac{1}{2} = \pm\sqrt{-\dfrac{1}{3}}$

$x = \dfrac{1}{2} \pm \dfrac{\sqrt{3}}{3}i$

$x = \dfrac{3}{6} \pm \dfrac{2\sqrt{3}}{6}i$

$x = \dfrac{3 \pm 2\sqrt{3}i}{6}$

97.

$36x^2 + 96x + 73 = 0$

$x^2 + \dfrac{8}{3}x + \dfrac{73}{36} = 0$ Multiply by $\dfrac{1}{36}$

$x^2 + \dfrac{8}{3}x + \dfrac{16}{9} = -\dfrac{73}{36} + \dfrac{16}{9}$

$\left(x + \dfrac{4}{3}\right)^2 = -\dfrac{9}{36}$

$x + \dfrac{4}{3} = \pm\sqrt{-\dfrac{1}{4}}$

$x = -\dfrac{4}{3} \pm \dfrac{1}{2}i$

$x = \dfrac{-8}{6} \pm \dfrac{3}{6}i$

$x = \dfrac{-8 + 3i}{6}$

Exercises 7.4

1.

$2x^2 - 11x + 5 = 0$

$a = 2, b = -11, c = 5$

$x = \dfrac{11 \pm \sqrt{11^2 - 4(2)(5)}}{2 \cdot 2}$

$x = \dfrac{11 \pm \sqrt{121 - 40}}{4}$

#1. continued

$x = \dfrac{11 \pm \sqrt{81}}{4}$

$x = \dfrac{11 \pm 9}{4}; \quad x = \dfrac{11 + 9}{4} = \dfrac{20}{4} = 5$

$\qquad\qquad x = \dfrac{11 - 9}{4} = \dfrac{2}{4} = \dfrac{1}{2} \quad \left\{5, \dfrac{1}{2}\right\}$

3.

$4x^2 - 12x + 9 = 0$

$a = 4, b = -12, c = 9$

$x = \dfrac{12 \pm \sqrt{12^2 - 4(4)(9)}}{2 \cdot 4}$

$x = \dfrac{12 \pm \sqrt{144 - 144}}{8}$

$x = \dfrac{12}{8} \quad \text{or} \quad \dfrac{3}{2} \qquad \left\{\dfrac{3}{2}\right\}$

5.

$x^2 - 9 = 0$

$a = 1, b = 0, c = -9$

$x = \dfrac{0 \pm \sqrt{0 - 4(1)(-9)}}{2 \cdot 1}$

$x = \dfrac{\pm\sqrt{36}}{2}$

$x = \pm\dfrac{6}{2} \quad \text{or} \quad \pm 3 \qquad \{3, -3\}$

7.

$2x^2 - 8x = 0$

$x^2 - 4x = 0$ Multiply by $\dfrac{1}{2}$

$a = 1, b = -4, c = 0$

$x = \dfrac{4 \pm \sqrt{(-4)^2 - 4(1)(0)}}{2 \cdot 1}$

$x = \dfrac{4 \pm \sqrt{16}}{2}$

$x = \dfrac{4 \pm 4}{2}; \quad x = \dfrac{4 + 4}{2} = 4$

$\qquad\qquad x = \dfrac{4 - 4}{2} = 0 \qquad \{0, 4\}$

9.

$$x^2 - 15x + 56 = 0$$

$a = 1, b = -15, c = 56$

$$x = \frac{15 \pm \sqrt{(-15)^2 - 4(1)(56)}}{2}$$

$$x = \frac{15 \pm \sqrt{225 - 224}}{2}$$

$$x = \frac{15 \pm 1}{2}; \quad x = \frac{15 + 1}{2} = 8$$

$$x = \frac{15 - 1}{2} = 7 \qquad \{7, 8\}$$

11.

$$6x^2 + 17x + 12 = 0$$

$a = 6, b = 17, c = 12$

$$x = \frac{-17 \pm \sqrt{17^2 - 4(6)(12)}}{2 \cdot 6}$$

$$x = \frac{-17 \pm \sqrt{289 - 288}}{12}$$

$$x = \frac{-17 \pm 1}{12}; \quad x = \frac{-17 + 1}{12} = -\frac{4}{3}$$

$$x = \frac{-17 - 1}{12} = -\frac{3}{2} \qquad \left\{-\frac{4}{3}, -\frac{3}{2}\right\}$$

13.

$$3x^2 - 4x - 20 = 0$$

$a = 3, b = -4, c = -20$

$$x = \frac{4 \pm \sqrt{(-4)^2 - 4(3)(-20)}}{2 \cdot 3}$$

$$x = \frac{4 \pm \sqrt{16 + 240}}{6}$$

$$x = \frac{4 \pm \sqrt{256}}{6}$$

$$x = \frac{4 \pm 16}{6}; \quad x = \frac{4 + 16}{6} = \frac{10}{3}$$

$$x = \frac{4 - 16}{6} = -2 \qquad \left\{\frac{10}{3}, -2\right\}$$

15.

$$3x^2 - 7x + 1 = 0$$

$a = 3, b = -7, c = 1$

#15. continued

$$x = \frac{7 \pm \sqrt{(-7)^2 - 4(3)(1)}}{2 \cdot 3}$$

$$x = \frac{7 \pm \sqrt{49 - 12}}{6}$$

$$x = \frac{7 \pm \sqrt{37}}{6} \qquad \left\{\frac{7 \pm \sqrt{37}}{6}\right\}$$

17.

$$2x^2 + 3x - 7 = 0$$

$a = 2, b = 3, c = -7$

$$x = \frac{-3 \pm \sqrt{(-3)^2 - 4(2)(-7)}}{2 \cdot 2}$$

$$x = \frac{-3 \pm \sqrt{9 + 56}}{4}$$

$$x = \frac{-3 \pm \sqrt{65}}{4} \qquad \left\{\frac{-3 \pm \sqrt{65}}{4}\right\}$$

19.

$$3x^2 + 7x - 1 = 0$$

$a = 3, b = 7, c = -1$

$$x = \frac{-7 \pm \sqrt{(-7)^2 - 4(3)(-1)}}{2 \cdot 3}$$

$$x = \frac{-7 \pm \sqrt{49 + 12}}{6}$$

$$x = \frac{-7 \pm \sqrt{61}}{6} \qquad \left\{\frac{-7 \pm \sqrt{61}}{6}\right\}$$

21.

$$x^2 - 6x + 7 = 0$$

$a = 1, b = -6, c = 7$

$$x = \frac{6 \pm \sqrt{(-6)^2 - 4(1)(7)}}{2 \cdot 1}$$

$$x = \frac{6 \pm \sqrt{36 - 28}}{2}$$

$$x = \frac{6 \pm \sqrt{8}}{2}$$

$$x = \frac{6 \pm 2\sqrt{2}}{2} = \frac{\cancel{2}(3 \pm \sqrt{2})}{\cancel{2}}$$

$$= 3 \pm \sqrt{2} \qquad \left\{3 \pm \sqrt{2}\right\}$$

23.

$2x^2 + 6x + 1 = 0$

$a = 2, b = 6, c = 1$

$x = \dfrac{-6 \pm \sqrt{6^2 - 4(2)(1)}}{2 \cdot 2}$

$x = \dfrac{-6 \pm \sqrt{36 - 8}}{4}$

$x = \dfrac{-6 \pm \sqrt{28}}{4}$

$x = \dfrac{-6 \pm 2\sqrt{7}}{4}$

$x = \dfrac{\cancel{2}(-3 \pm \sqrt{7})}{\cancel{2} \cdot 2} = \dfrac{-3 \pm \sqrt{7}}{2} \qquad \left\{ \dfrac{-3 \pm \sqrt{7}}{2} \right\}$

25.

$2x^2 + 5x + 7 = 0$

$a = 2, b = 5, c = 7$

$x = \dfrac{-5 \pm \sqrt{5^2 - 4(2)(7)}}{2 \cdot 2}$

$x = \dfrac{-5 \pm \sqrt{25 - 56}}{4}$

$x = \dfrac{-5 \pm \sqrt{-31}}{4}$

$x = \dfrac{-5 \pm \sqrt{31}i}{4} \qquad \left\{ \dfrac{-5 \pm \sqrt{31}i}{4} \right\}$

27.

$3x^2 - x + 1 = 0$

$a = 3, b = -1, c = 1$

$x = \dfrac{1 \pm \sqrt{(-1)^2 - 4(3)(1)}}{2 \cdot 3}$

$x = \dfrac{1 \pm \sqrt{1 - 12}}{6}$

$x = \dfrac{1 \pm \sqrt{-11}}{6}$

$x = \dfrac{1 \pm \sqrt{11}i}{6} \qquad \left\{ \dfrac{1 \pm \sqrt{11}i}{6} \right\}$

29.

$9x^2 + 25 = 0$

$a = 9, b = 0, c = 25$

$x = \dfrac{0 \pm \sqrt{0^2 - 4(9)(25)}}{2 \cdot 9}$

$x = \dfrac{\pm \sqrt{-900}}{18}$

$x = \dfrac{\pm 30i}{18} = \pm \dfrac{5}{3}i \qquad \left\{ \pm \dfrac{5}{3}i \right\}$

31.

$x^2 + 2x + 17 = 0$

$a = 1, b = 2, c = 17$

$x = \dfrac{-2 \pm \sqrt{2^2 - 4(1)(17)}}{2 \cdot 1}$

$x = \dfrac{-2 \pm \sqrt{4 - 68}}{2}$

$x = \dfrac{-2 \pm \sqrt{-64}}{2}$

$x = \dfrac{-2 \pm 8i}{2} = \dfrac{\cancel{2}(-1 \pm 4i)}{\cancel{2}} = -1 \pm 4i \qquad \{ -1 \pm 4i \}$

33.

$9x^2 + 6x + 5 = 0$

$a = 9, b = 6, c = 5$

$x = \dfrac{-6 \pm \sqrt{6^2 - 4(9)(5)}}{2 \cdot 9}$

$x = \dfrac{-6 \pm \sqrt{36 - 180}}{18}$

$x = \dfrac{-6 \pm \sqrt{-144}}{18}$

$x = \dfrac{-6 \pm 12i}{18}$

$x = \dfrac{\cancel{6}(-1 \pm 2i)}{\cancel{6} \cdot 3} = \dfrac{-1 \pm 2i}{3} \qquad \left\{ \dfrac{-1 \pm 2i}{3} \right\}$

35.

$6x^2 \left(1 + \dfrac{1}{6x} - \dfrac{1}{3x^2} \right) = 0 \cdot 6x^2$

$6x^2 + x - 2 = 0$

#35. continued

$a = 6, b = 1, c = -2$

$$x = \frac{-1 \pm \sqrt{1^2 - 4(6)(-2)}}{2 \cdot 6}$$

$$x = \frac{-1 \pm \sqrt{1 + 48}}{12}$$

$$x = \frac{-1 \pm \sqrt{49}}{12}$$

$$x = \frac{-1 \pm 7}{12}; \quad x = \frac{-1 + 7}{12} = \frac{1}{2}$$

$$x = \frac{-1 - 7}{12} = -\frac{2}{3} \qquad \left\{\frac{1}{2}, -\frac{2}{3}\right\}$$

37.

$$2x^2\left(\frac{1}{2} - \frac{1}{x} - \frac{1}{x^2}\right) = 0 \cdot 2x^2$$

$$x^2 - 2x - 2 = 0$$

$a = 1, b = -2, c = -2$

$$x = \frac{2 \pm \sqrt{(-2)^2 - 4(1)(-2)}}{2 \cdot 1}$$

$$x = \frac{2 \pm \sqrt{4 + 8}}{2}$$

$$x = \frac{2 \pm 2\sqrt{3}}{2} = \frac{\cancel{2}(1 \pm \sqrt{3})}{\cancel{2}} = 1 \pm \sqrt{3} \qquad \left\{1 \pm \sqrt{3}\right\}$$

39.

$$4x^2\left(\frac{1}{4} + \frac{2}{x} + \frac{5}{x^2}\right) = 0 \cdot 4x^2$$

$$x^2 + 8x + 20 = 0$$

$a = 1, \quad b = 8, \quad c = 20$

$$x = \frac{-8 \pm \sqrt{8^2 - 4(1)(20)}}{2 \cdot 1}$$

$$x = \frac{-8 \pm \sqrt{64 - 80}}{2}$$

$$x = \frac{-8 \pm \sqrt{-16}}{2}$$

$$x = \frac{-8 \pm 4i}{2} = \frac{\cancel{2}(-4 \pm 2i)}{\cancel{2}} = -4 \pm 2i \qquad \left\{-4 \pm 2i\right\}$$

41.

$2x^2 - x - 6 = 0$

$a = 2, b = -1, c = -6$

$b^2 - 4ac = (-1)^2 - 4(2)(-6) = 49$

Perfect square: Two rational roots

43.

$4x^2 - 12x + 9 = 0$

$a = 4, b = -12, c = 9$

$b^2 - 4ac = (-12)^2 - 4(4)(9) = 144 - 144 = 0$

One rational root

45.

$2x^2 - 4x + 5 = 0$

$a = 2, b = -4, c = 5$

$b^2 - 4ac = (-4)^2 - 4(2)(5) = -24$

Two imaginary solutions

47.

$4x^2 - 25x - 21 = 0$

$a = 4, b = -25, c = -21$

$b^2 - 4ac = (-25)^2 - 4(4)(21) = 289$

Perfect square: Two rational roots

49.

$25x^2 + 20x + 4 = 0$

$a = 25, b = 20, c = 4$

$b^2 - 4ac = 20^2 - 4(25)(4) = 400 - 400 = 0$

One rational root

51.

$4x^2 - 12x + m = 0$

$a = 4, b = -12, c = m$

$b^2 - 4ac = 0; \quad (-12)^2 - 4(4 \cdot m) = 0$

$$144 - 16m = 0$$

$$-16m = -144$$

$$m = 9$$

53.

$mx^2 - 8x + 1 = 0$

$a = m, b = -8, c = 1$

$b^2 - 4ac = 0; \quad (-8)^2 - 4(m \cdot 1) = 0$

$$64 - 4m = 0$$

$$-4m = -64$$

$$m = 16$$

55.

$9x^2 + mx + 1 = 0$

$a = 9, b = m, c = 1$

$b^2 - 4ac = 0; \quad m^2 - 4(9)(1) = 0$

$$m^2 - 36 = 0$$

$$m^2 = 36$$

$$m = \pm 6$$

57.

$x^2 + 2x + m = 0$

$a = 1, b = 2, c = m$

$b^2 - 4ac < 0; \quad 2^2 - 4(1)(m) < 0$

$$4 - 4m < 0$$

$$-4m < -4$$

$$m > 1$$

59.

$mx^2 + 2x - 4 = 0$

$a = m, b = 2, c = -4$

$b^2 - 4ac < 0; \quad 2^2 - 4(m(-4)) < 0$

$$4 + 16m < 0$$

$$16m < -4$$

$$m < -\frac{1}{4}$$

61.

The solution is in the answer section of the main text.

63.

The ones that can readily be factored.

Exercises 7.5

1.

$4x^2 - 9 = 0$ \qquad Placing in stardard form

$(2x + 3)(2x - 3) = 0$

$2x + 3 = 0 \quad \text{or} \quad 2x - 3 = 0$

$x = -\frac{3}{2} \qquad\qquad x = \frac{3}{2}$

Solution set: $\left\{ -\frac{3}{2}, \frac{3}{2} \right\}$

3.

$x^2 - 7x + 2 = 0$

$a = 1, b = -7, c = 2$

$x = \dfrac{7 \pm \sqrt{(-7)^2 - 4(1)(2)}}{2 \cdot 1}$

$x = \dfrac{7 \pm \sqrt{49 - 8}}{2}$

$x = \dfrac{7 \pm \sqrt{41}}{2}$ \qquad Solution set: $\left\{ \dfrac{7 \pm \sqrt{41}}{2} \right\}$

5.

$5x^2 - 2 = 0$

$5x^2 = 2$

$x^2 = \dfrac{2}{5}$

$x = \pm\sqrt{\dfrac{2}{5}}$

$x = \pm\dfrac{\sqrt{10}}{5}$ \qquad Solution set: $\left\{ \pm\dfrac{\sqrt{10}}{5} \right\}$

7.

$4x^2 - 12x - 16 = 0$ \qquad Placing in standard form

$x^2 - 3x - 4 = 0$ \qquad Multiply by $\dfrac{1}{4}$

$(x - 4)(x + 1) = 0$

$x - 4 = 0 \quad \text{or} \quad x + 1 = 0$

$x = 4 \qquad\qquad x = -1$

Solution set: $\{4, -1\}$

9.

$4x^2 - 20x + 29 = 0$ \qquad Placing in standard form

$a = 4, b = -20, c = 29$

$x = \dfrac{20 \pm \sqrt{(-20)^2 - 4(4)(29)}}{2 \cdot 4}$

$x = \dfrac{20 \pm \sqrt{400 - 464}}{8}$

$x = \dfrac{20 \pm \sqrt{-64}}{8}$

$x = \dfrac{20 \pm 8i}{8} = \dfrac{\cancel{4}(5 \pm 2i)}{\cancel{4} \cdot 2} = \dfrac{5 \pm 2i}{2}$

Solution set: $\left\{ \dfrac{5 \pm 2i}{2} \right\}$

11.

$3x^2 - 9 = 0$ Placing in standard form

$\quad 3x^2 = 9$

$\quad\ x^2 = 3$

$\qquad x = \pm\sqrt{3}$ Solution set: $\left\{-\sqrt{3}, \sqrt{3}\right\}$

13.

$6x^2 - 10x = 0$ Placing in standard form

$2x(3x - 5) = 0$

$2x = 0 \quad$ or $\quad 3x - 5 = 0$

$\quad x = 0 \qquad\qquad x = \dfrac{5}{3}$

Solution set: $\left\{0, \dfrac{5}{3}\right\}$

15.

$\quad 2x^2 - x - 15 = 0$ Placing in standard form

$(2x + 5)(x - 3) = 0$

$2x + 5 = 0 \quad$ or $\quad x - 3 = 0$

$\qquad x = -\dfrac{5}{2} \qquad\quad x = 3$

Solution set: $\left\{-\dfrac{5}{2}, 3\right\}$

17.

$\quad 18x^2 + 15x - 7 = 0$ Placing in standard form

$(6x + 7)(3x - 1) = 0$

$6x + 7 = 0 \qquad\quad 3x - 1 = 0$

$\qquad x = -\dfrac{7}{6} \qquad\qquad x = \dfrac{1}{3}$

Solution set: $\left\{-\dfrac{7}{6}, \dfrac{1}{3}\right\}$

19.

$4x^2 + 1 = 0$ Placing in standard form

$4x^2 = -1$

$x^2 = -\dfrac{1}{4}$

$x = \pm\sqrt{-\dfrac{1}{4}}$

$x = \pm\dfrac{1}{2}i$ Solution set: $\left\{\dfrac{1}{2}i, -\dfrac{1}{2}i\right\}$

21.

$16x^2 + 8x - 12 = 0$ Placing in standard form

$\quad 4x^2 + 2x - 3 = 0$ Multiply by $\dfrac{1}{4}$

$a = 4, b = 2, c = -3$

$x = \dfrac{-2 \pm \sqrt{2^2 - 4(4)(-3)}}{2 \cdot 4}$

$x = \dfrac{-2 \pm \sqrt{4 + 48}}{8}$

$x = \dfrac{-2 \pm \sqrt{52}}{8}$

$x = \dfrac{-2 \pm 2\sqrt{13}}{8} = \dfrac{\cancel{2}\left(1 \pm \sqrt{13}\right)}{\cancel{2} \cdot 4} = \dfrac{1 \pm \sqrt{13}}{4}$

Solution set: $\left\{\dfrac{1 \pm \sqrt{13}}{4}\right\}$

23.

$2x^2 + 3x - 3 = 0$

$a = 2, b = 3, c = -3$

$x = \dfrac{-3 \pm \sqrt{3^2 - 4(2)(-3)}}{2 \cdot 2}$

$x = \dfrac{-3 \pm \sqrt{9 + 24}}{4}$

$x = \dfrac{-3 \pm \sqrt{33}}{4}$ Solution set: $\left\{\dfrac{-3 \pm \sqrt{33}}{4}\right\}$

25.

$2x^2 - 6x + 3 = 0$

$a = 2, b = -6, c = 3$

$x = \dfrac{6 \pm \sqrt{(-6)^2 - 4(2)(3)}}{2 \cdot 2}$

$x = \dfrac{6 \pm \sqrt{36 - 24}}{4}$

$x = \dfrac{6 \pm \sqrt{12}}{4}$

$x = \dfrac{6 \pm 2\sqrt{3}}{4} = \dfrac{\cancel{2}\left(3 \pm \sqrt{3}\right)}{\cancel{2} \cdot 2} = \dfrac{3 \pm \sqrt{3}}{2}$

Solution set: $\left\{\dfrac{3 \pm \sqrt{3}}{2}\right\}$

27.

$9x^2 - 12x + 2 = 0$

$a = 9, b = -12, c = 2$

$x = \dfrac{12 \pm \sqrt{(-12)^2 - 4(9)(2)}}{2 \cdot 9}$

$x = \dfrac{12 \pm \sqrt{144 - 72}}{18}$

$x = \dfrac{10 \pm \sqrt{72}}{18}$

$x = \dfrac{12 \pm 6\sqrt{2}}{18} = \dfrac{\cancel{6}(2 \pm \sqrt{2})}{\cancel{6} \cdot 3} = \dfrac{2 \pm \sqrt{2}}{3}$

Solution set: $\left\{ \dfrac{2 \pm \sqrt{2}}{3} \right\}$

29.

$x^2 - 4x - 9 = 0$ Placing in standard form

$a = 1, b = -4, c = -9$

$x = \dfrac{4 \pm \sqrt{(-4)^2 - 4(1)(-9)}}{2 \cdot 1}$

$x = \dfrac{4 \pm \sqrt{16 + 36}}{2}$

$x = \dfrac{4 \pm \sqrt{52}}{2}$

$x = \dfrac{4 \pm 2\sqrt{13}}{2} = \dfrac{\cancel{2}(2 \pm \sqrt{13})}{\cancel{2}} = 2 \pm \sqrt{13}$

Solution set: $\left\{ 2 \pm \sqrt{13} \right\}$

31.

$6x^2 + 19x + 8 = 0$

$(3x + 8)(2x + 1) = 0$

$3x + 8 = 0$ or $2x + 1 = 0$

    $x = \dfrac{-8}{3}$ $x = -\dfrac{1}{2}$

Solution set: $\left\{ -\dfrac{8}{3}, -\dfrac{1}{2} \right\}$

33.

$x^2 + 6x + 13 = 0$ Placing in standard form

$a = 1, b = 6, c = 13$

$x = \dfrac{-6 \pm \sqrt{6^2 - 4(1)(13)}}{2 \cdot 1}$

#33. continued

$x = \dfrac{-6 \pm \sqrt{36 - 52}}{2}$

$x = \dfrac{-6 \pm \sqrt{-16}}{2}$

$x = \dfrac{-6 \pm 4i}{2} = \dfrac{\cancel{2}(-3 \pm 2i)}{\cancel{2}} = -3 \pm 2i$

Solution set: $\{-3 \pm 2i\}$

35.

$6x^2 + 3x = 0$ Placing in standard form

$3x(2x + 1) = 0$

$3x = 0$ or $2x + 1 = 0$

   $x = 0$ $x = -\dfrac{1}{2}$

Solution set: $\left\{ 0, -\dfrac{1}{2} \right\}$

37.

$x^2 - 16x - 57 = 0$

$(x - 19)(x + 3) = 0$

$x - 19 = 0$ or $x + 3 = 0$

    $x = 19$ $x = -3$

Solution set: $\{19, -3\}$

39.

$16x^2 + 8x - 5 = 0$ Placing in standard form

$a = 16, b = 8, c = -5$

$x = \dfrac{-8 \pm \sqrt{8^2 - 4(16)(-5)}}{2 \cdot 16}$

$x = \dfrac{-8 \pm \sqrt{64 + 320}}{32}$

$x = \dfrac{-8 \pm \sqrt{384}}{32}$

$x = \dfrac{-8 \pm 8\sqrt{6}}{32} = \dfrac{\cancel{8}(1 \pm \sqrt{6})}{\cancel{8} \cdot 4} = \dfrac{1 \pm \sqrt{6}}{4}$

Solution set: $\left\{ \dfrac{1 \pm \sqrt{6}}{4} \right\}$

41.

$4x^2 - 5 = 0$ Placing in standard form

$4x^2 = 5$

$x^2 = \dfrac{5}{4}$

$x = \pm\sqrt{\dfrac{5}{4}} = \pm\dfrac{\sqrt{5}}{2}$

Solution set: $\left\{\dfrac{\sqrt{5}}{2}, -\dfrac{\sqrt{5}}{2}\right\}$

43.

$x^2 - 8x + 17 = 0$

$a = 1, b = -8, c = 17$

$x = \dfrac{8 \pm \sqrt{(-8)^2 - 4(1)(17)}}{2 \cdot 1}$

$x = \dfrac{8 \pm \sqrt{64 - 68}}{2}$

$x = \dfrac{8 \pm \sqrt{-4}}{2}$

$x = \dfrac{8 \pm 2i}{2} = \dfrac{\cancel{2}(4 \pm i)}{\cancel{2}} = 4 \pm i$

Solution set: $\{4 \pm i\}$

Exercises 7.6

1.

Let x and $x + 1 =$ the positive integer.

Thus, $x(x + 1) = 42$

$x^2 + x - 42 = 0$

$(x + 7)(x - 6) = 0$

$x + 7 = 0$ or $x - 6 = 0$

$\quad x = -7 \qquad\quad x = 6$

$\qquad\qquad\qquad x + 1 = 6 + 1 = 7$

The integers are 6 and 7.

3.

Let $x =$ the number.

Then $\dfrac{1}{x} =$ reciprocal.

Thus, $\qquad x + \dfrac{1}{x} = 4$

$\qquad\qquad x^2 + 1 = 4x$ Multiply by 4

$\qquad x^2 - 4x + 1 = 0$ Placing in standard form

#3. continued

$a = 1, b = -4, c = 1$

$x = \dfrac{4 \pm \sqrt{(-4)^2 - 4(1)(1)}}{2 \cdot 1}$

$x = \dfrac{4 \pm \sqrt{16 - 4}}{2}$

$x = \dfrac{4 \pm 2\sqrt{3}}{2} = \dfrac{\cancel{2}(2 \pm \sqrt{3})}{\cancel{2}} = 2 \pm \sqrt{3}$

The numbers are $2 + \sqrt{3}$ and $2 - \sqrt{3}$.

5.

Let x and $x + 1 =$ the positive numbers.

Then x^2 and $(x + 1)^2 =$ their squares.

Thus, $x^2 + (x + 1)^2 = 61$

$\qquad 2x^2 + 2x - 60 = 0$ Placing in standard form

$\qquad\quad x^2 + x - 30 = 0$ Multiply by $\dfrac{1}{2}$

$\qquad\quad (x - 5)(x + 6) = 0$

$x - 5 = 0$ or $x + 6 = 0$

$\quad x = 5 \qquad\qquad x = -6$

$x + 1 = 5 + 1 = 6$

The numbers are 5 and 6.

7.

Let $x =$ the number.

Thus, $\qquad x^2 - 2x = 14$

$\qquad\qquad x^2 - 2x - 14 = 0$

$a = 1, b = -2, c = -14$

$x = \dfrac{2 \pm \sqrt{(-2)^2 - 4(1)(-14)}}{2 \cdot 1}$

$x = \dfrac{2 \pm \sqrt{4 + 56}}{2}$

$x = \dfrac{2 \pm \sqrt{60}}{2}$

$x = \dfrac{2 \pm 2\sqrt{15}}{2} = \dfrac{\cancel{2}(1 \pm \sqrt{15})}{\cancel{2}} = 1 \pm \sqrt{15}$

The numbers are $1 + \sqrt{15}$ and $1 - \sqrt{15}$.

9.

Let $x =$ length of one leg.

Then $x + 2 =$ length of the other leg.

#9. continued

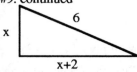

Thus, $x^2 + (x+2)^2 = 6^2$

$2x^2 + 4x - 32 = 0$ Placing in standard form

$x^2 + 2x - 16 = 0$ Multiply by $\frac{1}{2}$

$a = 1, b = 2, c = -16$

$x = \dfrac{-2 \pm \sqrt{2^2 - 4(1)(-16)}}{2 \cdot 1}$

$x = \dfrac{-2 \pm \sqrt{4 + 64}}{2}$

$x = \dfrac{-2 \pm \sqrt{68}}{2}$

$x = \dfrac{-2 \pm 2\sqrt{17}}{2} = \dfrac{\cancel{2}\left(-1 \pm \sqrt{7}\right)}{\cancel{2}} = -1 \pm \sqrt{7}$

Answer: Approximately 3.1 in and 5.1 in.

Note: $-1 - \sqrt{7}$ is extraneous.

11.

Let x = length of the height.

Then 2x - 1 = length of the base.

Thus, $A = \dfrac{1}{2} bh$

$6 = \dfrac{1}{2}(2x - 1) \cdot x$

$12 = x(2x - 1)$ Multiply by 2

$0 = 2x^2 - x - 12$ Placing in standard form

$a = 2, b = -1, c = -12$

$x = \dfrac{1 \pm \sqrt{(-1)^2 - 4(2)(-12)}}{2 \cdot 2}$

$x = \dfrac{1 \pm \sqrt{1 + 96}}{4}$

$x = \dfrac{1 \pm \sqrt{97}}{4}$

Answer: Height $\doteq$ 2.7 in   Base $\doteq$ 4.4 in.

Note: $\dfrac{1 - \sqrt{97}}{4}$ is extraneous.

13.

Let x = width of border.

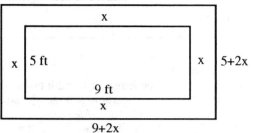

To find the solution, subtract the pool area from the total area.

Thus, $(9 + 2x)(5 + 2x) - 5 \cdot 9 = 40$

$4x^2 + 28x - 40 = 0$ Placing in standard form

$x^2 + 7x - 10 = 0$ Multiply by $\frac{1}{4}$

$a = 1, b = 7, c = -10$

$x = \dfrac{-7 \pm \sqrt{7^2 - 4(1)(-10)}}{2 \cdot 1}$

$x = \dfrac{-7 \pm \sqrt{49 + 40}}{2}$

$x = \dfrac{-7 \pm \sqrt{89}}{2}$

Answer: Approximately 1.2 ft.

Note: $\dfrac{-7 - \sqrt{89}}{2}$ is extraneous.

15.

Let x = width.

Then x + 3 = length.

Then x + 6 and x + 3 = the length and width of the picture frame.

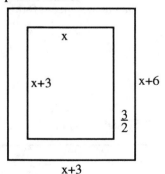

Thus, $(x + 6)(x + 3) = 108$

$x^2 + 9x - 90 = 0$

$(x - 6)(x + 15) = 0$

#15. continued

$x - 6 = 0$ or $x + 15 = 0$

$x = 6$ $x = -15$

Answer: 6 by 9 in.

17.

Let x = width of border.

Thus, 16+2x and 10+x = the length and width of the pool and border. To find the solution, subtract the pool area from the total area.

Thus, $(16 + 2x)(10 + x) - 16 \cdot 10 = 90$

$2x^2 + 36x - 90 = 0$ Placing in standard form

$x^2 + 18x - 45 = 0$ Multiply by $\frac{1}{2}$

$a = 1, b = 18, c = -45$

$x = \dfrac{-18 \pm \sqrt{18^2 - 4(1)(-45)}}{2 \cdot 1}$

$x = \dfrac{-18 \pm \sqrt{324 + 180}}{2}$

$x = \dfrac{-18 \pm \sqrt{504}}{2}$

$x = \dfrac{-18 \pm 6\sqrt{14}}{2} = \dfrac{2\left(-9 \pm 3\sqrt{14}\right)}{2} = -9 \pm 3\sqrt{14}$

Answer: Approximately 2.2 ft.

Note: $-9 - 3\sqrt{14}$ is extraneous.

19.

Let x = one pipe's time.

Then x + 1 = other pipe's time.

Thus, $\dfrac{1}{x} + \dfrac{1}{x+1} = \dfrac{1}{4}$

$4(x + 1) + 4x = x(x + 1)$ Multiply by $4x(x + 1)$

$x^2 - 7x - 4 = 0$ Placing in standard form

$a = 1, b = -7, c = -4$

$x = \dfrac{7 \pm \sqrt{(-7)^2 - 4(1)(-4)}}{2 \cdot 1}$

$x = \dfrac{7 \pm \sqrt{49 + 16}}{2} = \dfrac{7 \pm \sqrt{65}}{2}$

Answer: Approximately 7.53 and 8.53 hrs.

21.

Let x = old processing time.

Then x - 2 = new processing time.

#21. continued

Thus, $\dfrac{1}{x} + \dfrac{1}{x-2} = \dfrac{1}{6}$

$6(x - 2) + 6x = x(x - 2)$ Multiply by $6x(x - 2)$

$x^2 - 14x + 12 = 0$ Placing in standard form

$a = 1, b = -14, c = 12$

$x = \dfrac{14 \pm \sqrt{(-14)^2 - 4(1)(12)}}{2 \cdot 1}$

$x = \dfrac{14 \pm \sqrt{196 - 48}}{2}$

$x = \dfrac{14 \pm \sqrt{148}}{2}$

$x = \dfrac{14 \pm 2\sqrt{37}}{2} = \dfrac{2\left(7 \pm \sqrt{37}\right)}{2} = 7 \pm \sqrt{37}$

Answer: Approximately 13.1 das and 11.1 das.

Note: $7 - \sqrt{37}$ is extraneous.

23.

	Distance	÷ Rate	= Time
Up	4	x	$\dfrac{4}{x}$
Down	4	x + 1	$\dfrac{4}{x + 1}$

Thus, $\dfrac{4}{x} + \dfrac{4}{x+1} = 5$

$4(x + 1) + 4x = 5x(x + 1)$ Multiply by $x(x + 1)$

$5x^2 - 3x - 4 = 0$

$a = 5, b = -3, c = -4$

$x = \dfrac{3 \pm \sqrt{(-3)^2 - 4(5)(-4)}}{2 \cdot 5}$

$x = \dfrac{3 \pm \sqrt{9 + 80}}{10} = \dfrac{3 \pm \sqrt{89}}{10}$

Answer: Up = 1.2 mph, Down = 2.2 mph

Note: $\dfrac{3 - \sqrt{89}}{10}$ is extraneous

25.

	Distance	÷ Rate	= Time
Walk	3	x	$\dfrac{3}{x}$
Bike	3	x + 5	$\dfrac{3}{x + 5}$

#25. continued

Thus, $\dfrac{3}{x} + \dfrac{3}{x+5} = 1$

$3(x+5) + 3x = x(x+5)$ Multiply by $x(x+5)$

$x^2 - x - 15 = 0$

$a = 1,\ b = -1,\ c = -15$

$x = \dfrac{1 \pm \sqrt{(-1)^2 - 4(-15)(1)}}{2 \cdot 1}$

$x = \dfrac{1 \pm \sqrt{1+60}}{2} = \dfrac{1 \pm \sqrt{61}}{2}$

Answer: approximately 4.4 mph

Note: $\dfrac{1 - \sqrt{61}}{2}$ is extraneous

27.

$h = -16t^2 + 32t + 80$

Let $h = 0$

$0 = -16t^2 + 32t + 80$

$0 = t^2 - 2t - 5$ Multiply by $-\dfrac{1}{16}$

$a = 1,\ b = -2,\ c = -5$

$t = \dfrac{2 \pm \sqrt{4+20}}{2} = \dfrac{2 \pm 2\sqrt{6}}{2} = 1 \pm \sqrt{6}$

Answer: approximately 3.4 sec

Note: $1 - \sqrt{6}$ is extraneous

29.

Step 1: Find ground hit time.

$h = -16t^2 + 96t + 104$ Let $h = 0$

$0 = -16t^2 + 96t + 104$

$0 = t^2 - 6t - 6.5$

$a = 1,\ b = -6,\ c = -6.5$

$t = \dfrac{6 \pm \sqrt{36+26}}{2} = \dfrac{6 \pm \sqrt{62}}{2}$

Answer: Hits ground approximately 6.95 sec

Step 2: Find maximum height time and maximum height.

$h = -16t^2 + 96t + 104$

$h = -16(t^2 - 6t + 9) + 104 + 144$

$h = -16(t-3)^2 + 248$

Answer: Maximum height at 3 sec

Maximum height = 248 ft

Exercises 7.7

1.

$x^4 - 5x^2 + 4 = 0$ Let $u = x^2$

$u^2 - 5u + 4 = 0$

$(u-4)(u-1) = 0$

$u - 4 = 0$ or $u - 1 = 0$

$u = 4$ $u = 1$

$x^2 = 4$ $x^2 = 1$

$x = \pm 2$ $x = \pm 1$

3.

$4x^4 - 37x^2 + 9 = 0$ Let $u = x^2$

$4u^2 - 37u + 9 = 0$

$(4u-1)(u-9) = 0$

$4u - 1 = 0$ or $u - 9 = 0$

$u = \dfrac{1}{4}$ $u = 9$

$x^2 = \dfrac{1}{4}$ $x^2 = 9$

$x = \pm\dfrac{1}{2}$ $x = \pm 3$

5.

$x^4 + x^2 - 12 = 0$ Let $u = x^2$

$u^2 + u - 12 = 0$

$(u-3)(u+4) = 0$

$u - 3 = 0$ or $u + 4 = 0$

$u = 3$ $u = -4$

$x^2 = 3$ $x^2 = -4$

$x = \pm\sqrt{3}$ $x = \pm 2i$

7.

$2x^4 - 11x^2 + 5 = 0$ Let $u = x^2$

$2u^2 - 11u + 5 = 0$

$(2u-1)(u-5) = 0$

$2u - 1 = 0$ or $u - 5 = 0$

$u = \dfrac{1}{2}$ $u = 5$

$x^2 = \dfrac{1}{2}$ $x^2 = 5$

$x = \pm\dfrac{2\sqrt{2}}{2}$ $x = \pm\sqrt{5}$

9.

$3x^4 + 4x^2 + 1 = 0$ Let $u = x^2$

$3u^2 + 4u + 1 = 0$

$(3u + 1)(u + 1) = 0$

$3u + 1 = 0$ or $u + 1 = 0$

$u = -\dfrac{1}{3}$ $u = -1$

$x^2 = -\dfrac{1}{3}$ $x^2 = -1$

$x = \pm\dfrac{\sqrt{3}}{3}i$ $x = \pm i$

11.

$x - 5\sqrt{x} + 4 = 0$ Let $u = \sqrt{x}$

$u^2 - 5u + 4 = 0$

$(u - 4)(u - 1) = 0$

$u - 4 = 0$ or $u - 1 = 0$

$u = 4$ $u = 1$

$\sqrt{x} = 4$ $\sqrt{x} = 1$

$x = 16$ $x = 1$

13.

$2x - 11\sqrt{x} + 5 = 0$ Let $u = \sqrt{x}$

$2u^2 - 11u + 5 = 0$

$(2u - 1)(u - 5) = 0$

$2u - 1 = 0$ or $u - 5 = 0$

$u = \dfrac{1}{2}$ $u = 5$

$\sqrt{x} = \dfrac{1}{2}$ $\sqrt{x} = 5$

$x = \dfrac{1}{4}$ $x = 25$

15.

$2x - \sqrt{x} - 15 = 0$ Let $u = \sqrt{x}$

$2u^2 - u - 15 = 0$

$(2u + 5)(u - 3) = 0$

$2u + 5 = 0$ or $u - 3 = 0$

$u = -\dfrac{5}{2}$ $u = 3$

$\sqrt{x} = -\dfrac{5}{7}$ $\sqrt{x} = 3$

                 $x = 9$

Note: $\sqrt{x}$ cannot be negative.

17.

$2x + 7\sqrt{x} + 3 = 0$ Let $u = \sqrt{x}$

$2u^2 + 7u + 3 = 0$

$(2u + 1)(u + 3) = 0$

$2u + 1 = 0$ or $u + 3 = 0$

$u = -\dfrac{1}{2}$ $u = -3$

$\sqrt{x} = -\dfrac{1}{2}$ $\sqrt{x} = -3$

            $\varnothing$

Note: $\sqrt{x}$ cannot be negative.

19.

$(x + 1)^2 + (x + 1) - 20 = 0$ Let $u = (x + 1)$

$u^2 + u - 20 = 0$

$(u - 4)(u + 5) = 0$

$u - 4 = 0$ or $u + 5 = 0$

$u = 4$ $u = -5$

$x + 1 = 4$ $x + 1 = -5$

$x = 3$ $x = -6$

21.

$(x - 3)^2 - 4 = 0$ Let $u = x - 3$

$u^2 - 4 = 0$

$(u + 2)(u - 2) = 0$

$u + 2 = 0$ or $u - 2 = 0$

$u = -2$ $u = 2$

$x - 3 = -2$ $x - 3 = 2$

$x = 1$ $x = 5$

23.

$4(x - 1)^2 - 3(x - 1) - 10 = 0$ Let $u = x - 1$

$4u^2 - 3u - 10 = 0$

$(4u + 5)(u - 2) = 0$

$4u + 5 = 0$ or $u - 2 = 0$

$4(x - 1) + 5 = 0$ $(x - 1) - 2 = 0$

$x = -\dfrac{1}{4}$ $x = 3$

25.

$4(2x - 3)^2 + 8(2x - 3) + 3 = 0$ Let $u = (2x - 3)$

$4u^2 + 8u + 3 = 0$

$(2y + 3)(2u + 1) = 0$

#25. continued

$2u + 3 = 0$ or $2u + 1 = 0$

$u = -\dfrac{3}{2}$ $u = -\dfrac{1}{2}$

$2x - 3 = -\dfrac{3}{2}$ $2x - 3 = -\dfrac{1}{2}$

$2x = \dfrac{3}{2}$ $2x = \dfrac{5}{2}$

$x = \dfrac{3}{4}$ $x = \dfrac{5}{4}$

27.

$(3x - 4)^2 - 4(3x - 4) + 4 = 0$ Let $u = 3x - 4$

$u^2 - 4u + 4 = 0$

$(u - 2)(u - 2) = 0$

$u - 2 = 0$

$(3x - 4) - 2 = 0$

$x = 2$

29.

$x^{-2} + x^{-1} - 6 = 0$ Let $u = x^{-1}$

$u^2 + u - 6 = 0$

$(u - 2)(u + 3) = 0$

$u - 2 = 0$ or $u + 3 = 0$

$u = 2$ $u = -3$

$x^{-1} = 2$ $x^{-1} = -3$

$x = \dfrac{1}{2}$ $x = -\dfrac{1}{3}$

31.

$x^{-2} + 6x^{-1} + 5 = 0$ Let $u = x^{-1}$

$u^2 + 6u + 5 = 0$

$(u + 5)(u + 1) = 0$

$u + 5 = 0$ or $u + 1 = 0$

$u = -5$ $u = -1$

$x^{-1} = -5$ $x^{-1} = -1$

$x = -\dfrac{1}{5}$ $x = -1$

33.

$2x^{-2} + 13x^{-1} - 7 = 0$ Let $u = x^{-1}$

$2u^2 + 13u - 7 = 0$

$(2u - 1)(u + 7) = 0$

#33. continued

$2u - 1 = 0$ or $u + 7 = 0$

$u = \dfrac{1}{2}$ $u = -7$

$x^{-1} = \dfrac{1}{2}$ $x^{-1} = -7$

$x = 2$ $u = -\dfrac{1}{7}$

35.

$6x^{-2} - 7x^{-1} + 2 = 0$ Let $u = x^{-1}$

$6u^2 - 7u + 2 = 0$

$(3u - 2)(2u - 1) = 0$

$3u - 2 = 0$ or $2u - 1 = 0$

$u = \dfrac{2}{3}$ $u = \dfrac{1}{2}$

$x^{-1} = \dfrac{2}{3}$ $x^{-1} = \dfrac{1}{2}$

$x = \dfrac{3}{2}$ $x = 2$

37.

$x^{2/3} - x^{1/3} - 30 = 0$ Let $u = x^{1/3}$

$u^2 - u - 30 = 0$

$(u + 5)(u - 6) = 0$

$u + 5 = 0$ or $u - 6 = 0$

$u = -5$ $u = 6$

$x^{1/3} = -5$ $x^{1/3} = 6$

$x = -125$ $x = 216$

39.

$3x^{2/3} - 7x^{1/3} + 2 = 0$ Let $u = x^{1/3}$

$3u^2 - 7u + 2 = 0$

$(3u - 1)(u - 2) = 0$

$3u - 1 = 0$ or $u - 2 = 0$

$u = \dfrac{1}{3}$ $u = 2$

$x^{1/3} = \dfrac{1}{3}$ $x^{1/3} = 2$

$x = \dfrac{1}{27}$ $x = 8$

41.

$6x^{2/3} - 5x^{1/3} - 4 = 0$ Let $u = x^{1/3}$

$6u^2 - 5u - 4 = 0$

$(3u - 4)(2u + 1) = 0$

$3u - 4 = 0$ or $2u + 1 = 0$

$u = \dfrac{4}{3}$ $u = -\dfrac{1}{2}$

$x^{1/3} = \dfrac{4}{3}$ $x^{1/3} = -\dfrac{1}{2}$

$x = \dfrac{64}{27}$ $x = -\dfrac{1}{8}$

43.

$\left(\dfrac{x}{x+1}\right)^2 - 2\cdot\left(\dfrac{x}{x+1}\right) - 15 = 0$ Let $u = \dfrac{x}{x+1}$

$u^2 - 2u - 15 = 0$

$(u + 3)(u - 5) = 0$

$u + 3 = 0$ or $u - 5 = 0$

$u = -3$ $u = 5$

$\dfrac{x}{x+1} = -3$ $\dfrac{x}{x+1} = 5$

$x = -\dfrac{3}{4}$ $x = -\dfrac{5}{4}$

45.

$\left(\dfrac{3x+1}{x}\right)^2 + 6\cdot\left(\dfrac{3x+1}{x}\right) + 8 = 0$ Let $u = \dfrac{3x+1}{x}$

$u^2 + 6u + 8 = 0$

$(u + 2)(u + 4) = 0$

$u + 2 = 0$ or $u + 4 = 0$

$u = -2$ $u = -4$

$\dfrac{3x+1}{x} = -2$ $\dfrac{3x+1}{x} = -4$

$x = -\dfrac{1}{5}$ $x = -\dfrac{1}{7}$

47.

$6\cdot\left(\dfrac{2x}{x-3}\right)^2 - 5\cdot\left(\dfrac{2x}{x-3}\right) - 6 = 0$ Let $u = \dfrac{2x}{x-3}$

$6u^2 - 5u - 6 = 0$

$(2u - 3)(3u + 2) = 0$

#47. continued

$2u - 3 = 0$ or $3u + 2 = 0$

$u = \dfrac{3}{2}$ $u = -\dfrac{2}{3}$

$\dfrac{2x}{x-3} = \dfrac{3}{2}$ $\dfrac{2x}{x-3} = -\dfrac{2}{3}$

$x = -9$ $x = \dfrac{3}{4}$

49.

$4\cdot\left(\dfrac{x+1}{x-1}\right)^2 - 8\cdot\left(\dfrac{x+1}{x-1}\right) + 3 = 0$ Let $u = \dfrac{x+1}{x-1}$

$4u^2 - 8u + 3 = 0$

$(2u - 3)(2u - 1) = 0$

$2u - 3 = 0$ or $2u - 1 = 0$

$u = \dfrac{3}{2}$ $u = \dfrac{1}{2}$

$\dfrac{x+1}{x-1} = \dfrac{3}{2}$ $\dfrac{x+1}{x-1} = \dfrac{1}{2}$

$x = 5$ $x = -3$

51.

$x^4 - 8x^2 + 3 = 0$ Let $u = x^2$

$u^2 - 8u + 3 = 0$

$a = 1,\ b = -8,\ c = 3$

$u = \dfrac{8 \pm \sqrt{64 - 12}}{2}$

$u = \dfrac{8 \pm \sqrt{52}}{2}$

$u = \dfrac{8 \pm 2\sqrt{13}}{2} = 4 \pm \sqrt{13}$

$x^2 = 4 \pm \sqrt{13}$

$x = \pm\sqrt{4 \pm \sqrt{13}}$

53.

$x^3 + 7x^2 - x - 7 = 0$

$x^2(x + 7) - 1\cdot(x + 7) = 0$

$(x + 7)(x^2 - 1) = 0$

$(x + 7)(x + 1)(x - 1) = 0$

$x + 7 = 0$ or $x + 1 = 0$ or $x - 1 = 0$

$x = -7$ $x = -1$ $x = 1$

55.

$$3x^3 + 2x^2 - 12x - 8 = 0$$
$$x^2(3x + 2) - 4(3x + 2) = 0$$
$$(3x + 2)(x^2 - 4) = 0$$
$$(3x + 2)(x + 2)(x - 2) = 0$$
$$3x + 2 = 0 \quad \text{or} \quad x + 2 = 0 \quad \text{or} \quad x - 2 = 0$$
$$x = -\frac{3}{2} \qquad x = -2 \qquad x = 2$$

57.

$$4x^3 - 12x^2 - 9x + 27 = 0$$
$$4x^2(x - 3) - 9(x - 3) = 0$$
$$(x - 3)(4x^2 - 9) = 0$$
$$(x - 3)(2x + 3)(2x - 3) = 0$$
$$x - 3 = 0 \quad \text{or} \quad 2x + 3 = 0 \quad \text{or} \quad 2x - 3 = 0$$
$$x = 3 \qquad x = -\frac{3}{2} \qquad x = \frac{3}{2}$$

59.

$$9x^3 + 9x^2 - x - 1 = 0$$
$$9x^2(x + 1) - 1 \cdot (x + 1) = 0$$
$$(x + 1)(9x^2 - 1) = 0$$
$$(x + 1)(3x + 1)(3x - 1) = 0$$
$$x + 1 = 0 \quad \text{or} \quad 3x + 1 = 0 \quad \text{or} \quad 3x - 1 = 0$$
$$x = -1 \qquad x = -\frac{1}{3} \qquad x = \frac{1}{3}$$

61.

$$8x^3 - 12x^2 - 2x + 3 = 0$$
$$4x^2(2x - 3) - 1 \cdot (2x - 3) = 0$$
$$(2x - 3)(4x^2 - 1) = 0$$
$$(2x - 3)(2x + 1)(2x - 1) = 0$$
$$2x - 3 = 0 \quad \text{or} \quad 2x + 1 = 0 \quad \text{or} \quad 2x - 1 = 0$$
$$x = \frac{3}{2} \qquad x = -\frac{1}{2} \qquad x = \frac{1}{2}$$

63.

$$5x^3 - 2x^2 + 5x - 2 = 0$$
$$x^2(5x - 2) + (5x - 2) = 0$$
$$(5x - 2)(x^2 + 1) = 0$$

#63. continued

$$5x - 2 = 0 \quad \text{or} \quad x^2 + 1 = 0$$
$$x = \frac{2}{5} \qquad\qquad x^2 = -1$$
$$x = \pm i$$

65.

$$8x^3 - 12x^2 + 2x - 3 = 0$$
$$4x^2(2x - 3) + (2x - 3) = 0$$
$$(2x - 3)(4x^2 + 1) = 0$$
$$2x - 3 = 0 \quad \text{or} \quad 4x^2 + 1 = 0$$
$$x = \frac{3}{2} \qquad\qquad x^2 = -\frac{1}{4}$$
$$x = \pm \frac{1}{2}i$$

67.

$$5x^4 - 3x^3 - 5x + 3 = 0$$
$$x^3(5x - 3) - 1 \cdot (5x - 3) = 0$$
$$(5x - 3)(x^3 - 1) = 0$$
$$(5x - 3)(x - 1)(x^2 + x + 1) = 0$$
$$5x - 3 = 0 \quad \text{or} \quad x - 1 = 0 \quad \text{or} \quad x^2 + x + 1 = 0$$
$$x = \frac{3}{5} \qquad x = 1 \qquad x = \frac{-1 \pm \sqrt{3}i}{2}$$

Note: Solve $x^2 + x + 1 = 0$ using the quadratic formula.

69.

$$x^4 + 3x^3 + 27x + 81 = 0$$
$$x^3(x + 3) + 27(x + 3) = 0$$
$$(x + 3)(x^3 + 27) = 0$$
$$(x + 3)(x + 3)(x^2 - 3x + 9) = 0$$
$$x + 3 = 0 \quad \text{or} \quad x^2 - 3x + 9 = 0$$
$$x = -3 \qquad x = \frac{3 \pm 3\sqrt{3}i}{2}$$

Note: Solve $x^2 - 3x + 9 = 0$ using the quadratic formula.

71.
$$x^5 - 5x^4 - x + 5 = 0$$
$$x^4(x-5) - 1 \cdot (x-5) = 0$$
$$(x-5)(x^4 - 1) = 0$$
$$(x-5)(x^4 - 1) = 0$$
$$(x-5)(x^2 + 1)(x+1)(x-1) = 0$$
$$x = 5 \quad x = \pm i \quad x = -1 \quad x = 1$$

73.
$$x^7 + 3x^6 - x - 3 = 0$$
$$x^6(x+3) - 1 \cdot (x+3) = 0$$
$$(x+3)(x^6 - 1) = 0$$
$$(x+3)(x^3 + 1)(x^3 - 1) = 0$$
$$(x+3)(x+1)(x^2 - x + 1)(x-1)(x^2 + x + 1) = 0$$
$$x = 3 \quad x = -1 \quad x = \frac{1 \pm \sqrt{3}i}{2} \quad x = 1 \quad x = \frac{-1 \pm \sqrt{3}i}{2}$$

Exercises 7.8

The graphs are in the answer section of the main text.

1.
$$x^2 + x - 20 < 0$$
$(x-4)(x+5) < 0$ The product is negative.
$x - 4 = 0$ or $x + 5 = 0$
  $x = 4$ $x = -5$ Boundaries
Picking numbers from each of the three regions will show that
$$-5 < x < 4$$

3.
$$x^2 + 9x + 18 > 0$$
$(x+6)(x+3) > 0$ The product is positive.
$x + 6 = 0$ or $x + 3 = 0$
  $x = -6$ $x = -3$ Boundaries
Picking numbers from each of the three regions will show that
$$x < -6 \quad \text{or} \quad x > -3$$

The graphs are in the answer section of the main text.

5.
$$2x^2 + 3x > 0$$
$x(2x+3) > 0$ The product is positive.
$x = 0$ or $2x + 3 = 0$
$x = 0$ $x = -\dfrac{3}{2}$ Boundaries
Picking numbers from each of the three regions will show that
$$x < -\frac{3}{2} \quad \text{or} \quad x > 0$$

7.
$$x^2 + 10x + 16 \geq 0$$
$(x+2)(x+8) \geq 0$ The product is
                                   positive or zero
$x + 2 = 0$ or $x + 8 = 0$
  $x = -2$ $x = -8$ Boundaries
Picking numbers from each of the three regions will show that
$$x \leq -8 \quad \text{or} \quad x \geq -2$$

9.
$$x^2 - 7x + 6 \leq 0$$
$(x-6)(x-1) \leq 0$ The product is
                                   negative or zero
$x - 6 = 0$ or $x - 1 = 0$
  $x = 6$ $x = 1$ Boundaries
Picking numbers from each of the three regions will show that
$$1 \leq x \leq 6$$

11.
$$x^2 - 49 > 0$$
$(x+7)(x-7) > 0$ The product is positive
$x = -7, \quad x = 7$ Boundaries
Picking numbers from each of the three regions will show that
$$x < -7 \quad \text{or} \quad x > 7$$

13.

$$2x^2 - x - 6 < 0$$

$$(2x + 3)(x - 2) < 0 \qquad \text{The product is negative}$$

$$x = -\frac{3}{2}, \quad x = 2 \qquad \text{Boundaries}$$

Picking numbers from each of the three regions will show that

$$-\frac{3}{2} < x < 2$$

15.

$$8x^2 + 22x + 9 > 0$$

$$(4x + 9)(2x + 1) > 0 \qquad \text{The product is positive}$$

$$x = -\frac{9}{4}, \quad x = -\frac{1}{2} \qquad \text{Boundaries}$$

Picking numbers from each of the three regions will show that

$$x < -\frac{9}{4} \quad \text{or} \quad x > -\frac{1}{2}.$$

17.

$$x^2 + 2x + 1 \geq 0$$

$$(x + 1)(x + 1) \geq 0 \qquad \text{The product is positive or negative}$$

$$x = -1 \qquad \text{Boundary}$$

Picking numbers from each of the three regions will show that all real numbers are solutions

19.

$$4x^2 - 20x + 25 < 0$$

$$(2x - 5)(2x - 5) < 0 \qquad \text{The product is negative}$$

$$x = \frac{5}{2} \qquad \text{Boundary}$$

Picking numbers from each of the three regions will show that none of the real numbers are solutions.

21.

$$x^2 - 6x + 7 > 0$$

Using the quadratic formula will show that the boudaries are $3 - \sqrt{2}$ and $3 + \sqrt{2}$ or approximately 1.6 and 4.4.

Picking numbers from each of the three regions will show that

$$x < 3 - \sqrt{2} \quad \text{or} \quad x > 3 + \sqrt{2}$$

23.

$$x^2 + 2x - 5 \leq 0$$

Using the quadratic formula will show that the boundaries are $-1 + \sqrt{6}$ and $-1 - \sqrt{6}$ or approximately 1.4 and -3.4.

Picking numbers from each of the three regions will show that

$$-1 - \sqrt{6} \leq x \leq -1 + \sqrt{6}$$

25.

$$x^2 - x - 6 < 0$$

$$(x + 2)(x - 3) < 0 \qquad \text{The product is negative}$$

$$x = -2, \quad x = 3 \qquad \text{Boundaries}$$

Picking numbers from each of the three regions will show that

$$-2 < x < 3$$

27.

$$2x^2 + 11x + 12 \geq 0$$

$$(2x + 3)(x + 4) \geq 0 \qquad \text{The product is positive or zero}$$

$$x = -\frac{3}{2}, \quad x = -4 \qquad \text{Boundaries}$$

Picking numbers from each of the three regions will show that

$$x \leq -4 \quad \text{or} \quad x \geq -\frac{3}{2}$$

29.

$$4x^2 - 4x + 1 < 0$$

$$(2x - 1)(2x - 1) < 0 \qquad \text{The product is negative}$$

$$x = \frac{1}{2} \qquad \text{Boundary}$$

Picking numbers from each of the three regions will show that none of the real numbers are solutions

31.

$$16x^2 - 1 > 0$$

$$(4x + 1)(4x - 1) > 0 \qquad \text{The product is positive}$$

$$x = -\frac{1}{4}, \quad x = \frac{1}{4} \qquad \text{Boundaries}$$

The graphs are in the answer section of the main text.

#31. continued
Picking numbers from each of the three regions will show that
$$x < -\frac{1}{4} \quad \text{or} \quad x > \frac{1}{4}$$

33.
$-x^2 + 11x - 24 \leq 0$

$x^2 - 11x + 24 \geq 0$ Multiply by -1, reverse signs

$(x - 3)(x - 8) \geq 0$ The product is negative or zero

$x = 3, \quad x = 8$ Boundaries

Picking numbers from each of the three regions will show that
$$x \leq 3 \quad \text{or} \quad x \geq 8$$

35.
$-8x^2 + 18x - 9 > 0$

$8x^2 - 18x + 9 < 0$ Multiply by -1, reverse signs

$(4x - 3)(2x - 3) < 0$ The product is positive

$x = \frac{3}{4}, \quad x = \frac{3}{2}$ Boundaries

Picking numbers from each of the three regions will show that
$$\frac{3}{4} < x < \frac{3}{2}$$

37.
$-4x^2 + 12x + 7 \leq 0$

$4x^2 - 12x - 7 \geq 0$ Multiply by -1, reverse signs

$(2x + 1)(2x - 7) \geq 0$ The product is negative or zero

$x = -\frac{1}{2}, \quad x = \frac{7}{2}$ Boundaries

Picking numbers from each of the three regions will show that
$$x \leq -\frac{1}{2} \quad \text{or} \quad x \geq \frac{7}{2}$$

The graphs are in the answer section of the main text.

39.
$x^2 - 3x + 1 \geq 0$

Using the quadratic formula will show that the boundaries are $\dfrac{3 + \sqrt{5}}{2}$ and $\dfrac{3 - \sqrt{5}}{2}$ or approximately 2.6 and 0.38.

Picking numbers from each of the three regions will show that
$$x \leq \frac{3 - \sqrt{5}}{2} \quad \text{or} \quad x \geq \frac{3 + \sqrt{5}}{2}$$

41.
$$4x^2 - 1 > 0$$

$(2x + 1)(2x - 1) > 0$ The product is positive

$x = -\frac{1}{2}, \quad x = \frac{1}{2}$ Boundaries

Picking numbers from each of the three regions will show that
$$x < -\frac{1}{2} \quad \text{or} \quad x > \frac{1}{2}$$

43.
$$x^2 - 2x + 1 < 0$$

$(x - 1)(x - 1) < 0$ The product is negative

$x = 1$ Boundary

Picking numbers from each of the three regions will show that none of the real numbers are solutions.

45.
$9x^2 + 30x + 25 \geq 0$

$(3x + 5)(3x + 5) \geq 0$ The product is positive or zero

$x = -\frac{5}{3}$ Boundary

Picking numbers from each of the three regions will show that all of the real numbers are solutions.

47.
$x^2 + 1 > 0$

This inequality is true for all real numbers.

The graphs are in the answer section of the main text.

49.
$$x^2 - 2x + 5 \leq 0$$
Picking numbers from each of the three regions will show that none of the real numbers are solutions.

51.
$$x^2 + x - 6 < 0$$

$(x - 2)(x + 3) < 0$ The product is negative

$x = 2, \quad x = -3$ Boundaries

Picking numbers from each of the three regions will show that
$$-3 < x < 2$$

53.
$$6x^2 - 7x - 3 \geq 0$$

$(3x + 1)(2x - 3) \geq 0$ The product is positive or zero

$x = -\dfrac{1}{3}, \quad x = \dfrac{3}{2}$ Boundaries

Picking numbers from each of the three regions will show that
$$x \leq -\frac{1}{3} \quad \text{or} \quad x \geq \frac{3}{2}$$

55.
$$9x^2 - 16 > 0$$

$(3x + 4)(3x - 4) > 0$ The product is negative

$x = -\dfrac{4}{3}, \quad x = \dfrac{4}{3}$ Boundaries

Picking numbers from each of the three regions will show that
$$x < -\frac{4}{3} \quad \text{or} \quad x > \frac{4}{3}$$

57.
$$x^2 + 8x + 12 < 0$$

$(x + 6)(x + 2) < 0$ The product is negative

$x = -6, \quad x = -2$ Boundaries

Picking numbers from each of the three regions will show that
$$-6 < x < -2$$

The graphs are in the answer section of the main text.

59.
$$2x^2 - 8x - 4 \geq 0$$

$x^2 - 4x - 2 \geq 0$ The product is negative or zero

Using the quadratic formula will show that the boundaries are $2 + \sqrt{7}$ and $2 - \sqrt{7}$ or approximately 4.6 and -0.65.

Picking numbers from each of the three regions will show that
$$x \leq 2 - \sqrt{7} \quad \text{or} \quad x \geq 2 + \sqrt{7}$$

61.
$$2x^2 + 3x + 1 > 0$$

$(x + 1)(2x + 1) > 0$ The product is negative

$x = -1, \quad x = -\dfrac{1}{2}$ Boundaries

Picking numbers from each of the three regions will show that
$$x < -1 \quad \text{or} \quad x > -\frac{1}{2}$$

63.
$$x^2 - 8x + 15 \leq 0$$

$(x - 3)(x - 5) \leq 0$ The product is negative or zero

$x = 3, \quad x = 5$ Boundaries

Picking numbers from each of the three regions will show that
$$3 \leq x \leq 5$$

65.
$$2x^2 + x + 6 \geq 0$$

Picking numbers from each of the three regions will show that all real numbers are solutions.

67.
$$x^2 + 8x + 5 \leq 0$$

$(2x + 5)(x + 1) \leq 0$ The product is negative or zero

$x = -\dfrac{5}{2}, \quad x = -1$ Boundaries

| The graphs are in the answer section of the main text. |

#67. continued

Picking numbers from each of the three regions will show that

$$-\frac{5}{2} \le x \le -1$$

69.

$3x^2 - 4x + 1 > 0$

$(3x-1)(x-1) > 0$ The product is positive

$x = \frac{1}{3}, \quad x = 1$ Boundaries

Picking numbers from each of the three regions will show that

$$x < \frac{1}{3} \quad \text{or} \quad x > 1$$

71.

$x^2 + 4x + 4 < 0$

Picking numbers from each of the three regions will show that none of the real numbers are solutions.

73.

$25x^2 + 10x - 1 > 0$

Using the quadratic formula will show that the

boundaries are $\dfrac{-1+\sqrt{2}}{5}$ and $\dfrac{-1-\sqrt{2}}{5}$ or 0.083

or - 0.48.

Picking numbers from each of the three regions will show that

$$x < \frac{-1-\sqrt{2}}{5} \quad \text{or} \quad x > \frac{-1+\sqrt{2}}{5}$$

75.

$6x^2 - x - 12 \le 0$

$(3x+4)(2x-3) \le 0$ The product is negative or zero

$x = -\frac{4}{3}, \quad x = \frac{3}{2}$ Boundaries

Picking numbers from each of the three regions will show that

$$-\frac{4}{3} \le x \le \frac{3}{2}$$

| The graphs are in the answer section of the main text. |

77.

$3x^2 - 4x - 2 < 0$

Using the quadratic formula will show that the

boundaries are $\dfrac{2+\sqrt{10}}{3}$ and $\dfrac{2-\sqrt{10}}{3}$ or

approximately 1.7 and - 0.39.

Picking numbers from each of the three regions will show that

$$\frac{2-\sqrt{10}}{3} < x < \frac{2+\sqrt{10}}{3}$$

79.

$x^2 + 2x - 15 < 0$

$(x+5)(x-3) < 0$ The product is negative

$x = -5, \quad x = 3$ Boundaries

Picking numbers from each of the three regions will show that

$$-5 < x < 3$$

81.

$3x^2 - 2x + 1 < 0$

Picking numbers from each of the three regions will show that none of the real numbers are solutions.

83.

$2x^2 - 8x < 0$

$2x(x-4) < 0$ The product is negative

$x = 0, \quad x = 4$ Boundaries

Picking numbers from each of the three regions will show that

$$0 < x < 4$$

85.

$\dfrac{x-4}{x+1} < 0$ The quotient is negative

$x = 4, \quad x = -1$ Boundaries

Picking numbers from each of the three regions will show that

$$-1 < x < 4$$

The graphs are in the answer section of the main text.

87.

$$\frac{2x-3}{x-7}>0 \qquad \text{The quotient is positive}$$

$$x=\frac{3}{2}, \quad x=7 \quad \text{Boundaries}$$

Picking numbers from each of the three regions will show that

$$x<\frac{3}{2} \quad \text{or} \quad x>7$$

89.

$$\frac{2x+3}{x+1}\geq 0 \qquad \text{Quotient is positive or zero}$$

$$x=-\frac{3}{2}, \quad x=-1 \quad \text{Boundaries}$$

Picking numbers from each of the three regions will show that

$$x\leq-\frac{3}{2} \quad \text{or} \quad x>-1$$

91.

$$\frac{2x}{x-4}-1>0$$

$$\frac{x+4}{x-4}>0 \qquad \text{Quotient is positive}$$

$$x=-4, \quad x=4$$

Picking numbers from each of the three regions will show that

$$x<-4 \quad \text{or} \quad x>4$$

Note: $\dfrac{2x}{x-4}-1=\dfrac{2x}{x-4}-\dfrac{x-4}{x-4}$

$$=\frac{2x-x+4}{x-4}=\frac{x+4}{x-4}$$

93.

$$\frac{2x+3}{x-1}+1\leq 0$$

$$\frac{3x+2}{x-1}\leq 0 \qquad \text{Quotient is negative or zero}$$

$$x=-\frac{2}{3}, \quad x=1 \quad \text{Boundaries}$$

Picking numbers from each of the three regions will show that

$$-\frac{2}{3}\leq x<1$$

The graphs are in the answer section of the main text.

#93. continued

Note: $\dfrac{2x+3}{x-1}+1=\dfrac{2x+3}{x-1}+\dfrac{x-1}{x-1}$

$$=\frac{3x+2}{x-1}$$

95.

$$\frac{x+1}{x-1}-\frac{1}{2}<0$$

$$\frac{x+3}{2x-2}<0 \qquad \text{Quotient is negative}$$

$$x=-3, \quad x=1 \quad \text{Boundaries}$$

Picking numbers from each of the three regions will show that

$$-3<x<1$$

97.

$(x-1)(x-2)(x+3)>0$ The product is positive

$x=1, \quad x=2, \quad x=-3$ Boundaries

Picking numbers from each of the four regions will show that

$$-3<x<1 \quad \text{or} \quad x>2$$

99.

$(2x-1)(x-3)(x+1)\leq 0$ The product is negative or zero

$$x=\frac{1}{2}, \quad x=3, \quad x=-1 \quad \text{Boundaries}$$

Picking numbers from each of the four regions will show that

$$x\leq-1 \quad \text{or} \quad \frac{1}{2}\leq x\leq 3$$

101.

$(x-1)^2(x+3)\geq 0$ The product is positive or zero

$x=1, \quad x=-3$ Boundaries

Picking numbers from each of the three regions will show that

$$x\geq-3$$

Review Exercises

1.-11.

The graphs are in the answer section of the main text.

13.

$y = x^2 - 6x + 14$

$y = x^2 - 6x + 9 - 9 + 14$

$y = (x-3)^2 + 5$ Vertex: $(3, 5)$

15.

$y = -\left(x^2 + 14x + 49\right) - 47 + 49$

$y = -(x+7)^2 + 2$ Vertex: $(-7, 2)$

17.

$y = -3\left(x^2 - 2x + 1\right) + 4 + 3$

$y = -3(x-1)^2 + 7$ Vertex: $(1, 7)$

19.

$y = x^2 - \dfrac{4}{3}x + \dfrac{4}{9} + 3 - \dfrac{4}{9}$

$y = \left(x - \dfrac{2}{3}\right)^2 + \dfrac{23}{9}$ Vertex: $\left(\dfrac{2}{3}, \dfrac{23}{9}\right)$

21.

$y = 2\left(x^2 + x + \dfrac{1}{4}\right) + \dfrac{7}{2} - \dfrac{1}{2}$

$y = 2\left(x + \dfrac{1}{2}\right)^2 + 3$ Vertex: $\left(-\dfrac{1}{2}, 3\right)$

23.

$y = -2\left(x^2 - 5x + \dfrac{25}{4}\right) - 6 + \dfrac{25}{2}$

$y = -2\left(x - \dfrac{5}{2}\right)^2 + \dfrac{13}{2}$ Vertex: $\left(\dfrac{5}{2}, \dfrac{13}{2}\right)$

25.

$y = -2\left(x^2 - 180x + 8100\right) - 14,060 + 16,200$

$y = -2(x - 90)^2 + 2140$

Answer: 90 tires, profit = \$2140

27.

$x^2 = 16$

$\quad x = \pm 4$

29.

$x^2 = 121$

$\quad x = \pm 11$

31.

$9x^2 = -16$

$\quad x^2 = -\dfrac{16}{9}$

$\quad x = \pm\sqrt{\dfrac{-16}{9}}$

$\quad x = \pm\dfrac{4}{3}i$

33.

$3x^2 = -1$

$\quad x^2 = -\dfrac{1}{3}$

$\quad x = \pm\sqrt{-\dfrac{1}{3}}$

$\quad x = \pm\dfrac{\sqrt{3}}{3}i$

35.

$(3x + 1)^2 = 13$

$\quad 3x + 1 = \pm\sqrt{13}$

$\quad 3x = -1 \pm \sqrt{13}$

$\quad x = \dfrac{-1 \pm \sqrt{13}}{3}$

37.

$x^2 - 6x + 5 = 0$

$x^2 - 6x + 9 = -5 + 9$

$\quad (x - 3)^2 = 4$

$\quad\quad x - 3 = \pm 2$

$\quad\quad\quad x = 3 \pm 2; \quad\quad x = 3 + 2 = 5; \quad x = 3 - 2 = 1$

39.

$x^2 - 2x - 1 = 0$

$x^2 - 2x + 1 = 1 + 1$

$\quad (x - 1)^2 = 2$

$\quad\quad x - 1 = \pm\sqrt{2}$

$\quad\quad\quad x = 1 \pm \sqrt{2}$

41.

$$9x^2 - 12x + 20 = 0$$

$$x^2 - \frac{4}{3}x + \frac{20}{9} = 0 \qquad \text{Multiply by } \frac{1}{9}$$

$$x^2 - \frac{4}{3}x + \frac{4}{9} = -\frac{20}{9} + \frac{4}{9}$$

$$\left(x - \frac{2}{3}\right)^2 = -\frac{16}{9}$$

$$x - \frac{2}{3} = \pm\sqrt{-\frac{16}{9}}$$

$$x = \frac{2}{3} \pm \frac{4}{3}i \quad \text{or} \quad \frac{2 \pm 4i}{3}$$

43.

$$3x^2 + 8x + 3 = 0$$

$$x^2 + \frac{8}{3}x + 1 = 0 \qquad \text{Multiply by } \frac{1}{3}$$

$$x^2 + \frac{8}{3}x + \frac{16}{9} = -1 + \frac{16}{9}$$

$$\left(x + \frac{4}{3}\right)^2 = \frac{7}{9}$$

$$x + \frac{4}{3} = \pm\sqrt{\frac{7}{9}}$$

$$x = -\frac{4}{3} \pm \frac{\sqrt{7}}{3} \quad \text{or} \quad \frac{-4 \pm \sqrt{7}}{3}$$

45.

$$2x^2 - 2x - 1 = 0$$

$$x^2 - x + \frac{1}{4} = \frac{1}{2} + \frac{1}{4} \qquad \text{Multiply by } \frac{1}{2}$$

$$\left(x - \frac{1}{2}\right)^2 = \frac{3}{4}$$

$$x - \frac{1}{2} = \pm\sqrt{\frac{3}{4}}$$

$$x = \frac{1}{2} \pm \frac{\sqrt{3}}{2} \quad \text{or} \quad \frac{1 \pm \sqrt{3}}{2}$$

47.

$$3x^2 - 5x - 2 = 0$$

$$a = 3, b = -5, c = -2$$

$$x = \frac{5 \pm \sqrt{25 + 24}}{6}$$

$$x = \frac{5 \pm \sqrt{49}}{6}$$

#47. continued

$$x = \frac{5 \pm 7}{6}; \qquad x = \frac{5 + 7}{6} = 2; \quad x = \frac{5 - 7}{6} = -\frac{1}{3}$$

49.

$$5x^2 - 4 = 0$$

$$a = 5, b = 0, c = -4$$

$$x = \frac{\pm\sqrt{80}}{10}$$

$$x = \frac{\pm 4\sqrt{5}}{10} = \frac{\pm 2\sqrt{5}}{5}$$

51.

$$2x^2 + 5x - 18 = 0$$

$$a = 2, b = 5, c = -18$$

$$x = \frac{-5 \pm \sqrt{25 + 144}}{4}$$

$$x = \frac{-5 \pm \sqrt{169}}{4}$$

$$x = \frac{-5 \pm 13}{4}; \qquad x = \frac{-5 + 13}{4} = 2; \quad x = \frac{-5 - 13}{4} = -\frac{9}{2}$$

53.

$$x^2 + 6x + 7 = 0$$

$$a = 1, \ b = 6, \ c = 7$$

$$x = \frac{-6 \pm \sqrt{36 - 28}}{2}$$

$$x = \frac{-6 \pm \sqrt{8}}{2}$$

$$x = \frac{-6 \pm 2\sqrt{2}}{2} = -3 \pm \sqrt{2}$$

55.

$$x^2 - 6x + 10 = 0$$

$$a = 1, b = -6, c = 10$$

$$x = \frac{6 \pm \sqrt{36 - 40}}{2}$$

$$x = \frac{6 \pm \sqrt{-4}}{2}$$

$$x = \frac{6 \pm 2i}{2} = 3 \pm i$$

57.

$3 - \dfrac{11}{x} - \dfrac{4}{x^2} = 0$

$3x^2 - 11x - 4 = 0$ Multiply by x^2

$a = 3, b = -11, c = -4$

$x = \dfrac{11 \pm \sqrt{121 + 48}}{6}$

$x = \dfrac{11 \pm \sqrt{169}}{6}$

$x = \dfrac{11 \pm 13}{6};$ $x = \dfrac{11 + 13}{6} = 4$

$\phantom{x = \dfrac{11 \pm 13}{6};} \quad x = \dfrac{11 - 13}{6} = -\dfrac{1}{3}$

59.

$\dfrac{1}{13} - \dfrac{4}{13x} + \dfrac{1}{x^2} = 0$

$x^2 - 4x + 13 = 0$ Multiply by $13x^2$

$a = 1, b = -4, c = 13$

$x = \dfrac{4 \pm \sqrt{16 - 52}}{2}$

$x = \dfrac{4 \pm \sqrt{-36}}{2}$

$x = \dfrac{4 \pm 6i}{2} = 2 \pm 3i$

61.

$9x^2 - 12x + 4 = 0$

$a = 9, b = -12, c = 4$

$b^2 - 4ac = 144 - 144 = 0$

One rational solution

63.

$3x^2 - 5x + 1 = 0$

$a = 3, b = -5, c = 1$

$b^2 - 4ac = 25 - 12 = 13$

Two irrational roots

65.

Let x = one number.

Then 2 - x = the other number.

#65. continued

Thus, $x^2 + (2 - x)^2 = \dfrac{5}{2}$

$x^2 + 4 - 4x + x^2 = \dfrac{5}{2}$

$2x^2 - 4x + 4 = \dfrac{5}{2}$

$4x^2 - 8x + 8 = 5$

$4x^2 - 8x + 3 = 0$

$(2x - 1)(2x - 3) = 0$

$2x - 1 = 0$ or $2x - 3 = 0$

$x = \dfrac{1}{2}$ $x = \dfrac{3}{2}$

The numbers are $\dfrac{1}{2}$ and $\dfrac{3}{2}$.

67.

Let x = height of the building.

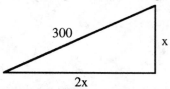

Use the Pythagorean Theorem.

$(2x)^2 + x^2 = 300^2$

$5x^2 = 90,000$

$x^2 = 18,000$

$x = \pm\sqrt{18000}$

$x = \pm 60\sqrt{5}$

Answer: $60\sqrt{5}$ or approximately 13.16 ft

69.

Let x = width of the border.

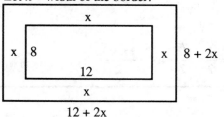

To find the solution, subtract the pool's area from the total area.

#69. continued

$(12 + 2x)(8 + 2x) - 8 \cdot 12 = 124$

$96 + 40x + 4x^2 - 96 = 124$

$4x^2 + 40x - 124 = 0$

$x^2 + 10x - 31 = 0$ Multiply by $\frac{1}{4}$

$a = 1, b = 10, c = -31$

$x = \dfrac{-10 \pm \sqrt{100 + 124}}{2}$

$x = \dfrac{-10 \pm 4\sqrt{14}}{2} = -5 \pm 2\sqrt{14}$

Answer: $-5 + 2\sqrt{14} \approx 2.5$ ft

71.

Let $h = 0$

$0 = -16t^2 + 48t + 80$

$0 = t^2 - 3t - 5$

$a = 1, b = -3, c = -5$

$t = \dfrac{3 \pm \sqrt{9 + 20}}{2}$

$t = \dfrac{3 \pm \sqrt{29}}{2}$

Answer: $\dfrac{3 + 29}{2} \approx 4.19$ sec

73.

$x^4 + 5x^2 - 36 = 0$ Let $u = x^2$

$u^2 + 5u - 36 = 0$

$a = 1, b = 5, c = -36$

$u = \dfrac{-5 \pm \sqrt{25 + 144}}{2}$

$u = \dfrac{-5 \pm \sqrt{169}}{2} = \dfrac{-5 \pm 13}{2}$

$u = \dfrac{-5 + 13}{2} = 4$

$u = \dfrac{-5 - 13}{2} = -9$

$x^2 = 4$ or $x^2 = -9$

$x = \pm 2$ $x = \pm 3i$

75.

$x - 4\sqrt{x} + 3 = 0$ Let $u = \sqrt{x}$

$u^2 - 4u + 3 = 0$

$(u - 3)(u - 1) = 0$

$u = 3$ or $u = 1$

$\sqrt{x} = 3$ $\sqrt{x} = 1$

$x = 9$ $x = 1$

77.

$4(x + 2)^2 + 9(x + 2) - 9 = 0$ Let $u = x + 2$

$4u^2 + 9u - 9 = 0$

$(4u - 3)(u + 3) = 0$

$4u - 3 = 0$ or $u + 3 = 0$

$4(x + 2) - 3 = 0$ $(x + 2) + 3 = 0$

$x = -\dfrac{5}{4}$ $x = -5$

79.

$x^{-2} + 9x^{-1} + 20 = 0$ Let $u = x$

$u^2 + 9u + 20 = 0$

$(u + 4)(u + 5) = 0$

$u + 4 = 0$ or $u + 5 = 0$

$u = -4$ $u = -5$

$x^{-1} = -4$ $x^{-1} = -5$

$x = -\dfrac{1}{4}$ $x = -\dfrac{1}{5}$

81.

$x^{2/3} - 3x^{1/3} - 10 = 0$ Let $u = x^{1/3}$

$u^2 - 3u - 10 = 0$

$(u + 2)(u - 5) = 0$

$u + 2 = 0$ or $u - 5 = 0$

$u = -2$ $u = 5$

$x^{1/3} = -2$ $x^{1/3} = 5$

$x = -8$ $x = 125$

83.

$\left(\dfrac{x}{x - 1}\right)^2 - 4 \cdot \left(\dfrac{x}{x - 1}\right) - 5 = 0$ Let $u = \dfrac{x}{x - 1}$

$u^2 - 4u - 5 = 0$

$(u - 5)(u + 1) = 0$

#83. continued

$u - 5 = 0$ or $u + 1 = 0$

$u = 5$ $u = -1$

$\dfrac{x}{x-1} = 5$ $\dfrac{x}{x-1} = -1$

$x = \dfrac{5}{4}$ $x = \dfrac{1}{2}$

85.

$$3x^3 + 4x^2 - 27x - 36 = 0$$

$$x^2(3x + 4) - 9(3x + 4) = 0$$

$$(3x + 4)(x + 3)(x - 3) = 0$$

$$x = -\dfrac{4}{3}, x = -3, x = 3$$

87.

$$4x^3 + 12x^2 - x - 3 = 0$$

$$4x^2(x + 3) - 1 \cdot (x + 3) = 0$$

$$(x + 3)(2x + 1)(2x - 1) = 0$$

$$x = -3, x = -\dfrac{1}{2}, x = \dfrac{1}{2}$$

89.

$$7x^4 - 6x^3 - 7x + 6 = 0$$

$$x^3(7x - 6) - 1 \cdot (7x - 6) = 0$$

$$(7x - 6)(x - 1)\left(x^2 + x + 1\right) = 0$$

$$x = \dfrac{6}{7}, x = 1, x = \dfrac{-1 \pm \sqrt{3}i}{2}$$

Note: Solve $x^2 + x + 1$ using the quadratic formula.

91.

$$2x^2 + 3x - 5 \geq 0$$

$(2x + 5)(x - 1) \geq 0$ The product is
                                  positive or zero

$x = -\dfrac{5}{2}, \quad x = 1$ Boundaries

Picking numbers from each of the three regions will show that

$$x \leq -\dfrac{5}{2} \quad \text{or} \quad x \geq 1$$

93.

$$x^2 - 4x + 13 \geq 0$$

Picking numbers from each of the three regions will show that all of the real numbers are solutions.

95.

$x^2 - 7x + 10 < 0$ Placing in standard form

$(x - 2)(x - 5) < 0$ The product is negative

$x = 2, \quad x = 5$ Boundaries

Picking numbers from each of the three regions will show that

$$2 < x < 5$$

97.

$x^2 + 2x - 4 > 0$ Placing in standard form

Using the quadratic formula will show that the boundaries are $-1 + \sqrt{5}$ and $-1 - \sqrt{5}$ or approximately 1.23 and -3.24.

Picking numbers from each of the three regions will show that

$$x < -1 - \sqrt{5} \quad \text{or} \quad x > -1 + \sqrt{5}$$

99.

$x^2 - 10x + 7 < 0$ Placing in standard form

Using the quadratic formula will show that the boundaries are $5 + 3\sqrt{2}$ and $5 - 3\sqrt{2}$ or approximately 9.24 and 0.76.

Picking numbers from each of the three regions will show that

$$5 - 3\sqrt{2} < x < 5 + 3\sqrt{2}$$

101.

$9x^2 + 24x + 16 \leq 0$ Placing in standard form

$(3x + 4)^2 \leq 0$ The product is
                            positive or zero

$x = -\dfrac{4}{3}$ Boundary

No real number but $-\dfrac{4}{3}$ is a solution since $(3x + 4)^2$ is positive or zero.

103.

$\dfrac{2x-3}{x+4} \geq 0$ Quotient is positive or zero

$x = \dfrac{3}{2}, \quad x = -4$ Boundaries

Picking numbers from each of the three regions will show that

$$x < -4 \quad \text{or} \quad x \geq \dfrac{3}{2}$$

105.

$\dfrac{4x}{x-2} - 1 < 0$

$\dfrac{3x+2}{x-2} < 0$ Quotient is negative

$x = -\dfrac{2}{3}, \quad x = 2$ Boundaries

Picking numbers from each of the three regions will show that

$$-\dfrac{2}{3} < x < 2$$

107.

$(x+3)(x-4)(x-7) \geq 0$ Product is positive or zero

$x = -3, \, x = 4, \, x = 7$ Boundaries

Picking numbers from each of the four regions will show that

$$-3 \leq x \leq 4 \quad \text{or} \quad x \geq 7$$

Chapter 7 Test Solutions

1.

The graph is in the answer section of the main text.

3.

$(2x+7)^2 = 16$

$2x + 7 = \pm 4$

$2x = -7 \pm 4$

$x = \dfrac{-7 \pm 4}{2}; \quad x = \dfrac{-7+4}{2} = -\dfrac{3}{2}$

$\phantom{x = \dfrac{-7 \pm 4}{2}; \quad} x = \dfrac{-7-4}{2} = -\dfrac{11}{2}$

5.

$4x^2 + 12x + 1 = 0$

$a = 4, \, b = 12, \, c = 1$

$x = \dfrac{-12 \pm \sqrt{144 - 16}}{2 \cdot 4}$

$x = \dfrac{-12 \pm \sqrt{128}}{8}$

$x = \dfrac{-12 \pm 8\sqrt{2}}{8} = \dfrac{-3 \pm 2\sqrt{2}}{2}$

7.

$2x^4 - 7x^2 - 72 = 0$ Let $u = x^2$

$2u^2 - 7u - 72 = 0$

$(2u+9)(u-8) = 0$

$2u + 9 = 0 \quad \text{or} \quad u - 8 = 0$

$\quad u = -\dfrac{9}{2} \qquad\qquad u = 8$

$\quad x^2 = -\dfrac{9}{2} \qquad\qquad x^2 = 8$

$\quad x = \pm\sqrt{-\dfrac{9}{2}} \qquad x = \pm 2\sqrt{2}$

$\quad x = \pm\dfrac{3\sqrt{2}i}{2}$

9.

$27x^3 + 45x^2 - 3x - 5 = 0$

$9x^2(3x+5) - 1 \cdot (3x+5) = 0$

$(3x+5)(3x+1)(3x-1) = 0$

$x = -\dfrac{5}{3}, \, x = -\dfrac{1}{3}, \, x = \dfrac{1}{3}$

11.

$2x^2 - 5x - 12 \geq 0$

$(2x+3)(x-4) \geq 0$ The product is positive or zero

$x = -\dfrac{3}{2}, \quad x = 4$ Boundaries

Picking numbers from each of the three regions will show that

$$x \leq -\dfrac{3}{2} \quad \text{or} \quad x \geq 4$$

13.

$$\frac{1}{x-4} - 1 \geq 0$$

$$\frac{-x+5}{x-4} \geq 0 \qquad \text{Quotient is positive or zero}$$

$x = 5 \quad \text{or} \quad x = 4 \qquad \text{Boundaries}$

Picking numbers from each of the three regions will show that

$$4 < x \leq 5$$

15.

Let x = width of moat.

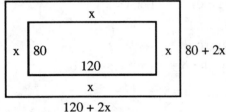

$120 + 2x$

To find the solution, subtract the castle's area from the total area.

$$(120 + 2x)(80 + 2x) - 80 \cdot 120 = 1600$$

$$9600 + 400x + 4x^2 - 9600 = 1600$$

$$4x^2 + 400x - 1600 = 0$$

$$x^2 + 100x - 400 = 0$$

$a = 1, b = 100, c = -400$

$$x = \frac{-100 \pm \sqrt{10000 + 1600}}{2}$$

$$x = \frac{-100 \pm \sqrt{11600}}{2}$$

$$x = \frac{-100 \pm 20\sqrt{29}}{2} = -50 \pm 10\sqrt{29}$$

Answer: $-50 + 10\sqrt{29} \approx 3.85$ yd

Test Your Memory

1.

$$\frac{\left(2^{-1}x^3y^{-4}\right)^2}{\left(4^0 x^{-3}y^{-1}\right)^{-3}} = \frac{2^{-2}x^6 y^{-8}}{x^9 y^3} \qquad \text{Note:} \quad 4^0 = 1$$

$$= \frac{1}{2^2 x^{9-6} y^{3+8}}$$

$$= \frac{1}{4x^3 y^{11}}$$

3.

$$\frac{4x(x-6)(x-1)}{(x+1)(x-1)(5x+8)} \cdot \frac{(x+6)(x+1)}{12x^3(x+6)(x-6)}$$

$$= \frac{4x}{12x^3(5x+8)}$$

$$= \frac{4x}{3(4x)x^2(5x+8)} = \frac{1}{3x^2(5x+8)}$$

5.

$$\frac{-\left(15x^{-2} - x^{-1} - 2\right)}{-\left(21x^{-2} + 4x^{-1} - 1\right)} \qquad \text{Let } u = x^{-1}$$

$$= \frac{15u^2 - u - 2}{21u^2 + 4u - 1} = \frac{(5u-2)(3u+1)}{(3u+1)(7u-1)}$$

$$= \frac{5x^{-1} - 2}{7x^{-1} - 1} = \frac{\left(\frac{5}{x} - 2\right)x}{\left(\frac{7}{x} - 1\right)x}$$

$$= \frac{5 - 2x}{7 - x} \quad \text{or} \quad \frac{2x - 5}{x - 7}$$

7.

$$\sqrt[3]{135} - 5\sqrt[3]{40} = \sqrt[3]{27 \cdot 5} - 5\sqrt[3]{8 \cdot 5}$$

$$= 3\sqrt[3]{5} - 10\sqrt[3]{5}$$

$$= -7\sqrt[3]{5}$$

9.

$$\frac{\sqrt{x} + \sqrt{y}}{\sqrt{x} - \sqrt{y}} \cdot \frac{\sqrt{x} + \sqrt{y}}{\sqrt{x} + \sqrt{y}} = \frac{x + 2\sqrt{xy} + y}{x - y}$$

11.

$$y = 2x^2 + 4x - 1$$

$$y = 2\left(x^2 + 2x + 1\right) - 1 - 2$$

$$y = 2(x+1)^2 - 3$$

Vertex: $(-1, -3)$

13.

$$g(-1) = 2(1)^2 - 7 = -5$$

15.

$$8f(1) - 2g(2) = 8(3 \cdot 1 + 2) - 2\left(2 \cdot 2^2 - 7\right)$$
$$= 8(5) - 2(1)$$
$$= 40 - 2 = 38$$

17.

$(6, -2), \quad (-9, 4)$

$$m = \frac{4 + 2}{-9 - 6} = \frac{6}{-15} = -\frac{2}{5}$$

$$y + 2 = -\frac{2}{5}(x - 6)$$

$$y + 2 = -\frac{2}{5}x + \frac{12}{5}$$

$$y = -\frac{2}{5}x + \frac{12}{5} - 2$$

$$y = -\frac{2}{5}x + \frac{2}{5}$$

19.

$(-1, -6), \quad m = -\frac{2}{3}$

$$y + 6 = -\frac{2}{3}(x + 1)$$

$$y + 6 = -\frac{2}{3}x - \frac{2}{3}$$

$$y = -\frac{2}{3}x - \frac{2}{3} - 6$$

$$y = -\frac{2}{3}x - \frac{20}{3}$$

21.

$$\frac{2}{x + 3} = \frac{5}{x - 1}$$
$$2x - 2 = 5x + 15 \qquad \text{Cross multiply}$$
$$-17 = 3x$$
$$-\frac{17}{3} = x \qquad \left\{-\frac{17}{3}\right\}$$

23.

$$\frac{x + 3}{x - 1} + \frac{3}{x + 2} = \frac{5x + 13}{(x - 1)(x + 2)}$$
$$\text{LCD} = (x - 1)(x + 2)$$

$$\frac{(x - 1)(x + 2)}{1}\left(\frac{x + 3}{x - 1} + \frac{3}{x + 2}\right)$$
$$= \frac{5x + 13}{(x - 1)(x + 2)} \cdot \frac{(x - 1)(x + 2)}{1}$$

#23. continued

$$(x + 2)(x + 3) + 3(x - 1) = 5x + 13$$
$$x^2 + 5x + 6 + 3x - 3 = 5x + 13$$
$$x^2 + 3x - 10 = 0$$
$$(x + 5)(x - 2) = 0$$
$$x + 5 = 0 \quad \text{or} \quad x - 2 = 0$$
$$x = -5 \qquad \qquad x = 2$$
$$\{-5, 2\}$$

25.

$$\sqrt{3x - 2} + 4 = 0$$
$$\sqrt{3x - 2} = -4$$
$$\varnothing$$

27.

$$\sqrt{x + 2} - \sqrt{x - 3} = 1$$
$$\left(\sqrt{x + 2}\right)^2 = \left(\sqrt{x - 3} + 1\right)^2$$
$$x + 2 = (x - 3) + 2\sqrt{x - 3} + 1$$
$$4 = 2\sqrt{x - 3}$$
$$(2)^2 = \left(\sqrt{x - 3}\right)^2 \qquad \text{Multiply by } \frac{1}{2}$$
$$4 = x - 3$$
$$7 = x \qquad \{7\}$$

29.

$$2\left(\frac{x}{x + 1}\right)^2 - 5\left(\frac{x}{x + 1}\right) - 3 = 0 \qquad \text{Let } u = \frac{x}{x + 1}$$
$$2u^2 - 5u - 3 = 0$$
$$(2u + 1)(u - 3) = 0$$
$$2u + 1 = 0 \quad \text{or} \quad u - 3 = 0$$
$$u = -\frac{1}{2} \qquad\qquad u = 3$$
$$\frac{x}{x + 1} = -\frac{1}{2} \qquad\qquad \frac{x}{x + 1} = 3$$
$$2x = -x - 1 \qquad\qquad 3x + 3 = x$$
$$x = -\frac{1}{3} \qquad\qquad x = -\frac{3}{2}$$
$$\left\{-\frac{1}{3}, -\frac{3}{2}\right\}$$

31.
$$9x^4 - 37x^2 + 4 = 0 \qquad \text{Let } u = x^2$$
$$9u^2 - 37u + 4 = 0$$
$$(9u - 1)(u - 4) = 0$$
$$9u - 1 = 0 \quad \text{or} \quad u - 4 = 0$$
$$u = \frac{1}{9} \qquad\qquad u = 4$$
$$x^2 = \frac{1}{9} \qquad\qquad x^2 = 4$$
$$x = \pm\frac{1}{3} \qquad\qquad x = \pm 2$$
$$\left\{\pm\frac{1}{3}, \pm 2\right\}$$

33.
$$(3x - 1)^2 = 25$$
$$3x - 1 = \pm 5$$
$$3x = 1 \pm 5$$
$$x = \frac{1 \pm 5}{3}; \quad x = \frac{1 + 5}{3} = 2$$
$$x = \frac{1 - 5}{3} = -\frac{4}{3}$$
$$\left\{2, -\frac{4}{3}\right\}$$

35.
$$x^2 - 4x - 2 = 0$$
$$a = 1, b = -4, c = -2$$
$$x = \frac{4 \pm \sqrt{16 + 8}}{2}$$
$$x = \frac{4 \pm \sqrt{24}}{2} = \frac{4 \pm 2\sqrt{6}}{2} = 2 \pm \sqrt{6}$$
$$\left\{2 \pm \sqrt{6}\right\}$$

37.
$$2 - 3x \geq -10$$
$$-3x \geq -12$$
$$x \leq \frac{12}{3}$$
$$x \leq 4$$
The graph is in the answer section of the main text.

39.
$$x^2 - 2x - 8 < 0$$
$$(x + 2)(x - 4) < 0 \qquad \text{The product is negative}$$
$$x = -2, \quad x = 4 \qquad \text{Boundaries}$$
Picking numbers from each of the three regions will show that
$$-2 < x < 4$$
The graph is in the answer section of the main text.

41.
$$\frac{x - 2}{x + 6} > 0 \qquad \text{Quotient is positive}$$
$$x = 2, \quad x = -6 \qquad \text{Boundaries}$$
Picking numbers from each of the three regions will show that
$$x < -6 \quad \text{or} \quad x > 2$$
The graph is in the answer section of the main text.

43.
The graph is in the answer section of the main text.

45.
Let x = number of quarters.
Let $2x$ = number of nickels.
Let $x + 5$ = number of dimes.
Thus, $25x + 5(2x) + 10(x + 5) = 320$
$$25x + 10x + 10x + 50 = 320$$
$$45x = 270$$
$$x = 6$$
$$2x = 2 \cdot 6 = 12$$
$$x + 5 = 6 + 5 = 11$$
Answer: 6 quarters, 12 nickels, 11 dimes

47.
$$S = ke^2$$

Find k	Find S
$54 = k \cdot 3^2$	$S = 6 \cdot 6^2$
$6 = k$	$S = 216$

Answer: 216 sq in

49.
Let x = width of border.

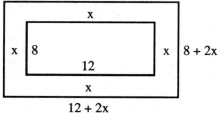

$12 + 2x$

To solve, subtract the pool's area from the total area.

$$(12 + 2x)(8 + 2x) - 8 \cdot 12 = 120$$

$$96 + 40x + 4x^2 - 96 = 120$$

$$4x^2 + 40x - 120 = 0$$

$$x^2 + 10x - 30 = 0$$

$$a = 1, b = 10, c = -30$$

$$x = \frac{-10 \pm \sqrt{100 + 120}}{2}$$

$$x = \frac{-10 \pm \sqrt{220}}{2}$$

$$x = \frac{-10 \pm 2\sqrt{55}}{2} = -5 \pm \sqrt{55}$$

Answer: $-5 + \sqrt{55} \approx 2.4$ ft

Chapter 7 Study Guide

Self-Test Exercises

I. Place the following quadratic functions in standard form, if necessary. Also find the vertex and sketch the graph of each.

1. $y = \dfrac{1}{2}(x-2)^2 + 2$
2. $y = x^2 + 6x + 5$
3. $y = -3(x+4)^2 + 1$
4. $y = 3x^2 + 4x + 3$

II. Find the solutions of the following quadratic equations using the indicated method.

5. $(3x-4)^2 = 25$ Extraction of roots
6. $x^2 + 6x + 8 = 63$ Completing the square

7. $3x^2 + 4 = 8x$ Quadratic formula
8. $(x-1)^2 = 3$ Extraction of roots

9. $-2x^2 + 14x + 4 = 0$ Completing the square
10. $3x^2 - 4x + 5 = 0$ Quadratic formula

III. Find the solutions of the following equations.

11. $6x^4 + x^2 - 1 = 0$
12. $4x^{-2} - x^{-1} - 5 = 0$
13. $2\left(\dfrac{x+1}{x}\right)^2 - 7\left(\dfrac{x+1}{x}\right) - 9 = 0$
14. $9x^3 + 9x^2 - x - 1 = 0$

IV. Find the solutions of the following inequalities, and graph each solution on a number line.

15. $x^2 + x - 2 \geq 0$
16. $3x^2 + 7x + 2 < 0$
17. $\dfrac{x-4}{x+5} > 0$
18. $\dfrac{1}{x+3} \leq 0$

V. Find the solutions of the following problems. For the problems that have irrational solutions, use a decimal approximation of the solution.

19. Mike plans to build a fence next to his barn to corral his pigs. He has 400 feet of fencing, but wishes the enclosed area to be 15,000 sq ft. What are the possible dimentions for the area?

20. The length of a rectangle is twice the width. If the area of the rectangle is 256 square meters, find the dimentions.

The worked-out solutions begin on the next page.

Self-Test Solutions

1.

$$y = \frac{1}{2}(x-2)^2 + 2$$

Vertex: $(2,2)$

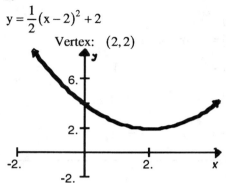

2.

$$y = x^2 + 6x + 5$$
$$y = x^2 + 6x + 9 + 5 - 9$$
$$y = (x+3)^2 - 4$$

Vertex: $(-3,-4)$

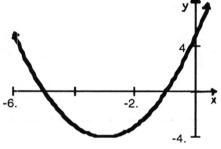

3.

$$y = -3(x+4)^2 + 1$$

Vertex: $(-4,1)$

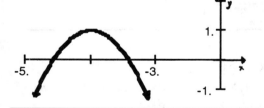

4.

$$y = 3x^2 + 4x + 3$$
$$y = 3\left(x^2 + \frac{4}{3}x + \frac{4}{9}\right) + 3 - \frac{4}{3}$$
$$y = 3\left(x + \frac{2}{3}\right)^2 + \frac{5}{3}$$

Vertex: $\left(-\frac{2}{3}, \frac{5}{3}\right)$

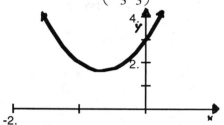

5.

$$(3x-4)^2 = 25$$
$$3x - 4 = \pm 5$$
$$3x = 4 \pm 5$$
$$x = \frac{4 \pm 5}{3}; \quad x = \frac{4+5}{3} = 3$$
$$x = \frac{4-5}{3} = -\frac{1}{3}$$

6.

$$x^2 + 6x + 8 = 63$$
$$x^2 + 6x - 55 = 0$$
$$x^2 + 6x + 9 = 55 + 9$$
$$(x+3)^2 = 63$$
$$x + 3 = \pm\sqrt{63}$$
$$x = -3 \pm 3\sqrt{3}$$

7.

$$3x^2 - 8x + 4 = 0$$
$$a = 3, b = -8, c = 4$$
$$x = \frac{8 \pm \sqrt{64 - 48}}{6}$$
$$x = \frac{8 \pm \sqrt{16}}{6}$$
$$x = \frac{8 \pm 4}{6}; \quad x = \frac{8+4}{6} = 2$$
$$x = \frac{8-4}{6} = \frac{2}{3}$$

8.

$$(x-1)^2 = 3$$
$$x - 1 = \pm\sqrt{3}$$
$$x = 1 \pm \sqrt{3}$$

9.

$$-2x^2 + 14x + 4 = 0$$
$$x^2 - 7x - 2 = 0 \qquad \text{Multiply by } -\frac{1}{2}$$
$$x^2 - 7x + \frac{49}{4} = 2 + \frac{49}{4}$$
$$\left(x - \frac{7}{2}\right)^2 = \frac{57}{4}$$
$$x - \frac{7}{2} = \frac{\pm\sqrt{57}}{2}$$
$$x = \frac{7}{2} \pm \frac{\sqrt{57}}{2} = \frac{7 \pm \sqrt{57}}{2}$$

10.

$$3x^2 - 4x + 5 = 0$$
$$a = 3, b = -4, c = 5$$
$$x = \frac{4 \pm \sqrt{16 - 60}}{6}$$
$$x = \frac{4 \pm \sqrt{-44}}{6}$$
$$x = \frac{4 \pm 2\sqrt{11}i}{6} = \frac{2 \pm \sqrt{11}i}{3}$$

11.

$$6x^4 + x^2 - 1 = 0 \qquad \text{Let } u = x^2$$
$$6u^2 + u - 1 = 0$$
$$(2u + 1)(3u - 1) = 0$$
$$2u + 1 = 0 \quad \text{or} \quad 3u - 1 = 0$$
$$u = -\frac{1}{2} \qquad\qquad u = \frac{1}{3}$$
$$x^2 = -\frac{1}{2} \qquad\qquad x^2 = \frac{1}{3}$$
$$x = \pm\frac{\sqrt{2}}{2}i \qquad\quad x = \pm\frac{\sqrt{3}}{3}$$

12.

$$4x^{-2} - x^{-1} - 5 = 0 \qquad \text{Let } u = x^{-1}$$
$$4u^2 - u - 5 = 0$$
$$(4u - 5)(u + 1) = 0$$
$$4u - 5 = 0 \quad \text{or} \quad u + 1 = 0$$
$$u = \frac{5}{4} \qquad\qquad u = -1$$
$$x^{-1} = \frac{5}{4} \qquad\qquad x^{-1} = -1$$
$$x = \frac{4}{5} \qquad\qquad x = -1$$

13.

$$2\left(\frac{x+1}{x}\right)^2 - 7\left(\frac{x+1}{x}\right) - 9 = 0 \qquad \text{Let } u = \frac{x+1}{x}$$
$$2u^2 - 7u - 9 = 0$$
$$(2u - 9)(u + 1) = 0$$
$$2u - 9 = 0 \quad \text{or} \quad u + 1 = 0$$
$$u = \frac{9}{2} \qquad\qquad u = -1$$
$$\frac{x+1}{x} = \frac{9}{2} \qquad\quad \frac{x+1}{x} = -1$$
$$2x + 2 = 9x \qquad\quad x + 1 = -x$$
$$x = \frac{2}{7} \qquad\qquad x = -\frac{1}{2}$$

14.

$$9x^3 + 9x^2 - x - 1 = 0$$
$$9x^2(x + 1) - 1(x + 1) = 0$$
$$(x + 1)(3x + 1)(3x - 1) = 0$$
$$x = 1, x = -\frac{1}{3}, x = \frac{1}{3}$$

15.

$$x^2 + x - 2 \geq 0$$
$$(x + 2)(x - 1) \geq 0 \qquad \text{Product is positive or zero}$$
$$x = -2, \quad x = 1 \qquad \text{Borders}$$

Picking numbers from each of the three regions will show that

$$x \leq -2 \quad \text{or} \quad x \geq 1$$

16.

$3x^2 + 7x + 2 < 0$

$(3x + 1)(x + 2) < 0$ The product is negative

$x = -\dfrac{1}{3}, \quad x = -2$ Boundaries

Picking numbers from each of the three regions will show that

$$-2 < x < -\dfrac{1}{3}$$

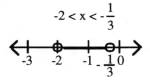

17.

$\dfrac{x - 4}{x + 5} > 0$ Quotient is positive

$x = 4, \quad x = -5$ Boundaries

Picking numbers from each of the three regions will show that

$$x < -5 \quad \text{or} \quad x > 4$$

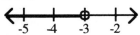

18.

$\dfrac{1}{x + 3} \le 0$ Quotient is negative or zero

$x = -3$ Boundary

$x < -3$

19.

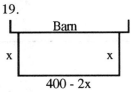

The equation is

$x(400 - 2x) = 15,000$

$400x - 2x^2 = 15,000$

$0 = 2x^2 - 400x + 15,000$

$0 = x^2 - 200x + 7,500$

$0 = (x - 50)(x - 150)$

#19. continued

There are two possibilities

     $x = 50$ $x = 150$

$400 - 2x = 400 - 2 \cdot 50$ $400 - 2x = 400 - 2 \cdot 150$

            $= 300$ $= 100$

Answer: 50 ft by 300 ft or 150 ft by 100 ft

20.

Let x = width.

Let $2x$ = length.

So, $x(2x) = 256$

$2x^2 = 256$

$x^2 = 128$

$x = \pm\sqrt{128} = \pm 8\sqrt{2}$

Answer: $8\sqrt{2} \approx 11.3$ meters

          $16\sqrt{2} \approx 22.6$ meters

CHAPTER 8 CONIC SECTIONS

Solutions To Text Exercises

Exercises 8.1

1.-13.
The graphs are in the answer section of the main text.

15.

$x = y^2 - 8y + 12$

$x = y^2 - 8y + 16 + 12 - 16$

$x = (y-4)^2 - 4$ Vertex: $(-4, -4)$

17.

$x = y^2 - 3y + 1$

$x = y^2 - 3y + \dfrac{9}{4} + 1 - \dfrac{9}{4}$

$x = \left(y - \dfrac{3}{2}\right)^2 - \dfrac{5}{4}$ Vertex: $\left(-\dfrac{5}{4}, -\dfrac{3}{2}\right)$

19.

$x = 4y^2 - 24y + 32$

$x = 4\left(y^2 - 6y + 9\right) + 32 - 36$

$x = 4(y-3)^2 - 4$ Vertex: $(-4, 3)$

21.

$x = 2y^2 - y + 1$

$x = 2\left(y^2 - \dfrac{1}{2}y + \dfrac{1}{16}\right) + 1 - \dfrac{1}{8}$

$x = 2\left(y - \dfrac{1}{4}\right)^2 + \dfrac{7}{8}$ Vertex: $\left(\dfrac{7}{8}, \dfrac{1}{4}\right)$

23.

$x = -y^2 + 10y - 24$

$x = -\left(y^2 - 10y + 25\right) - 24 + 25$

$x = -(y-5)^2 + 1$ Vertex: $(1, 5)$

25.

$3x = -2y^2 - 8y + 1$

$x = -\dfrac{2}{3}y^2 - \dfrac{8}{3}y + \dfrac{1}{3}$

$x = -\dfrac{2}{3}\left(y^2 + 4y + 4\right) + \dfrac{1}{3} + \dfrac{8}{3}$

$x = -\dfrac{2}{3}(y+2)^2 + 3$ Vertex: $(3, -2)$

Exercises 8.2

1.-21.
The graphs are in the answer section of the main text.

23.

$x^2 + y^2 + 4x - 5y - 2 = 0$

$x^2 + 4x + 4 + y^2 - 5y + \dfrac{25}{4} = 2 + 4 + \dfrac{25}{4}$

$(x+2)^2 + \left(y - \dfrac{5}{2}\right)^2 = \dfrac{49}{4}$

Center: $\left(-2, \dfrac{5}{2}\right)$ Radius: $\dfrac{7}{2}$

25.

$x^2 + y^2 + 2x - 12y + 36 = 0$

$x^2 + 2x + 1 + y^2 - 12y + 36 = -36 + 1 + 36$

$(x+1)^2 + (y-6)^2 = 1$

Center: $(-1, 6)$ Radius: 1

27.

$x^2 + y^2 + \dfrac{2}{3}y = 0$

$x^2 + y^2 + \dfrac{2}{3}y + \dfrac{1}{9} = \dfrac{1}{9}$

$x^2 + \left(y + \dfrac{1}{3}\right)^2 = \dfrac{1}{9}$

Center: $\left(0, -\dfrac{1}{3}\right)$ Radius: $\dfrac{1}{3}$

29.

$$4x^2 + 4y^2 - 24x + 8y + 31 = 0$$

$$4(x^2 - 6x + 9) + 4(y^2 + 2y + 1) = -31 + 36 + 4$$

$$4(x - 3)^2 + 4(y + 1)^2 = 9$$

$$(x - 3)^2 + (y + 1)^2 = \frac{9}{4}$$

Center: $(3, -1)$ Radius: $\frac{3}{2}$

Exercises 8.3

1.-15.
The graphs are in the answer section of the main text.

17.

$$9x^2 + 25y^2 - 18x - 150y + 9 = 0$$

$$9x^2 - 18x + 25y^2 - 150y = -9$$

$$9(x^2 - 2x) + 25(y^2 - 6y) = -9$$

$$9(x^2 - 2x + 1) + 25(y^2 - 6y + 9) = -9 + 9 + 225$$

$$9(x - 1)^2 + 25(y - 3)^2 = 225$$

$$\frac{(x - 1)^2}{25} + \frac{(y - 3)^2}{9} = 1$$

Center: $(1, 3)$ $a = 5, b = 3$

19.

$$9x^2 + 4y^2 - 24x + 8y - 16 = 0$$

$$9x^2 - 24x + 4y^2 + 8y = 16$$

$$9\left(x^2 - \frac{8}{3}x + \frac{16}{9}\right) + 4(y^2 + 2y + 1) = 16 + 16 + 4$$

$$9\left(x - \frac{4}{3}\right)^2 + 4(y + 1)^2 = 36$$

$$\frac{\left(x - \frac{4}{3}\right)^2}{4} + \frac{(y + 1)^2}{9} = 1$$

Center: $\left(\frac{4}{3}, -1\right)$ $a = 2, b = 3$

21.

$$64x^2 + 225y^2 - 448x - 300y + 484 = 0$$

$$64\left(x^2 - 7x + \frac{49}{4}\right) + 225\left(y^2 - \frac{4}{3}y + \frac{4}{9}\right)$$

$$= -484 + 784 + 100$$

#21. continued

$$64\left(x - \frac{7}{2}\right)^2 + 225\left(y - \frac{2}{3}\right)^2 = 400$$

$$\frac{\left(x - \frac{7}{2}\right)^2}{\frac{25}{4}} + \frac{\left(y - \frac{2}{3}\right)^2}{\frac{16}{9}} = 1$$

Center: $\left(\frac{7}{2}, \frac{2}{3}\right)$ $a = \frac{5}{2}, b = \frac{4}{3}$

Exercises 8.4

1.-17.
The graphs are in the answer section of the main text.

19.

$$x^2 - 36y^2 - 2x + 288y - 539 = 0$$

$$x^2 - 2x + 1 - 36(y^2 - 8y + 16) = 539 + 1 - 576$$

$$(x - 1)^2 - 36(y - 4)^2 = -36$$

$$\frac{(y - 4)^2}{1} - \frac{(x - 1)}{36} = 1$$

Center: $(1, 4)$ Vertices: $(1, 5), (1, 3)$

21.

$$9x^2 - 16y^2 - 24x + 24y + 151 = 0$$

$$9\left(x^2 - \frac{8}{3}x + \frac{16}{9}\right) - 16\left(y^2 - \frac{3}{2}y + \frac{9}{16}\right) = -151 + 16 - 9$$

$$9\left(x - \frac{4}{3}\right)^2 - 16\left(y - \frac{3}{4}\right)^2 = -144$$

$$\frac{\left(y - \frac{3}{4}\right)^2}{9} - \frac{\left(x - \frac{4}{3}\right)^2}{16} = 1$$

Center: $\left(\frac{4}{3}, \frac{3}{4}\right)$ Vertices: $\left(\frac{4}{3}, \frac{15}{4}\right), \left(\frac{4}{3}, \frac{-9}{4}\right)$

Exercises 8.5

1.-27.
The graphs are in the answer section of the main text.

Review Exercises

1.-5.
The graphs are in the answer section of the main text.

7.

$x = y^2 - 2y - 1$

$x = y^2 - 2y + 1 - 1 - 1$

$x = (y-1)^2 - 2$

Vertex: $(-2,1)$

9.

$x = \dfrac{3}{2}y^2 + 6y + 3$

$x = \dfrac{3}{2}(y^2 + 4y + 4) + 3 - 6$

$x = \dfrac{3}{2}(y+2)^2 - 3$

Vertex: $(-3,-2)$

11.-17.
The graphs are in the answer section of the main text.

19.

$$x^2 + y^2 - 6x - 2y + 8 = 0$$

$$(x^2 - 6x + 9) + (y^2 - 2y + 1) = -8 + 9 + 1$$

$$(x-3)^2 + (y-1)^2 = 2$$

Center: $(3,1)$ Radius: $\sqrt{2}$

21.-27.
The graphs are in the answer section of the main text.

29.

$$25x^2 + 4y^2 + 150x + 8y + 129 = 0$$

$$25(x^2 + 6x + 9) + 4(y^2 + 2y + 1) = -129 + 225 + 4$$

$$25(x+3)^2 + 4(y+1)^2 = 100$$

$$\frac{(x+3)^2}{4} + \frac{(y+1)^2}{25} = 1$$

Center: $(-3,-1)$ $a = 2, b = 5$

31.-37.
The graphs are in the answer section of the main text.

39.

$$x^2 - 8y^2 + 6x + 32y - 39 = 0$$

$$(x^2 + 6x + 9) - 8(y^2 - 4y + 4) = 39 + 9 - 32$$

$$(x+3)^2 - 8(y-2)^2 = 16$$

$$\frac{(x+3)^2}{16} - \frac{(y-2)^2}{2} = 1$$

Center: $(-3,2)$ Vertices: $(1,2), (-7,2)$

41.-53.
The graphs are in the answer section of the main text.

Chapter 8 Test Solutions

1.-3.
The graphs are in the answer section of the main text.

5.

$x = -y^2 - 8y - 14$

$x = -(y^2 + 8y + 16) - 14 + 16$

$x = -(y+4)^2 + 2$

Vertex: $(2,-4)$

7.

$$2x^2 - 8y^2 - 20y + \frac{39}{2} = 0$$

$$x^2 - 4y^2 - 10y + \frac{39}{4} = 0$$

$$x^2 - 4\left(y^2 + \frac{5}{2}y + \frac{25}{16}\right) = -\frac{39}{4} - \frac{25}{4}$$

$$x^2 - 4\left(y + \frac{5}{4}\right)^2 = -16$$

$$\frac{\left(y + \frac{5}{4}\right)^2}{4} - \frac{x^2}{16} = 1$$

Center: $\left(0, -\dfrac{5}{4}\right)$ Vertices: $\left(0, \dfrac{3}{4}\right), \left(0, -\dfrac{13}{4}\right)$

9.-11.
The graphs and answers are in the answer section of the main text.

Test Your Memory

1.

$$\frac{\left(6^0 x^{-2} y\right)^4}{\left(3x^4 y^{-2}\right)^{-3}} = \frac{x^{-8} y^4}{3^{-3} x^{-12} y^6} \qquad \text{Note:} \quad 6^0 = 1$$

$$= \frac{3^3 x^{-8+12}}{y^{6-4}}$$

$$= \frac{27x^4}{y^2}$$

3.

$$\frac{(a+b)(3x+y)}{(a+b)(2x-y)} \cdot \frac{(2x-y)(3x-y)}{(3x+y)(3x-y)} = 1$$

5.

$$4x\sqrt{12xy} - 6\sqrt{3x^3 y} - \sqrt{27x^3 y}$$

$$= 4x\sqrt{4 \cdot 3xy} - 6\sqrt{x^2 \cdot 3xy} - \sqrt{9x^2 \cdot 3xy}$$

$$= 8x\sqrt{3xy} - 6x\sqrt{3xy} - 3x\sqrt{3xy}$$

$$= -x\sqrt{3xy}$$

7.

$$\sqrt{\frac{5x}{9y^3} \cdot \frac{y}{y}} = \sqrt{\frac{5xy}{9y^4}} = \frac{\sqrt{5xy}}{3y^2}$$

9.

$$f(3) = \frac{5}{3-1} = \frac{5}{2}$$

11.

$$[g(2)]^2 = \left(2^3 + 1\right)^2 = (8+1)^2 = 81$$

13.

$$(2,6), \quad (-3,-9)$$

$$m = \frac{-9-6}{-3-2} = \frac{-15}{-5} = 3$$

$$y - 6 = 3(x-2)$$

$$y - 6 = 3x - 6$$

$$y = 3x$$

15.

$$(-1,3); \quad m = -\frac{5}{2}$$

$$y - 3 = -\frac{5}{2}(x+1)$$

$$y - 3 = -\frac{5}{2}x - \frac{5}{2}$$

$$y = -\frac{5}{2}x - \frac{5}{2} + 3$$

$$y = -\frac{5}{2}x + \frac{1}{2}$$

17.-21.

The answers and graphs are in the answer section of the main text.

23.

$$9x^2 + y^2 - 18x + 6y + 9 = 0$$

$$9\left(x^2 - 2x + 1\right) + \left(y^2 + 6y + 9\right) = -9 + 9 + 9$$

$$9(x-1)^2 + (y+3)^2 = 9$$

$$\frac{(x-1)^2}{1} + \frac{(y+3)^2}{9} = 1$$

25.

$$x^2 - 6x + 13 = 0$$

$$a = 1, b = -6, c = 13$$

$$x = \frac{6 \pm \sqrt{36 - 52}}{2}$$

$$x = \frac{6 \pm \sqrt{-16}}{2}$$

$$x = \frac{6 \pm 4i}{2} = 3 \pm 2i$$

$$\{3 \pm 2i\}$$

27.

$$\frac{4x+1}{x-2} = \frac{5}{x+5}$$

$$4x^2 + 21x + 5 = 5x - 10$$

$$4x^2 + 16x + 15 = 0$$

$$(2x+3)(2x+5) = 0$$

$$2x + 3 = 0 \quad \text{or} \quad 2x + 5 = 0$$

$$x = -\frac{3}{2} \qquad x = -\frac{5}{2}$$

#27. continued
$$\left\{-\frac{5}{2}, -\frac{3}{2}\right\}$$

29.
$$\frac{1}{4}(2x-1) + \frac{1}{3}(x+4) = 1$$
$$12\left(\frac{1}{4}(2x-1) + \frac{1}{3}(x+4)\right) = 12 \cdot 1$$
$$3(2x-1) + 4(x+4) = 12$$
$$6x - 3 + 4x + 16 = 12$$
$$10x + 13 = 12$$
$$10x = -1$$
$$x = -\frac{1}{10}$$
$$\left\{-\frac{1}{10}\right\}$$

31.
$$|5x - 1| = 13$$
$$5x - 1 = 13 \quad \text{or} \quad 5x - 1 = -13$$
$$5x = 14 \qquad\qquad 5x = -12$$
$$x = \frac{14}{5} \qquad\qquad x = -\frac{12}{5}$$
$$\left\{\frac{14}{5}, -\frac{12}{5}\right\}$$

33.
$$x^4 - 20x^2 + 64 = 0 \qquad \text{Let } u = x^2$$
$$u^2 - 20u + 64 = 0$$
$$(u - 4)(u - 16) = 0$$
$$u - 4 = 0 \quad \text{or} \quad u - 16 = 0$$
$$u = 4 \qquad\qquad u = 16$$
$$x^2 = 4 \qquad\qquad x^2 = 16$$
$$x = \pm 2 \qquad\qquad x = \pm 4$$
$$\{2, -2, -4, 4\}$$

35.
$$\frac{(x+2)(x-3)}{1}\left(\frac{x-4}{x+2} + \frac{3}{x-3}\right)$$
$$= \left(\frac{5x}{(x+2)(x-3)}\right)\frac{(x+2)(x-3)}{1}$$
$$\text{LCD} = (x+2)(x-3)$$

#35. continued
$$(x-3)(x-4) + 3(x+2) = 5x$$
$$x^2 - 7x + 12 + 3x + 6 = 5x$$
$$x^2 - 9x + 18 = 0$$
$$(x-3)(x-6) = 0$$
$$x - 3 = 0 \quad \text{or} \quad x - 6 = 0$$
$$x = 3 \qquad\qquad x = 6$$
$$\{6\} \quad \text{3 does not check.}$$

37.
$$\left(\sqrt{2x+1}\right)^2 = \left(3 - \sqrt{x+4}\right)^2$$
$$2x + 1 = 9 - 6\sqrt{x+4} + (x+4)$$
$$2x + 1 = 9 - 6\sqrt{x+4} + x + 4$$
$$(x-12)^2 = \left(-6\sqrt{x+4}\right)^2$$
$$x^2 - 24x + 144 = 36(x+4)$$
$$x^2 - 24x + 144 = 36x + 144$$
$$x^2 - 60x = 0$$
$$x(x - 60) = 0$$
$$x = 0 \quad \text{or} \quad x - 60 = 0$$
$$x = 60$$
$$\{0\} \quad \text{60 does not check.}$$

39.
$$2x + 1 \le -3$$
$$2x \le -4$$
$$x \le -2$$
The graph is in the answer section of the main text.

41.
$$2x^2 - 5x - 3 > 0$$
$$(2x+1)(x-3) > 0 \qquad \text{The product is positive}$$
$$x = -\frac{1}{2}, \quad x = 3 \qquad \text{Boundaries}$$

Picking numbers from each of the three regions will show that

$$x < -\frac{1}{2} \quad \text{or} \quad x > 3$$

The graph is in the answer section of the main text.

43.
The graph is in the answer section of the main text.

45.

Let x = height.

Set $\frac{1}{2}x + 9$ = base.

Thus $A = \frac{1}{2}bh$

$22 = \frac{1}{2}\left(\frac{1}{2}x + 9\right)x$ Multiply by 2

$44 = \left(\frac{1}{2}x + 9\right)x$

$44 = \frac{1}{2}x^2 + 9x$

$88 = x^2 + 18x$ Multiply by 2

$0 = x^2 + 18x - 88$

$0 = (x - 4)(x + 22)$

$x = 4$ or $x = -22$

$\frac{1}{2}x + 9 = 11$

Answer: Height: 4 in   Base: 11 in

47.

	Time · Rate =		Distance
Biff	$x + \frac{1}{2}$	48	$48\left(x + \frac{1}{2}\right)$
Buffy	x	60	$60x$

Distance is the same.

$60x = 48\left(x + \frac{1}{2}\right)$

$60x = 48x + 24$

$12x = 24$

$x = 2$

Answer: 2 hrs

49.

$h = -16t^2 + 96t + 16$

$0 = -16t^2 + 96t + 16$

$0 = t^2 - 6t - 1$

$a = 1, b = -6, c = -1$

$t = \frac{6 \pm \sqrt{36 + 4}}{2}$

$t = \frac{6 \pm \sqrt{40}}{2}$

$t = \frac{6 \pm 2\sqrt{10}}{2} = 3 \pm \sqrt{10}$

Answer: $3 + \sqrt{10}$ or approximately 6.2 sec.

Chapter 8 Study Guide

Self-Test Exercises

I.   Graph the following equations.

1. $\dfrac{x^2}{9} - \dfrac{y^2}{4} = 1$

2. $x = -\dfrac{1}{3}(y-3)^2 - 2$

3. $x^2 + y^2 = 20$

4. $x^2 + 2y^2 = 8$

5. $x = -y^2 - 8y - 7$

6. $x^2 + y^2 + 10y - 75 = 0$

7. $4x^2 + 25y^2 - 8x - 100y + 4 = 0$

8. $9x^2 - 4y^2 - 36x + 24y - 36 = 0$

II.   Find the equation of the following circle.

9. Center = C(-4,1) and passing through (2,-5)

10. The endpoints of a diameter are (7,3) and (-1,-3)

III.   Graph the following inequalities.

11. $y < x^2 - x$

12. $y > x^2 - 6x + 8$

IV.   Graph the following system of inequalities.

13. $\begin{cases} \dfrac{y^2}{25} - \dfrac{x^2}{4} < 1 \\[2mm] \dfrac{x^2}{9} + \dfrac{y^2}{4} < 1 \end{cases}$

14. $\begin{cases} \dfrac{x^2}{25} + \dfrac{y^2}{1} \geq 1 \\[2mm] \dfrac{x^2}{4} - \dfrac{y^2}{4} > 1 \end{cases}$

The worked-out solutions begin on the next page.

Self-Test Solutions

1.

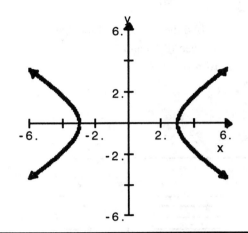

2.

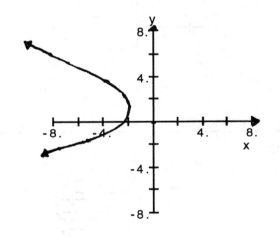

3.

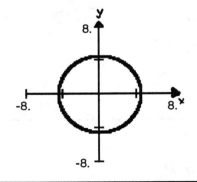

4.

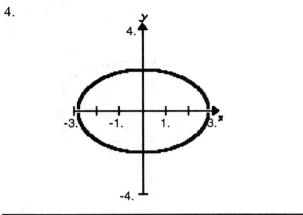

5.

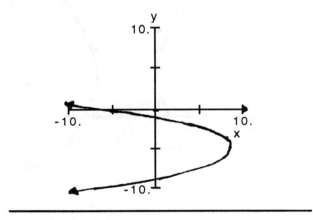

6.

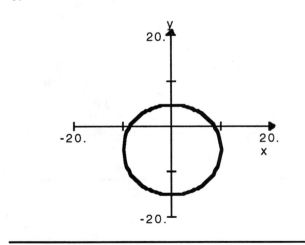

7.

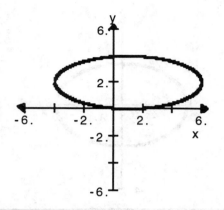

8.

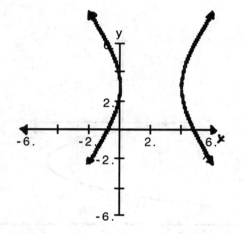

9.

$$(x-h)^2 + (y-k) = r^2$$
$$C(-4,1) \text{ and } (x,y) = (2,-5)$$
$$(2+4)^2 + (-5-1)^2 = r^2$$
$$6^2 + (-6)^2 = r^2$$
$$36 + 36 = r^2$$
$$72 = r^2$$
Thus $(x+4)^2 + (y-1)^2 = 72$

10.
Step 1: Find the center using the midpoint formula.
$$M\left(\frac{7-1}{2}, \frac{3-3}{2}\right) = M(3,0)$$
Step 2: To find the radius, find the distance between $(3,0)$ and $(7,3)$.
$$r^2 = (3-7)^2 + (0-3)^2$$
$$r^2 = 16 + 9$$
$$r^2 = 25$$
Thus, $(x-3)^2 + y^2 = 25$

11.

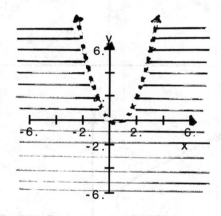

12.

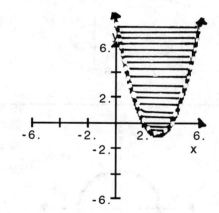

13.

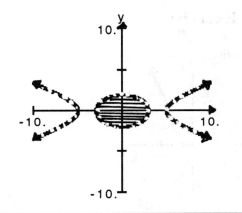

14.

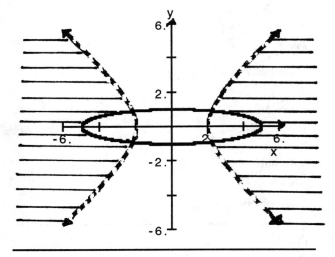

CHAPTER 9   SYSTEMS OF EQUATIONS

Solutions To Text Exercises

Exercises 9.1

1.
$x + 3y = 11$
$2x - y = -6$

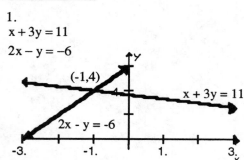

$(-1, 4)$;   Independent

3.
$-2x + 3y = 4$
$x - 4y = -2$

$(-2, 0)$;   Independent

5.
$x + 4y = 6$
$x + 4y = 0$

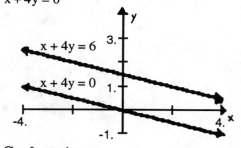

$\varnothing$;   Inconsistent

7.
$3x - y = 1$
$-6x + 2y = -2$

#7. continued

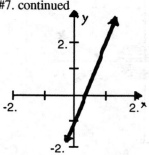

$\{(x, y) \mid 3x - y = 1\}$;   Dependent

9.
(1) $2x - 3y = 7$
(2) $y = x - 2$
Substitute x - 2 for y in (1).
$2x - 3(x - 2) = 7$
$2x - 3x + 6 = 7$
$-x = 1$
$x = -1$
Substitute -1 for x in (2).
$y = -1 - 2 = -3$
$(-1, -3)$;   Independent

11.
(1) $-2x + 5y = 11$
(2) $x = y - 2$
Substitute y - 2 for x in (1).
$-2(y - 2) + 5y = 11$
$-2y + 4 + 5y = 11$
$3y = 7$
$y = \dfrac{7}{3}$
Substitute $\dfrac{7}{3}$ for y in (2).
$x = \dfrac{7}{3} - 2 = \dfrac{1}{3}$
$\left(\dfrac{1}{3}, \dfrac{7}{3}\right)$;   Independent

13.
(1) $x - 2y = 5$
(2) $x = 2y + 11$
Substitute $2y + 11$ for x in (1).
$2y + 11 - 2y = 5$
$11 = 5$
$\varnothing$; Inconsistent

15.
(1) $x - y = 11$
(2) $2x + 3y = 5$
Solve (1) for x.
(3) $x = y + 11$
Substitute $y + 11$ for x in (2).
$2(y + 11) + 3y = 5$
$2y + 22 + 3y = 5$
$5y = -17$
$y = -\dfrac{17}{5}$
Substitute $-\dfrac{17}{5}$ for y in (3).
$y = -\dfrac{17}{5} + 11 = \dfrac{38}{5}$
$\left(\dfrac{38}{5}, -\dfrac{17}{5}\right)$; Independent

17.
(1) $3x - 5y = 8$
(2) $2x - y = -1$
Solve (2) for y.
(3) $y = 2x + 1$
Substitute $2x + 1$ for y in (1).
$3x - 5(2x + 1) = 8$
$3x - 10x - 5 = 8$
$-7x = 13$
$x = -\dfrac{13}{7}$
Substitute $-\dfrac{13}{7}$ for x in (3).
$y = 2\left(-\dfrac{13}{7}\right) + 1 = \dfrac{-19}{7}$
$\left(-\dfrac{13}{7}, -\dfrac{19}{7}\right)$; Independent

19.
(1) $2x - 3y = 4$
(2) $-3x + 4y = -1$
Solve (1) for x.
$2x = 3y + 4$
$x = \dfrac{3}{2}y + 2$
Substitute $\dfrac{3}{2}y + 2$ for x in (2).
$-3\left(\dfrac{3}{2}y + 2\right) + 4y = -1$
$-\dfrac{9}{2}y - 6 + 4y = -1$
$-9y - 12 + 8y = -2 \qquad$ Multiply by 2
$-y = 10$
$y = -10$
Substitute -10 for y in (2).
$-3x + 4(-10) = -1$
$-3x - 40 = -1$
$-3x = 39$
$x = -13$
$(-13, -10)$; Independent

21.
(1) $2x - 6y = 14$
(2) $5x - 15y = 35$
Solve (1) for x.
$2x = 6y + 14$
$x = 3y + 7$
Substitute $3y + 7$ for x in (2).
$5(3y + 7) - 15y = 35$
$15y + 35 - 15y = 35$
$35 = 35$
Dependent

23.
(1) $-3x + y = 7$
(2) $\underline{3x + 2y = 1}$
$3y = 8$
$y = \dfrac{8}{3}$
Substitute $\dfrac{8}{3}$ for y in (2).

241

#23. continued

$$3x + 2\left(\frac{8}{3}\right) = 1$$

$$3x + \frac{16}{3} = 1$$

$$9x + 16 = 3 \qquad \text{Multiply by 3}$$

$$9x = -13$$

$$x = -\frac{13}{9}$$

$\left(-\frac{13}{9}, \frac{8}{3}\right)$; Independent

25.

(1) $x - 5y = -2$

(2) $3x - 5y = 3$

Multiply (1) by -1.

(3) $-x + 5y = 2$

(2) $\underline{3x - 5y = 3}$

$$2x \qquad = 5$$

$$x = \frac{5}{2}$$

Substitute $\frac{5}{2}$ for x in (1).

$$\frac{5}{2} - 5y = -2$$

$$5 - 10y = -4 \qquad \text{Multiply by 2}$$

$$-10y = -9$$

$$y = \frac{9}{10}$$

$\left(\frac{5}{2}, \frac{9}{10}\right)$; Independent

27.

(1) $2x - 3y = -1$

(2) $x - 2y = 1$

Multiply (2) by -2.

(1) $2x - 3y = -1$

(3) $\underline{-2x + 4y = -2}$

$$y = -3$$

Substitute -3 for y in (2).

$$x - 2(-3) = 1$$

$$x + 6 = 1$$

$$x = -5$$

$(-5, -3)$; Independent

29.

(1) $2x - y = 7$

(2) $6x - 3y = 11$

Multiply (1) by -3.

(3) $-6x + 3y = -21$

$$\underline{6x - 3y = 11}$$

$$0 = -10$$

$\varnothing$ Inconsistent

31.

(1) $2x + 3y = 7$

(2) $-4x + 2y = 1$

Multiply (1) by 2.

(3) $4x + 6y = 14$

(2) $\underline{-4x + 2y = 1}$

$$8y = 15$$

$$y = \frac{15}{8}$$

Substitute $\frac{15}{8}$ for y in (2).

$$-4x + 2\left(\frac{15}{8}\right) = 1$$

$$-4x + \frac{15}{4} = 1$$

$$-16x + 15 = 4 \qquad \text{Multiply by 4}$$

$$-16x = -11$$

$$x = \frac{11}{6}$$

$\left(\frac{11}{6}, \frac{15}{8}\right)$; Independent

33.

(1) $-5x - 2y = 6$

(2) $2x - 5y = -2$

To eliminate the x's, multiply (1) by 2 and (2) by 5.

(3) $-10x - 4y = 12$

(4) $\underline{10x - 25y = -10}$

$$-29y = 2$$

$$y = -\frac{2}{29}$$

Substitute $-\frac{2}{29}$ for y in (1).

#33. continued

$$-5x - 2\left(-\frac{2}{29}\right) = 6$$

$$-5x + \frac{4}{29} = 6$$

$$-5x = 6 - \frac{4}{29}$$

$$-5x = \frac{170}{29}$$

$$x = -\frac{34}{29}$$

$\left(-\frac{34}{29}, -\frac{2}{29}\right)$; Independent

35.

(1) $2x - 4y = 10$

(2) $-3x + 6y = -15$

To eliminate the x' s, multiply (1) by 3 and (2) by 2.

(3) $6x - 12y = 30$

(4) $\underline{-6x + 12y = -30}$

$0 = 0$

Dependent

37.

(1) $\frac{1}{8}x - \frac{3}{4}y = -\frac{13}{4}$

(2) $\frac{2}{3}x + \frac{1}{3}y = 0$

To eliminate the fractions, multiply (1) by 8 and (2) by 3.

(3) $x - 6y = -26$

(4) $2x + y = 0$

To eliminate the y' s, multiply (4) by 6.

(3) $x - 6y = -26$

(5) $\underline{12x + 6y = 0}$

$13x = -26$

$x = -2$

Substitute - 2 for x in (2).

$$\frac{2}{3}(-2) + \frac{1}{3}y = 0$$

$$\frac{-4}{3} + \frac{1}{3}y = 0$$

$$-4 + y = 0 \qquad \text{Multiply by 3}$$

$$y = 4$$

$(-2, 4)$; Independent

39.

(1) $\frac{1}{3}x - \frac{1}{2}y = \frac{2}{3}$

#39. continued

(2) $\frac{1}{4}x - \frac{3}{8}y = \frac{1}{2}$

To eliminate the fractions, multiply (1) by 6 and (2) by 8.

(3) $2x - 3y = 4$

(4) $\underline{2x - 3y = 4}$

$0 = 0$

Dependent

41.

(1) $\frac{1}{2}x - \frac{1}{8}y = 3$

(2) $\frac{2}{3}x + \frac{1}{6}y = -\frac{1}{2}$

To eliminate the fractions, multiply (1) by 8 and (2) by 6.

(3) $4x - y = 24$

(4) $\underline{4x + y = -3}$

$8x = 21$

$x = \frac{21}{8}$

Substitute $\frac{21}{8}$ for x in (2).

$$\frac{2}{3} \cdot \frac{21}{8} + \frac{1}{6}y = -\frac{1}{2}$$

$$\frac{7}{4} + \frac{1}{6}y = -\frac{1}{2}$$

$$21 + 2y = -6 \qquad \text{Multiply by 12}$$

$$2y = -27$$

$$y = -\frac{27}{2}$$

$\left(\frac{21}{8}, -\frac{27}{2}\right)$; Independent

43.

(1) $\frac{1}{x} + \frac{2}{y} = 5$ Let $u = \frac{1}{x}$, $v = \frac{1}{y}$

(2) $\frac{1}{x} - \frac{2}{y} = 1$

(3) $u + 2v = 5$

(4) $\underline{u - 2v = 1}$

$2u = 6$

$u = 3;$ $\frac{1}{x} = 3;$ $x = \frac{1}{3}$

Substitute 3 for u in (3).

#43. continued

$3 + 2v = 5$

$2v = 2$

$v = 1; \quad \dfrac{1}{y} = 1; \quad y = 1$

$\left(\dfrac{1}{3}, 1\right);$ Independent

45.

(1) $-\dfrac{3}{x} + \dfrac{2}{y} = \dfrac{1}{2}$ \qquad Let $u = \dfrac{1}{x}, \, v = \dfrac{1}{y}$

(2) $\dfrac{2}{x} - \dfrac{1}{y} = -\dfrac{5}{12}$

(3) $-3u + 2v = \dfrac{1}{2}$

(4) $2u - v = -\dfrac{5}{12}$

Mulitply (3) by 2 and (4) by 12.

(5) $-6u + 4v = 1$

(6) $24u - 12v = -5$

To eliminate the u' s, multiply (5) by 4.

(7) $-24u + 16v = 4$

(6) $\dfrac{24u - 12v = -5}{4v = -1}$

$v = -\dfrac{1}{4}; \quad \dfrac{1}{y} = -\dfrac{1}{4}; \quad y = -4$

Substitute $-\dfrac{1}{4}$ for v in (3).

$-3u + 2\left(-\dfrac{1}{4}\right) = \dfrac{1}{2}$

$-3u - \dfrac{1}{2} = \dfrac{1}{2}$

$-3u = 1$

$u = -\dfrac{1}{3}; \quad \dfrac{1}{x} = -\dfrac{1}{3}; \quad x = -3$

$(-3, -4);$ Independent

47.

(1) $2u - 3v = -2$ \qquad Let $u = \dfrac{1}{x}, \, v = \dfrac{1}{y}$

(2) $-3u + 4v = -5$

Step 1:

$6u - 9v = -6$ \quad Multiply (1) by 3

$\dfrac{-6u + 8v = -10}{-v = -16}$ \quad Multiply (2) by 2

$v = 16$

$y = \dfrac{1}{16}$

#47. continued

Step 2:

(1) $2u - 3(16) = -2$

$u = 23$

$x = \dfrac{1}{23}$

$\left(\dfrac{1}{23}, \dfrac{1}{16}\right)$ Independent

Exercises 9.2

1.

(1) $2x + y - 3z = -4$

(2) $x - y + 4z = 13$

(3) $-x + 3y + z = 4$

(2) $x - y + 4z = 13$

(3) $\dfrac{-x + 3y + z = 4}{}$

(4) $\quad 2y + 5z = 17$

Multiply (3) by 2 and combine with (1).

(1) $\quad 2x + y - 3z = -4$

(3a) $\dfrac{-2x + 6y + 2z = 8}{}$

(5) $\quad 7y - z = 4$

Multiply (5) by 5 and combine with (4).

(4) $2y + 5z = 17$

$\dfrac{35y - 5z = 20}{37y = 37}$

$y = 1$

Substitute 1 for y in (5).

$7 \cdot 1 - z = 4$

$z = 3$

Substitute 1 for y and 3 for z in (2).

$x - 1 + 12 = 13$

$x = 2$

$(2, 1, 3);$ Independent

3.

(1) $5x + y + z = 6$

(2) $-2x + 3y - 4z = -17$

(3) $4x - 2y - z = -7$

#3. continued
(1) $\quad 5x + y + z = 6$
(3) $\quad \underline{4x - 2y - z = -7}$
(4) $\quad 9x - y \quad = -1$
Multiply (1) by 4 and combine with (2).
(1a) $20x + 4y + 4z = 24$
(2) $\quad \underline{-2x + 3y - 4z = -17}$
(5) $\quad 18x + 7y \quad = 7$
Multiply (4) by 7 and combine with (5).
(4a) $63x - 7y = -7$
(5) $\quad \underline{18x + 7y = 7}$
$\quad\quad 81x \quad = 0$
$\quad\quad\quad x = 0$
Substitute 0 for x in (4).
$9 \cdot 0 - y = -1$
$\quad\quad -y = 1$
Substitute 0 for x and 1 for y in (1).
$5 \cdot 0 + 1 + z = 6$
$\quad\quad\quad z = 5$
$(0, 1, 5); \quad$ Independent

5.
(1) $\quad 2x + 3y - z = -12$
(2) $\quad 4x - 5y + z = 10$
(3) $\quad -6x + 7y + z = -4$

(1) $\quad 2x + 3y - z = -12$
(2) $\quad \underline{4x - 5y + z = 10}$
(4) $\quad 6x - 2y \quad = -2$

(4a) $3x - y \quad = -1 \quad$ Multiply by $\dfrac{1}{2}$

Multiply (3) by -1 and combine with (2).
(2) $\quad 4x - 5y + z = 10$
(3a) $\quad \underline{6x - 7y - z = 4}$
(5) $\quad 10x - 12y = 14$

(5a) $5x - 6y = 7 \quad$ Multiply by $\dfrac{1}{2}$

Multiply (4a) by -6 and combine with (5a).
(4b) $-18x + 6y = 6$
(5a) $\quad \underline{5x - 6y = 7}$
$\quad\quad -13x \quad = 13$
$\quad\quad\quad\quad x = -1$
Substitute -1 for x in (4a).
$3(-1) - y = -1$
$\quad\quad\quad -y = -2$

#5. continued
Substitute -1 for x and -2 for y in (2).
$4(-1) - 5(-2) + z = 10$
$\quad\quad -4 + 10 + z = 10$
$\quad\quad\quad\quad\quad z = 4$
$(-1, -2, 4); \quad$ Independent

7.
(1) $\quad -2x + 3y + 2z = 16$
(2) $\quad 3x - 3y + z = -14$
(3) $\quad 3x - 2y - 5z = -16$
Multiply (2) by -2 and combine with (1).
(1) $\quad -2x + 3y + 2z = 16$
(2a) $\quad \underline{-6x + 6y - 2z = 28}$
(5) $\quad -8x + 9y \quad = 44$
Multiply (2) by 5 and combine with (3).
(2b) $15x - 15y + 5z = -70$
(3) $\quad \underline{3x - 2y - 5z = -16}$
(6) $\quad 18x - 17y \quad = -86$
Multiply (5) by 9 and (6) by 4.
(5a) $-72x + 81y = 396$
(6a) $\quad \underline{72x - 68y = -344}$
$\quad\quad\quad 13y = 52$
$\quad\quad\quad\quad y = 4$
Substitite 4 for y in (5).
$-8x + 9 \cdot 4 = 44$
$\quad\quad -8x = 8$
$\quad\quad\quad x = -1$
Substitute 4 for y and -1 for x in (2).
$3(-1) - 3 \cdot 4 + z = -14$
$\quad\quad -3 - 12 + z = -14$
$\quad\quad\quad\quad\quad z = 1$
$(-1, 4, 1); \quad$ Independent

9.
(1) $\quad 2x - 3y - 2z = -12$
(2) $\quad 3x + 4y + 4z = 20$
(3) $\quad 2x + 5y - 2z = 4$

#9. continued
Multiply (3) by -1 and combine with (1).

(1) $\quad 2x - 3y - 2z = -12$

(3a) $\underline{-2x - 5y + 2z = -4}$

(4) $\qquad -8y \qquad = -16$

$\qquad y = 2$

Substitute 2 for y in (1) and (2).

(1a) $\quad 2x - 3 \cdot 2 - 2z = -12$

(1a) $\quad 2x \qquad - 2z = -6$

(2a) $\quad 3x + 4 \cdot 2 + 4z = 20$

(2a) $\quad 3x \qquad + 4z = 12$

Multiply (1a) by 2 and combine with (2a).

(1b) $\quad 4x - 4z = -12$

(2a) $\quad \underline{3x + 4z = \quad 12}$

$\qquad 7x \qquad = 0$

$\qquad x = 0$

Substitute 0 for x in (2a).

$3 \cdot 0 + 4z = 12$

$\qquad z = 3$

$(0, 2, 3); \quad$ Independent

11.

(1) $\quad 5x + y - z = -2$

(2) $\quad x + 3y + 4z = 8$

(3) $\quad 3x - 3y - z = -7$

(2) $\quad x + 3y + 4z = 8$

(3) $\quad \underline{3x - 3y - \quad z = -7}$

(4) $\quad 4x \qquad + 3z = 1$

Multiply (1) by -3 and combine with (2).

(1a) $\quad -15x - 3y + 3z = 6$

(2) $\quad \underline{\quad x + 3y + 4z = 8}$

$\qquad -14x \qquad + 7z = 14$

(5) $\quad - 2x \qquad + z = 2 \quad$ Multiply by $\frac{1}{7}$

Multiply (5) by 2 and combine with (4).

(4) $\quad 4x + 3z = 1$

(5a) $\quad \underline{-4x + 2z = 4}$

$\qquad 5z = 5$

$\qquad z = 1$

#11. continued
Substitute 1 for z in (5).

$-2x + 1 = 2$

$\quad - 2x = 1$

$\qquad x = -\frac{1}{2}$

Substitute $-\frac{1}{2}$ for x and 1 for z in (1).

$5\left(-\frac{1}{2}\right) + y - 1 = -2$

$\qquad -\frac{5}{2} + y - 1 = -2$

$\qquad y - \frac{7}{2} = -2$

$\qquad y = -2 + \frac{7}{2}$

$\qquad y = \frac{3}{2}$

$\left(-\frac{1}{2}, \frac{3}{2}, 1\right); \quad$ Independent

---

13.

(1) $\quad x - 2y + 2z = 0$

(2) $\quad - 2x + y + 4z = 2$

(3) $\quad 7x - 2y - 2z = 0$

Multiply (2) by 2 and combine with (1).

(1) $\qquad x - 2y + 2z \ = 0$

(2a) $\quad \underline{-4x + 2y + 8z \ = 4}$

(4) $\quad -3x \qquad + 10z = 4$

Combine (3) with (2a).

(3) $\qquad 7x - 2y - 2z = 0$

(2a) $\quad \underline{-4x + 2y + 8z = 4}$

(5) $\qquad 3x \qquad + 6z = 4$

To eliminate the z's, multiply (4) by 3
and (5) by -5.

(4a) $\quad -9x + 30z = 12$

(5a) $\quad \underline{-15x - 30z = -20}$

$\qquad -24x \qquad = -8$

$\qquad x = \frac{1}{3}$

246

#13. continued

Substitute $\frac{1}{3}$ for x in (4).

$$-3 \cdot \frac{1}{3} + 10z = 4$$
$$-1 + 10z = 4$$
$$10z = 5$$
$$z = \frac{1}{2}$$

Substitute $\frac{1}{3}$ for x and $\frac{1}{2}$ for z in (2).

$$-2 \cdot \frac{1}{3} + y + 4 \cdot \frac{1}{2} = 2$$
$$-\frac{2}{3} + y + 2 = 2$$
$$y = \frac{2}{3}$$

$\left(\frac{1}{3}, \frac{2}{3}, \frac{1}{2}\right)$;   Independent

---

15.

(1) $2x - 4y + 3z = 11$

(2) $x + 3y - 5z = -22$

(3) $2y - z = -3$

Multiply (2) by -2 and combine with (1).

(1) $2x - 4y + 3z = 11$

(2a) $\underline{-2x - 6y + 10z = 44}$

(4) $-10y + 13z = 55$

Multiply (3) by 5 and combine with (4).

(3a) $10y - 5z = -15$

(4) $\underline{-10y + 13z = 55}$

$$8z = 40$$
$$z = 5$$

Substitute 5 for z in (3).

$$2y - 5 = -3$$
$$2y = 2$$
$$y = 1$$

Substitute 1 for y and 5 for z in (2).

$$x + 3 \cdot 1 - 5 \cdot 5 = -22$$
$$x + 3 - 25 = -22$$
$$x = 0$$

$(0, 1, 5)$;   Independent

17.

(1) $4x - 5y - 4z = 0$

(2) $2x + 3y \quad\quad = -1$

(3) $\quad\quad -y + z = 2$

Multiply (3) by 4 and combine with (1).

(1) $4x - 5y - 4z = 0$

(3a) $\underline{\quad -4y + 4z = 8}$

(4) $4x - 9y \quad\quad = 8$

Multiply (2) by -2 and combine with (4).

(2a) $-4x - 6y = 2$

(4) $\underline{\quad 4x - 9y = 8}$

$$-15y = 10$$
$$y = -\frac{2}{3}$$

Substitute $-\frac{2}{3}$ for y in (3).

$$-\left(-\frac{2}{3}\right) + z = 2$$
$$z = \frac{4}{3}$$

Substitute $-\frac{2}{3}$ for y in (2).

$$2x + 3\left(-\frac{2}{3}\right) = -1$$
$$2x - 2 = -1$$
$$x = \frac{1}{2}$$

$\left(\frac{1}{2}, -\frac{2}{3}, \frac{4}{3}\right)$;   Independent

---

19.

(1) $2x - y = 0$

(2) $3x + 4z = -15$

(3) $3y - 2z = 0$

Solve (1) for x.

(1a) $x = \frac{1}{2}y$

Substitute $\frac{1}{2}y$ for x in (2).

(2a) $3\left(\frac{1}{2}y\right) + 4z = -15$

(2a) $\frac{3}{2}y + 4z = -15$

(2a) $3y + 8z = -30$ Multiply by 2

#19. continued

Multiply (3) by -1 and combine with (2a).

(3a) $-3y + 2z = 0$

(2a) $\underline{3y + 8z = -30}$

$\qquad 10z = -30$

$\qquad z = -3$

Substitute -3 for z in (2).

$3x + 4(-3) = -15$

$\quad 3x - 12 = -15$

$\qquad x = -1$

Substitute -1 for x in (1).

$2(-1) - y = 0$

$\qquad y = -2$

$(-1, -2, -3);$ Independent

21.

(1) $3x - 4y + z = -11$

(2) $\qquad 3y - 4z = -7$

(3) $\qquad z = 4$

Substitute 4 for z in (2).

$3y - 4 \cdot 4 = -7$

$\qquad y = 3$

Substitute 4 for z and 3 for y in (1).

$3x - 4 \cdot 3 + 4 = -11$

$\quad 3x - 12 + 4 = -11$

$\qquad x = -1$

$(-1, 3, 4);$ Independent

23.

(1) $2x - 4y + 6z = -2$

(2) $x - 2y + 3z = -1$

(3) $-3x + 6y - 9z = 3$

Multiply (1) by $\frac{1}{2}$ and (3) by $-\frac{1}{3}$.

(1a) $x - 2y + 3z = -1$

(3a) $x - 2y + 3z = -1$

The system is dependent.

25.

(1) $x + 5y + 3z = 11$

(2) $-x + 6y + 8z = 22$

(3) $2x + 7y + 3z = 13$

(1) $x + 5y + 3z = 11$

(2) $\underline{-x + 6y + 8z = 22}$

(4) $\qquad 11y + 11z = 33$

(4) $\qquad y + z = 3$ Multiply by $\frac{1}{11}$

Multiply (2) by 2 and combine with (3).

(2a) $-2x + 12y + 16z = 44$

(3) $\underline{2x + 7y + 3z = 13}$

(5) $\qquad 19y + 19z = 57$

(5) $\qquad y + z = 3$

Note: (4) and (5).   The system is dependent.

27.

(1) $-6x + 3y - 9z = -1$

(2) $2x - y + 3z = 7$

(3) $4x - 2y + 6z = 0$

Multiply (2) by 3 and combine with (1).

(1) $-6x + 3y - 9z = -1$

$\quad \underline{6x - 3y + 9z = 21}$

$\qquad 0 = 20$

The system is inconsistent.

29.

(1) $x + 2y - 2z = 7$

(2) $9x - 6y + 3z = 7$

(3) $3x - 2y + z = 1$

(1) $x + 2y - 2z = 7$

(3) $\underline{3x - 2y + z = 1}$

(4) $4x - z = 8$

Multiply (1) by 3 and combine with (2).

(1a) $3x + 6y - 6z = 21$

(2) $\underline{9x - 6y + 3z = 7}$

(5) $12x - 3z = 28$

(5) $4x - z = \dfrac{28}{3}$

Note: (4) and (5).   The system is inconsistent.

31.

(1) $\quad x - 2y + z + \ w = 4$

(2) $\quad 2x + \ y - z + 2w = -1$

(3) $\quad -x + 3y + 2z - \ w = 5$

(4) $\quad 3x - \ y + 4z - 3w = 3$

(1) $\quad x - 2y + \ z + w = 4$

(3) $\quad \underline{-x + 3y + 2z - w = 5}$

(5) $\qquad y + 3z \quad = 9$

Multiply (3) by 2 and combine with (2).

(2) $\quad 2x + \ y - \ z + 2w = -1$

(3a) $\quad \underline{-2x + 6y + 4z - 2w = 10}$

(6) $\qquad 7y + 3z \qquad = 9$

Multiply (5) by -7 and combine with (6).

(5a) $\quad -7y - 21z = -63$

$\qquad \underline{7y + \ 3z = 9}$

$\qquad\quad -18z = -54$

$\qquad\qquad z = 3$

Substitute 3 for z in (5).

$y + 3 \cdot 3 = 9$

$\qquad y = 0$

Substitute 0 for y and 3 for z in (1) and (4) and combine.

(1a) $\quad x + \ w = 1$

(4) $\quad 3x - 3w = -9$

Multiply (1a) by 3.

(1b) $\quad 3x + 3w = 3$

(4) $\quad \underline{3x - 3w = -9}$

$\qquad 6x = -6$

$\qquad\ x = -1$

Substitute -1 for x, 0 for y, and 3 for z in (1).

$-1 - 2 \cdot 0 + 3 + w = 4$

$\qquad 2 + w = 4$

$\qquad\quad w = 2$

$(-1, 0, 3, 2);$ Independent

Exercises 9.3

1.
Let x and y = the numbers.

Then, $\quad x + y = 25$

$\qquad\qquad x = 5y + 1$

(1) $\quad x + y = 25$

(2) $\quad x - 5y = 1$

Step 1: (1a) $\quad -x - \ y = -25$

(2) $\qquad \underline{x - 5y = 1}$

$\qquad\qquad -6y = -24$

$\qquad\qquad\quad y = 4$

Step 2: (1) $\quad x + y = 25$

$\qquad\qquad x + 4 = 25$

$\qquad\qquad\quad x = 21$

Answer: 4 and 21

3.
Let x and y = two numbers.

Then, (1) $\quad x + y = 6$

(2) $\quad x - y = 10$

Step 1: (1) $\quad x + y = 6$

(2) $\quad \underline{x - y = 10}$

$\qquad\quad 2x = 16$

$\qquad\quad\ x = 8$

Step 2: (1) $\quad 8 + y = 6$

$\qquad\qquad y = -2$

Answer: 8 and -2

5.
Let x and y = two numbers.

Then, (1) $\quad x - y = 10$

(2) $\quad 2x - y = 14$

Step 1: (1a) $\quad -x + y = -10$

(2) $\quad \underline{2x - y = 14}$

$\qquad\qquad x = 4$

Step 2: (1) $\quad 4 - y = 10$

$\qquad\qquad y = -6$

Answer: 4 and -6

7.
Let x = number of nickels.
Let y = number of dimes.
Let 5x = value of the nickels.
Let 10y = value of the dimes.

#7. continued

Then, (1) $x + y = 40$

(2) $5x + 10y = 280$

Step 1: (1a) $-5x - 5y = -200$

(2) $\underline{5x + 10y = 280}$

$5y = 80$

$y = 16$

Step 2: (1) $x + 16 = 40$

$x = 24$

Answer: 24 nickels, 16 dimes

9.

Let x = number of 20¢ stamps.
Let y = number of 32¢ stamps.
Let 20x = value of 20¢ stamps.
Let 32y = value of 32¢ stamps.

Then, (1) $x + y = 21$

(2) $20x + 32y = 528$

Step 1: (1a) $-20x - 20y = -420$

(2) $\underline{20x + 32y = 528}$

$12y = 108$

$y = 9$

Step 2: (1) $x + 9 = 21$

$x = 12$

Answer: 12–20¢ stamps, 9–32¢ stamps

11.

Let x = number of adult tickets sold.
Let y = number of child tickets sold.
Let 350x = value of adult tickets.
Let 125y = value of child tickets.

Then, (1) $x + y = 37$

(2) $350x + 125y = 7325$

Step 1: (1a) $-350x - 350y = -12950$

(2) $\underline{350x + 125y = 7325}$

$-225y = -5625$

$y = 25$

Step 2: (1) $x + 25 = 37$

$x = 12$

Answer: 12 adult tickets, 25 child tickets

13.

Let x = number of nickels.
Let y = number of dimes.
Let z = number of quarters.

#13. continued

Then, 5x = value of nickels

10y = values of dimes

25z = values of quarters

Thus, (1) $x + y + z = 34$

(2) $5x + 10y + 25z = 450$

(2a) $x + 2y + 5z = 90$ Multiply by $\frac{1}{5}$

(3) $x = z + 2$ Subtract z

(3a) $x - z = 2$

Step 1: (1) $x + y + z = 34$

(3a) $\underline{x \quad\;\; - z = \quad 2}$

(4) $2x + y \quad\;\; = 36$

Step 2: (2a) $x + 2y + 5z = 90$

(3b) $\underline{5x \quad\quad - 5z = 10}$ Multiply by 5

(5) $6x + 2y \quad\;\; = 100$

Step 3: (4a) $-4x - 2y = -72$ Multiply by -2

(5) $\underline{6x + 2y = 100}$

$2x \quad\;\; = 28$

$x = 14$

Step 4: (4) $2 \cdot 14 + y = 36$

$y = 8$

Step 5: (1) $14 + 8 + z = 34$

$z = 12$

Answer: 14 nickels, 8 dimes, 12 quarters.

15.

Let x = cost of a pencil.
Let y = cost of an eraser.

Thus, (1) $4x + y = 0.62$

(2) $3x + 7y = 0.84$

Step 1: (1a) $-28x - 7y = -4.34$ Multiply by -7

(2) $\underline{3x + 7y = 0.84}$

$-25x \quad\;\; = -3.5$

$x = 0.14$

Step 2: $4(0.14) + y = 0.62$

$0.56 + y = 0.62$

$y = 0.06$

Answer: Pencil = 14¢, eraser = 6¢.

17.

Let x = cost of a basketball.
Let y = cost of a football.

#17. continued

Thus　(1)　$3x + 2y = 60$

　　　(2)　$2x + 5y = 73$

Step 1:　(1a)　$6x + 4y = 120$　Multiply by 2

　　　(2a)　$\underline{-6x - 15y = -219}$　Multiply by - 3

　　　　　　　$-11y = -99$

　　　　　　　　　$y = 9$

Step 2:　(2)　$2x + 5 \cdot 9 = 73$

　　　　　$2x + 45 = 73$

　　　　　　$2x = 28$

　　　　　　　$x = 14$

Answer:　Basketball = $14, football = $9.

19.

Let x = cost of a saw.

Let y = cost of a hammer.

Let z = cost of a screwdiver.

Thus　(1)　$x + y + z = 36$　(Larry)

　　　(2)　$2x + 2y + 3z = 76$　(Moe)

　　　(3)　$x + 2y + 11z = 86$　(Curley)

Step 1:　(1)　$x + y + z = 36$

　　　(3a)　$\underline{-x - 2y - 11z = -86}$　Multiply by - 1

　　　(4)　　$-y - 10z = -50$

Step 2:　(1a)　$-2x - 2y - 2z = -72$　Multiply by - 2

　　　(2)　　$\underline{2x + 2y + 3z = 76}$

　　　　　　　　　$z = 4$

Step 3:　(4)　$-y - 10 \cdot 4 = -50$

　　　　　$-y - 40 = -50$

　　　　　$-y = -10$

　　　　　　$y = 10$

Step 4:　(1)　$x + 10 + 4 = 36$

　　　　　　$x = 22$

Answer:　Saw = $22, hammer = $10,

　　　　screwdiver = $4.

21.

Let x = pounds of Mr. Peanut.

Let y = pounds of King Peanut.

Thus,　(1)　$x + y = 20$

　　　(2)　$2x + 7y = 4 \cdot 20$

Step 1:　(1a)　$-2x - 2y = -40$　Multiply by - 2

　　　(2)　　$\underline{2x + 7y = 80}$

　　　　　　　$5y = 40$

　　　　　　　$y = 8$

#21. continued

Step 2:　$x + 8 = 20$

　　　　$x = 12$

Answer:　12 lb of Mr. Peanut, 8 lb King Peanut.

23.

Let x = amount invested at 9%.

Let y = amount invested at 11%.

Thus,　(1)　$x + y = 7000$

　　　(2)　$0.09x + 0.11y = 750$

　　　(2a)　$9x + 11y = 75,000$

　　　　　　　　　　　Multiply by 100

Step 1:　(1a)　$-9x - 9y = -63,000$

　　　(2a)　$\underline{9x + 11y = 75,000}$

　　　　　　　$2y = 12,000$

　　　　　　　$y = 6,000$

Step 2:

　　　(1)　$x + 6,000 = 7,000$

　　　　　$x = 1,000$

Answer:　$1,000 at 9%, $6,000 at 11%.

25.

Let x = gal. of Lone Star

Let y = gal. of Big Time

Thus,　(1)　$x + y = 10$

　　　(2)　$0.74x + 0.58y = 0.68(10)$

　　　(2a)　$74x + 58y = 680$

　　　　　　　　　　Multiply by 100

Step 1:　$-74x - 74y = -740$　Multiply by - 74

　　　$\underline{74x + 58y = 680}$

　　　　　$-16y = -60$

　　　　　$y = 3.75$

Step 2:　(1)　$x + 3.75 = 10$

　　　　　$x = 6.25$

Answer:　6.25 gal. of Lone Star, 3.75 gal. of Big
Time.

27.

　　　(1)　$\begin{aligned} 5 &= a \cdot 1^2 + b \cdot 1 + c \\ 5 &= a + b + c \end{aligned}$　For $(1, 5)$

　　　(2)　$\begin{aligned} 14 &= a(-2)^2 + b(-2) + c \\ 14 &= 4a - 2b + c \end{aligned}$　For $(-2, 14)$

　　　(3)　$\begin{aligned} 7 &= a(-1)^2 + b(-1) + c \\ 7 &= a - b + c \end{aligned}$　For $(-1, 7)$

#27. continued

Thus, (1) $a + b + c = 5$

(2) $4a - 2b + c = 14$

(3) $a - b + c = 7$

Step 1: (1) $a + b + c = 5$

(3) $\underline{a - b + c = 7}$

$2a \quad + 2c = 12$

(4) $a \quad + c = 6$ Multiply by $\frac{1}{2}$

Step 2: (1a) $2a + 2b + 2c = 10$ Multiply by 2

(2) $\underline{4a - 2b \quad + c = 14}$

$6a \quad\quad + 3c = 24$

(5) $2a \quad\quad + c = 8$

Step 3: (4) $a + c = 6$

(5a) $\underline{-2a - c = -8}$ Multiply by -1

$-a \quad = -2$

$a = 2$

Step 4: (4) $2 + c = 6$

$c = 4$

(1) $2 + b + 4 = 5$

$b + 6 = 5$

$b = -1$

Answer: $a = 2,\ b = -1,\ c = 4$

29.

Let u = the units digit.

t = the tens digit.

Thus, (1) $t + u = 9$

$10u + t = 10t + u + 9$

$9u - 9t = 9$

(2) $u - t = 1$

Step 1: (1) $t + u = 9$

(2) $\underline{-t + u = 1}$

$2u = 10$

$u = 5$

Step 2: (1) $t + 5 = 9$

$t = 4$

Answer: 45

31.

$(-3)^2 + (-1)^2 + D(-3) + E(-1) + F = 0$ For $(-3, -1)$

$9 + 1 - 3D - E + F = 0$

(1) $-3D - E + F = -10$

#31. continued

$(-1)^2 + 3^2 + D(-1) + E \cdot 3 + F = 0$ For $(-1, 3)$

$1 + 9 - D + 3E + F = 0$

(2) $-D + 3E + F = -10$

$5^2 + (-5)^2 + D \cdot 5 + E(-5) + F = 0$ For $(5, -5)$

$25 + 25 + 5D - 5E + F = 0$

(3) $5D - 5E + F = -50$

Thus, (1) $-3D - E + F = -10$

(2) $-D + 3E + F = -10$

(3) $5D - 5E + F = -50$

Step 1: (1) $-3D - E + F = -10$

(2a) $\underline{D - 3E - F = 10}$ Multiply by -1

(4) $-2D - 4E \quad = 0$

Step 2: (2) $-D + 3E + F = -10$

(3a) $\underline{-5D + 5E - F = 50}$

(5) $-6D + 8E \quad = 40$

Step 3: (4a) $-4D - 8E = 0$ Multiply by 2

(5) $\underline{-6D + 8E = 40}$

$-10D \quad = 40$

$D \quad = -4$

Step 4: (4) $-2(-4) - 4E = 0$

$-4E = -8$

$E = 2$

Step 5: (1) $-3(-4) - 2 + F = -10$

$12 - 2 + F = -10$

$F = -20$

Answer: $x^2 + y^2 - 4x + 2y - 20 = 0$

Exercises 9.4

1.

$x^2 + y^2 = 4$

$x + y = 2$

$y = -x + 2$ Solve for y

Substitute and solve.

$x^2 + (-x + 2)^2 = 4$

$x^2 + x^2 - 4x + 4 = 4$

$2x^2 - 4x = 0$

$2x(x - 2) = 0$

#1. continued

$2x = 0 \quad \text{or} \quad x - 2 = 0$

$x = 0 \qquad\qquad x = 2$

$y = -x + 2 \qquad y = -x + 2$

$y = 0 + 2 = 2 \qquad y = -2 + 2$

$\qquad\qquad\qquad\quad y = 0$

$\{(0,2),(2,0)\}$

3.

$x^2 + y^2 = 4$

$x - 2y = -2$

$\quad x = 2y - 2 \qquad$ Solve for x

Substitute and solve.

$(2y - 2)^2 + y^2 = 4$

$4y^2 - 8y + 4 + y^2 = 4$

$\qquad 5y^2 - 8y = 0$

$\qquad y(5y - 8) = 0$

$y = 0 \quad \text{or} \quad 5y - 8 = 0$

$\qquad\qquad\qquad y = \dfrac{8}{5}$

$x = 2y - 2 \qquad x = 2y - 2$

$x = 2 \cdot 0 - 2 \qquad x = 2\left(\dfrac{8}{5}\right) - 2$

$x = -2 \qquad\qquad x = \dfrac{6}{5}$

$\left\{(-2,0),\left(\dfrac{6}{5},\dfrac{8}{5}\right)\right\}$

5.

$x^2 - y = 0$

$4x - y = 4$

$\quad y = 4x - 4 \qquad$ Solve for y

Substitute and solve.

$x^2 - (4x - 4) = 0$

$x^2 - 4x + 4 = 0$

$(x - 2)(x - 2) = 0$

$\qquad x - 2 = 0$

$\qquad\qquad x = 2$

$y = 4x - 4$

$y = 4 \cdot 2 - 4$

$y = 4$

$\{(2,4)\}$

7.

$x^2 - y = 1$

$4x - y = 6$

$\quad y = 4x - 6 \qquad$ Solve for y

Substitute and solve.

$x^2 - (4x - 6) = 1$

$x^2 - 4x + 6 = 1$

$x^2 - 4x + 5 = 0$

No real solution

9.

$-x - y^2 = 0$

$x - 4y = -4$

$\quad x = 4y - 4 \qquad$ Solve for x

Substitute and solve.

$-(4y - 4) + y^2 = 0$

$-4y + 4 + y^2 = 0$

$y^2 - 4y + 4 = 0$

$(y - 2)(y - 2) = 0$

$\qquad y - 2 = 0$

$\qquad\qquad y = 2$

$x = 4y - 4$

$x = 4 \cdot 2 - 4$

$x = 4$

$\{(4,2)\}$

11.

$2x + y = -6$

$xy = 4$

$\quad y = -2x - 6 \qquad$ Solve for y

Substitute and solve.

$x(-2x - 6) = 4$

$-2x^2 - 6x - 4 = 0$

$x^2 + 3x + 2 = 0 \quad$ Multiply by $-\dfrac{1}{2}$

$(x + 2)(x + 1) = 0$

$x + 2 = 0 \quad \text{or} \quad x + 1 = 0$

$x = -2 \qquad\qquad x = -1$

$y = -2x - 6 \qquad y = -2x - 6$

$y = -2(-2) - 6 \quad y = -2(-1) - 6$

$y = -2 \qquad\qquad y = -4$

$\{(-2,-2),(-1,-4)\}$

13.

$$xy = 1$$
$$3x - 2y = 5$$
$$y = \frac{3}{2}x - \frac{5}{2} \qquad \text{Solve for y}$$

Substitute and solve.

$$x\left(\frac{3}{2}x - \frac{5}{2}\right) = 1$$

$$\frac{3}{2}x^2 - \frac{5}{2}x - 1 = 0$$

$$3x^2 - 5x - 2 = 0 \qquad \text{Multiply by 2}$$

$$(3x + 1)(x - 2) = 0$$

$$3x + 1 = 0 \quad \text{or} \quad x - 2 = 0$$

$$x = -\frac{1}{3} \qquad\qquad x = 2$$

$$y = \frac{3}{2}x - \frac{5}{2} \qquad y = \frac{3}{2}x - \frac{5}{2}$$

$$y = \frac{3}{2}\left(-\frac{1}{3}\right) - \frac{5}{2} \qquad y = \frac{3}{2} \cdot 2 - \frac{5}{2}$$

$$y = -3 \qquad\qquad y = \frac{1}{2}$$

$$\left\{\left(-\frac{1}{3}, -3\right), \left(2, \frac{1}{2}\right)\right\}$$

15.

$$x^2 + 25y^2 = 25$$
$$x - 5y = -1$$
$$x = 5y - 1 \qquad \text{Solve for x}$$

Substitute and solve.

$$(5y - 1)^2 + 25y^2 = 25$$

$$25y^2 - 10y + 1 + 25y^2 = 25$$

$$50y^2 - 10y - 24 = 0$$

$$25y^2 - 5y - 12 = 0 \qquad \text{Multiply by } \frac{1}{2}$$

$$(5y + 3)(5y - 4) = 0$$

$$5y + 3 = 0 \quad \text{or} \quad 5y - 4 = 0$$

$$y = -\frac{3}{5} \qquad\qquad y = \frac{4}{5}$$

$$x = 5y - 1 \qquad\qquad x = 5y - 1$$

$$x = 5\left(-\frac{3}{5}\right) - 1 \qquad x = 5\left(\frac{4}{5}\right) - 1$$

$$x = -4 \qquad\qquad x = 3$$

$$\left\{\left(-4, -\frac{3}{5}\right), \left(3, \frac{4}{5}\right)\right\}$$

17.

$$2x^2 + 3y^2 = 4$$
$$x - y = -2$$
$$x = y - 2 \qquad \text{Solve for x}$$

Substitute and solve.

$$2(y - 2)^2 + 3y^2 = 4$$

$$2(y^2 - 4y + 4) + 3y^2 = 4$$

$$2y^2 - 8y + 8 + 3y^2 = 4$$

$$5y^2 - 8y + 4 = 0$$

No real solutions

19.

$$x^2 - 2y^2 = 17$$
$$x - 2y = -1$$
$$x = 2y - 1 \qquad \text{Solve for x}$$

Substitute and solve.

$$(2y - 1)^2 - 2y^2 = 17$$

$$4y^2 - 4y + 1 - 2y^2 = 17$$

$$2y^2 - 4y - 16 = 0$$

$$y^2 - 2y - 8 = 0 \qquad \text{Multiply by } \frac{1}{2}$$

$$(y - 4)(y + 2) = 0$$

$$y - 4 = 0 \quad \text{or} \quad y + 2 = 0$$

$$y = 4 \qquad\qquad y = -2$$

$$x = 2y - 1 \qquad x = 2y - 1$$

$$x = 2 \cdot 4 - 1 \qquad x = 2(-2) - 1$$

$$x = 7 \qquad\qquad x = -5$$

$$\{(7, 4), (-5, -2)\}$$

21.

$$4y^2 - x^2 = 4$$
$$3x - 2y = 2$$
$$y = \frac{3}{2}x - 1 \qquad \text{Solve for y}$$

Substitute and solve.

$$4\left(\frac{3}{2}x - 1\right)^2 - x^2 = 4$$

$$4\left(\frac{9}{4}x^2 - 3x + 1\right) - x^2 = 4$$

$$9x^2 - 12x + 4 - x^2 = 4$$

$$8x^2 - 12x = 0$$

$$2x^2 - 3x = 0 \qquad \text{Multiply by } \frac{1}{4}$$

$$x(2x - 3) = 0$$

#21. continued

$x = 0$ or $2x - 3 = 0$

$$x = \frac{3}{2}$$

$y = \frac{3}{2}x - 1$ $y = \frac{3}{2}x - 1$

$y = \frac{3}{2} \cdot 0 - 1$ $y = \frac{3}{2} \cdot \frac{3}{2} - 1$

$y = -1$ $y = \frac{5}{4}$

$$\left\{ (0, -1), \left(\frac{3}{2}, \frac{5}{4} \right) \right\}$$

23.

(1) $x^2 + y^2 = 12$

(2) $3x^2 - 4y^2 = 8$

(1a) $4x^2 + 4y^2 = 48$ Multiply by 4

(2) $\underline{3x^2 - 4y^2 = 8}$

$7x^2 = 56$

$x^2 = 8$

$x = \pm 2\sqrt{2}$

$x = 2\sqrt{2}$ $x = -2\sqrt{2}$

(1) $\left(2\sqrt{2}\right)^2 + y^2 = 12$ $\left(-2\sqrt{2}\right)^2 + y^2 = 12$

$8 + y^2 = 12$ $y = \pm 2$

$y^2 = 4$

$y = \pm 2$

$$\left\{ \left(2\sqrt{2}, 2\right), \left(2\sqrt{2}, -2\right), \left(-2\sqrt{2}, 2\right), \left(-2\sqrt{2}, -2\right), \right\}$$

25.

(1) $5x^2 + 2y^2 = 7$

(2) $x^2 + y^2 = 2$

(1) $5x^2 + 2y^2 = 7$

(2a) $\underline{-2x^2 - 2y^2 = -4}$ Multiply (2) by -2

$3x^2 = 3$

$x = \pm 1$

$x = 1$ $x = -1$

(2) $1^2 + y^2 = 2$ $(-1)^2 + y^2 = 2$

$y^2 = 1$ $y = \pm 1$

$y = \pm 1$

$$\{(1,1), (1,-1), (-1,1), (-1,-1)\}$$

27.

(1) $-7x^2 + 2y^2 = 13$

(2) $2x^2 + 3y^2 = 7$

(1a) $-14x^2 + 4y^2 = 26$

$\underline{14x^2 + 21y^2 = 49}$

$25y^2 = 75$

$y^2 = 3$

$y = \pm\sqrt{3}$

$y = \sqrt{3}$

(2) $2x^2 + 3\left(\sqrt{3}\right)^2 = 7$

$2x^2 + 9 = 7$

$2x^2 = -2$

$x^2 = -1$

$x = i$

No real solutions.

29.

(1) $x^2 + y^2 = 2$

(2) $3x - 2y^2 = -5$

$y^2 = \frac{3}{2}x + \frac{5}{2}$ Solve for y^2

Substitute and solve

$x^2 + \frac{3}{2}x + \frac{5}{2} = 2$

$2x^2 + 3x + 5 = 4$ Multiply by 2

$2x^2 + 3x + 1 = 0$

$(2x + 1)(x + 1) = 0$

$2x + 1 = 0$ or $x + 1 = 0$

$x = -\frac{1}{2}$ $x = -1$

(1) $x^2 + y^2 = 2$ $x^2 + y^2 = 2$

$\frac{1}{4} + y^2 = 2$ $1 + y^2 = 2$

$y^2 = \frac{7}{4}$ $y^2 = 1$

$y^2 = \pm\frac{\sqrt{7}}{2}$ $y = \pm 1$

$$\left\{ \left(-\frac{1}{2}, \frac{\sqrt{7}}{2} \right), \left(-\frac{1}{2}, -\frac{\sqrt{7}}{2} \right), (-1,1), (-1,-1) \right\}$$

31.

(1) $x^2 + 4y^2 = 4$

(2) $x^2 - y = -1$

$\qquad x^2 = y - 1 \qquad$ Solve for x^2

Substitute and solve.

$y - 1 + 4y^2 = 4$

$4y^2 + y - 5 = 0$

$(4y + 5)(y - 1) = 0$

$4y + 5 = 0 \quad$ or $\quad y - 1 = 0$

$\qquad y = -\dfrac{5}{4} \qquad\qquad y = 1$

(2) $x^2 + \dfrac{5}{4} = -1 \qquad x^2 - 1 = -1$

$\qquad x^2 = \dfrac{-9}{4} \qquad\qquad x^2 = 0$

$\qquad x = \pm\dfrac{3}{2}i \qquad\qquad x = 0$

$\{(0,1)\}$

33.

(1) $x^2 - 4y^2 = -16$

(2) $2x - y^2 = 0$

$\qquad y^2 = 2x \qquad$ Solve for y^2

Substitute and solve

$\qquad x^2 - 4(2x) = -16$

$\qquad x^2 - 8x + 16 = 0$

$\qquad (x - 4)(x - 4) = 0$

$\qquad x - 4 = 0$

$\qquad\qquad x = 4$

(2) $2 \cdot 4 - y^2 = 0$

$\qquad\qquad y^2 = 8$

$\qquad\qquad y = \pm 2\sqrt{2}$

$\left\{\left(4, 2\sqrt{2}\right), \left(4, -2\sqrt{2}\right)\right\}$

35.

(1) $x^2 + y^2 = 5$

(2) $x^2 + 2xy + y^2 = 9$

#35. continued

(1a) $-x^2 \qquad\quad -y^2 = -5 \qquad$ Multiply (1) by -1

(2) $\quad\underline{x^2 + 2xy + y^2 = 9}$

$\qquad\quad 2xy \qquad = 4$

$\qquad\quad xy \qquad = 2$

$\qquad\qquad\qquad y = \dfrac{2}{x}$

Substitute and solve

(1) $x^2 + \left(\dfrac{2}{x}\right)^2 = 5$

$\qquad x^2 + \dfrac{4}{x^2} = 5$

$\qquad x^4 + 4 = 5x^2 \qquad$ Multiply by x^2

$\qquad x^4 - 5x^2 + 4 = 0$

$\left(x^2 - 4\right)\left(x^2 - 1\right) = 0$

$x^2 - 4 = 0 \quad$ or $\quad x^2 - 1 = 0$

$\qquad x = \pm 2 \qquad\qquad x = \pm 1$

(1) $\quad 4 + y^2 = 5 \qquad 1 + y^2 = 5$

$\qquad\quad y^2 = 1 \qquad\qquad y^2 = 4$

$\qquad\quad y = \pm 1 \qquad\qquad y = \pm 2$

A check will show that the solution set is

$\{(2,1), (-2,-1), (1,2), (-1,-2)\}$

37.

(1) $4x^2 + y^2 = 37$

(2) $4x^2 - 2xy + y^2 = 31$

(1a) $-4x^2 \qquad\quad -y^2 = -37 \qquad$ Multiply (-1) by -1

(2) $\quad\underline{4x^2 - 2xy + y^2 = 31}$

$\qquad\qquad -2xy = -6$

$\qquad\qquad xy = 3$

$\qquad\qquad\quad y = \dfrac{3}{x}$

Substitute and solve

(1) $\quad 4x^2 + \left(\dfrac{3}{x}\right)^2 = 37$

$\qquad\quad 4x^2 + \dfrac{9}{x^2} = 37$

$\qquad\qquad 4x^4 + 9 = 37x^2 \quad$ Multiply by x^2

$\qquad 4x^4 - 37x^2 + 9 = 0$

$\left(4x^2 - 1\right)\left(x^2 - 9\right) = 0$

256

#37. continued

$$4x^2 - 1 = 0 \quad \text{or} \quad x^2 - 9 = 0$$

$$x^2 = \frac{1}{4} \qquad\qquad x^2 = 9$$

$$x = \pm\frac{1}{2} \qquad\qquad x = \pm 3$$

(1) $\quad 4 \cdot \frac{1}{4} + y^2 = 37 \qquad 4 \cdot 9 + y^2 = 37$

$$y^2 = 36 \qquad\qquad y^2 = 1$$

$$y = \pm 6 \qquad\qquad y = \pm 1$$

A check will show that the solution set is

$$\left\{\left(\frac{1}{2}, 6\right), \left(-\frac{1}{2}, -6\right), (3, 1), (-3, -1)\right\}$$

39.

(1) $\quad y = \sqrt{2x + 1}$

(2) $\quad 2x + y = 5$

Substitute and solve

(2) $\quad 2x + \sqrt{2x + 1} = 5$

$$\left(\sqrt{2x + 1}\right)^2 = (5 - 2x)^2$$

$$2x + 1 = 25 - 20x + 4x^2$$

$$0 = 4x^2 - 22x + 24$$

$$0 = 2x^2 - 11x + 12 \quad \text{Multiply by } \frac{1}{2}$$

$$0 = (2x - 3)(x - 4)$$

$$2x - 3 = 0 \quad \text{or} \quad x - 4 = 0$$

$$x = \frac{3}{2} \qquad\qquad x = 4$$

A check will show that the solution set is

$$\left\{\left(\frac{3}{2}, 2\right)\right\}$$

41.

Let x and y = two numbers.

Thus, (1) $\quad x^2 + y^2 = 65$

(2) $\quad x + y = 11$

$$y = 11 - x \qquad \text{Solve for y}$$

Substitute and solve

$$x^2 + (11 - x)^2 = 65$$

$$x^2 + 121 - 22x + x^2 = 65$$

$$2x^2 - 22x + 56 = 0$$

$$x^2 - 11x + 28 = 0$$

$$(x - 4)(x - 7) = 0$$

#41. continued

$$x - 4 = 0 \quad \text{or} \quad x - 7 = 0$$

$$x = 4 \qquad\qquad x = 7$$

(2) $\quad 4 + y = 11 \qquad 7 + y = 11$

$$y = 7 \qquad\qquad y = 4$$

Answer: 4 and 7

43.

Let x = width

$\quad$ y = length

Thus, (1) $\quad 2x + 2y = 32$

$\quad\quad$ (2) $\quad\quad xy = 48$

$$y = \frac{48}{x} \qquad \text{Solve for y}$$

Substitute and solve

(1) $\quad 2x + 2\left(\frac{48}{x}\right) = 32$

$$2x + \frac{96}{x} = 32$$

$$2x^2 + 96 = 32x \quad \text{Multiply by x}$$

$$x^2 + 48 = 16x \quad \text{Multiply by } \frac{1}{2}$$

$$x^2 - 16x + 48 = 0$$

$$(x - 4)(x - 12) = 0$$

$$x - 4 = 0 \quad \text{or} \quad x - 12 = 0$$

$$x = 4 \qquad\qquad x = 12$$

(2) $\quad 4 \cdot y = 48 \qquad 12 \cdot y = 48$

$$y = 12 \qquad\qquad y = 4$$

Answer: 4 by 12 ft

45.

Let x and y = legs of a right triangle.

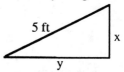

Thus, (1) $\quad x^2 + y^2 = 25$

$\quad\quad$ (2) $\quad 6 = \frac{1}{2}xy$

$$y = \frac{12}{x} \qquad \text{Solve for y}$$

#45. continued
Substitute and solve

(1) $\quad x^2 + \left(\dfrac{12}{x}\right)^2 = 25$

$\qquad x^2 + \dfrac{144}{x^2} = 25 \qquad$ Multiply by x^2

$\qquad x^4 + 144 = 25x^2$

$x^4 - 25x^2 + 144 = 0$

$\left(x^2 - 9\right)\left(x^2 - 16\right) = 0$

$x^2 - 9 = 0 \qquad$ or $\qquad x^2 - 16 = 0$

$\qquad x = \pm 3 \qquad\qquad\qquad x = \pm 4$

$\qquad y = \dfrac{12}{3} = 4 \qquad\qquad y = \dfrac{12}{4} = 3$

Measurement is not negative.

Answer: 3 and 4 ft

47.

(1) $\quad \dfrac{\left(2\sqrt{3}\right)^2}{a^2} + \dfrac{(-1)^2}{b^2} = 1$

(1) $\quad \dfrac{12}{a^2} + \dfrac{1}{b^2} = 1$

(2) $\quad \dfrac{2^2}{a^2} + \dfrac{\left(\sqrt{3}\right)^2}{b^2} = 1$

(2) $\quad \dfrac{4}{a^2} + \dfrac{3}{b^2} = 1$

(1) $\quad 12b^2 + a^2 = a^2b^2 \qquad$ Multiply by a^2b^2

(2) $\quad 4b^2 + 3a^2 = a^2b^2 \qquad$ Multiply by a^2b^2

(1) $\qquad 12b^2 + \ a^2 = \ a^2b^2$

(2a) $\dfrac{-12b^2 - 9a^2 = -3a^2b^2}{}$

$\qquad\qquad -8a^2 = -2a^2b^2$

$\qquad\qquad 4a^2 = a^2b^2 \qquad$ Multiply by $-\dfrac{1}{2}$

$\qquad 4a^2 - a^2b^2 = 0 \qquad$ Subtract a^2b^2

$\qquad a^2\left(4 - b^2\right) = 0 \qquad$ Factor

$\quad a^2 = 0 \ $ or $\ 4 - b^2 = 0$

$\qquad\qquad\qquad\qquad b^2 = 4$

#47. continued

(1) $\quad 12 \cdot 4 + a^2 = a^2 \cdot 4$

$\qquad 48 + a^2 = 4a^2$

$\qquad\qquad 48 = 3a^2$

$\qquad\qquad 16 = a^2$

Answer: $\dfrac{x^2}{16} + \dfrac{y^2}{4} = 1$

Review Exercises

1.
$2x - y = 5$
$x + 3y = -8$

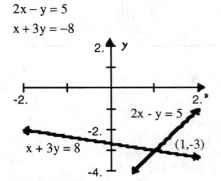

$(1, -3)$; Independent

3.
$x - 2y = 4$
$2x - 4y = 0$

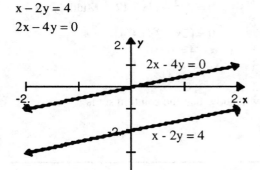

$\varnothing$; Inconsistent

5.
(1) $\quad 3x + 2y = 19$
(2) $\quad y = x + 2$

Substitute and solve
(1) $\quad 3x + 2(x + 2) = 19$
$\qquad 3x + 2x + 4 = 19$
$\qquad\qquad 5x = 15$
$\qquad\qquad x = 3$

#5. continued
(2) $y = x + 2$
$y = 3 + 2$
$y = 5$
$(3, 5)$; Independent

7.
(1) $3x - 9y = -21$
(2) $x = 3y - 7$
Substitute and solve
(1) $3(3y - 7) - 9y = -21$
$9y - 21 - 9y = -21$
$-21 = -21$
$\{(x, y) | x = 3y - 7\}$; Dependent

9.
(1) $2x + 4y = -11$
(2) $x + y = -3$
$y = -x - 3$ Solve for y

(1) $2x + 4(-x - 3) = -11$
$2x - 4x - 12 = -11$
$-2x = 1$
$x = -\dfrac{1}{2}$
(2) $y = -x - 3$
$y = \dfrac{1}{2} - 3$
$y = -\dfrac{5}{2}$
$\left(-\dfrac{1}{2}, -\dfrac{5}{2}\right)$; Independent

11.
(1) $2x + y = 8$
(2) $\underline{3x - y = 12}$
$5x \quad = 20$
$x = 4$
(1) $2 \cdot 4 + y = 8$
$y = 0$
$(4, 0)$; Independent

13.
(1) $2x + 8y = 7$
(2) $x + 4y = 1$

(1) $2x + 8y = 7$
(2a) $\underline{-2x - 8y = -2}$ Multiply (2) by -2
$0 = 5$
$\varnothing$; Inconsistent

15.
(1) $-\dfrac{1}{2}x + \dfrac{1}{3}y = \dfrac{1}{6}$
(2) $\dfrac{1}{4}x - \dfrac{1}{2}y = \dfrac{1}{4}$

(1a) $-3x + 2y = 1$ Multiply (1) by 6
(2a) $\underline{x - 2y = 1}$ Multiply (2) by 4
$-2x \quad = 2$
$x = -1$
(1a) $-3(-1) + 2y = 1$
$2y = -2$
$y = -1$
$(-1, -1)$; Independent

17.
(1) $x - 3y + z = 8$
(2) $2x + y - z = 0$
(3) $-x - y + 4z = 11$

(1) $x - 3y + \quad z = 8$
(3) $\underline{-x - \quad y + 4z = 11}$
(4) $\quad -4y + 5z = 19$

(2) $2x + \quad y - \quad z = 0$
(3a) $\underline{-2x - 2y + 8z = 22}$ Multiply (3) by 2
(5) $\quad - y + 7z = 22$
Solve for z
(4) $-4y + 5z = 19$
(5a) $\underline{4y - 28z = -88}$ Multiply (5) by -4
$-23z = -69$
$z = 3$

#17. continued
Solve for y

(5) $\quad -y + 7\cdot 3 = 22$

$\qquad\qquad -y = 1$

$\qquad\qquad y = -1$

Solve for x

(2) $\quad 2x + (-1) - 3 = 0$

$\qquad\qquad 2x - 4 = 0$

$\qquad\qquad 2x = 4$

$\qquad\qquad x = 2$

$(2, -1, 3);\quad$ Independent

19.

$(1)\quad x - 3y + 2z = 10$

$(2)\quad 5x + 2y + z = 3$

$(3)\quad -2x - y + 3z = 1$

$(1)\qquad x - 3y + 2z = 10$

$(2a)\quad \underline{-10x - 4y - 2z = -6}$

$(4)\quad -9x - 7y \qquad = 4$

$(3)\qquad -2x - y + 3z = 1$

$(2b)\quad \underline{-15x - 6y - 3z = -9}$

$(5)\quad -17x - 7y \qquad = -8$

Solve for x.

$(4)\quad -9x - 7y = 4$

$(5a)\quad \underline{17x + 7y = 8}\qquad$ Multiply (5) by -1

$\qquad\qquad 8x = 12$

$\qquad\qquad x = \dfrac{3}{2}$

Solve for y.

$(4)\quad -9\left(\dfrac{3}{2}\right) - 7y = 4$

$\qquad\quad \dfrac{-27}{2} - 7y = 4$

$\qquad\qquad -7y = \dfrac{35}{2}$

$\qquad\qquad y = -\dfrac{5}{2}$

#19. continued
Solve for z.

$(2)\quad 5\left(\dfrac{3}{2}\right) + 2\left(-\dfrac{5}{2}\right) + z = 3$

$\qquad\quad \dfrac{15}{2} - 5 + z = 3$

$\qquad\qquad z = -\dfrac{15}{2} + 5 + 3$

$\qquad\qquad z = \dfrac{1}{2}$

$\left(\dfrac{3}{2}, -\dfrac{5}{2}, \dfrac{1}{2}\right);\quad$ Independent

---

21.

$(1)\quad -2x - 6y + 2z = 8$

$(2)\quad 3x + 9y - 3z = -12$

$(3)\quad x + 3y - z = -4$

$(1)\quad -2x - 6y + 2z = \;\;8$

$(3)\quad \underline{2x + 6y - 2z = -8}$

$\qquad\qquad\qquad 0 = 0$

Dependent

23.

$(1)\quad x + 4y - z = 0$

$(2)\quad -x - y + 2z = -4$

$(3)\quad 3x + y + 3z = 5$

$(1)\quad x + 4y - \;z = 0$

$(2)\quad \underline{-x - \;y + 2z = -4}$

$(4)\qquad\quad 3y + \;z = -4$

$(3)\quad 3x + \;y + 3z = 5$

$(2a)\quad \underline{-3x - 3y + 6z = -12}\qquad$ Multiply (2) by 3

$(5)\qquad\quad -2y + 9z = -7$

Solve for y.

$(5)\quad -2y + 9z = -7$

$(4a)\quad \underline{-27y - 9z = 36}\qquad$ Multiply (4) by -9

$\qquad\qquad -29y = 29$

$\qquad\qquad\quad y = -1$

Solve for z.

$(4)\quad 3(-1) + z = -4$

$\qquad\qquad z = -1$

#23. continued
Solve for x.

(1) $\quad x + 4(-1) - (-1) = 0$

$\qquad x - 4 + 1 \qquad = 0$

$\qquad\qquad\qquad x = 3$

$(3, -1, -1);$ Independent

25.

(1) $\quad -3x + 2y + z = 5$

(2) $\quad x - 2y + z = 1$

(3) $\quad 2x - y - z = -4$

(1) $\quad -3x + 2y + z = 5$

(3) $\quad \underline{2x - \ y - z = -4}$

(4) $\quad - \ x + \ y \quad\ = 1$

(2) $\quad x - 2y + z = 1$

(3) $\quad \underline{2x - \ y - z = -4}$

(5) $\quad 3x - 3y \quad\ = -3$

(6) $\quad - x + \ y \quad\ = 1 \qquad$ Multiply (5) by $-\dfrac{1}{3}$

Note: (4) and (5) are the same.

The system is dependent.

27.
Let x and y = the number.

Thus, (1) $\quad x + y = 19$

$\qquad$ (2) $\quad x = y + 5$

Substitute and solve

(1) $\quad y + 5 + y = 19$

$\qquad\ 2y + 5 = 19$

$\qquad\qquad 2y = 14$

$\qquad\qquad\ y = 7$

Solve for x.

(1) $\quad x + 7 = 19$

$\qquad\quad x = 12$

Answer: 7 and 12

29.
Let x = number of adult tickets

$\quad$ y = number of child tickets

$\quad$ 4x = value of adult tickets

$\quad$ 1.5y = value of child tickets

#29. continued
Thus, (1) $\quad x + y = 32$

$\qquad$ (2) $\quad 4x + 1.5y = 103$

$\qquad\qquad\qquad y = 32 - x \qquad$ Solve (1) for y.

Substitute and solve

$4x + 1.5(32 - x) = 103$

$\quad 4x + 48 - 1.5x = 103$

$\qquad\qquad\ 2.5x = 55$

$\qquad\qquad\quad x = 22$

Solve for y.

(1) $\quad 22 + y = 32$

$\qquad\qquad y = 10$

Answer: 22 adult tickets, 10 child tickets

31.
Let x = amount of "Apple of Your Eye"

$\quad$ y = amount of "Generic"

(1) $\quad x + y = 40$

(2) $\quad 150x + 70y = 120(40)$

$\qquad\qquad\qquad y = 40 - x \qquad\qquad$ Solve (1) for y

Substitute and solve

(2) $\quad 150x + 70(40 - x) = 4800$

$\qquad 150x + 2800 - 70x = 4800$

$\qquad\qquad\qquad\quad 80x = 2000$

$\qquad\qquad\qquad\qquad x = 25$

Solve for y.

(1) $\quad 25 + y = 40$

$\qquad\qquad y = 15$

Answer: 25 qt of "Apple of Your Eye",

$\qquad\quad$ 15 qt of "Generic"

33.
Let x = amount of Big D toothpaste

y = amount of Big G toothpaste

Thus, (1) $\quad x + y = 12$

$\qquad$ (2) $\quad .42x + .24y = .36(12)$

$\qquad\qquad\qquad 42x + 24y = 36(12)$

$\qquad\qquad\qquad\qquad y = 12 - x \qquad$ Solve (1) for y

Substitute and solve

(2) $\quad 42x + 24(12 - x) = 432$

$\qquad 42x + 288 - 24x = 432$

$\qquad\qquad\qquad 18x = 144$

$\qquad\qquad\qquad\ x = 8$

#33. continued
Solve for y.

(1) $8 + y = 12$

   $y = 4$

Answer: 8 oz of Big D, 4 oz of Big G

35.
Let x = football unit cost

y = basketball unit cost

z = soccer ball unit cost

Thus,   (1) $x + y + z = 47$

   (2) $2x + y + 2z = 80$

   (3) $2x + 3y + z = 93$

(1a) $-2x - 2y - 2z = -94$ Multiply (1) by -2

(2) $\underline{2x + y + 2z = 80}$

   $-y = -14$

   $y = 14$

(1) $x + 14 + z = 47$

(4) $x + z = 33$

(3) $2x + 3(14) + z = 93$

   $2x + 42 + z = 93$

(5) $2x + z = 51$

(5) $2x + z = 51$

(4a) $\underline{-x - z = -33}$

   $x = 18$

Solve for z.

(1) $18 + 14 + z = 47$

   $32 + z = 47$

   $z = 15$

Answer: football = $18;   basketball = $14;

   soccer ball = $15

37.
$x^2 + y = 3$

$\underline{x - y = 3}$

$x^2 + x = 6$

$x^2 + x - 6 = 0$

$(x - 2)(x + 3) = 0$

#37. continued
$x - 2 = 0$ or $x + 3 = 0$

   $x = 2$ $x = -3$

$x - y = 3$ $x - y = 3$

$2 - y = 3$ $-3 - y = 3$

   $y = -1$ $y = -6$

$\{(2, -1), (-3, -6)\}$

39.
(1) $2x - y = 3$

(2) $xy = 20$

   $y = \dfrac{20}{x}$ Solve for y

Substitute and solve

   $2x - \dfrac{20}{x} = 3$

   $2x^2 - 20 = 3x$ Multiply by x

$2x^2 - 3x - 20 = 0$

$(2x + 5)(x - 4) = 0$

$2x + 5 = 0$ or $x - 4 = 0$

   $x = -\dfrac{5}{2}$ $x = 4$

$y = \dfrac{20}{-\dfrac{5}{2}}$ $y = \dfrac{20}{4}$

$y = -8$ $y = 5$

$\left\{\left(-\dfrac{5}{2}, -8\right), (4, 5)\right\}$

41.
(1) $y^2 - 9x^2 = 9$

(2) $x - y = 3$

   $x = y + 3$ Solve for (2) for y

(1) $y^2 - 9(y + 3)^2 = 9$

   $y^2 - 9\left(y^2 + 6y + 9\right) = 9$

   $y^2 - 9y^2 - 54y - 81 = 9$

   $-8y^2 - 54y = 90$

   $4y^2 + 27y = -45$ Multiply by $-\dfrac{1}{2}$

   $4y^2 + 27y + 45 = 0$

   $(4y + 15)(y + 3) = 0$

#41. continued

$4y + 15 = 0$ or $y + 3 = 0$

$$y = -\frac{15}{4} \qquad\qquad y = -3$$

$$x + \frac{15}{4} = 3 \qquad x + 3 = 3$$

$$x = -\frac{3}{4} \qquad\qquad x = 0$$

$$\left\{ \left(-\frac{3}{4}, -\frac{15}{4}\right), (0, -3) \right\}$$

43.

(1) $2x^2 - 3y^2 = 8$

(2) $3x^2 + 4y^2 = 29$

(1a) $8x^2 - 12y^2 = 32$ Multiply (1) by 4

(2a) $\underline{9x^2 + 12y^2 = 87}$

$\qquad 17x^2 \qquad\quad = 119$

$$x^2 = 7$$

$$x = \pm\sqrt{7}$$

  Solve for y

$2 \cdot 7 - 3y^2 = 8$

$\quad -3y^2 = -6$

$\quad\quad y^2 = 2$

$\quad\quad y = \pm\sqrt{2}$

$$\left\{ \left(\sqrt{7}, \sqrt{2}\right), \left(\sqrt{7}, -\sqrt{2}\right), \left(-\sqrt{7}, \sqrt{2}\right), \left(-\sqrt{7}, -\sqrt{2}\right) \right\}$$

45.

(1) $x^2 - 4y^2 = 5$

(2) $x^2 - y = 8$

$\qquad x^2 = y + 8$

(1) $y + 8 - 4y^2 = 5$

$\qquad\qquad 0 = 4y^2 - y - 3$

$\qquad\qquad 0 = (4y + 3)(y - 1)$

$4y + 3 = 0$ or $y - 1 = 0$

$\quad y = -\frac{3}{4} \qquad\quad y = 1$

$x^2 = -\frac{3}{4} + 8 \qquad x^2 = 1 + 8$

$x^2 = \frac{29}{4} \qquad\qquad x^2 = 9$

$x = \pm\dfrac{\sqrt{29}}{2} \qquad\quad x = \pm 3$

#45. continued

$$\left\{ \left(\frac{\sqrt{29}}{2}, -\frac{3}{4}\right), \left(-\frac{\sqrt{29}}{2}, -\frac{3}{4}\right), (3, 1), (-3, 1) \right\}$$

47.

(1) $x^2 + y^2 = 20$

(2) $x^2 + 5xy + y^2 = 60$

(1a) $-x^2 \qquad\quad - y^2 = -20$

(2a) $\underline{x^2 + 5xy + y^2 = \quad 60}$

$\qquad 5xy \qquad\quad = \quad 40$

$\qquad\qquad xy = 8$

$\qquad\qquad y = \dfrac{8}{x}$ \qquad Solve for y

Substitute and solve

(1) $x^2 + \left(\dfrac{8}{x}\right)^2 = 20$

$\qquad x^2 + \dfrac{64}{x^2} = 20$

$\qquad x^4 + 64 = 20x^2$ \qquad Multiply by x^2

$x^4 - 20x^2 + 64 = 0$

$\left(x^2 - 4\right)\left(x^2 - 16\right) = 0$

$x^2 - 4 = 0$ or $x^2 - 16 = 0$

$x^2 = 4 \qquad\qquad x^2 = 16$

$x = \pm 2 \qquad\qquad x = \pm 4$

$y = \dfrac{8}{2} = 4 \qquad\quad y = \dfrac{8}{4} = 2$

$y = \dfrac{8}{-2} = -4 \qquad y = \dfrac{8}{-4} = -2$

$$\left\{ (4, 2), (-4, -2), (2, 4), (-2, -4) \right\}$$

49.

Let x = width

Let y = length

Thus, $2x + 2y = 36$

$\qquad\qquad xy = 72$

$$y = \frac{72}{x} \qquad \text{Solve for y}$$

263

#49. continued
Substitute and solve

$$2x + 2\left(\frac{72}{x}\right) = 36$$

$$2x + \frac{144}{x} = 36$$

$$2x^2 + 144 = 36x$$

$$2x^2 - 36x + 144 = 0$$

$$x^2 - 18x + 72 = 0 \qquad \text{Multiply by } \frac{1}{2}$$

$$(x - 6)(x - 12) = 0$$

$$x - 6 = 0 \quad \text{or} \quad x - 12 = 0$$

$$x = 6 \qquad\qquad x = 12$$

$$y = \frac{72}{6} = 12 \qquad y = \frac{72}{12} = 6$$

Answer: 6 ft. by 12 ft.

Chapter 1 Test Solutions

1.
(1) $2x - 3y = 5$
(2) $y = 2x + 2$

Substitute and solve
(1) $2x - 3(2x + 2) = 5$

$$2x - 6x - 6 = 5$$

$$-4x = 11$$

$$x = -\frac{11}{4}$$

(2) $y = 2\left(-\frac{11}{4}\right) + 2$

$$y = -\frac{11}{2} + 2$$

$$y = -\frac{7}{2}$$

$$\left(-\frac{11}{4}, -\frac{7}{2}\right)$$

3.
(1) $3x - 6y = 11$
(2) $x - 2y = 5$

$$x = 2y + 5 \qquad \text{Solve for } x$$

(1) $3(2y + 5) - 6y = 11$

$$6y + 15 - 6y = 11$$

$$15 = 11$$

#3. continued
$\varnothing$; Inconsistent

5.
(1) $x - 2y + 3z = 7$
(2) $2x + y + z = 4$
(3) $-3x + 2y + 3z = 19$

(1) $x - 2y + 3z = 7$
(3) $\underline{-3x + 2y + 3z = 19}$
$\quad\ -2x \qquad\ \ + 6z = 26$

(4) $\quad x \qquad - 3z = -13 \qquad \text{Multiply by } -\frac{1}{2}$

(1) $x - 2y + 3z = 7$
(2a) $\underline{4x + 2y + 2z = 8} \qquad \text{Multiply (2) by 2}$
$\quad\ 5x \qquad\ + 5z = 15$

(5) $\quad x \qquad\ \ + z = 3$

(4) $\quad x - 3z = -13$
(5a) $\underline{-x -\ z = -3} \qquad \text{Multiply (5) by } -1$
$\qquad\ - 4z = -16$
$\qquad\quad\ z = 4$

(5) $x + 4 = 3$
$\quad\ \ x = -1$

(2) $2(-1) + y + 4 = 4$
$\quad\ -2 + y + 4 = 4$
$\qquad\qquad y = 2$

$(-1, 2, 4)$

7.
(1) $x^2 + y^2 = 1$
(2) $x - 2y = -1$
$x = 2y - 1 \qquad \text{Solve for x.}$
Substitute and solve.

$$(2y - 1)^2 + y^2 = 1$$

$$4y^2 - 4y + 1 + y^2 = 1$$

$$5y^2 - 4y = 0$$

$$y(5y - 4) = 0$$

#7. continued

$y = 0$ or $5y - 4 = 0$

$x = 2y - 1$ $\quad y = \dfrac{4}{5}$

$x = 2 \cdot 0 - 1$

$x = -1$ $\qquad x = 2y - 1$

$\qquad\qquad x = 2 \cdot \dfrac{4}{5} - 1$

$\qquad\qquad x = \dfrac{8}{5} - 1$

$\qquad\qquad x = \dfrac{3}{5}$

$\left\{ (-1, 0), \left(\dfrac{3}{5}, \dfrac{4}{5} \right) \right\}$

9.

(1) $\quad x^2 + y^2 = 29$

(2) $\quad x^2 + 14xy + y^2 = 169$

(1a) $\quad -x^2 \qquad\quad -y^2 = -29$

(2) $\quad \dfrac{x^2 + 14xy + y^2 = 169}{}$

$\qquad\quad 14xy \qquad\quad = 140$

$\qquad\qquad xy \qquad\quad = 10$

$\qquad\qquad y = \dfrac{10}{x}$

Substitute and solve

(1) $\quad x^2 + \left(\dfrac{10}{x} \right)^2 = 29$

$\qquad x^2 + \dfrac{100}{x^2} = 29$

$\qquad x^4 + 100 = 29x^2$

$x^4 - 29x^2 + 100 = 0$

$\left(x^2 - 25 \right)\left(x^2 - 4 \right) = 0$

$x^2 - 25 = 0$ or $x^2 - 4 = 0$

$\qquad x^2 = 25 \qquad\quad x^2 = 4$

$\qquad x = \pm 5 \qquad\quad x = \pm 2$

$y = \dfrac{10}{5} = 2 \qquad y = \dfrac{10}{2} = 5$

$y = \dfrac{10}{-5} = -2 \qquad y = \dfrac{10}{-2} = -5$

$\{ (5, 2), (-5, -2), (2, 5), (-2, -5) \}$

11.

Let x = amount of "Garden Grow"

y = amount of "This Might Work"

Thus, (1) $\quad x + y = 30$

$\qquad\quad$ (2) $\quad 170x + 50y = 90(30)$

$\qquad\quad$ (2) $\quad 17x + 5y = 270 \qquad$ Multiply by $\dfrac{1}{10}$

$\qquad\quad$ (1) $\quad -5x - 5y = -150$

$\qquad\quad$ (2) $\quad \dfrac{17x + 5y = 270}{}$

$\qquad\qquad\quad 12x = 120$

$\qquad\qquad\quad\; x = 10$

$\qquad\quad$ (1) $\quad 10 + y = 30$

$\qquad\qquad\qquad\; y = 20$

Answer: 10 lb "Garden Grow",

$\qquad\qquad$ 20 lb "This Might Grow"

Test Your Memory

1.

$\left(\dfrac{10x^{-9}y^4}{6x^{-2}y^{10}} \right)^2 = \left(\dfrac{\cancel{2} \cdot 5}{\cancel{2} \cdot 3x^{-2+9}y^{10-4}} \right)^2$

$= \left(\dfrac{5}{3x^7 y^6} \right)^2$

$= \dfrac{25}{9x^{14} y^{12}}$

3.

$x\sqrt{48x} - 3\sqrt{12x^3} - \sqrt{300x^3}$

$= x\sqrt{16 \cdot 3x} - 3\sqrt{4x^2 \cdot 3x} - \sqrt{100x^2 \cdot 3x}$

$= 4x\sqrt{3x} - 6x\sqrt{3x} - 10x\sqrt{3x}$

$= -12x\sqrt{3x}$

5.

$\dfrac{x+4}{x+4} \cdot \dfrac{x-4}{(x-6)(x-2)} + \dfrac{x-11}{(x+4)(x-6)} \cdot \dfrac{x-2}{x-2}$

$\qquad\qquad\qquad LCD = (x-6)(x-2)(x+4)$

$= \dfrac{(x+4)(x-4)}{(x+4)(x-6)(x-2)} + \dfrac{(x-11)(x-2)}{(x+4)(x-6)(x-2)}$

$= \dfrac{x^2 - 16 + x^2 - 13x + 22}{(x+4)(x-6)(x-2)}$

#5. continued

$$= \frac{2x^2 - 13x + 6}{(x+4)(x-6)(x-2)}$$

$$= \frac{(2x-1)\cancel{(x-6)}}{(x+4)\cancel{(x-6)}(x-2)}$$

$$= \frac{2x-1}{(x+4)(x-2)}$$

7.

$$f(0) = \frac{0}{0-2} = 0$$

9.

$$[g(-4)]^3 = \left(\frac{3}{-4-1}\right)^3 = \left(-\frac{3}{5}\right)^3 = -\frac{27}{125}$$

11.

$(10,1), (-2,5)$

$$m = \frac{5-1}{-2-10} = \frac{4}{-12} = -\frac{1}{3}$$

$$y - 1 = -\frac{1}{3}(x-10)$$

$$y - 1 = -\frac{1}{3}x + \frac{10}{3}$$

$$y = -\frac{1}{3}x + \frac{13}{3}$$

13.

$$x^2 + (y+4)^2 = 36$$

15.-19.

The graphs are in the answer section of the main text.

21.

$$x^2(3x+1) - 4(3x+1) = 0$$

$$(3x+1)(x+2)(x-2) = 0$$

$$3x + 1 = 0 \quad \text{or} \quad x + 2 = 0 \quad \text{or} \quad x - 2 = 0$$

$$x = -\frac{1}{3} \qquad x = -2 \qquad x = 2$$

$$\left\{-\frac{1}{3}, -2, 2\right\}$$

23.

$$\frac{x+3}{1}\left(1 - \frac{3x+14}{x+3}\right) = \left(\frac{4x+7}{x+3}\right)\frac{x+3}{1} \qquad LCD = x+3$$

$$x + 3 - (3x+14) = 4x + 7$$

$$x + 3 - 3x - 14 = 4x + 7$$

$$-2x - 11 = 4x + 7$$

$$-18 = 6x$$

$$-3 = x$$

$\varnothing$ since $x + 3 \neq 0$ and $x \neq -3$

25.

$$\sqrt[3]{2x-1} + 3 = 1$$

$$\left(\sqrt[3]{2x-1}\right)^3 = (-2)^3$$

$$2x - 1 = -8$$

$$2x = -7$$

$$x = -\frac{7}{2} \qquad \left\{-\frac{7}{2}\right\}$$

27.

$$2x^2 - 2x + 5 = 0$$

$$a = 2, b = -2, c = 5$$

$$x = \frac{2 \pm \sqrt{4-40}}{2\cdot 2}$$

$$x = \frac{2 \pm \sqrt{-36}}{4}$$

$$x = \frac{2 \pm 6i}{4} = \frac{1 \pm 3i}{2}$$

$$\left\{\frac{1 \pm 3i}{2}\right\}$$

29.

$$\left(\sqrt{2x+11}\right)^2 = \left(2 - \sqrt{x+6}\right)^2$$

$$2x + 11 = 4 - 4\sqrt{x+6} + x + 6$$

$$2x + 11 = 10 - 4\sqrt{x+6} + x$$

$$(x+1)^2 = \left(-4\sqrt{x+6}\right)^2$$

$$x^2 + 2x + 1 = 16(x+6)$$

$$x^2 + 2x + 1 = 16x + 96$$

$$x^2 - 14x - 95 = 0$$

$$(x+5)(x-19) = 0$$

$$x + 5 = 0 \quad \text{or} \quad x - 19 = 0$$

$$x = -5 \qquad x = 19$$

#29. continued

$\{-5\}$ 19 does not check.

31.

(1) $-6x + 4y = -11$

(2) $x - y = 2$

(1) $-6x + 4y = -11$

(2a) $\underline{6x - 6y = 12}$ Multiply (2) by 6

      $-2y = 1$

      $y = -\dfrac{1}{2}$

(2) $x - \left(-\dfrac{1}{2}\right) = 2$

      $x + \dfrac{1}{2} = 2$

      $x = 2 - \dfrac{1}{2}$

      $x = \dfrac{3}{2}$

$\left\{\left(\dfrac{3}{2}, -\dfrac{1}{2}\right)\right\}$ Independent

---

33.

(1) $2x + 3y = -5$

(2) $3x - 4y = -16$

(1a) $8x + 12y = -20$ Multiply (1) by 4

(2a) $\underline{9x - 12y = -48}$ Multiply (2) by 3

    $17x = -68$

       $x = -4$

Solve for y

(1) $2(-4) + 3y = -5$

     $-8 + 3y = -5$

       $3y = 3$

        $y = 1$

$\{(-4, 1)\}$ Independent

35.

(1) $x + 2y - z = -4$

(2) $2x - y + 3z = 7$

(3) $-x + y - 4z = -13$

#35. continued

(1) $x + 2y - z = -4$

(3) $\underline{-x + y - 4z = -13}$

(4) $3y - 5z = -17$

(2) $2x - y + 3z = 7$

(3a) $\underline{-2x + 2y - 8z = -26}$ Multiply (3) by 2

(5) $y - 5z = -19$

(4) $3y - 5z = -17$

(5a) $\underline{-3y + 15z = 57}$

       $10z = 40$

        $z = 4$

Solve for y

(5) $y - 5 \cdot 4 = -19$

     $y - 20 = -19$

       $y = 1$

Solve for x

(1) $x + 2 \cdot 1 - 4 = -4$

     $x + 2 - 4 = -4$

        $x = -2$

$\{(-2, 1, 4)\}$ Independent

37.

(1) $x^2 + y^2 = 17$

(2) $x^2 + 3xy + y^2 = 29$

(1a) $-x^2 - y^2 = -17$ Multiply (1) by -1

(2) $\underline{x^2 + 3xy + y^2 = 29}$

      $3xy = 12$

       $xy = 4$

        $y = \dfrac{4}{x}$ Solve for y

Substitute and solve

(1) $x^2 + \left(\dfrac{4}{x}\right)^2 = 17$

      $x^2 + \dfrac{16}{x^2} = 17$

      $x^4 + 16 = 17x^2$ Multiply by x^2

   $x^4 - 17x^2 + 16 = 0$

$\left(x^2 - 16\right)\left(x^2 - 1\right) = 0$

#37. continued

$$x^2 - 16 = 0 \quad \text{or} \quad x^2 - 1 = 0$$
$$x^2 = 16 \qquad\qquad x^2 = 1$$
$$x = \pm 4 \qquad\qquad x = \pm 1$$
$$y = \frac{4}{4} = 1 \qquad\qquad y = \frac{4}{1} = 4$$
$$y = \frac{4}{-4} = -1 \qquad y = \frac{4}{-1} = -1$$

A check shows that the solution set is:

$$\{(4,1),(-4,-1),(1,4),(-1,-4)\}$$

39.-41.

The graphs are in the answer section of the main text.

43.

Let x = number of quarters.

Let 22 − x = number of nickels.

Thus, $\quad 25x + 5(22 - x) = 250$
$$25x + 110 - 5x = 250$$
$$20x = 140$$
$$x = 7$$
$$22 - x = 22 - 7 = 15$$

Answer: 7 quarters; 15 nickels

45.

T = hk	Find T
60 = 2k	T = 3 · 30
30 = k	T = 90

Answer: 90° F

47.

Let x = speed in still water.

	Distance	=	Rate	· Time
Up	12		x - 1	$\frac{12}{x-1}$
Down	18		x + 1	$\frac{18}{x+1}$

Time is the same.

$$\frac{12}{x-1} = \frac{18}{x+1}$$
$$12x + 12 = 18x - 18 \qquad \text{Cross multiply}$$
$$30 = 6x$$
$$5 = x$$

Answer: 5 mph

49.

Let x = ballpoint per unit cost.

Let y = pencil per unit cost.

Thus,

(1) $\quad 3x + 2y = 171$

(2) $\quad 5x + y = 257$

(1) $\qquad\quad 3x + 2y = \;\;\;171$

(2a) $\quad \underline{-10x - 2y = -514} \qquad$ Multiply (2) by -2

$\qquad\qquad -7x \qquad\;\; = -343$

$\qquad\qquad\qquad x = 49$

Solve for y

(2) $\quad 5 \cdot 49 + y = 257$

$\qquad\quad 245 + y = 257$

$\qquad\qquad\quad y = 12$

Answer: ballpoint pen = 49¢

$\qquad\qquad\;$ pencil = 12¢

Chapter 9 Study Guide

Self-Test Exercises

I. Using the substitution or elimination method, determine the solutions of the following linear systems of equations. Identify the system of as independent, inconsistent, or dependent.

1. $3x + 2y = -2$
   $y = 4x + 6$

2. $y = 2x + 4$
   $2x + 2y = 8$

3. $18x + 30y = 6$
   $6x + 10y = 2$

4. $3x + 6y = 26$
   $2x + 4y = 18$

5. $2x + 3y = 4$
   $x + 2y = -6$

6. $9x - 8y = -2$
   $3x + 2y = 3$

7. $\frac{1}{3}x - \frac{1}{2}y = \frac{1}{30}$
   $\frac{3}{2}x + \frac{1}{4}y = \frac{13}{20}$

8. $4x + y = 1$
   $3x - 3y = 7$

9. $x + 8y - 6z = -47$
   $3x - 2y + 7z = 13$
   $7x - 9y - 9z = -3$

10. $2x - 3y + 3z = -15$
    $3x + 2y - 5z = 19$
    $5x - 4y - 2z = -2$

II. Determine the solutions of the following systems of equations.

11. $2x^2 - y^2 = -8$
    $x - y = 6$

12. $2x^2 - 2y^2 = 8$
    $x + y = 4$

13. $4x^2 + 3y^2 = 39$
    $-2x^2 + 5y^2 = -13$

14. $3x^2 + 4y^2 = 16$
    $x^2 + 2y^2 = 6$

15. $x^2 + y^2 = 5$
    $x^2 + 4xy + y^2 = 13$

16. $x^2 + y^2 = 13$
    $x^2 + 3xy + y^2 = 31$

III. Find the solutions to the following problems.

17. The sum of two numbers is 25. One number is 2 less than twice the other number. Find the numbers.

18. James bought 3 hammers and 1 chisel for $17 at Acme Hardware. A week later he bought 4 hammers and 1 chisel for $22. What is the cost of each tool?

19. Johney's Fish Inc. sells a 20 lb mixture of crab and shrimp for $76.95. Crab cost $4.25 a pound and shrimp cost $3.10 a pound. Find how many pounds of each were used.

20. The perimeter of a rectangle is 46 m. The rectangle has an area of 112 m. Find the dimensions of the rectangle.

The worked-out solutions begin on the next page.

Self-Test Solutions

1.

(1) $\quad 3x + 2y = -2$

(2) $\quad y = 4x + 6$

Substitute and solve

(1) $\quad 3x + 2(4x + 6) = -2$

$\qquad 3x + 8x + 12 = -2$

$\qquad 11x = -14$

$\qquad x = -\dfrac{14}{11}$

(2) $\quad y = 4x + 6 = 4\left(\dfrac{-14}{11}\right) + 6$

$\qquad y = \dfrac{-56}{11} + \dfrac{66}{11}$

$\qquad y = \dfrac{10}{11} \qquad \left(-\dfrac{14}{11}, \dfrac{10}{11}\right)$

2.

(1) $\quad y = 2x + 4$

(2) $\quad 2x + 2y = 8$

Substitute and solve

(2) $\quad 2x + 2(2x + 4) = 8$

$\qquad 2x + 4x + 8 = 8$

$\qquad 6x = 0$

$\qquad x = 0$

(1) $\quad y = 2x + 4 = 2 \cdot 0 + 4$

$\qquad y = 4 \qquad (0, 4)$

3.

(1) $\quad 18x + 30y = 6$

(2) $\quad 6x + 10y = 2$

(1a) $\quad 6x + 10y = 2 \qquad$ Multiply (1) by $\dfrac{1}{3}$

Note: (1) is the same as (1a).

The system is dependent.

4.

(1) $\quad 3x + 6y = 26$

(2) $\quad 2x + 4y = 18$

#4. continued

(1a) $\quad 6x + 12y = 52 \qquad$ Multiply (1) by 2

(2a) $\quad \underline{-6x - 12y = -54}$

$\qquad\qquad 0 = -2$

The system is inconsistent.

5.

(1) $\quad 2x + 3y = 4$

(2) $\quad x + 2y = -6$

(1) $\quad 2x + 3y = 4$

(2a) $\quad \underline{-2x - 4y = 12} \qquad$ Multiply (2) by -2

$\qquad\qquad -y = 16$

$\qquad\qquad y = -16$

Solve for x.

(2) $\quad x + 2(-16) = -6$

$\qquad x - 32 = -6$

$\qquad x = 26 \qquad (26, -16)$

6.

(1) $\quad 9x - 8y = -2$

(2) $\quad 3x + 2y = 3$

(1) $\quad 9x - 8y = -2$

(2a) $\quad \underline{12x + 8y = 12} \qquad$ Multiply (2) by 4

$\qquad\quad 21x = 10$

$\qquad\quad x = \dfrac{10}{21}$

Solve for y.

(2) $\quad 3\left(\dfrac{10}{21}\right) + 2y = 3$

$\qquad \dfrac{10}{7} + 2y = 3$

$\qquad 2y = 3 - \dfrac{10}{7}$

$\qquad 2y = \dfrac{11}{7}$

$\qquad y = \dfrac{11}{14} \qquad \left(\dfrac{10}{21}, \dfrac{11}{14}\right)$

7.

(1) $\dfrac{1}{3}x - \dfrac{1}{2}y = \dfrac{1}{30}$

(2) $\dfrac{3}{2}x + \dfrac{1}{4}y = \dfrac{13}{20}$

(1a) $10x - 15y = 1$ Multiply (1) by 30

(2a) $30x + 5y = 13$ Multiply (2) by 20

(1a) $10x - 15y = 1$

(2b) $\underline{90x + 15y = 39}$ Multiply (2a) by 3

$\qquad 100x = 40$

$\qquad\quad x = \dfrac{4}{10} = \dfrac{2}{5}$

Solve for y.

(2a) $30\left(\dfrac{2}{5}\right) + 5y = 13$

$\qquad 12 + 5y = 13$

$\qquad\quad 5y = 1$

$\qquad\quad y = \dfrac{1}{5}$ $\left(\dfrac{2}{5}, \dfrac{1}{5}\right)$

8.

(1) $4x + y = 1$

(2) $3x - 3y = 7$

(1) $\qquad y = -4x + 1$ Solve (1) for y.

Substitute and solve

(2) $3x - 3(-4x + 1) = 7$

$\qquad 3x + 12x - 3 = 7$

$\qquad\qquad 15x = 10$

$\qquad\qquad\quad x = \dfrac{2}{3}$

Solve for y.

(1) $y = -4\left(\dfrac{2}{3}\right) + 1$

$\quad y = -\dfrac{8}{3} + 1$

$\quad y = -\dfrac{5}{3}$

$\left(\dfrac{2}{3}, -\dfrac{5}{3}\right)$

9.

(1) $x + 8y - 6z = -47$

(2) $3x - 2y + 7z = 13$

(3) $7x - 9y - 9z = -3$

(1a) $-3x - 24y + 18z = 141$ Multiply (1) by -3

(2) $\underline{\quad 3x - 2y + 7z = 13}$

(4) $\qquad -26y + 25z = 154$

(1b) $-7x - 56y + 42z = 329$ Multiply (1) by -7

(3) $\underline{\quad 7x - 9y - 9z = -3}$

(5) $\qquad -65y + 33z = 326$

(4a) $-130y + 125z = 770$ Multiply (4) by 5

(5a) $\underline{\quad 130y - 66z = -652}$ Multiply (5) by -2

$\qquad\qquad 59z = 118$

$\qquad\qquad\quad z = 2$

Solve for y.

(4) $-26y + 25 \cdot 2 = 154$

$\qquad -26y = 104$

$\qquad\quad y = -4$

Solve for x.

(1) $x + 8(-4) - 6 \cdot 2 = -47$

$\quad x - 32 - 12 = -47$

$\qquad\qquad x = -3$ $(-3, -4, 2)$

10.

(1) $2x - 3y + 3z = -15$

(2) $3x + 2y - 5z = 19$

(3) $5x - 4y - 2z = -2$

(2a) $6x + 4y - 10z = 38$ Multiply (2) by 2

(3) $\underline{\quad 5x - 4y - 2z = -2}$

(4) $11x \qquad\quad - 12z = 36$

(1a) $4x - 6y + 6z = -30$ Multiply (1) by 2

(2a) $\underline{\quad 9x + 6y - 15z = 57}$ Multiply (2) by 3

(5) $13x \qquad\quad - 9z = 27$

#10. continued

(4a) $\quad 33x - 36z = 108$ $\qquad$ Multiply (4) by 3

(5a) $\quad \underline{-52x + 36z = -108}$ $\qquad$ Multiply (5) by -4

$\qquad -19x = 0$

$\qquad \quad x = 0$

Solve for z.

(5) $\quad 13 \cdot 0 - 9z = 27$

$\qquad \qquad x = -3$

Solve for y.

(2) $\quad 3 \cdot 0 + 2y - 5(-3) = 19$

$\qquad \qquad 2y + 15 = 19$

$\qquad \qquad 2y = 4$

$\qquad \qquad y = 2 \quad (0, 2, -3)$

11.

(1) $\quad 2x^2 - y^2 = -8$

(2) $\quad x - y = 6$

$\qquad x = y + 6$ $\qquad$ Solve for x.

(1) $\qquad 2(y + 6)^2 - y^2 = -8$

$\qquad 2(y^2 + 12y + 36) - y^2 = -8$

$\qquad 2y^2 + 24y + 72 - y^2 = -8$

$\qquad \quad y^2 + 24y + 80 = 0$

$\qquad \quad (y + 4)(y + 20) = 0$

$y + 4 = 0 \quad$ or $\quad y + 20 = 0$

$\quad y = -4 \qquad \qquad y = -20$

(2) $\quad x = y + 6 = -4 + 6 \qquad x = y + 6 = -20 + 6$

$\qquad \qquad = 2 \qquad \qquad \qquad = -14$

$\{(2, -4), (-20, -14)\}$

12.

(1) $\quad 2x^2 - 2y^2 = 8$

(2) $\quad x + y = 4$

$\qquad y = -x + 4$ $\qquad$ Solve for y.

(1) $\qquad 2x^2 - 2(-x + 4)^2 = 8$

$\qquad 2x^2 - 2(x^2 - 8x + 16) = 8$

$\qquad 2x^2 - 2x^2 - 16x - 32 = 8$

$\qquad \qquad -16x = 40$

$\qquad \qquad x = -\dfrac{5}{2}$

#12. continued

Solve for y.

(2) $\quad y = -\left(-\dfrac{5}{2}\right) + 4$

$\qquad y = \dfrac{3}{2} \qquad \qquad \left(-\dfrac{5}{2}, \dfrac{3}{2}\right)$

13.

(1) $\qquad 4x^2 + 3y^2 = 39$

(2) $\quad -2x^2 + 5y^2 = -13$

(1) $\qquad 4x^2 + 3y^2 = 39$

(2a) $\quad \underline{-4x^2 + 10y^2 = -26}$ $\qquad$ Multiply (2) by 2

$\qquad \qquad 13y^2 = 13$

$\qquad \qquad y^2 = 1$

$\qquad \qquad y = \pm 1$

Solve for x.

(1) $\quad 4x^2 + 3 \cdot 1 = 39$

$\qquad \quad 4x^2 = 36$

$\qquad \quad x^2 = 9$

$\qquad \quad x = \pm 3$

$\{(3, 1), (-3, 1), (3, -1), (-3, -1)\}$

14.

(1) $\quad 3x^2 + 4y^2 = 16$

(2) $\quad x^2 + 2y^2 = 6$

(1) $\qquad 3x^2 + 4y^2 = 16$

(2a) $\quad \underline{-3x^2 - 6y^2 = -18}$ $\qquad$ Multiply (2) by -3

$\qquad \qquad -2y^2 = -2$

$\qquad \qquad y^2 = 1$

$\qquad \qquad y = \pm 1$

Solve for x.

(2) $\quad x^2 + 2 \cdot 1 = 6$

$\qquad \quad x^2 = 4$

$\qquad \quad x = \pm 2$

$\{(2, 1), (-2, 1), (2, -1), (-2, -1)\}$

15.

(1) $x^2 + y^2 = 5$

(2) $x^2 + 4xy + y^2 = 13$

(1a) $\quad -x^2 \quad\quad -y^2 = -5 \qquad$ Multiply (1) by -1

(2) $\quad \underline{x^2 + 4xy + y^2 = 13}$

$\qquad\qquad 4xy \quad\quad = 8$

$\qquad\qquad\quad xy = 2$

$\qquad\qquad\quad\; y = \dfrac{2}{x}$

Solve for x.

(1) $\qquad x^2 + \left(\dfrac{2}{x}\right)^2 = 5$

$\qquad\qquad x^2 + \dfrac{4}{x^2} = 5$

$\qquad\qquad x^4 + 4 = 5x^2$

$\qquad\quad x^4 - 5x^2 + 4 = 0$

$\qquad \left(x^2 - 4\right)\left(x^2 - 1\right) = 0$

$\quad x^2 - 4 = 0 \quad$ or $\quad x^2 - 1 = 0$

$\qquad x^2 = 4 \qquad\qquad x^2 = 1$

$\qquad x = \pm 2 \qquad\qquad x = \pm 1$

$\; 2^2 + y^2 = 5 \qquad 1^2 + y^2 = 5$

$\qquad y^2 = 1 \qquad\qquad y^2 = 4$

$\qquad y = \pm 1 \qquad\qquad y = \pm 2$

After checking, the solution set is:

$\qquad \{(2,1), (-2,-1), (1,2), (-1,-2)\}$

16.

(1) $x^2 + y^2 = 13$

(2) $x^2 + 3xy + y^2 = 31$

(1a) $\quad -x^2 \quad\quad -y^2 = -6 \qquad$ Multiply (1) by -1

(2) $\quad \underline{x^2 + 3xy + y^2 = 24}$

$\qquad\qquad 3xy \quad\quad = 18$

$\qquad\qquad\quad xy = 6$

$\qquad\qquad\quad\; y = \dfrac{6}{x}$

#16. continued

(1) $\qquad x^2 + \left(\dfrac{6}{x}\right)^2 = 13$

$\qquad\qquad x^2 + \dfrac{36}{x^2} = 13$

$\qquad\qquad x^4 + 36 = 13x^2 \qquad$ Multiply by x^2

$\qquad\quad x^4 - 13x^2 + 36 = 0$

$\qquad \left(x^2 - 4\right)\left(x^2 - 9\right) = 0$

$\quad x^2 - 4 = 0 \quad$ or $\quad x^2 - 9 = 0$

$\qquad x^2 = 4 \qquad\qquad x^2 = 9$

$\qquad x = \pm 2 \qquad\qquad x = \pm 3$

(1) $\; 2^2 + y^2 = 13 \qquad 3^2 + y^2 = 13$

$\qquad y^2 = 9 \qquad\qquad y^2 = 4$

$\qquad y = \pm 3 \qquad\qquad y = \pm 2$

Answer checking, the solution set is:

$\qquad \{(2,3), (-2,-3), (3,2), (-3,-2)\}$

17.

Let x and y = two numbers.

Thus, (1) $\; x + y = 25$

$\qquad$ (2) $\qquad x = 2y - 2$

Substitute and solve.

(1) $\; 2y - 2 + y = 25$

$\qquad\qquad 3y = 27$

$\qquad\qquad\; y = 9$

Solve for x.

(1) $\; x + 9 = 25$

$\qquad x = 16$

Answer: 16 and 9

18.

Let x = cost of a hammer.

Let y = cost of a chisel.

Thus,

(1) $\; 3x + y = 17$

(2) $\; 4x + y = 22$

(1) $\qquad 3x + y = \;\; 17$

(2a) $\; \underline{-4x - y = -22} \qquad$ Multiply (2) by -1

$\qquad -x \quad\;\; = -5$

$\qquad\quad x = 5$

#18. continued
Solve for y.

(1) $3 \cdot 5 + y = 17$

$y = 2$

Answer: hammer = \$5; chisel = \$2

19.
Let x = pounds of crab.
Let y = pounds of shrimp.
Thus,

(1) $x + y = 20$

(2) $425x + 310y = 7695$

(1) $y = -x + 20$ Solve (1) for y

Substitute and solve.

(2) $425x + 310(-x + 20) = 7695$

$425x - 310x + 6200 = 7695$

$115x = 1495$

$x = 13$

Solve for y.

(1) $13 + y = 20$

$y = 7$

Answer: 13 lb crab, 7 lb shrimp

20.
Let x = width.
Let y = length.
Thus,

(1) $2x + 2y = 46$

(2) $xy = 112$

(1a) $x + y = 23$ Multiply (1) by $\dfrac{1}{2}$

(2a) $y = \dfrac{112}{x}$ Solve (2) for y

Substitute and solve.

(1a) $x + \dfrac{112}{x} = 23$

$x^2 + 112 = 23x$ Multiply by x

$x^2 - 23x + 112 = 0$

$(x - 7)(x - 16) = 0$

$x - 7 = 0$ or $x - 16 = 0$

$x = 7$ $x = 16$

(2a) $y = \dfrac{112}{7} = 16$ $y = \dfrac{112}{16} = 7$

Answer: 16 m by 7 m

CHAPTER 10 EXPONENTIAL AND LOGARITHMIC FUNCTIONS

Solutions To Text Exercises

1.

$f(x) = 3x + 1$

$y = 3x + 1 \xrightarrow{\text{Inverse}} x = 3y + 1$ Interchange x and y

$\qquad x - 1 = 3y$ Solve for y

$\qquad \frac{1}{3}(x - 1) = 3y\left(\frac{1}{3}\right)$

$\qquad \frac{1}{3}x - \frac{1}{3} = y$

$\qquad\qquad$ or

$\qquad f^{-1}(x) = \frac{1}{3}x - \frac{1}{3}$

3.

$f(x) = 3 - 2x$

$y = 3 - 2x \xrightarrow{\text{Inverse}} x = 3 - 2y$ Interchange x and y

$\qquad 3 - x = -2y$ Solve for y

$\qquad 3 - x = 2y$

$\qquad \frac{1}{2}(3 - x) = 2y\left(\frac{1}{2}\right)$

$\qquad \frac{3}{2}x - \frac{1}{2} = y$

$\qquad\qquad$ or

$\qquad f^{-1}(x) = -\frac{1}{2}x + \frac{3}{2}$

5.

$f(x) = \frac{x + 3}{4}$

$y = \frac{x + 3}{4} \xrightarrow{\text{Inverse}} x = \frac{y + 3}{4}$ Interchange x and y

$\qquad (4)x = \frac{y + 3}{4}(4)$ Solve for y

$\qquad 4x = y + 3$

$\qquad 4x - 3 = y$

$\qquad\qquad$ or

$\qquad f^{-1}(x) = 4x - 3$

7.

$f(x) = \frac{2x - 5}{3}$

$y = \frac{2x - 5}{3} \xrightarrow{\text{Inverse}} x = \frac{2y - 5}{3}$

$\qquad\qquad$ Interchange x and y

$\qquad (3)x = \frac{2y - 5}{3}(3)$

$\qquad\qquad$ Solve for y

$\qquad 3x = 2y - 5$

$\qquad 3x + 5 = 2y$

$\qquad \frac{1}{2}(3x + 5) = 2y\left(\frac{1}{2}\right)$

$\qquad \frac{3}{2}x + \frac{5}{2} = y$

$\qquad\qquad$ or

$\qquad f^{-1}(x) = \frac{3}{2}x + \frac{5}{2}$

9.

$f(x) = x^2 - 3$

$y = x^2 - 3 \xrightarrow{\text{Inverse}} x = y^2 - 3$ Interchange x and y

The inverse is not a function since two y values exist for one x.

11.

$f(x) = \frac{1}{2}x^2 + 2$

$y = \frac{1}{2}x^2 + 2 \xrightarrow{\text{Inverse}} x = \frac{1}{2}y^2 + 2$

$\qquad\qquad$ Interchange x and y

The inverse is not a function since two y values exist for one x.

13.

$f(x) = 5 - 2x^2$

$y = 5 - 2x^2 \xrightarrow{\text{Inverse}} x = 5 - 2y^2$

$\qquad\qquad$ Interchange x and y

The inverse is not a function since two y values exist for one x.

15.

$$f(x) = \frac{6 - x^2}{4}$$

$$y = \frac{6 - x^2}{4} \xrightarrow{\text{Inverse}} x = \frac{6 - y^2}{4}$$

Interchange x and y

The inverse is not a function since two y values exist for one x.

17.

$$f(x) = \sqrt{x + 2} \qquad \text{Note: } f(x) \geq 0$$

$$y = \sqrt{x + 2} \xrightarrow{\text{Inverse}} x = \sqrt{y + 2}$$

Interchange x and y

$$(x)^2 = \left(\sqrt{y + 2}\right)^2$$

Solve for y

$$x^2 = y + 2$$

$$x^2 - 2 = y$$

or

$$f^{-1}(x) = x^2 - 2, \, x \geq 0$$

19.

$$f(x) = \sqrt{x - 1} \qquad \text{Note: } f(x) \geq 0$$

$$y = \sqrt{x - 1} \xrightarrow{\text{Inverse}} x = \sqrt{y - 1}$$

Interchange x and y

$$(x)^2 = \left(\sqrt{y - 1}\right)^2$$

Solve for y

$$x^2 = y - 1$$

$$x^2 + 1 = y \qquad \text{where } x \geq 0$$

or

$$f^{-1}(x) = x^2 + 1, \, x \geq 0$$

21.

$$f(x) = \sqrt{5 - 2x} \qquad \text{Note: } f(x) \geq 0$$

$$y = \sqrt{5 - 2x} \xrightarrow{\text{Inverse}} x = \sqrt{5 - 2y}$$

Interchange x and y

$$(x)^2 = \left(\sqrt{5 - 2y}\right)^2$$

Solve for y

$$x^2 = 5 - 2y$$

$$x^2 - 5 = -2y$$

#21. continued

$$-\frac{1}{2}\left(x^2 - 5\right) = -2y\left(-\frac{1}{2}\right)$$

$$-\frac{1}{2}x^2 + \frac{5}{2} = y$$

$$\frac{-x^2 + 5}{2} = y \qquad \text{where } x \geq 0$$

or

$$f^{-1}(x) = \frac{-x^2 + 5}{2}, \, x \geq 0$$

23.

$$f(x) = \sqrt{x} + 1 \qquad \text{Note: } f(x) \geq 1$$

$$y = \sqrt{x} + 1 \xrightarrow{\text{Inverse}} x = \sqrt{y} + 1$$

Interchange x and y

$$x - 1 = \sqrt{y}$$

Solve for y

$$(x - 1)^2 = \left(\sqrt{y}\right)^2$$

$$(x - 1)^2 = y \qquad \text{where } x \geq 1$$

or

$$f^{-1}(x) = (x - 1)^2, \, x \geq 1$$

25.

$$f(x) = |x + 1|$$

$$y = |x + 1| \xrightarrow{\text{Inverse}} x = |y + 1| \qquad \text{Interchange x and y}$$

$$x = y + 1$$

The inverse is not a function since two y values exist for one x.

27.

$$f(x) = |x - 2|$$

$$y = |x - 2| \xrightarrow{\text{Inverse}} x = |y - 2| \qquad \text{Interchange x and y}$$

$$x = y - 2$$

The inverse is not a function since two y values exist for one x.

The graphs are in the answer section of the main text.

29.

$f(x) = |x| + 2$

$y = |x| + 2 \xrightarrow{\text{Inverse}} x = |y| + 2$ Interchange x and y

$$x = y + 2$$

The inverse is not a function since two y values exist for one x.

31.

$f(x) = x^3$

$y = x^3 \xrightarrow{\text{Inverse}} x = y^3$ Interchange x and y

$$\sqrt[3]{x} = \sqrt[3]{y^3}$$ Solve for y

$$\sqrt[3]{x} = y$$

or

$$f^{-1}(x) = \sqrt[3]{x}$$

33.

$f(x) = \sqrt[3]{x+1}$

$y = \sqrt[3]{x+1} \xrightarrow{\text{Inverse}} x = \sqrt[3]{y+1}$

Interchange x and y

$$(x)^3 = \left(\sqrt[3]{y+1}\right)^3$$

Solve for y

$$x^3 = y + 1$$

$$x^3 - 1 = y$$

or

$$f^{-1}(x) = x^3 - 1$$

35.

$f(x) = x^5$

Let $f(a) = f(b)$.

$a^5 = b^5$

By inspection, a = b for any set of values. Therefore f is one to one.

$y = x^5 \xrightarrow{\text{Inverse}} x = y^5$ Interchange x and y

$$\sqrt[5]{x} = y$$ Solve for y

or

$$f^{-1}(x) = \sqrt[5]{x}$$

37.

$f(x) = \dfrac{3}{x+2}$

Let $f(a) = f(b)$.

$$\frac{3}{a+2} = \frac{3}{b+2}$$

$$3(b+2) = 3(a+2)$$

$$\left(\frac{1}{3}\right)(3)(b+2) = \left(\frac{1}{3}\right)(3)(a+2)$$

$$b + 2 = a + 2$$

$$b = a \qquad \text{f is one to one.}$$

Therefore,

$y = \dfrac{3}{x+2} \xrightarrow{\text{Inverse}} x = \dfrac{3}{y+2}$

Interchange x and y

$$(y+2)x = \frac{3}{y+2}(y+2)$$

Solve for y

$$(y+2)x = 3$$

$$\frac{1}{x}(y+2)x = 3\left(\frac{1}{x}\right)$$

$$y + 2 = \frac{3}{x}$$

$$y = \frac{3}{x} - 2$$

$$y = \frac{3}{x} - 2\left(\frac{x}{x}\right)$$

$$y = \frac{3 - 2x}{x}$$

or

$$f^{-1}(x) = \frac{3 - 2x}{x}$$

39.

$f(x) = \dfrac{x}{x-4}$

Let $f(a) = f(b)$.

$$\frac{a}{a-4} = \frac{b}{b-4}$$

$$a(b-4) = b(a-4)$$

$$ab - 4a = ba - 4b$$

$$-4a = -4b$$

$$a = b \qquad \text{f is one to one.}$$

Therefore,

#39. continued

$$y = \frac{x}{x-4} \xrightarrow{\text{Inverse}} x = \frac{y}{y-4}$$

Interchange x and y

$$x(y-4) = y$$

Solve for y

$$xy - 4x = y$$
$$-4x = y - xy$$
$$4x = -y + xy$$
$$4x = y(x-1)$$
$$\frac{4x}{x-1} = y$$

or

$$f^{-1}(x) = \frac{4x}{x-1}$$

41.

$$f(x) = \frac{1}{x^2 - 4}$$

Let $f(a) = f(b)$.

$$\frac{1}{a^2 - 4} = \frac{1}{b^2 - 4}$$
$$b^2 - 4 = a^2 - 4$$
$$b^2 = a^2$$
$$b^2 - a^2 = 0$$
$$(a+b)(a-b) = 0$$
$$a + b = 0 \quad \text{or} \quad a - b = 0$$
$$a = -b \qquad a = b$$

f is not one to one since $a = -b$.

43.

$$f(x) = |2x + 3|$$

Let $f(a) = f(b)$.

$$|2a + 3| = |2b + 3|$$

Two possibilities:

$$2a + 3 = -(2b+3) \quad \text{or} \quad 2a + 3 = 2b + 3$$
$$2a + 3 = -2b - 3 \qquad\qquad 2a = 2b$$
$$2a = -2b - 6 \qquad \left(\frac{1}{2}\right)2a = 2b\left(\frac{1}{2}\right)$$
$$\left(\frac{1}{2}\right)2a = (-2b-6)\frac{1}{2} \qquad a = b$$
$$a = -b - 3$$

f is not one to one since one possibility is $a = -b - 3$.

45.

$$f(x) = x^2 + 4$$

Let $f(a) = f(b)$.

$$a^2 + 4 = b^2 + 4$$
$$a^2 = b^2$$
$$a^2 - b^2 = 0$$

Two possibilities:

$$a = b \quad \text{or} \quad a = -b$$

f is not one to one since one possibility is $a = -b$.

47.

$$f(x) = \sqrt{x+2}$$

Let $f(a) = f(b)$.

$$\sqrt{a+2} = \sqrt{b+2}$$
$$\left(\sqrt{a+2}\right)^2 = \left(\sqrt{b+2}\right)^2$$
$$a + 2 = b + 2$$
$$a = b \qquad \text{f is one to one.}$$

Therefore,

$$y = \sqrt{x+2} \xrightarrow{\text{Inverse}} x = \sqrt{y+2}$$

Interchange x and y

$$x^2 = \left(\sqrt{y+2}\right)^2$$

Solve for y

$$x^2 = y + 2$$
$$x^2 - 2 = y \qquad x \geq 0$$

or

$$f^{-1}(x) = x^2 - 2, \, x \geq 0$$

49.

$$f(x) = \sqrt[3]{x-5}$$

Let $f(a) = f(b)$.

$$\sqrt[3]{a-5} = \sqrt[3]{b-5}$$
$$\left(\sqrt[3]{a-5}\right)^3 = \left(\sqrt[3]{b-5}\right)^3$$
$$a - 5 = b - 5$$
$$a = b \qquad \text{f is one to one.}$$

Therefore,

$$y = \sqrt[3]{x-5} \xrightarrow{\text{Inverse}} x = \sqrt[3]{y-5}$$

Interchange x and y

$$(x)^3 = \left(\sqrt[3]{y-5}\right)^3$$

Solve for y

#49. continued

$$x^3 = y - 5$$
$$x^3 + 5 = y$$

or

$$f^{-1}(x) = x^3 + 5$$

51.

$f(x) = 2x + 1$

$g(x) = 3 - x$

a. $(f \circ g)(2) = 2(3 - x) + 1$

$$= 2(1) + 1$$
$$= 3$$

b. $(f \circ g)(-3) = 2(3 - x) + 1$

$$= 2(6) + 1$$
$$= 13$$

c. $(f \circ g)(x) = f[g(x)]$

$$= f[(3 - x)]$$

Substitute $(3 - x)$ for $g(x)$

$$= 2(3 - x) + 1$$
$$= 6 - 2x + 1$$
$$= 7 - 2x$$

d. $(g \circ f)(3) = 3 - (2x + 1)$

$$= 3 - (2(3) + 1)$$
$$= -4$$

e. $(g \circ f)(-1) = 3 - (2x + 1)$

$$= 3 - (2(-1) + 1)$$
$$= 4$$

f. $(g \circ f)(x) = g[f(x)]$

$$= g[2x + 1]$$

Substitute $(2x + 1)$ for $f(x)$

$$= 3 - (2x + 1)$$
$$= 3 - 2x - 1 = 2 - 2x$$

53.

$f(x) = x^2 + 1$

$g(x) = x - 5$

a. $(f \circ g)(2) = (x - 5)^2 + 1$

$$= (2 - 5)^2 + 1$$
$$= (-3)^2 + 1$$
$$= 10$$

#53. continued

b. $(f \circ g)(-3) = (x - 5)^2 + 1$

$$= (-3 - 5)^2 + 1$$
$$= (-8)^2 + 1$$
$$= 65$$

c. $(f \circ g)(x) = f[g(x)]$

$$= f[(x - 5)]$$

Substitute $(x - 5)$ for $g(x)$

$$= (x - 5)^2 + 1$$
$$= x^2 - 10x + 25 + 1$$
$$= x^2 - 10x + 26$$

d. $(g \circ f)(3) = (x^2 + 1) - 5$

$$= (3^2 + 1) - 5$$
$$= 5$$

e. $(g \circ f)(-1) = (x^2 + 1) - 5$

$$= ((-1)^2 + 1) - 5$$
$$= (1 + 1) - 5$$
$$= -3$$

f. $(g \circ f)(x) = g[f(x)]$

$$= g[x^2 + 1]$$

Substitute $(x^2 + 1)$ for $f(x)$

$$= (x^2 + 1) - 5 = x^2 - 4$$

55.

$f(x) = 4$

$g(x) = 2x$

a. $(f \circ g)(2) = f[g(x)]$

$$= f[2x]$$
$$= f[2(2)]$$
$$= f(4)$$
$$= 4 \qquad \text{For any value of } x, f(x)$$
$$\text{always equals 4}$$

b. $(f \circ g)(-3) = f[g(x)]$

$$= f[2x]$$
$$= f[2(-3)]$$
$$= f(-6) = 4$$

c. $(f \circ g)(x) = f[g(x)]$

$$= f[2x] = 4$$

#55. continued

d. $(g \circ f)(3) = g[f(x)]$

$= g[4] = 2x$

$= 2(4) = 8$

e. $(g \circ f)(-1) = g[f(x)]$

$= g[4] = 2x$

$= 2(4) = 8$

f. $(g \circ f)(x) = g[f(x)]$

$= g[4] = 2x$

$= 2(4) = 8$

57.

$f(x) = 3x + 5$

$g(x) = x - \dfrac{5}{3}$

$(f \circ g)(x) = f[g(x)]$

$= f\left[x - \dfrac{5}{3}\right]$

$= 3\left(x - \dfrac{5}{3}\right) + 5$

$= 3x - 5 + 5 = 3x$

$(g \circ f)(x) = g[f(x)]$

$= g[3x + 5]$

$= (3x + 5) - \dfrac{5}{3}$

$= 3x + \dfrac{15}{3} - \dfrac{5}{3}$

$= 3x + \dfrac{10}{3}$

Since $3x \neq 3x + \dfrac{10}{3}$, an inverse does not exist.

59.

$f(x) = \dfrac{2x + 3}{4}$

$g(x) = \dfrac{4x - 3}{2}$

$(f \circ g)(x) = f[g(x)]$

$= f\left[\dfrac{4x - 3}{2}\right]$

$= \dfrac{2\left(\dfrac{4x - 3}{2}\right) + 3}{4}$

#59. continued

$= \dfrac{4x - 3 + 3}{4}$

$= \dfrac{4x}{4} = x$

$(g \circ f)(x) = g[f(x)]$

$= g\left[\dfrac{2x + 3}{4}\right]$

$= \dfrac{4\left(\dfrac{2x + 3}{4}\right) - 3}{2}$

$= \dfrac{2x + 3 - 3}{2}$

$= \dfrac{2x}{2} = x$

Since $x = x$, there are inverses.

61.

$f(x) = \dfrac{2}{x + 3}$

$g(x) = \dfrac{2 - 3x}{x}$

$(f \circ g)(x) = f[g(x)]$

$= f\left[\dfrac{2 - 3x}{x}\right]$

$= \dfrac{2}{\left(\dfrac{2 - 3x}{x}\right) + 3}$

$= \dfrac{2}{\left(\dfrac{2 - 3x}{x}\right) + 3} \cdot \left(\dfrac{x}{x}\right)$

$= \dfrac{2x}{2 - 3x + 3x}$

$= \dfrac{2x}{2} = x$

$(g \circ f)(x) = g[f(x)]$

$= g\left[\dfrac{2}{x + 3}\right]$

$= \dfrac{2 - 3\left(\dfrac{2}{x + 3}\right)}{\left(\dfrac{2}{x + 3}\right)}$

$= \dfrac{2 - 3\left(\dfrac{2}{x + 3}\right)}{\left(\dfrac{2}{x + 3}\right)} \cdot \dfrac{(x + 3)}{(x + 3)}$

#61. continued

$$= \frac{2(x+3)-6}{2}$$

$$= x + 3 - 3 = x$$

Since $x = x$, there are inverses.

63.

$$f(x) = \frac{1}{x-2}$$

$$g(x) = \frac{2+x}{x}$$

$$(f \circ g)(x) = f[g(x)]$$

$$= f\left[\frac{2+x}{x}\right]$$

$$= \frac{1}{\left(\frac{2+x}{x}\right)-2}$$

$$= \frac{1}{\left(\frac{2+x}{x}\right)-2} \cdot \left(\frac{x}{x}\right)$$

$$= \frac{x}{2+x-2x} = \frac{x}{2-x}$$

$$(g \circ f)(x) = g[f(x)]$$

$$= g\left[\frac{1}{x-2}\right]$$

$$= \frac{2+\left(\frac{1}{x-2}\right)}{\left(\frac{1}{x-2}\right)}$$

$$= \frac{2+\left(\frac{1}{x-2}\right)}{\left(\frac{1}{x-2}\right)} \cdot \frac{(x-2)}{(x-2)}$$

$$= \frac{2(x-2)+1}{1}$$

$$= 2x - 4 + 1 = 2x - 3$$

Since $\frac{x}{2-x} \neq 2x - 3$, an inverse does not exist.

Exercises 10.2

1.-19.

The graphs are in the answer section of the main text.

21.

$$y = 5^{-2} = \frac{1}{5^2} = \frac{1}{25}$$

23.

$$y = 27^{2/3} = \left(27^{1/3}\right)^2 = \left(\sqrt[3]{27}\right)^2 = (3)^2 = 9$$

25.

$$y = 25^{-3/2} = \frac{1}{25^{3/2}} = \frac{1}{\left(25^{1/2}\right)^3} = \frac{1}{5^3} = \frac{1}{125}$$

27.

$$16 = a^2$$

$$(16)^{1/2} = \left(a^2\right)^{1/2}$$

$$\sqrt{16} = a$$

$$4 = a$$

29.

$$3 = a^{1/3}$$

$$3^3 = \left(a^{1/3}\right)^3$$

$$27 = a$$

31.

$$36 = a^{2/3}$$

$$(36)^{3/2} = \left(a^{2/3}\right)^{3/2}$$

$$\left(36^{1/2}\right)^3 = a$$

$$6^3 = a$$

$$216 = a$$

33.

$$5 = a^{-1/3}$$

$$(5)^{-3} = \left(a^{-1/3}\right)^{-3}$$

$$\frac{1}{5^3} = a$$

$$\frac{1}{125} = a$$

35.

$$\frac{1}{16} = a^{-4/3}$$

$$\left(\frac{1}{16}\right)^{-3/4} = \left(a^{-4/3}\right)^{-3/4}$$

$$16^{3/4} = a$$

$$\left(16^{1/4}\right)^3 = a$$

$$\left(\sqrt[4]{16}\right)^3 = a$$

$$2^3 = a$$

$$8 = a$$

37.

$$8 = 2^2$$

$$2^3 = 2^x \qquad \text{Write 8 as a power of 2}$$

$$3 = x \qquad \text{Equate the exponents}$$

39.

$$\frac{1}{4} = 2^x$$

$$4^{-1} = 2^x$$

$$2^{-2} = 2^x \qquad \text{Write 4 as a power of 2}$$

$$-2 = x \qquad \text{Equate the exponents}$$

41.

$$\frac{4}{9} = \left(\frac{2}{3}\right)^x$$

$$\frac{2^2}{3^2} = \left(\frac{2}{3}\right)^x \qquad \text{Write 4 and 9 as powers of 2 and 3}$$

$$\left(\frac{2}{3}\right)^2 = \left(\frac{2}{3}\right)^x$$

$$2 = x \qquad \text{Equate the exponents}$$

43.

$$\frac{1}{4} = 8^x$$

$$4^{-1} = 8^x$$

$$2^{-2} = 8^x \qquad \text{Write 4 as a power of 2}$$

$$\left(8^{1/3}\right)^{-2} = 8^x \qquad \text{Write 2 as a power of } \frac{1}{3}$$

#43. continued

$$8^{-2/3} = 8^x$$

$$-\frac{2}{3} = x \qquad \text{Equate exponents}$$

45.

$$\frac{1}{25} = 5^{x-1}$$

$$25^{-1} = 5^{x-1}$$

$$\left(5^2\right)^{-1} = 5^{x-1} \qquad \text{Write 25 as a power of 2}$$

$$5^{-2} = 5^{x-1}$$

$$-2 = x - 1 \qquad \text{Equate exponents}$$

$$-1 = x$$

47.

$$8 = 4^{x/3}$$

$$(2)^3 = \left(2^2\right)^{x/3} \qquad \text{Express 8 and 4 as powers of 2}$$

$$(2)^3 = (2)^{2x/3}$$

$$3 = \frac{2x}{3} \qquad \text{Equate exponents}$$

$$\left(\frac{3}{2}\right)3 = \frac{2}{3}x\left(\frac{3}{2}\right)$$

$$\frac{9}{2} = x$$

49.

$$\frac{1}{9} = 4\left(6^{3x}\right)$$

$$\left(\frac{1}{4}\right)\frac{1}{9} = 4\left(6^{3x}\right)\left(\frac{1}{4}\right)$$

$$\frac{1}{36} = 6^{3x}$$

$$(36)^{-1} = 6^{3x}$$

$$\left(6^2\right)^{-1} = 6^{3x} \qquad \text{Write 36 as a power of 2}$$

$$6^{-2} = 6^{3x}$$

$$-2 = 3x \qquad \text{Equate exponents}$$

$$-\frac{2}{3} = x$$

51.

1940	$P = 6.4(1.02)^0 = 6,400,000$
1950	$P = 6.4(1.02)^{10} = 7,801,564$
1960	$P = 6.4(1.02)^{20} = 9,510,063$
1970	$P = 6.4(1.02)^{30} = 11,592,714$
1980	$P = 6.4(1.02)^{40} = 14,131,454$
2000	$P = 6.4(1.02)^{60} = 20,998,597$

53.

After 1 yr the account will have a balance of

$P = 5000(1.02)^{4t}$

$= 5000(1.02)^{4(1)}$

$= \$5412.16$

After 5 yrs...

$P = 5000(1.02)^{4(5)}$

$= 5000(1.02)^{20}$

$= \$7429.74$

We are asked to calculate the number of years it will take to double the account, i.e., $2 \times \$5000 = \$10,000$.

Thus, $\$10000 = 5000(1.02)^{4t}$

$\dfrac{10000}{5000} = (1.02)^{4t}$

$2 = (1.02)^{4t}$

$(1.02)^{35} = (1.02)^{4t}$ Write 2 as a power of 35

$35 = 4t$ Equate exponents

$8.75 = t$ or

$t = 9$ yrs $\big($round off to nearest year$\big)$

55.

Amount of money after 1 year is

$P = 20,100\big(1.005^{12t} - 1\big)$

$= 20,100\big(1.005^{12(1)} - 1\big)$

$= \$1239.72$

After 5 years...

$P = 20,100\big(1.005^{12(5)} - 1\big)$

$= 20,100\big(1.005^{60} - 1\big)$

$= \$7011.89$

57.

The sandwich temperature at 8:00 am, after 60 minutes of elapsed time is:

$T = 70 + (40 - 70)e^{-.015t}$

$= 70 + (40 - 70)e^{-.015(60)}$

$= 57.8°F$

At 9:00 am, after 120 minutes of elapsed time, the temperature is:

$T = 70 + (40 - 70)e^{-.015t}$

$= 70 + (40 - 70)e^{-.015(120)}$

$= 65°F$

Exercises 10.3

1.

$5^{-3} = \dfrac{1}{125}$

$\log_5\left(\dfrac{1}{125}\right) = -3$

3.

$10^4 = 10000$

$\log_{10}(10000) = 4$

5.

$6^1 = 6$

$\log_6(6) = 1$

7.

$y = 3^x$

$x = \log_3(y)$

9.

$y = 3^{x+2}$

$x + 2 = \log_3(y)$

11.

$P = e^{2t}$

$2t = \ln(P)$

13.

$M = (1.02)^{12t}$

$12t = \log_{1.02}(M)$

15.
$$Q = Q_0 a^{kt}$$
$$\frac{Q}{Q_0} = a^{kt}$$
$$kt = \log_a\left(\frac{Q}{Q_0}\right)$$

17.
$$\log_8 64 = 2$$
$$8^2 = 64$$

19.
$$\log_9 3 = \frac{1}{2}$$
$$9^{\frac{1}{2}} = 3$$

21.
$$\log_{10} 10 = 1$$
$$10^1 = 10$$

23.
$$y = \log_5 x$$
$$5^y = x$$

25.
$$y = \log_5(2x+1)$$
$$5^y = 2x+1$$

27.
$$\log_{27} 9 = x$$
$$27^x = 9$$

29.
$$4 = \log_3(1-3x)$$
$$3^4 = 1-3x$$

31.
$$\log_x 16 = 2$$
$$x^2 = 16$$

33.
$$\log(x+1) = 4$$
$$\log_{10}(x+1) = 4$$
$$10^4 = x+1$$

35.
$$y = \ln x^2$$
$$y = \ln_e x^2$$
$$e^y = x^2$$

37.-45.
The graphs are in the answer section of the main text.

Exercises 10.4

1.
$$\log_b x^2 y^5$$
$$= \log_b x^2 + \log_b y^5$$
$$= 2\log_b x + 5\log_b y$$

3.
$$\log_b \frac{(xy)^3}{\sqrt{z}}$$
$$= \log_b (xy)^3 - \log_b(\sqrt{z})$$
$$= \log_b (xy)^3 - \log_b(z)^{\frac{1}{2}}$$
$$= 3\log_b (xy) - \frac{1}{2}\log_b(z)$$
$$= 3\log_b x + 3\log_b y - \frac{1}{2}\log_b(z)$$

5.
$$\log_b \frac{x^5}{y^2 z^4}$$
$$= \log_b x^5 - \log_b y^2 z^4$$
$$= \log_b x^5 - \left(\log_b y^2 + \log_b z^4\right)$$
$$= 5\log_b x - \left(2\log_b y + 4\log_b z\right)$$
$$= 5\log_b x - 2\log_b y - 4\log_b z$$

7.

$$\log_b \sqrt{\frac{x}{y^3}}$$

$$= \log_b \left(\frac{x}{y^3}\right)^{\frac{1}{2}}$$

$$= \frac{1}{2} \log_b \left(\frac{x}{y^3}\right)$$

$$= \frac{1}{2} \left(\log_b x - \log_b y^3\right)$$

$$= \frac{1}{2} \left(\log_b x - 3 \log_b y\right)$$

$$= \frac{1}{2} \log_b x - \frac{3}{2} \log_b y$$

9.

$$\log_b \frac{\sqrt{x}}{\sqrt[3]{yz^2}}$$

$$= \log_b \frac{x^{\frac{1}{2}}}{\left(yz^2\right)^{\frac{1}{3}}}$$

$$= \log_b x^{\frac{1}{2}} - \log_b \left(yz^2\right)^{\frac{1}{3}}$$

$$= \frac{1}{2} \log_b x - \frac{1}{3} \log_b \left(yz^2\right)$$

$$= \frac{1}{2} \log_b x - \frac{1}{3} \left(\log_b y + \log_b z^2\right)$$

$$= \frac{1}{2} \log_b x - \frac{1}{3} \left(\log_b y + 2 \log_b z\right)$$

$$= \frac{1}{2} \log_b x - \frac{1}{3} \log_b y - \frac{2}{3} \log_b z$$

11.

$$\log_b x\left(y^{2t}\right)$$

$$= \log_b x + \log_b y^{2t}$$

$$= \log_b x + 2t \log_b y$$

13.

$$2 \log_b x + \frac{1}{2} \log_b y + \log_b z$$

$$= \log_b x^2 + \log_b y^{\frac{1}{2}} + \log_b z$$

$$= \log_b \left(x^2\right)\left(y^{\frac{1}{2}}\right)(z)$$

$$= \log_b x^2 z \sqrt{y}$$

15.

$$\frac{1}{4} \log_b x + \frac{3}{4} \log_b y - \log_b z$$

$$= \log_b x^{\frac{1}{4}} + \log_b 4^{\frac{3}{4}} - \log_b z$$

$$= \log_b \left(x^{\frac{1}{4}}\right)\left(y^{\frac{3}{4}}\right) - \log_b z$$

$$= \log_b \frac{x^{\frac{1}{4}} y^{\frac{3}{4}}}{z}$$

$$= \log_b \frac{\left(xy^3\right)^{\frac{1}{4}}}{z}$$

$$= \log_b \frac{\sqrt[4]{xy^3}}{z}$$

17.

$$\frac{1}{2} \log_b x - 2 \log_b y - \frac{2}{3} \log_b z$$

$$= \log_b x^{\frac{1}{2}} - \log_b y^2 - \log_b z^{\frac{2}{3}}$$

$$= \log_b x^{\frac{1}{2}} - \left(\log_b y^2 + \log_b z^{\frac{2}{3}}\right)$$

$$= \log_b x^{\frac{1}{2}} - \log_b y^2 z^{\frac{2}{3}}$$

$$= \log_b \frac{x^{\frac{1}{2}}}{y^2 z^{\frac{2}{3}}}$$

$$= \log_b \frac{\sqrt{x}}{y^2 \sqrt[3]{z^2}}$$

19.
$$\log_6 35$$

$$= \frac{\log 35}{\log 6} \approx 1.9842775$$

21.
$$\log_7 .0419$$

$$= \frac{\log .0419}{\log 7} \approx -1.63032679$$

23.
$$\log_{\frac{1}{3}} 4$$

$$= \frac{\log 4}{\log \frac{1}{3}} \approx -1.2618595$$

25.
$$6^{\log_6 2} = 2$$

27.
$$4^{3\log_4 5}$$
$$= 4^{\log_4 5^3}$$
$$= 5^3 = 125$$

29.
$$\sqrt{5^{\log_5 4}}$$
$$= \sqrt{4} = 2$$

31.
$$\log_9 9^3$$
$$= 3\log_9 9$$
$$= 3(1) = 3$$

33.
$$\log_7 \sqrt[3]{7}$$
$$= \log_7 7^{\frac{1}{3}}$$
$$= \frac{1}{3}\log_7 7$$
$$= \frac{1}{3}(1) = \frac{1}{3}$$

35.
$$\log_2 8$$
$$= \log_2 2^3 \qquad \text{Express 8 as a power of 2}$$
$$= 3\log_2 2$$
$$= 3(1) = 3$$

37.
$$\log_3 \frac{1}{9}$$
$$= \log_3 9^{-1}$$
$$= \log_3 \left(3^2\right)^{-1} \qquad \text{Express 9 as a power of 2}$$
$$= \log_3 3^{-2}$$
$$= -2\log_3 3$$
$$= -2(1) = -2$$

39.
$$\log_4 8$$
$$= \log_4 2^3 \qquad \text{Express 8 as a power of 2}$$
$$= \log_4 \left((4)^{\frac{1}{2}}\right)^3 \qquad \text{Express 2 as a power of 4}$$
$$= \log_4 4^{\frac{3}{2}}$$
$$= \frac{3}{2}\log_4 4$$
$$= \frac{3}{2}(1) = \frac{3}{2}$$

41.
$$\log_{16} \frac{1}{8}$$
$$= \log_{16} 8^{-1}$$
$$= \log_{16} \left(2^3\right)^{-1} \qquad \text{Express 8 as a power of 2}$$
$$= \log_{16} \left(\left(16^{\frac{1}{4}}\right)^3\right)^{-1} \qquad \text{Express 2 as a power of 16}$$
$$= \log_{16} 16^{-\frac{3}{4}}$$
$$= -\frac{3}{4}\log_{16} 16$$
$$= -\frac{3}{4}(1) = -\frac{3}{4}$$

43.
$$\log_b 15$$
$$= \log_b (3)(5)$$
$$= \log_b (3) + \log_b (5)$$
$$= 1.4 + 2.0 = 3.4$$

45.
$$\log_b 45$$
$$= \log_b (3)(5)(3)$$
$$= \log_b (3) + \log_b (5) + \log_b (3)$$
$$= 1.4 + 2.0 + 1.4 = 4.8$$

47.
$$\log_b 49$$
$$= \log_b (7)^2$$
$$= 2\log_b 7$$
$$= 2(2.4) = 4.8$$

49.

$\log_b \sqrt{3}$

$= \log_b 3^{1/2}$

$= \dfrac{1}{2} \log_b 3$

$= \dfrac{1}{2}(1.4) = 0.7$

51.

$\log_b \dfrac{7}{5}$

$= \log_b 7 - \log_b 5$

$= 2.4 - 2.0 = 0.4$

53.

$\log_b \dfrac{9}{5}$

$= \log_b 9 - \log_b 5$

$= \log_b 3^2 - \log_b 5$

$= 2(\log_b 3) - \log_b 5$

$= 2(1.4) - 2.0$

$= 2.8 - 2.0 = 0.8$

55.

$\log_b \sqrt[3]{\dfrac{15}{7}}$

$= \log_b \left(\dfrac{15}{7}\right)^{1/3}$

$= \dfrac{1}{3} \log_b \left(\dfrac{15}{7}\right)$

$= \dfrac{1}{3}(\log_b 15 - \log_b 7)$

$= \dfrac{1}{3}(\log_b (3)(5) - \log_b 7)$

$= \dfrac{1}{3}(\log_b (3) + \log_b (5) - \log_b 7)$

$= \dfrac{1}{3}(1.4 + 2.0 - 2.4)$

$= \dfrac{1}{3}(1.0) = \dfrac{1}{3}$

57.

$\log_b \dfrac{5\sqrt[4]{75}}{63}$

$= \log_b \dfrac{5(75)^{1/4}}{63}$

$= \log_b 5(75)^{1/4} - \log_b 63$

$= \log_b 5 + \log_b 75^{1/4} - \log_b 63$

$= \log_b 5 + \dfrac{1}{4} \log_b 75 - \log_b 63$

$= \log_b 5 + \dfrac{1}{4} \log_b (3)\left(5^2\right) - \log_b \left(3^2\right)(7)$

$= \log_b 5 + \dfrac{1}{4}\left(\log_b 3 + 2\log_b 5\right) - \left(2\log_b 3 + \log_b 7\right)$

$= 2.0 + \dfrac{1}{4}\left(1.4 + 2(2.0)\right) - \left(2(1.4) + 2.4\right)$

$= 2.0 + \dfrac{1}{4}(5.4) - 5.2$

$= \dfrac{1}{4}(5.4) - 3.2 = -1.85$

59.

Alaska Earthquake:

$M = \dfrac{\log E - 11.4}{1.5}$

$= \dfrac{\log(1.41)(10)^{24} - 11.4}{1.5}$

$= \dfrac{\log(1.41) + 24\log 10 - 11.4}{1.5} = 8.5$

San Francisco Earthquake:

$M = \dfrac{\log E - 11.4}{1.5}$

$= \dfrac{\log(5.96)(10)^{23} - 11.4}{1.5}$

$= \dfrac{\log 5.96 + 23\log 10 - 11.4}{1.5} = 8.25$

61.

The proof is in the answer section of the main text.

63.

$2\ln|x + 3| + \dfrac{1}{2}\ln|2x - 1| - \ln\left|x^2 - x - 3\right|$

$= \ln(x + 3)^2 + \ln(2x - 1)^{1/2} - \ln\left|x^2 - x - 3\right|$

#63. continued

Note: The quantities $(x + 3)^2$ and $(2x - 1)^{\frac{1}{2}}$ are always positive by definition. Therefore the absolute value can be removed.

$$= \ln \frac{(x + 3)^2 \sqrt{2x - 1}}{\left| x^2 - x - 3 \right|}$$

Exercises 10.5

1.

$$5 = 11^x$$

$\log 5 = \log 11^x$ Take the log of both sides

$\log 5 = x \log 11$

$$\frac{\log 5}{\log 11} = x$$

$$x \approx 0.6711877$$

3.

$$25 = 4^x$$

$\log 25 = \log 4^x$ Take the log of both sides

$\log 25 = x \log 4$

$$\frac{\log 25}{\log 4} = x$$

$$x \approx 2.3219281$$

5.

$$6^{x+3} = 2.5$$

$\log 6^{x+3} = \log 2.5$ Take the log of both sides

$(x + 3) \log 6 = \log 2.5$

$$x + 3 = \frac{\log 2.5}{\log 6}$$

$$x = \frac{\log 2.5}{\log 6} - 3$$

$$x \approx -2.4886084$$

7.

$$2^{5x-2} = 3.5^x$$

$\log 2^{5x-2} = \log 3.5^x$ Take the log of both sides

$(5x - 2) \log 2 = x \log 3.5$

$5x \log 2 - 2 \log 2 = x \log 3.5$

$-2 \log 2 = x \log 3.5 - 5x \log 2$

$-2 \log 2 = x(\log 3.5 - 5 \log 2)$

#7. continued

$$\frac{-2 \log 2}{\log 3.5 - 5 \log 2} = x$$

$$x = \frac{-\log 2^2}{\log 3.5 - \log 2^5}$$

$$= \frac{-\log 4}{\log \frac{3.5}{32}}$$

$$\approx 0.6264398$$

9.

$$2.54^{3-x} = 7^{x+2}$$

$\log 2.54^{3-x} = \log 7^{x+2}$

            Take the log of both sides

$3 - x(\log 2.54) = (x + 2)(\log 7)$

$3(\log 2.54) - x(\log 2.54) = x \log 7 + 2 \log 7$

$3 \log 2.54 - 2 \log 7 = x \log 2.54 + x \log 7$

$3 \log 2.54 - \log 7^2 = x(\log 2.54 + \log 7)$

$$\frac{3 \log 2.54 - \log 49}{\log 7 + \log 2.54} = x$$

$$x \approx -0.3805767$$

11.

$$3 = 4(5.2)^t$$

$$\frac{3}{4} = (5.2)^t$$

$\log \frac{3}{4} = \log(5.2)^t$ Take the log of both sides

$$\log \frac{3}{4} = t \log(5.2)$$

$$\frac{\log \frac{3}{4}}{\log 5.2} = t$$

$$t \approx -0.1744946$$

13.

$$200 = 125(2.45)^{2t}$$

$$\frac{200}{125} = 2.45^{2t}$$

$\log \frac{200}{125} = \log 2.45^{2t}$ Take the log of both sides

$$\log \frac{8}{5} = 2t \log 2.45$$

#13. continued

$$\frac{\log \frac{8}{5}}{2 \log 2.45} = t$$

$$t \approx 0.2622530$$

15.
$$\log_x 9 = -2$$

$\quad x^{-2} = 9 \qquad$ Change from log to exponential form

$$\frac{1}{x^2} = 9$$

$$\frac{1}{9} = x^2$$

$$\sqrt{\frac{1}{9}} = x$$

$$x = \frac{1}{3}$$

17.
$$\log_x 4 = \frac{2}{5}$$

$\left(x^{2/5}\right)^5 = (4)^5 \qquad$ Change from log to exponential form

$$x^2 = 4^5$$

$$\sqrt{x^2} = \sqrt{1024}$$

$$x = 32$$

19.
$$\log_3\left(\frac{1}{27}\right) = x$$

$$3^x = \frac{1}{27}$$

$\qquad$ Change from log to exponential form

$$3^x = 3^{-3}$$

$$x = -3$$

21.
$$\log_8 16 = x$$

$$8^x = 16$$

$\left(2^3\right)^x = 2^4 \qquad$ Rewrite with the same base

$$2^{3x} = 2^4$$

$3x = 4 \qquad$ Equate powers

$$x = \frac{4}{3}$$

23.
$$\log_6 x = -3$$

$\quad 6^{-3} = x \qquad$ Change from log to exponential form

$$\frac{1}{6^3} = x$$

$$x = \frac{1}{216}$$

25.
$$\log x = 5$$

$\log_{10} x = 5 \qquad$ Change from log to exponential form

$$10^5 = x$$

$$x = 100,000$$

27.
$$\log_3 x + \log_3(x+6) = 3$$

$$\log_3 x(x+6) = 3$$

$3^3 = x(x+6) \qquad$ Change from log to exponential form

$$27 = x^2 + 6x$$

$$x^2 + 6x - 27 = 0$$

$$(x+9)(x-3) = 0$$

$$x = -9 \quad \text{or} \quad x = 3$$

Checking the solutions, we have

$x = -9$

$$\log_3(-9) + \log_3(-9+6) = 3$$

$\qquad \uparrow \qquad\qquad \uparrow$

Both these terms are undefined
since the log of a negative number
doesn't exist.

$x = 3$

$$\log_3(3) + \log_3(9) = 3$$

$$1 + \log_3(3)^2$$

$$1 + 2(1) = 3$$

Thus, $x = 3$ is the only valid solution.

29.

$$\log_4(x+12) - \log_4(x-3) = 2$$

$$\log_4 \frac{(x+12)}{(x-3)} = 2$$

$$4^2 = \frac{x+12}{x-3}$$

Change from log to
exponential form

$$16 = \frac{x+12}{x-3}$$

$$16(x-3) = x+12$$

$$16x - 48 = x + 12$$

$$16x - x = 12 + 48$$

$$15x = 60$$

$$x = 4$$

31.

$$2\log_4 x = 3$$

$$\log_4 x = \frac{3}{2}$$

$$4^{3/2} = x$$

$$x = 8$$

33.

$$\log_2 5 + \log_2(x-3) = \log_2(x+5)$$

$$\log_2 5(x-3) = \log_2(x+5)$$

$$5(x-3) = x+5 \qquad \text{Drop the logs}$$

$$5x - 15 = x + 5$$

$$5x - x = 5 + 15$$

$$4x = 20$$

$$x = 5$$

35.

$$\log(x+1) = \log(x+9) - \log x$$

$$\log(x+1) = \log\left(\frac{x+9}{x}\right)$$

$$x + 1 = \frac{x+9}{x} \qquad \text{Drop the logs}$$

$$x(x+1) = x + 9$$

$$x^2 + x = x + 9$$

$$x^2 + x - x - 9 = 0$$

$$x^2 - 9 = 0$$

$$(x+3)(x-3) = 0$$

#35. continued

$$x = 3 \quad \text{or} \quad x = -3$$

Check the solutions:

$$x = 3$$

$$\log(3+1) = \log(3+9) - \log 3$$

$$\log 4 = \log 12 - \log 3$$

$$\log 4 = \log \frac{12}{3}$$

$$4 = 4$$

$$x = -3$$

$$\log(-3+1) = \log(-3+9) - \log(-3)$$

$$\uparrow \qquad\qquad\qquad \uparrow$$

Both these terms are undefined since the
log of a negative number doesn't exist.
Thus, $x = 3$ is the only valid solution.

37.

$$\frac{1}{2}\log_4(2x+1) = \log_4 3$$

$$\log_4(2x+1)^{1/2} = \log_4 3$$

$$\sqrt{2x+1} = 3 \qquad \text{Drop the logs}$$

$$\left(\sqrt{2x+1}\right)^2 = (3)^2$$

$$2x + 1 = 9$$

$$2x = 8$$

$$x = 4$$

39.

Using the equation,

$$M = D\left(1 + \frac{r}{n}\right)^{nt}, \quad \text{where } n = 2, \ r = 0.10, \ D = \$10,000$$

and $M = \$15,000$, we can calculate the time t.
Substituting, we have,

$$15,000 = 10,000\left(1 + \frac{0.10}{2}\right)^{2t}$$

$$\frac{15000}{10000} = \left(1 + \frac{0.10}{2}\right)^{2t}$$

$$= (1 + 0.05)^{2t}$$

$$\log\frac{3}{2} = \log(1.05)^{2t} \qquad \text{Take log of both sides}$$

$$\log\frac{3}{2} = 2t\log(1.05)$$

#39. continued

$$\frac{\log\frac{3}{2}}{\log 1.05} = 2t$$

$$\frac{\log 1.5}{2\log 1.05} = t$$

$$t = 4.15 \quad \text{or} \quad 4 \text{ yrs 2 mos}$$

If M = \$20,000, we have

$$20,000 = 10,000\left(1 + \frac{0.10}{2}\right)^{2t}$$

$$\frac{20,000}{10,000} = (1 + 0.05)^{2t}$$

$$2 = (1.05)^{2t}$$

$$\log 2 = \log(1.05)^{2t} \qquad \text{Take the log of both sides}$$

$$\log 2 = 2t\log(1.05)$$

$$\frac{\log 2}{\log 1.05} = 2t$$

$$\frac{\log 2}{2\log 1.05} = t$$

$$t = 7.1 \text{ yrs or 7 yrs 6 mos}$$

41.

The population of Texas can be predicted with the equation

$$P = 6.4(1.02)^t$$

If $P = 25 \times 10^6$, we can solve for t or time:

$$25 = 6.4(1.02)^t$$

$$\frac{25}{6.4} = (1.02)^t$$

$$\log\left(\frac{25}{6.4}\right) = t\log(1.02) \qquad \text{Take the log of both sides}$$

$$\frac{\log\left(\frac{25}{6.4}\right)}{\log(1.02)} = t$$

$$t = 68.8 \text{ yrs}$$

The year that this happens will be:

$$68.8 + 1940 = 2008.8 \quad \text{or} \quad 2009$$

Review Exercises

The graphs are in the answer section of the main text.

1.
$$f(x) = 3x - 4$$

$$y = 3x - 4 \xrightarrow{\text{Inverse}} x = 3y - 4 \qquad \text{Interchange x and y}$$

$$x + 4 = 3y \qquad \text{Solve for y}$$

$$\left(\frac{1}{3}\right)x + 4 = 3y\left(\frac{1}{3}\right)$$

$$\frac{1}{3}x + \frac{4}{3} = y \quad \text{or} \quad f^{-1}(x) = \frac{1}{3}x + \frac{4}{3}$$

3.
$$f(x) = \frac{2x + 4}{3}$$

$$y = \frac{2x + 4}{3} \xrightarrow{\text{Inverse}} x = \frac{2y + 4}{3} \qquad \text{Interchange x and y}$$

$$3x = 2y + 4 \quad \text{Solve for y}$$

$$3x - 4 = 2y$$

$$\frac{1}{2}(3x - 4) = 2y\left(\frac{1}{2}\right)$$

$$\frac{3}{2}x - 2 = y \quad \text{or} \quad f^{-1}(x) = \frac{3x}{2} - 2$$

5.
$$f(x) = 3 - \frac{1}{2}x^2$$

$$y = 3 - \frac{1}{2}x^2 \xrightarrow{\text{Inverse}} x = 3 - \frac{1}{2}y^2$$

Interchange x and y

The inverse is not a function since two y values exist for one x.

7.
$$f(x) = \sqrt{3 - x} \qquad \text{Note: } f(x) \geq 0$$

$$y = \sqrt{3 - x} \xrightarrow{\text{Inverse}} x = \sqrt{3 - y}$$

Interchange x and y

$$(x)^2 = \left(\sqrt{3 - y}\right)^2 \quad \text{Solve for y}$$

$$x^2 = 3 - y$$

$$-x^2 = -3 + y$$

$$3 - x^2 = y \quad \text{where } x \geq 0$$

$$\text{or} \quad f^{-1}(x) = 3 - x^2, \quad x \geq 0$$

9.

$$f(x) = \frac{2}{x+1}$$

Let $f(a) = f(b)$

$$\frac{2}{a+1} = \frac{2}{b+1}$$

$$2(b+1) = 2(a+1)$$

$$\left(\frac{1}{2}\right)2(b+1) = 2(a+1)\left(\frac{1}{2}\right)$$

$$b+1 = a+1$$

$$b = a \qquad \text{f is one to one.}$$

Therefore

$$y = \frac{2}{x+1} \xrightarrow{\text{Inverse}} x = \frac{2}{y+1} \qquad \text{Substitute x and y}$$

$$(y+1)x = 2$$

$$\frac{1}{x}(y+1)x = 2\left(\frac{1}{x}\right)$$

$$y+1 = \frac{2}{x}$$

$$y = \frac{2}{x} - 1$$

$$y = \frac{2}{x} - \frac{x}{x}$$

$$y = \frac{2-x}{x} \quad \text{or} \quad f^{-1}(x) = \frac{2-x}{x}$$

11.

$$f(x) = |x+4|$$

Let $f(a) = f(b)$

$$|a+4| = |b+4|$$

Two possibilities:

$$a+4 = -(b+4) \quad \text{or} \quad a+4 = b+4$$

$$a+4 = -b-4 \qquad\qquad a = b$$

$$a = -b-8$$

f is <u>not</u> one to one since one possibility is a = -b - 8.

13.

$$f(x) = \sqrt{3x+1}$$

Let $f(a) = f(b)$

$$\sqrt{3a+1} = \sqrt{3b+1}$$

$$\left(\sqrt{3a+1}\right)^2 = \left(\sqrt{3b+1}\right)^2$$

$$3a+1 = 3b+1$$

$$3a = 3b$$

$$a = b \qquad \text{f is one to one.}$$

#13. continued

Therefore,

$$y = \sqrt{3x+1} \xrightarrow{\text{Inverse}} x = \sqrt{3y+1} \quad \text{Substitute x and y}$$

$$x^2 = \left(\sqrt{3y+1}\right)^2 \qquad\qquad \text{Solve for y}$$

$$x^2 = 3y+1 \qquad\qquad x \geq 0$$

$$x^2 - 1 = 3y \qquad\qquad x \geq 0$$

$$\frac{x^2-1}{3} = y \qquad\qquad x \geq 0$$

$$\text{or} \quad f^{-1}(x) = \frac{1}{3}x^2 - \frac{1}{3}, \quad \text{for } x \geq 0$$

15.

$$f(x) = 4x - 7$$

$$g(x) = x + 5$$

a. $(f \circ g)(-2) = 4(x+5) - 7$

$$= 4(-2+5) - 7$$

$$= 4(3) - 7$$

$$= 5$$

b. $(f \circ g)(x) = f[g(x)]$

$$= f[x+5]$$

$$= 4(x+5) - 7$$

$$= 4x + 20 - 7$$

$$= 4x + 13$$

c. $(g \circ f)(4) = (4x-7) + 5$

$$= \left(4(4) - 7\right) + 5$$

$$= 16 - 7 + 5$$

$$= 14$$

d. $(g \circ f)(x) = g[f(x)]$

$$= g[4x - 7]$$

$$= (4x-7) + 5$$

$$= 4x - 2$$

17.

$$f(x) = \frac{1}{4}x + 3$$

$$g(x) = 4x - 12$$

$$(f \circ g)(x) = f[g(x)]$$

$$= f[4x - 12]$$

$$= \frac{1}{4}(4x - 12) + 3$$

$$= x - 3 + 3 = x$$

#17. continued

$(g \circ f)(x) = g[f(x)]$

$\quad = g\left[\dfrac{1}{4}x + 3\right]$

$\quad = 4\left(\dfrac{1}{4}x + 3\right) - 12$

$\quad = x + 12 - 12 = x$

Since $x = x$ there are inverses.

19.

$f(x) = \dfrac{x}{x + 4}$

$g(x) = \dfrac{4x}{1 - x}$

$(f \circ g)(x) = f[g(x)]$

$\quad = f\left[\dfrac{4x}{1 - x}\right]$

$\quad = \dfrac{\left(\dfrac{4x}{1 - x}\right)}{\left(\dfrac{4x}{1 - x}\right) + 4}$

$\quad = \dfrac{\left(\dfrac{4x}{1 - x}\right)}{\left(\dfrac{4x}{1 - x}\right) + 4} \cdot \dfrac{(1 - x)}{(1 - x)}$

$\quad = \dfrac{4x}{4x + 4(1 - x)}$

$\quad = \dfrac{4x}{4x + 4 - 4x}$

$\quad = \dfrac{4x}{4} = x$

$(g \circ f)(x) = g[f(x)]$

$\quad = g\left[\dfrac{x}{x + 4}\right]$

$\quad = \dfrac{4\left(\dfrac{x}{x + 4}\right)}{1 - \left(\dfrac{x}{x + 4}\right)}$

$\quad = \dfrac{4\left(\dfrac{x}{x + 4}\right)}{1 - \left(\dfrac{x}{x + 4}\right)} \cdot \dfrac{(x + 4)}{(x + 4)}$

$\quad = \dfrac{4x}{(x + 4) - x}$

$\quad = \dfrac{4x}{4} = x$

Since $x = x$ there are inverses.

21.-25.

The graphs are in the answer section of the main text.

27.

$y = 8^{-\frac{2}{3}} = \dfrac{1}{8^{\frac{2}{3}}} = \dfrac{1}{\left(8^{\frac{1}{3}}\right)^2} = \dfrac{1}{\left(\sqrt[3]{8}\right)^2} = \dfrac{1}{(2)^2} = \dfrac{1}{4}$

29.

$64 = a^{\frac{3}{4}}$

$(64)^{\frac{4}{3}} = \left(a^{\frac{3}{4}}\right)^{\frac{4}{3}}$

$\left(64^{\frac{1}{3}}\right)^4 = a$

$\left(\sqrt[3]{64}\right)^4 = a$

$(4)^4 = a$

$256 = a$

31.

$\dfrac{1}{9} = 3^x$

$(9)^{-1} = 3^x$

$\left(3^2\right)^{-1} = 3^x \qquad$ Write 9 as a power of 3

$3^{-2} = 3^x$

$-2 = x \qquad$ Equate the exponents

33.

$\dfrac{3}{4} = \left(\dfrac{16}{4}\right)^{3x+1}$

$\left(\dfrac{4}{3}\right)^{-1} = \left(\dfrac{16}{9}\right)^{3x+1}$

$\left(\dfrac{4}{3}\right)^{-1} = \left(\dfrac{4^2}{3^2}\right)^{3x+1} \qquad$ Write 16 and 9 as powers

of 4 and 3

$\left(\dfrac{4}{3}\right)^{-1} = \left(\dfrac{4}{3}\right)^{2(3x+1)}$

$-1 = 2(3x + 1) \qquad$ Equate exponents

$-1 = 6x + 2$

$-1 - 2 = 6x$

$-3 = 6x$

$-\dfrac{1}{2} = x$

35.

The number of aphids is represented by the equation:

$$A = 10(1.12)^t$$

On April 15, 14 days later, there are:

$$A = 10(1.12)^{14}$$
$$= 48.87 \approx 49 \text{ aphids}$$

On May 1, 30 days later, there are:

$$A = 10(1.12)^{30}$$
$$= 299.60 \approx 300 \text{ aphids}$$

37.

$$4^{1/2} = 2$$
$$\log_4 2 = \frac{1}{2}$$

39.

$$y = 2^{x-5}$$
$$\log_2 y = x - 5$$

41.

$$B = e^{kt}$$
$$\log_e B = kt$$
$$\text{or}$$
$$\ln B = kt$$

43.

$$\log_3 3 = 1$$
$$3^1 = 3$$

45.

$$y = \log_3(x + 2)$$
$$3^y = x + 2$$

47.

$$2t = \log_7 100$$
$$7^{2t} = 100$$

49.-51.

The graphs are in the answer section of the main text.

53.

The number of days, t, that it takes an instructor to learn N names is given by the equation:

$$t = 2 - 12 \ln\left(1 - \frac{N}{40}\right) \qquad N < 40$$

For 25 names, we have,

$$t = 2 - 12 \ln\left(1 - \frac{25}{40}\right)$$
$$= 2 - 12 \ln\left(\frac{15}{40}\right)$$
$$= 2 - 12 \ln\left(\frac{3}{8}\right)$$
$$= 13.77 \approx 14 \text{ days}$$

For 35 names, we have,

$$t = 2 - 12 \ln\left(1 - \frac{35}{40}\right)$$
$$= 2 - 12 \ln\left(\frac{5}{40}\right)$$
$$= 2 - 12 \ln\left(\frac{1}{8}\right)$$
$$= 26.95 \approx 27 \text{ days}$$

55.

$$\log_b \sqrt{\frac{x^3}{y^5}} = \log_b\left(\frac{x^3}{y^5}\right)^{1/2}$$
$$= \log_b\left(\frac{x^{3/2}}{y^{5/2}}\right)$$
$$= \log_b x^{3/2} - \log_b y^{5/2}$$
$$= \frac{3}{2}\log_b x - \frac{5}{2}\log_b y$$

57.

$$\log_b D(e^{rt}) = \log_b D + \log_b e^{rt} = \log_b D + rt \log_b e$$

59.

$$\frac{3}{2}\log_b x - \frac{1}{2}\log_b y - \log_b z$$
$$= \log_b x^{3/2} - \log_b y^{1/2} - \log_b z$$
$$= \log_b x^{3/2} - \left(\log_b y^{1/2} + \log_b z\right)$$

#59. continued

$$= \log_b \frac{x^{3/2}}{zy^{1/2}}$$

$$= \log_b = \frac{\sqrt{x^3}}{z\sqrt{y}}$$

61.

$$\log_5 24 = \frac{\log 24}{\log 5} = 1.9746359$$

63.

$$\log_5 \sqrt[3]{5} = \log_5 5^{1/3} = \frac{1}{3}\log_5 5 = \frac{1}{3}(1) = \frac{1}{3}$$

65.

$$\log_2 \frac{1}{8} = \log_2 8^{-1}$$

$$= \log_2 \left(2^3\right)^{-1}$$

$$= \log_2 2^{-3} = -3\log_2 2 = -3(1) = -3$$

67.

$$\log_b 63 = \log_b (3)(3)(7)$$

$$= \log_b(3) + \log_b(3) + \log_b(7)$$

$$= 1.4 + 1.4 + 2.4 = 5.2$$

69.

$$\log_b \sqrt{7} = \log_b 7^{1/2} = \frac{1}{2}\log_b 7 = \frac{1}{2}(2.4) = 1.2$$

71.

The formula for pH as a function of ion hydrogen concentration is given by the equation:

$$pH = -\log\left[H^+\right]$$

To find the ion hydrogen concentration we must solve for H^+. Thus,

$$-pH = \log_{10}\left[H^+\right]$$

$$10^{-pH} = H^+$$

For vinegar, pH = 2.5

$$10^{-2.5} \approx 0.003162$$

$$\approx 3.2\times10^{-3}$$

#71. continued

For human spinal fluid, pH = 7.4

$$10^{-7.4} \approx 3.981072\times10^{-8}$$

$$\approx 4.0\times10^{-8}$$

For egg whites, pH = 8.0

$$10^{-8} = 1\times10^{-8}$$

73.

$$7.5 = 3^{x-4}$$

$$\log 7.5 = \log 3^{x-4} \qquad \text{Take the log of both sides}$$

$$\log 7.5 = (x-4)\log 3$$

$$\frac{\log 7.5}{\log 3} = x - 4$$

$$\frac{\log 7.5}{\log 3} + 4 = x$$

$$x \approx 5.8340438$$

75.

$$250 = 100(2)^{0.03t}$$

$$\frac{250}{100} = 2^{0.03t}$$

$$\log\frac{250}{100} = \log 2^{0.03t} \qquad \text{Take the log of both sides}$$

$$\log 2.5 = 0.03t\log 2$$

$$\frac{\log 2.5}{\log 2.0} = 0.03t$$

$$\frac{\log 2.5}{0.03\log 2} = t$$

$$t \approx 44.0642698$$

77.

$$\log_9 \frac{1}{81} = x$$

$$9^x = \frac{1}{81}$$

$$9^x = 81^{-1}$$

$$9^x = \left(9^2\right)^{-1}$$

$$9^x = 9^{-2}$$

$$x = -2 \qquad \text{Equate exponents}$$

79.

$\log_7 x = -2$

$7^{-2} = x$

$\dfrac{1}{7^2} = x$

$x = \dfrac{1}{49}$

81.

$\log_3(4x+1) - \log_3(x-4) = 1$

$\log_3\left(\dfrac{4x+1}{x-4}\right) = 1$

$3^1 = \dfrac{4x+1}{x-4}$

$3(x-4) = 4x+1$

$3x - 12 = 4x + 1$

$-12 - 1 = 4x - 3x$

$-13 = x$

Since $(4x+1)$ and $(x-4)$ are negative when $x = -13$, there is no solution since logs of negative numbers are undefined.

83.

$\log_5 3 + \log_5(x+7) = \log_5(13-5x)$

$\log_5 3(x+7) = \log_5(13-5x)$ Drop the logs

$3(x+7) = 13 - 5x$

$3x + 21 = 13 - 5x$

$3x + 5x = 13 - 21$

$8x = -8$

$x = -1$

85.

$2\log_6(5x+1) = \log_6(13-3x)$

$\log_6(5x+1)^2 = \log_6(13-3x)$

Drop the logs

$(5x+1)^2 = 13 - 3x$

$25x^2 + 10x + 1 = 13 - 3x$

$25x^2 + 10x + 3x - 1 - 13 = 0$

$25x^2 + 13x - 12 = 0$

#85. continued

Using the quadratic formula, we have, $a = 25$, $b = 13$, $c = -12$,

$\dfrac{-b \pm \sqrt{b^2 - 4ac}}{2a} = \dfrac{-13 \pm \sqrt{13^2 - 4(25)(-12)}}{2(25)}$

$= \dfrac{-13 \pm 37}{50}$

Therefore,

$x = \dfrac{24}{50} = \dfrac{12}{25}$ or $x = \dfrac{-50}{50} = -1$

The only solutions is $x = \dfrac{12}{25}$, since $x = -1$ produces a negative log, i.e., $(5x+1) = -4$, which is undefined.

Chapter 10 Test Solutions

1.

$f(x) = 3x - 6$

$y = 3x - 6 \xrightarrow{\text{Inverse}} x = 3y - 6$ Interchange x and y

$x + 6 = 3y$ Solve for y

$\dfrac{x+6}{3} = y$

$y = \dfrac{1}{3}x + 2$

or

$f^{-1}(x) = \dfrac{1}{3}x + 2$

The graph is in the answer section of the main text.

3.

The graphs are in the answer section of the main text.

5.

$A = A_0 r^{rt}$

$\dfrac{A}{A_0} = r^{rt}$

$\log_r \dfrac{A}{A_0} = \log_r r^{rt}$ Take the log, base r,

             of both sides

$\log_r \dfrac{A}{A_0} = \log_r \left(r^r\right)^t$

$\log_r \dfrac{A}{A_0} = rt \log_r r$

$\log_r \dfrac{A}{A_0} = rt(1)$

$\log_r \dfrac{A}{A_0} = rt$

7.

$$2^{\log_2 9} = 9$$

9.

$$\log_8 \frac{1}{4} = \log_8 (4)^{-1}$$

$$= \log_8 \left(\left(2^3\right)^{2/3} \right)^{-1}$$

$$= \log_8 8^{-2/3}$$

$$= -\frac{2}{3} \log_8 8$$

$$= -\frac{2}{3}(1) = -\frac{2}{3}$$

11.

$$9 = a^{-1/2}$$

$$9^2 = \left(a^{-1/2} \right)^2$$

$$9^2 = a^{-1}$$

$$9^2 = \frac{1}{a}$$

$$\frac{1}{9^2} = a$$

$$a = \frac{1}{81}$$

13.

$$7^{3-x} = 1.5$$

$$\log 7^{3-x} = \log 1.5 \qquad \text{Take the log of both sides}$$

$$(3-x)\log 7 = \log 1.5$$

$$3 - x = \frac{\log 1.5}{\log 7}$$

$$-x = \frac{\log 1.5}{\log 7} - 3$$

$$x = 3 - \frac{\log 1.5}{\log 7}$$

$$\approx 2.7916322$$

15.

$$\log 7 - \log(2x - 3) = \log(x - 4)$$

$$\log \frac{7}{(2x - 3)} = \log(x - 4)$$

$$\frac{7}{2x - 3} = x - 4 \qquad \text{Drop the logs}$$

$$7 = (x - 4)(2x - 3)$$

$$7 = 2x^2 - 11x + 12$$

$$2x^2 - 11x + 5 = 0$$

Using the quadratic formula, we have, a = 2, b = -11, c = 5,

$$\frac{-b \pm \sqrt{b^2 - 4ac}}{2a} = \frac{11 \pm \sqrt{(-11)^2 - 4(2)(5)}}{2(2)}$$

$$= \frac{11 \pm \sqrt{121 - 40}}{4}$$

$$= \frac{11 \pm \sqrt{81}}{4}$$

$$= \frac{11 \pm 9}{4}$$

Therefore,

$$x = \frac{20}{4} = 5 \quad \text{or} \quad x = \frac{2}{4} = \frac{1}{2}$$

The only solution is x = 5, since $x = \frac{1}{2}$ produces negative logs, i.e., $(2x - 3) = -2$ and $(x - 4) = -\frac{7}{2}$, which are both undefined.

Test Your Memory

1.

$$\left(\frac{8x^{-4}y^5}{12x^8y} \right)^{-3} = \left(\frac{2 \cdot 4y^{5-1}}{3 \cdot 4x^{8+4}} \right)^{-3} = \left(\frac{2y^4}{3x^{12}} \right)^{-3}$$

$$= \frac{2^{-3}y^{-12}}{3^{-3}x^{-36}}$$

$$= \frac{27x^{36}}{8y^{12}}$$

3.

$$4x\sqrt[3]{48} - \sqrt[3]{162x^3}$$

$$= 4x\sqrt[3]{8 \cdot 6} - \sqrt[3]{27x^3 \cdot 6}$$

$$= 8x\sqrt[3]{6} - 3x\sqrt[3]{6}$$

$$= 5x\sqrt[3]{6}$$

5.

$$\frac{2ax + 2bx - 3a - 3b}{2x^2 - 11x + 12} \cdot \frac{x^2 - 16}{a^3 + b^3}$$

$$= \frac{(2x - 3)(a + b)}{(2x - 3)(x - 4)} \cdot \frac{(x + 4)(x - 4)}{(a + b)(a^2 - ab + b^2)}$$

$$= \frac{x + 4}{a^2 - ab + b^2}$$

7.

$$f(-2) = 2(-2) - 5 = -4 - 5 = -9$$

9.

$$y = 2x - 5 \xrightarrow{\text{Inverse}} x = 2y - 5$$

$$x + 5 = 2y$$

$$\frac{1}{2}x + \frac{5}{2} = y \quad \text{or}$$

$$f^{-1}(x) = \frac{1}{2}x + \frac{5}{2}$$

$$= \frac{x + 5}{2}$$

11.

$$(g \circ f)(x) = (2x - 5)^2 + 4$$

$$= 4x^2 - 20x + 25 + 4$$

$$= 4x^2 - 20x + 29$$

13.

$$x - 2y = 13 \qquad \text{To find the slope, solve for y}$$

$$-2y = -x + 13$$

$$y = \frac{1}{2}x - \frac{13}{2} \qquad m = \frac{1}{2}$$

Perpendicular slope is -2.

Thus, $(4, -2)$, $m = -2$

$$y + 2 = -2(x - 4)$$

$$y + 2 = -2x + 8$$

$$y = -2x + 6$$

15.

Step 1: Find the radius.

The radius is the distance between $(1, -7)$ and $(4, -3)$.

$$r^2 = (4 - 1)^2 + (-3 + 7)^2$$

$$r^2 = 9 + 16 = 25; \quad r = 25$$

#15. continued

Step 2: State the equation.

Let $(h, k) = (1, -7)$

$$(x - h)^2 + (y - k)^2 = r^2$$

$$(x - 1)^2 + (y + 7)^2 = 25$$

17.

The graph is in the answer section of the main text.

19.

$$9x^2 + 4y^2 - 36x + 8y + 4 = 0$$

$$9x^2 - 36x + 4y^2 + 8y = 0 - 4$$

$$9(x^2 - 4x) + 4(y^2 + 2y) = 0 - 4$$

$$9(x^2 - 4x + 4) + 4(y^2 + 2y + 1) = 36 + 4 - 4$$

$$9(x - 2)^2 + 4(y + 1)^2 = 36$$

$$\frac{(x - 2)^2}{4} + \frac{(y + 1)^2}{9} = 1$$

21.

The graph is in the answer section of the main text.

23.

$$\log_3 \sqrt[3]{3} = \log_3 3^{\frac{1}{3}}$$

$$= \frac{1}{3}\log_3 3$$

$$= \frac{1}{3} \cdot 1 = \frac{1}{3}$$

25.

$$2\left(\frac{2x}{x + 1}\right)^2 + 7\left(\frac{2x}{x + 1}\right) - 4 = 0 \qquad \text{Let } u = \frac{2x}{x + 1}$$

$$2u^2 + 7u - 4 = 0$$

$$(2u - 1)(u + 4) = 0$$

$$2u - 1 = 0 \quad \text{or} \quad u + 4 = 0$$

$$u = \frac{1}{2} \qquad\qquad u = -4$$

$$\frac{2x}{x + 1} = \frac{1}{2} \qquad \frac{2x}{x + 1} = \frac{-4}{1}$$

$$4x = x + 1 \qquad 2x = -4x - 4$$

$$3x = 1 \qquad\qquad 6x = -4$$

$$x = \frac{1}{3} \qquad\qquad x = -\frac{2}{3}$$

#25. continued

Solution set: $\left\{\dfrac{1}{3}, -\dfrac{2}{3}\right\}$

27.

$$\left(\sqrt{3x}\right)^2 = \left(1 + \sqrt{x+1}\right)^2$$

$$3x = 1 + 2\sqrt{x+1} + x + 1$$

$$2x - 2 = 2\sqrt{x+1}$$

$$(x-1)^2 = \left(\sqrt{x+1}\right)^2 \qquad \text{Multiply by } \dfrac{1}{2}$$

$$x^2 - 2x + 1 = x + 1$$

$$x^2 - 3x = 0$$

$$x(x-3) = 0$$

$$x = 0 \quad \text{or} \quad x - 3 = 0$$

$$x = 3$$

Solution set: $\{3\}$ 0 does not check.

29.

$$4x^4 - 29x^2 + 25 = 0$$

$$\left(4x^2 - 25\right)\left(x^2 - 1\right) = 0$$

$$(2x+5)(2x-5)(x+1)(x-1) = 0$$

$$2x + 5 = 0 \quad \text{or} \quad 2x - 5 = 0 \quad \text{or} \quad x + 1 = 0$$

$$x = -\dfrac{5}{2} \qquad\qquad x = \dfrac{5}{2} \qquad\qquad x = -1$$

$$\text{or} \quad x - 1 = 0$$

$$x = 1$$

Solution set: $\left\{1, -1, \dfrac{5}{2}, -\dfrac{5}{2}\right\}$

31.

$$5x^2 - 6x + 2 = 0$$

$$a = 5, \quad b = -6, \quad c = 2$$

$$x = \dfrac{6 \pm \sqrt{36 - 40}}{2 \cdot 5}$$

$$x = \dfrac{6 \pm \sqrt{-4}}{10}$$

$$x = \dfrac{6 \pm 2i}{10} = \dfrac{3 \pm i}{5}$$

Solution set: $\left\{\dfrac{3 \pm i}{5}\right\}$

33.

$$\dfrac{1}{125} = 25^{x+3}$$

$$125^{-1} = 25^{x+3}$$

$$\left(5^3\right)^{-1} = \left(5^2\right)^{x+3}$$

$$5^{-3} = 5^{2x+6}$$

Therefore,

$$-3 = 2x + 6$$

$$-9 = 2x$$

$$-\dfrac{9}{2} = x$$

Solution set: $\left\{-\dfrac{9}{2}\right\}$

35.

$$\log(x+5) + \log(x-1) = \log(6x+19)$$

$$\log(x+5)(x-1) = \log(6x+19)$$

Therefore,

$$(x+5)(x-1) = 6x + 19$$

$$x^2 + 4x - 5 = 6x + 19$$

$$x^2 - 2x - 24 = 0$$

$$(x+4)(x-6) = 0$$

$$x + 4 = 0 \quad \text{or} \quad x - 6 = 0$$

$$x = -4 \qquad\qquad x = 6$$

Solution set: $\{6\}$ -4 does not check.

37.

$$\log_2 3x - \log_2(x-1) = \log_2 2^2$$

$$\log_2 \dfrac{3x}{x-1} = \log_2 4$$

Therefore,

$$\dfrac{3x}{x-1} = \dfrac{4}{1}$$

$$4x - 4 = 3x$$

$$x = 4$$

Solution set: $\{4\}$

39.

(1) $\quad 2x + 3y = -1$

(2) $\quad 3x + 7y = -2$

#39. continued

To eliminate the x' s, multiply (1) by 3 and
(2) by - 2.

(1a) $\quad 6x + 9y = -3$

(2a) $\quad \underline{-6x - 14y = 4}$

$\qquad -5y = 1$

$\qquad y = -\dfrac{1}{5}$

Substitute $-\dfrac{1}{5}$ for y in (1).

$2x + 3\left(-\dfrac{1}{5}\right) = -1$

$\qquad 2x - \dfrac{3}{5} = -1$

$\qquad 2x = -1 + \dfrac{3}{5}$

$\qquad 2x = -\dfrac{2}{5}$

$\qquad x = -\dfrac{1}{5}$

Solution set: $\left\{\left(-\dfrac{1}{5}, -\dfrac{1}{5}\right)\right\}$; Independent

---

41.

(1) $\quad 3x - 2y = 1$

(2) $\quad y = 3x + 4$

Substitute $3x + 4$ for y in (1).

(1) $\quad 3x - 2(3x + 4) = 1$

$\qquad 3x - 6x - 8 = 1$

$\qquad -3x = 9$

$\qquad x = -3$

Substitute - 3 for x in (2).

(2) $\quad y = 3(-3) + 4 = -5$

Solution set: $\{(-3, -5)\}$; Independent

43.

(1) $\quad x^2 + y^2 = 25$

(2) $\quad x^2 = y + 13 \qquad$ Solve for x^2

Substitute $y + 13$ for x^2 in (1).

(1) $\quad (y + 13) + y^2 = 25$

$\qquad y^2 + y - 12 = 0$

$\qquad (y - 3)(y + 4) = 0$

#43. continued

$y - 3 = 0 \quad$ or $\quad y + 4 = 0$

$\quad y = 3 \quad y = -4$

(2) $\quad x^2 = 3 + 13 \qquad x^2 = -4 + 13$

$\qquad x^2 = 16 \qquad\qquad x^2 = 9$

$\qquad x = \pm 4 \qquad\qquad x = \pm 3$

Solution set: $\{(4, 3)(-4, 3)(3, -4)(-3, -4)\}$

45.

Let x = width.

Let 3x - 3 = length.

Thus, $\quad A = L \cdot W$

$\qquad x(3x - 3) = 90$

$3x^2 - 3x - 90 = 0$

$\quad x^2 - x - 30 = 0 \qquad$ Multiply by $\dfrac{1}{3}$

$(x + 5)(x - 6) = 0$

$x + 5 = 0 \quad$ or $\quad x - 6 = 0$

$\quad \cancel{x = -5} \qquad\qquad x = 6$

$\qquad\qquad 3x - 3 = 3 \cdot 6 - 3 = 15$

Answer: 6 ft by 15 ft

47.

Let x = Vanna' s boat in still water.

	D	÷	R	=	T
Up	6		x - 3		$\dfrac{6}{x - 3}$
Down	6		x + 3		$\dfrac{6}{x + 3}$

The total trip took $\dfrac{3}{2}$ hours.

Thus, $\qquad \dfrac{6}{x - 3} + \dfrac{6}{x + 3} = \dfrac{3}{2}$

$\qquad\qquad\qquad\qquad$ LCD $= 2(x - 3)(x + 3)$

$2(x - 3)(x + 3)\left(\dfrac{6}{x - 3} + \dfrac{6}{x + 3}\right) = \left(\dfrac{3}{2}\right)2(x - 3)(x + 3)$

$\qquad 12(x + 3) + 12(x - 3) = 3(x - 3)(x + 3)$

$\qquad 12x + 36 + 12x - 36 = 3(x^2 - 9)$

$\qquad\qquad\qquad\qquad 24x = 3x^2 - 27$

$\qquad\qquad\qquad\qquad 0 = 3x^2 - 24x - 27$

$\qquad\qquad\qquad\qquad 0 = x^2 - 8x - 9$

$\qquad\qquad\qquad\qquad$ Multiply by $\dfrac{1}{3}$

$\qquad\qquad\qquad\qquad 0 = (x - 9)(x + 1)$

#47. continued

$x - 9 = 0$ or $x + 1 = 0$

$x = 9$ $\cancel{x = -1}$

Answer: 9 mph

49.

Let x = pounds of Jumbo Delux.

Let y = pounds of Small Fry.

Thus,   (1) $x + y = 10$

        (2) $300x + 50y = 200(10)$

        (2a) $6x + y = 40$

Multiply (2) by $\dfrac{1}{50}$

Multiply (1) by -1 and combine with (2a).

(1a) $-x - y = -10$

(2a) $\underline{6x + y = 40}$

        $5x = 30$

        $x = 6$

Substitute 6 for x in (1).

(1) $6 + y = 10$

        $y = 4$

Answer: 6 lb of Jumbo Delux, 4 lb of Small Fry

Chapter 10 Study Guide

Self-Test Exercises

I. Inverse functions and composition of functions.

1. Find the equation that defines the inverse of $f(x) = 2x + 5$, and write it in the form $y = f^{-1}(x)$. Also, graph both equations on the same coordinate plane.

2. Determine whether $f(x) = \dfrac{x}{3x+2}$ and $g(x) = \dfrac{-2x}{3x-1}$ are inverses by finding $(f \circ g)(x)$ and $(g \circ f)(x)$.

II. Graph the following functions.

3. $f(x) = 2^{3-x}$

4. $f(x) = \log_2(x+1)$

5. $f(x) = \log_{1/2}(x-2)$

III. Change to the indicated form.

6. $B = C^{an}$, logarithmic

7. $y = \log_5(x+4)$, exponential

8. $y = \log_7 x^2(x-2)$, exponential

9. $y = 3^{3-x}$, logarithmic

IV. Simplify the following.

10. $3^{\log_3 15}$

11. $\log_3 \sqrt{3}$

12. $\log_9 \dfrac{1}{3}$

13. Use the properties of logarithms to express
$\left[2\log_a x + \log_a(x-4)\right] + \log_a(x-2)$ as a single logarithm.

V. Solve the following equations.

14. $\dfrac{1}{4} = 16^{-x}$

15. $64 = (x+3)^3$

16. $81 = 3^{2x-3}$

17. $1.63^{x+1} = 25$

18. $\log_7(x+4) + \log_7(x-4) = 2$

19. $\log_3(x+4) - \log_3(x+1) = \log_3 x$

VI. Application

20. The amount of money, A, that a principal P will be worth after n years at interest rate r, compounded annually, is given by the formula $A = P(1+r)^n$. Maude puts $1200 in a money market account offering 5% interest compounded annually for 10 years. Find the amount accumulated.

21. Donald found that his forgetfulness between tests grew greater between certain periods of time. He discovered that the formula $f(t) = 86 - 9\log_{10}(1+t)$, where t is time in months after a certain test is given will help him to minimize his forgetfulness. How many months can be expected to elapse before the average score will be 68?

The worked-out solutions begin on the following page.

Self-Test Solutions

1.

$f(x) = 2x + 5$

$y = 2x + 5 \xrightarrow{\text{Inverse}} x = 2y + 5$ Interchange x and y

$\qquad\qquad x - 5 = 2y$ Solve for y

$\qquad\qquad \dfrac{1}{2}x - \dfrac{5}{2} = y$ Multiply by $\dfrac{1}{2}$

$\qquad\qquad$ or $f(x) = \dfrac{1}{2}x - \dfrac{5}{2}$

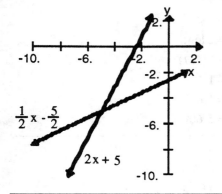

2.

$f(x) = \dfrac{x}{3x + 2}, \quad g(x) = \dfrac{-2x}{3x - 1}$

$(f \circ g)(x) = \dfrac{\dfrac{-2x}{3x - 1}}{3\left(\dfrac{-2x}{3x - 1}\right) + 2} = \dfrac{\dfrac{-2x}{3x - 1}}{\dfrac{-2}{3x - 1}}$

$\qquad\qquad\qquad = x$

$(g \circ f)(x) = \dfrac{-2\left(\dfrac{x}{3x + 2}\right)}{3\left(\dfrac{x}{3x + 2}\right) - 1} = \dfrac{\dfrac{-2x}{3x + 2}}{\dfrac{-2}{3x + 2}}$

$\qquad\qquad\qquad = x$

$f(x)$ and $g(x)$ are inverses to each other.

3.

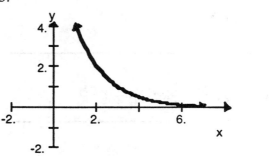

4.

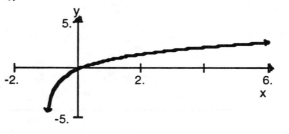

5.

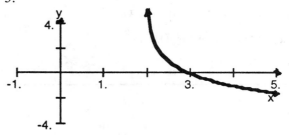

6.

$B = C^{an}; \quad \log_c B = an$

7.

$y = \log_5(x + 4); \quad 5^y = x + 4$

8.

$y = \log_7 x^2(x - 2); \quad 7^y = x^2(x - 2)$

9.

$y = 3^{3-x} = \log_3 y = 3 - x$

10.

$3^{\log_3 15} = x$

$\log_3 x = \log_3 15$

$\qquad x = 15$

11.

$\log_3 \sqrt{3} = \log_3 3^{\frac{1}{2}} = \frac{1}{2} \log_3 3 = \frac{1}{2} \cdot 1 = \frac{1}{2}$

12.

$\log_9 \dfrac{1}{3} = x$

$\qquad 9^x = \dfrac{1}{3}$

$\qquad 3^{2x} = 3^{-1}$

$\qquad 2x = -1$

$\qquad x = -\dfrac{1}{2}$

13.

$\left[2\log_a x + \log_a(x-4) \right] + \log_a(x-2)$

$= \left[\log_a x^2 + \log_a(x-4) \right] + \log_a(x-2)$

$= \left[\log_a x^2(x-4) \right] + \log_a(x-2)$

$= \log_a x^2(x-4)(x-2)$

14.

$\dfrac{1}{4} = 16^{-x}$

$\dfrac{1}{2^2} = 2^{-4x}$

$2^{-2} = 2^{-4x}$

$-2 = -4x$

$\dfrac{1}{2} = x$

15.

$64 = (x+3)^3$

$4^3 = (x+3)^3$

$4 = (x+3)$ Cube root of both sides

$1 = x$

16.

$81 = 3^{2x-3}$

$3^4 = 3^{2x-3}$

$4 = 2x - 3$

$7 = 2x$

$\dfrac{7}{2} = x$

17.

$1.63^{x+1} = 25$

$\log 1.63^{x+1} = \log 5^2$

$(x+1)\log 1.63 = 2\log 5$

$(x+1)(0.212) = 2(0.7)$

$0.212x + 0.212 = 1.4$

$0.212x = 1.188$

$x = 5.60$

18.

$\log_7(x+4) + \log_7(x-4) = 2$

$\log_7(x^2 - 16) = \log_7 7^2$

$x^2 - 16 = 49$

$x^2 = 65$

$x = \pm\sqrt{65}$

19.

$\log_3(x+4) - \log_3(x+1) = \log_3 x$

$\log_3 \dfrac{x+4}{x+1} = \log_3 x$

$\dfrac{x+4}{x+4} = x$

$x^2 + x = x + 4$

$x^2 - 4 = 0$

$(x+2)(x-2) = 0$

$x + 2 = 0 \quad \text{or} \quad x - 2 = 0$

$x = -2 \qquad\qquad x = 2$

Solution set: $\{2\}$

-2 does not check since $\log_3 -2$ is not a real number.

20.

$$A = P(1+r)^n$$

$$A = 1200(1+0.05)^{10}$$

$$A = 1200(1.05)^{10}$$

$$\log A = \log 1200(1.05)^{10}$$

$$= \log 1200 + \log(1.05)^{10}$$

$$= \log 1200 + 10 \log 1.05$$

$$= 3.07918 + 10(0.021189)$$

$$= 3.07918 + 0.21189$$

$$= 3.29107$$

$$A = \text{anti} \log 3.29107$$

$$A \approx 1950$$

Answer: $1950

21.

$$f(t) = 86 - 9 \log_{10}(1+t)$$

$$68 = 86 - 9 \log_{10}(1+t)$$

$$-18 = -9 \log_{10}(1+t)$$

$$2 = \log_{10}(1+t)$$

$$2 \log_{10} 10 = \log_{10}(1+t)$$

$$\log_{10} 10^2 = \log_{10}(1+t)$$

$$100 = 1+t$$

$$99 = t$$

Answer: 99 weeks

CHAPTER 11 SEQUENCES AND SERIES

Solutions To Text Exercises

1.

$a_n = 3n + 2$

$a_1 = 3(1) + 2 = 5$

$a_2 = 3(2) + 2 = 8$

$a_3 = 3(3) + 2 = 11$

$a_4 = 3(4) + 2 = 14$

$a_5 = 3(5) + 2 = 17$

The first five terms are 5, 8, 11, 14, 17.

3.

$a_n = 4n - 1$

$a_1 = 4(1) - 1 = 3$

$a_2 = 4(2) - 1 = 7$

$a_3 = 4(3) - 1 = 11$

$a_4 = 4(4) - 1 = 15$

$a_5 = 4(5) - 1 = 19$

The first five terms are 3, 7, 11, 15, 19.

5.

$a_n = 3^n$

$a_1 = 3^1 = 3$

$a_2 = 3^2 = 9$

$a_3 = 3^3 = 27$

$a_4 = 3^4 = 81$

$a_5 = 3^5 = 243$

The first five terms are 3, 9, 27, 81, 243.

7.

$a_n = \left(\dfrac{2}{3}\right)^n$

$a_1 = \left(\dfrac{2}{3}\right)^1 = \dfrac{2}{3}$

$a_2 = \left(\dfrac{2}{3}\right)^2 = \dfrac{4}{9}$

$a_3 = \left(\dfrac{2}{3}\right)^3 = \dfrac{8}{27}$

#7. continued

$a_4 = \left(\dfrac{2}{3}\right)^4 = \dfrac{16}{81}$

$a_5 = \left(\dfrac{2}{3}\right)^5 = \dfrac{32}{243}$

The first five terms are $\dfrac{2}{3}, \dfrac{4}{9}, \dfrac{8}{27}, \dfrac{16}{81}, \dfrac{32}{243}$.

9.

$a_n = \dfrac{1}{n}$

$a_1 = \dfrac{1}{1} = 1$

$a_2 = \dfrac{1}{2}$

$a_3 = \dfrac{1}{3}$

$a_4 = \dfrac{1}{4}$

$a_5 = \dfrac{1}{5}$

The first five terms are $1, \dfrac{1}{2}, \dfrac{1}{3}, \dfrac{1}{4}, \dfrac{1}{5}$.

11.

$a_n = n^2 - 4$

$a_1 = (1)^2 - 4 = -3$

$a_2 = (2)^2 - 4 = 0$

$a_3 = (3)^2 - 4 = 5$

$a_4 = (4)^2 - 4 = 12$

$a_5 = (5)^2 - 4 = 21$

The first five terms are -3, 0, 5, 12, 21.

13.

$a_n = n^2 + n + 1$

$a_1 = (1)^2 + (1) + 1 = 3$

$a_2 = (2)^2 + (2) + 1 = 7$

$a_3 = (3)^2 + (3) + 1 = 13$

#13. continued

$a_4 = (4)^2 + (4) + 1 = 21$

$a_5 = (5)^2 + (5) + 1 = 31$

The first five terms are 3, 7, 13, 21, 31.

15.

$a_n = (-1)^n (2n + 1)$

$a_1 = (-1)^1 (2(1) + 1) = (-1)(3) = -3$

$a_2 = (-1)^2 (2(2) + 1) = (1)(5) = 5$

$a_3 = (-1)^3 (2(3) + 1) = (-1)(7) = -7$

$a_4 = (-1)^4 (2(4) + 1) = (1)(9) = 9$

$a_5 = (-1)^5 (2(5) + 1) = (-1)(11) = -11$

The first five terms are -3, 5, -7, 9, -11.

17.

$a_n = \dfrac{(-1)^{n+1}}{n}$

$a_1 = \dfrac{(-1)^{1+1}}{1} = \dfrac{(-1)^2}{1} = 1$

$a_2 = \dfrac{(-1)^{2+1}}{2} = \dfrac{(-1)^3}{2} = -\dfrac{1}{2}$

$a_3 = \dfrac{(-1)^{3+1}}{3} = \dfrac{(-1)^4}{3} = \dfrac{1}{3}$

$a_4 = \dfrac{(-1)^{4+1}}{4} = \dfrac{(-1)^5}{4} = -\dfrac{1}{4}$

$a_5 = \dfrac{(-1)^{5+1}}{5} = \dfrac{(-1)^6}{5} = \dfrac{1}{5}$

The first five terms are $1, \ -\dfrac{1}{2}, \dfrac{1}{3}, -\dfrac{1}{4}, \dfrac{1}{5}$

19.

$a_n = 5n + 7$

$a_6 = 5(6) + 7 = 37$

21.

$a_n = \left(\dfrac{9}{5}\right)^n$

$a_3 = \left(\dfrac{9}{5}\right)^3 = \dfrac{729}{125}$

23.

$a_n = \dfrac{2n}{n^3 + 1}$

$a_9 = \dfrac{2(9)}{9^3 + 1} = \dfrac{18}{730} = \dfrac{9}{365}$

25.

$a_n = (-1)^n \left(3n + \dfrac{n^2}{n+1} \right)$

$a_5 = (-1)^5 \left(3(5) + \dfrac{5^2}{5+1} \right)$

$= (-1)\left(15 + \dfrac{25}{6} \right)$

$= (-1)\left(15 + 4\dfrac{1}{6} \right) = -19\dfrac{1}{6} = -\dfrac{115}{6}$

27.

Find S_5 when $a_n = 3n + 2$

$S_5 = a_1 + a_2 + a_3 + a_4 + a_5$

$\quad = 5 + 8 + 11 + 14 + 17 \qquad$ From Exercise 1

$\quad = 55$

29.

Find S_5 when $a_n = 4n - 1$

$S_5 = a_1 + a_2 + a_3 + a_4 + a_5$

$\quad = 3 + 7 + 11 + 15 + 19 \qquad$ From Exercise 3

$\quad = 55$

31.

Find S_5 when $a_n = 3^n$

$S_5 = a_1 + a_2 + a_3 + a_4 + a_5$

$\quad = 3 + 9 + 27 + 81 + 243 \qquad$ From Exercise 5

$\quad = 363$

33.

Find S_5 when $a_n = \left(\dfrac{2}{3}\right)^n$

$S_5 = a_1 + a_2 + a_3 + a_4 + a_5$

$\quad = \dfrac{2}{3} + \dfrac{4}{9} + \dfrac{8}{27} + \dfrac{16}{81} + \dfrac{32}{243} \quad$ From Exercise 7

$\quad = \dfrac{162}{243} + \dfrac{108}{243} + \dfrac{72}{243} + \dfrac{48}{243} + \dfrac{32}{243} = \dfrac{442}{243}$

35.

Find S_5 when $a_n = \dfrac{1}{n}$

$S_5 = a_1 + a_2 + a_3 + a_4 + a_5$

$= 1 + \dfrac{1}{2} + \dfrac{1}{3} + \dfrac{1}{4} + \dfrac{1}{5}$ From Exercise 9

$= \dfrac{60}{60} + \dfrac{30}{60} + \dfrac{20}{60} + \dfrac{15}{60} + \dfrac{12}{60} = \dfrac{137}{60}$

37.

Find S_5 when $a_n = n^2 - 4$

$S_5 = a_1 + a_2 + a_3 + a_4 + a_5$

$= (-3) + 0 + 5 + 12 + 21$ From Exercise 11

$= 35$

39.

Find S_5 when $a_n = n^2 + n + 1$

$S_5 = a_1 + a_2 + a_3 + a_4 + a_5$

$= 3 + 7 + 13 + 21 + 31$ From Exercise 13

$= 75$

41.

Find S_5 when $a_n = (-1)^n (2n + 1)$

$S_5 = a_1 + a_2 + a_3 + a_4 + a_5$

$= (-3) + 5 + (-7) + 9 + (-11)$ From Exercise 15

$= -7$

43.

Find S_5 when $a_n = \dfrac{(-1)^{n+1}}{n}$

$S_5 = a_1 + a_2 + a_3 + a_4 + a_5$

$= 1 + \left(-\dfrac{1}{2}\right) + \dfrac{1}{3} + \left(-\dfrac{1}{4}\right) + \dfrac{1}{5}$ From Exercise 17

$= \dfrac{60}{60} + \left(-\dfrac{30}{60}\right) + \dfrac{20}{60} + \left(-\dfrac{15}{60}\right) + \dfrac{12}{60}$

$= \dfrac{47}{60}$

45.

$\displaystyle\sum_{i=1}^{8} (3i - 7) = (3(1) - 7) + (3(2) - 7) + (3(3) - 7)$

$\qquad + (3(4) - 7) + (3(5) - 7) + (3(6) - 7)$

$\qquad + (3(7) - 7) + (3(8) - 7)$

#45. continued

$= (-4) + (-1) + (2) + (5) + (8) + (11)$

$\quad + (14) + (17)$

$= 52$

47.

$\displaystyle\sum_{i=1}^{3} (2i^2 + i + 3) = (2(1)^2 + 1 + 3) + (2(2)^2 + 2 + 3)$

$\qquad\qquad + (2(3)^2 + 3 + 3)$

$= (2 + 1 + 3) + (8 + 2 + 3) + (18 + 3 + 3)$

$= 6 + 13 + 24 = 43$

49.

$\displaystyle\sum_{i=1}^{4} \left(\dfrac{2}{3}\right)^i = \left(\dfrac{2}{3}\right)^1 + \left(\dfrac{2}{3}\right)^2 + \left(\dfrac{2}{3}\right)^3 + \left(\dfrac{2}{3}\right)^4$

$= \dfrac{2}{3} + \dfrac{4}{9} + \dfrac{8}{27} + \dfrac{16}{81}$

$= \dfrac{54}{81} + \dfrac{36}{81} + \dfrac{24}{81} + \dfrac{16}{81}$

$= \dfrac{130}{81}$

51.

$\displaystyle\sum_{i=3}^{7} 2i(i + 4) = 2(3)(3 + 4) + 2(4)(4 + 4) + 2(5)(5 + 4)$

$\qquad + 2(6)(6 + 4) + 2(7)(7 + 4)$

$= 6(7) + 8(8) + 10(9) + 12(10) + 14(11)$

$= 42 + 64 + 90 + 120 + 154 = 470$

53.

$\displaystyle\sum_{i=2}^{6} (-1)^i (i) = (-1)^2 (2) + (-1)^3 (3) + (-1)^4 (4)$

$\qquad + (-1)^5 (5) + (-1)^6 (6)$

$= (1)(2) + (-1)(3) + (1)(4)$

$\qquad + (-1)(5) + (1)(6)$

$= 2 + (-3) + 4 + (-5) + 6 = 4$

55.

$$\sum_{i=6}^{7}(-1)^{i+1}\left(\frac{1}{2i}\right) = (-1)^{6+1}\left(\frac{1}{2(6)}\right) + (-1)^{7+1}\left(\frac{1}{2(7)}\right)$$

$$= (-1)^{7}\left(\frac{1}{12}\right) + (-1)^{8}\left(\frac{1}{14}\right)$$

$$= -\frac{1}{12} + \frac{1}{14}$$

$$= -\frac{7}{84} + \frac{6}{84} = -\frac{1}{84}$$

57.

By inspection, the difference between terms is $(2-1)=(3-2)=1$.

Thus, $1+2+3+4+5+6+7 = \sum_{i=0}^{6}(i+1)$

or $1+2+3+4+5+6+7 = \sum_{i=1}^{7}(i)$

59.

The difference between terms is $(9-7)=(11-9)=2$

Thus, $7+9+11+13+15+17 = \sum_{i=1}^{6}2i+5$

or $7+9+11+13+15+17 = \sum_{i=4}^{9}2i-1$

61.

The difference between terms is not equal, $(9-3) \neq (27-9)$.

The terms, however, are all divisible by 3.

Thus, $3+9+27+81+243 = \sum_{i=1}^{5}3^{i}$

63.

The difference in terms decrease by $(80-40)=40$, $(40-20)=20$, and $(20-10)=10$. Therefore, the general term must contain $\left(\frac{1}{2}\right)^{i}$.

Thus, $80+40+20+10 = \sum_{i=1}^{4}80\left(\frac{1}{2}\right)^{i-1}$

65.

The terms alternate by positive and negative.

Therefore, the general term must contain $(-1)^{i+1}$.

The absolute value of the terms increase by i^{3}.

Thus, $1-8+27-64+125 = \sum_{i=1}^{5}(-1)^{i+1}(i)^{3}$

67.

Each term decreases by a factor of $\frac{1}{3}$.

Thus, $\frac{1}{3}+\frac{1}{9}+\frac{1}{27}+\cdots = \sum_{i=1}^{\infty}\frac{1}{3^{i}}$

Since there is no end term, this is an infinite series.

69.

The terms increase by $(5-2)=(8-5)=3$.

Thus, $2+5+8+\cdots = \sum_{i=0}^{\infty}(3i+2)$

or $2+5+8+\cdots = \sum_{i=1}^{\infty}(3i-1)$

71.

The sequence is denoted by $a_{n}=2^{n}$.

The prize money is calculated by the eighth term of the sequence.

Therefore, $a_{8}=2^{8}=256$.

Answer: $256

Exercises 11.2

1.

Common difference = 2nd term - 1st term

$$d = a_{2} - a_{1}$$
$$d = 8-6$$
$$d = 2$$

3.

Common difference = 2nd term - 1st term

$$d = a_{2} - a_{1}$$
$$d = 0-\frac{1}{2}$$
$$d = -\frac{1}{2}$$

5.

Common difference = 2nd term - 1st term

$$d = a_2 - a_1$$
$$d = 7 - 7$$
$$d = 0$$

7.

From the formula $a_n = a_1 + (n-1)d$,

if $a_1 = 8$ and $d = 7$, then,

$$a_9 = a_1 + (9-1)d$$
$$a_9 = a_1 + 8d$$
$$a_9 = 8 + 8(7) = 8 + 56 = 64$$

9.

Given sequence 6, 8, 10, 12, 14, $\cdots$ the common

difference is $d = (8-6) = 2$ and the first term is $a_1 = 6$.

The 18th term is:
$$a_{18} = a_1 + (n-1)d$$
$$= a_1 + (18-1)d$$
$$= a_1 + 17d$$
$$= 6 + 17(2) = 6 + 34 = 40$$

11.

Given sequence $\frac{1}{2}, 0, -\frac{1}{2}, -1, -\frac{3}{2}, \cdots$ the common

difference is $d = \left(0 - \frac{1}{2}\right) = -\frac{1}{2}$ and the first term is

$a_1 = \frac{1}{2}$. The 30th term is:

$$a_{30} = a_1 + (n-1)d$$
$$= a_1 + (30-1)d$$
$$= a_1 + 29d$$
$$= \frac{1}{2} + 29\left(-\frac{1}{2}\right) = \frac{1}{2} - \frac{29}{2} = -\frac{28}{2} = -14$$

13.

Given sequence 7, 7, 7, 7, 7, $\cdots$ the common

difference is $d = (7-7) = 0$ and the first term is

$a_1 = 7$. The 99th term is

$$a_{99} = a_1 + (n-1)d$$
$$= a_1 + (99-1)d$$
$$= a_1 + 98d$$
$$= 7 + 98(0) = 7 + 0 = 7$$

15.

Given the sequence 7, 9, 11, 13, $\cdots$

$d = (a_2 - a_1) = (9-7) = 2$ and $a_1 = 7$.

The general term is

$$a_n = a_1 + (n-1)d$$
$$= 7 + (n-1)2$$
$$= 7 + 2n - 2$$
$$= 2n + 5$$

17.

Given the sequence -2, -5, -8, -11, $\cdots$

$d = (a_2 - a_1) = (-5 - (-2)) = -3$ and $a_1 = -2$.

The general term is

$$a_n = a_1 + (n-1)d$$
$$= -2 + (n-1)(-3)$$
$$= -2 + (-3n) + 3$$
$$= -3n + 1$$

19.

Given the sequence $-\frac{5}{2}, -2, -\frac{3}{2}, -1, \cdots$

$$d = (a_2 - a_1) = \left(-2 - \left(-\frac{5}{2}\right)\right) = \left(2\frac{1}{2} - 2\right) = \frac{1}{2}$$

and $a_1 = -\frac{5}{2}$.

The general term is

$$a_n = a_1 + (n-1)d$$
$$= -\frac{5}{2} + (n-1)\left(\frac{1}{2}\right)$$
$$= -\frac{5}{2} + \left(\frac{1}{2}n\right) - \frac{1}{2}$$
$$= -\frac{5}{2} + \frac{1}{2}n - \frac{1}{2}$$
$$= \frac{1}{2}n - 3$$

21.

Given the sequence 1, -3, -7, -11, $\cdots$

$d = (a_2 - a_1) = (-3 - 1) = -4$ and $a_1 = 1$.

The general term is

#21. continued
$$a_n = a_1 + (n-1)d$$
$$= 1 + (n-1)(-4)$$
$$= 1 + (-4n) + 4$$
$$= 1 - 4n + 4$$
$$= -4n + 5$$

23.

Given the sequence $-\dfrac{7}{4}, -\dfrac{13}{4}, -\dfrac{19}{4}, -\dfrac{25}{4}, \ldots$

$$d = (a_2 - a_1) = \left(-\frac{13}{4} - \left(-\frac{7}{4}\right)\right) = \frac{7}{4} - \frac{13}{4} = -\frac{6}{4}$$

and $a_1 = -\dfrac{7}{4}$.

The general term is
$$a_n = a_1 + (n-1)d$$
$$= -\frac{7}{4} + (n-1)\left(-\frac{6}{4}\right)$$
$$= -\frac{7}{4} + \left(-\frac{6}{4}n\right) + \frac{6}{4}$$
$$= -\frac{7}{4} - \frac{6}{4}n + \frac{6}{4}$$
$$= -\frac{1}{4} - \frac{6}{4}n = -\frac{3}{2}n - \frac{1}{4}$$

25.

If $a_7 = 22$ and $a_{12} = 37$, then a_1 and d can be found by the general term formula.

$a_7 = a_1 + (7-1)d$ so $a_7 = a_1 + 6d$

$a_{12} = a_1 + (12-1)d$ so $a_{12} = a_1 + 11d$

Substituting for a_7 and a_{12}, yields

(1) $22 = a_1 + 6d$

(2) $37 = a_1 + 11d$

(1a) $-22 = -a_1 - 6d$ Multiply by -1

(2) $\underline{37 = a_1 + 11d}$

      $15 = 5d$

(3) $3 = d$ Multiply by $\dfrac{1}{5}$

Substitute 3 for d in (1).

(1) $22 + a_1 + 6 \cdot 3$

      $22 = a_1 + 18$

      $4 = a_1$

Answer: $a_1 = 4, \quad d = 3$

27.

If $a_9 = -\dfrac{37}{2}$ and $a_{16} = -\dfrac{65}{2}$, then a_1 and d can be found by the general term formula.

$a_9 = a_1 + (9-1)d$ so $a_9 = a_1 + 8d$

$a_{16} = a_1 + (16-1)d$ so $a_{16} = a_1 + 15d$

Substituting for a_9 and a_{16} yields

(1) $-\dfrac{37}{2} = a_1 + 8d$

(2) $-\dfrac{65}{2} = a_1 + 15d$

(1a) $\dfrac{37}{2} = -a_1 - 8d$ Multiply by -1

(2) $-\dfrac{65}{2} = a_1 + 15d$

      _____

      $-14 = \quad 7d$

(3) $-2 = \quad d$

Substitute -2 for d in (1).

(1) $-\dfrac{37}{2} = a_1 + 8(-2)$

      $-\dfrac{37}{2} = a_1 - 16$

      $-\dfrac{37}{2} + \dfrac{32}{2} = a_1$

      $-\dfrac{5}{2} = a_1$

Answer: $a_1 = -\dfrac{5}{2}, \quad d = -2$

29.

If $a_5 = \dfrac{23}{6}$ and $a_9 = \dfrac{13}{2}$, a_1 and d can be found by the general term formula.

$a_5 = a_1 + (5-1)d$ so $a_5 = a_1 + 4d$

$a_9 = a_1 + (9-1)d$ so $a_9 = a_1 + 8d$

Substituting for a_5 and a_9, yields

(1) $\dfrac{23}{6} = a_1 + 4d$

(2) $\dfrac{13}{2} = a_1 + 8d$

#29. continued

(1a) $\quad -\dfrac{23}{6} = -a_1 - 4d \qquad$ Multiply by -1

(2) $\quad \dfrac{13}{2} = a_1 + 8d$

$\qquad \dfrac{16}{6} = 4d$

$\qquad \dfrac{4}{6} = d \qquad$ Multiply by $\dfrac{1}{4}$

$\qquad \dfrac{2}{3} = d$

Substituting $\dfrac{2}{3}$ for d in (1).

(1) $\quad \dfrac{23}{6} = a_1 + 4\left(\dfrac{2}{3}\right)$

$\qquad \dfrac{23}{6} = a_1 + \dfrac{8}{3}$

$\qquad \dfrac{23}{6} - \dfrac{16}{6} = a_1$

$\qquad \dfrac{23}{6} - \dfrac{16}{6} = a_1$

$\qquad \dfrac{7}{6} = a_1$

Answer: $\quad d = \dfrac{2}{3}, \quad a_1 = \dfrac{7}{6}$

31.

Since we are given n, a_1 and a_n, we apply equation (3) where $n = 10, a_1 = 1, a_n = 27$.

$S_n = \dfrac{n}{2}\left(a_1 + a_n\right)$

$S_{10} = \dfrac{10}{2}(1 + 27)$

$\quad = \dfrac{10}{2} + \dfrac{270}{2}$

$\quad = \dfrac{280}{2} = 140$

33.

Given the sequence $-7, -2, 3, 8, \cdots$

$d = \left(a_2 - a_1\right) = (-2 - (-7)) = (7 - 2) = 5$ and $a_1 = -7$,

S_{19} can be calculated from equation (4).

#33. continued

$S_n = \dfrac{n}{2}\left(2a_1 + (n-1)d\right)$

$S_{19} = \dfrac{19}{2}\left(2(-7) + (19 - 1)(5)\right)$

$\quad = \dfrac{19}{2}(-14 + 90)$

$\quad = \dfrac{19}{2}(76)$

$\quad = 722$

35.

Given the sequence $-\dfrac{8}{3}, -\dfrac{10}{3}, -4, -\dfrac{14}{3}, \cdots$

$d = \left(a_2 - a_1\right) = \left(-\dfrac{10}{3} - \left(-\dfrac{8}{3}\right)\right) = \left(\dfrac{8}{3} - \dfrac{10}{3}\right) = -\dfrac{2}{3}$

and $a_1 = -\dfrac{8}{3}$,

S_{25} can be calculated from equation (4).

$S_n = \dfrac{n}{2}\left(2a_1 + (n-1)d\right)$

$S_{25} = \dfrac{25}{2}\left(2\left(-\dfrac{8}{3}\right) + (25 - 1)\left(-\dfrac{2}{3}\right)\right)$

$\quad = \dfrac{25}{2}\left(-\dfrac{16}{3} + \left(-\dfrac{48}{3}\right)\right)$

$\quad = \dfrac{25}{2}\left(-\dfrac{64}{3}\right)$

$\quad = -\dfrac{1600}{6} = -\dfrac{800}{3}$

37.

Given $a_1 = -\dfrac{15}{2}$ and $d = -\dfrac{1}{2}$ we find S_{23} from equation (4).

$S_n = \dfrac{n}{2}\left(2a_1 + (n-1)d\right)$

$S_{23} = \dfrac{23}{2}\left(2\left(-\dfrac{15}{2}\right) + (23 - 1)\left(-\dfrac{1}{2}\right)\right)$

$\quad = \dfrac{23}{2}\left(-\dfrac{30}{2} + \left(-\dfrac{22}{2}\right)\right)$

$\quad = \dfrac{23}{2}\left(-\dfrac{52}{2}\right) = -\dfrac{1196}{4}$

$\quad = -299$

39.

Since $(2i + 7)$ is first degree i, then $\displaystyle\sum_{i=1}^{15}(2i + 7)$ is an arithmetic series.

Thus, $a_1 = (2(1) + 7) = 9$

and

$a_{15} = (2(15) + 7) = 37$

From equation (3), we have

$$s_n = \frac{n}{2}(a_1 + a_n)$$

$$s_{15} = \frac{15}{2}(9 + 37) = \frac{15}{2}(46) = \frac{690}{2} = 345$$

41.

Since $(7 - 3i)$ is first degree i, then $\displaystyle\sum_{i=1}^{42}(7 - 3i)$ is an arithmetic series.

Thus, $a_1 = (7 - 3(1)) = 4$

and

$a_{42} = (7 - 3(42)) = -119$

From equation (3), we have

$$s_n = \frac{n}{2}(a_1 + a_n)$$

$$s_{42} = \frac{42}{2}(4 + (-119))$$

$$= \frac{42}{2}(-115) = \frac{-4830}{2} = -2415$$

43.

Since $\left(\dfrac{1}{2}i - \dfrac{3}{4}\right)$ is first degree i, then $\displaystyle\sum_{i=1}^{48}\left(\dfrac{1}{2}i - \dfrac{3}{4}\right)$ is an arithmetic series.

Thus, $a_1 = \left(\dfrac{1}{2}(1) - \dfrac{3}{4}\right) = -\dfrac{1}{4}$

and

$a_{48} = \left(\dfrac{1}{2}(48) - \dfrac{3}{4}\right)$

$$= \left(24 - \dfrac{3}{4}\right) = 23\dfrac{1}{4} = \dfrac{93}{4}$$

#43. continued

From equation (3), we have

$$s_n = \frac{n}{2}(a_1 + a_n)$$

$$s_{48} = \frac{48}{2}\left(-\frac{1}{4} + \frac{93}{4}\right)$$

$$= 24\left(\frac{92}{4}\right) = 6(92) = 552$$

45.

Since $\left(\dfrac{1}{5}i - \dfrac{2}{3}\right)$ is first degree i, then $\displaystyle\sum_{i=1}^{11}\left(\dfrac{1}{5}i - \dfrac{2}{3}\right)$ is an arithmetic series.

Thus, $a_1 = \left(\dfrac{1}{5}(1) - \dfrac{2}{3}\right)$

$$= \frac{1}{5} - \frac{2}{3} = \frac{3}{15} - \frac{10}{15} = -\frac{7}{15}$$

and

$a_{11} = \left(\dfrac{1}{5}(11) - \dfrac{2}{3}\right)$

$$= \frac{11}{5} - \frac{2}{3} = \frac{33}{15} - \frac{10}{15} = \frac{23}{15}$$

From equation (3), we have

$$s_n = \frac{n}{2}(a_1 + a_n)$$

$$s_{11} = \frac{11}{2}\left(-\frac{7}{15} + \frac{23}{15}\right)$$

$$= \frac{11}{2}\left(\frac{16}{15}\right) = \frac{176}{30} = \frac{88}{15}$$

Exercises 11.3

1.

The common ratio of $4, 12, 36, 108, \cdots$ is

$$r = \frac{a_2}{a_1} = \frac{12}{4} = 3$$

(Any two consecutive terms can be used.)

3.

The common ratio of $9, 6, 4, \dfrac{8}{3}, \cdots$ is

$$r = \frac{a_3}{a_2} = \frac{4}{6} = \frac{2}{3}$$

(Any two consecutive terms can be used.)

5.

The common ratio of $-\dfrac{1}{2}, \dfrac{3}{4}, -\dfrac{9}{8}, \dfrac{27}{16}, \cdots$ is

$$r = \frac{a_2}{a_1} = \frac{3}{4} \div \left(-\frac{1}{2}\right) = \frac{3}{4} \cdot -2 = -\frac{3}{2}$$

(Any two consecutive terms can be used.)

7.

Given $a_1 = 6$ and $r = 2$,

$$\begin{aligned}
a_5 &= a_1 r^{n-1} \qquad \text{General Term Formula} \\
&= (6)(2)^{5-1} \\
&= 6(2)^4 \\
&= 6(16) = 96
\end{aligned}$$

9.

Given the geometric sequence $4, 12, 36, 108, \cdots$,

the common ratio is $r = \dfrac{a_2}{a_1} = \dfrac{12}{4} = 3$ and the first

term is $a_1 = 4$. The 6th term is

$$\begin{aligned}
a_5 &= a_1 r^{n-1} \qquad \text{General Term Formula} \\
&= 4(3)^5 \\
&= 4(243) = 972
\end{aligned}$$

11.

Given the geometric sequence $9, 6, 4, \dfrac{8}{3}, \cdots$, the

common ratio is $r = \dfrac{a_2}{a_1} = \dfrac{6}{9} = \dfrac{2}{3}$ and the first term

is $a_1 = 9$.

The 7th term is

$$\begin{aligned}
a_7 &= a_1 r^{n-1} \qquad \text{General term formula} \\
&= (9)\left(\frac{2}{3}\right)^{7-1} \\
&= 9\left(\frac{2}{3}\right)^6 \\
&= 9\left(\frac{64}{729}\right) = \frac{64}{81}
\end{aligned}$$

13.

Given the geometric sequence $-\dfrac{1}{2}, \dfrac{3}{4}, -\dfrac{9}{8}, \dfrac{27}{16}, \cdots$,

the common ratio is

$$r = \frac{a_2}{a_1} = \frac{3}{4} \div \left(-\frac{1}{2}\right) = \frac{3}{4} \cdot \left(-\frac{2}{1}\right) = -\frac{3}{2}$$

and the first term is $a_1 = -\dfrac{1}{2}$.

The 8th term is

$$\begin{aligned}
a_8 &= a_1 r^{n-1} \qquad \text{General term formula} \\
&= \left(-\frac{1}{2}\right)\left(-\frac{3}{2}\right)^{8-1} \\
&= \left(-\frac{1}{2}\right)\left(-\frac{3}{2}\right)^7 \\
&= \left(-\frac{1}{2}\right)\left(-\frac{2187}{128}\right) = \frac{2187}{256}
\end{aligned}$$

15.

Given the geometric sequence $2, 4, 8, 16, \cdots$, the

common ratio is $\dfrac{a_2}{a_1} = \dfrac{4}{2} = 2$ and the first term is

$a_1 = 2$. Therefore, the general term is

$$\begin{aligned}
a_n &= a_1 r^{n-1} \\
a_n &= 2^1 (2)^{n-1} \\
a_n &= 2^n
\end{aligned}$$

17.

Given the geometric sequence

$\dfrac{25}{4}, \dfrac{125}{8}, \dfrac{625}{16}, \dfrac{3125}{32}, \cdots$, the common ratio is

$\dfrac{a_2}{a_1} = \dfrac{125}{8} \div \dfrac{25}{4} = \dfrac{125}{8} \cdot \dfrac{4}{25} = \dfrac{500}{200} = \dfrac{5}{2}$ the first term is

$a_1 = \dfrac{25}{4}$. Therefore, the general term is

$$\begin{aligned}
a_n &= a_1 r^{n-1} \\
&= \frac{25}{4}\left(\frac{5}{2}\right)^{n-1} = \left(\frac{5}{2}\right)^2\left(\frac{5}{2}\right)^{n-1} \\
&= \left(\frac{5}{2}\right)^{n+1}
\end{aligned}$$

19.

Given the geometric sequence $-6,\ 18,\ -54,\ 162,\ \cdots,$

the common ratio is $\dfrac{a_2}{a_1} = \dfrac{18}{-6} = -3$ and the first term

is $a_1 = -6$. Therefore, the general term is

$$a_n = a_1 r^{n-1}$$
$$= -6(-3)^{n-1} = (-6)(-3)^{-1}(-3)^n$$
$$= (-6)\left(-\dfrac{1}{3}\right)(-3)^n$$
$$= 2(-3)^n$$

21.

Given the geometric sequence $16,\ 12,\ 9,\ \dfrac{27}{4},\ \cdots,$

the common ratio is $\dfrac{a_2}{a_1} = \dfrac{12}{16} = \dfrac{3}{4}$ and the first term

is $a_1 = 16$. Therefore, the general term is

$$a_n = a_1 r^{n-1}$$
$$= 16\left(\dfrac{3}{4}\right)^{n-1}$$

23.

Given the geometric sequence $\dfrac{7}{8},\dfrac{49}{16},\dfrac{343}{32},\dfrac{2401}{64},\ \cdots,$

the common ratio is $\dfrac{a_2}{a_1} = \dfrac{49}{16} \div \dfrac{7}{8} = \dfrac{49}{16}\cdot\dfrac{8}{7} = \dfrac{7}{2}$ and

the first term is $\dfrac{7}{8}$. Therefore, the general term is

$$a_n = a_1 r^{n-1}$$
$$= \dfrac{7}{8}\left(\dfrac{7}{2}\right)^{n-1} = \dfrac{7}{8}\dfrac{(7)^{n-1}}{(2)^{n-1}}$$
$$= \dfrac{(7)^1}{(2)^3}\dfrac{(7)^{n-1}}{(2)^{n-1}} = \dfrac{7^n}{(2)^{n+2}}$$

25.

The indicated sum S_6 given $a_1 = -3$ and $r = 2$ is

$$S_n = \dfrac{a_1 - a_1 r^n}{1-r}$$
$$S_6 = \dfrac{-3-(-3)(2)^6}{1-2}$$
$$= \dfrac{-3-(-3)(64)}{-1}$$
$$= \dfrac{-3+192}{-1} = -189$$

27.

S_5 in the geometric sequence $8,\ 12,\ 18,\ 27,\ \cdots,$ can

be found from the common ratio $r = \dfrac{a_2}{a_1} = \dfrac{12}{8} = \dfrac{3}{2}$

and $a_1 = 8$.

Thus, $\quad S_n = \dfrac{a_1 - a_1 r^n}{1-r}$

$$S_5 = \dfrac{8-8\left(\dfrac{3}{2}\right)^5}{1-\dfrac{3}{2}}$$
$$= \dfrac{8-8\left(\dfrac{243}{32}\right)}{-\dfrac{1}{2}}$$
$$= \left(8-\dfrac{243}{4}\right)(-2)$$
$$= -16+\dfrac{243}{2}$$
$$= -\dfrac{32}{2}+\dfrac{243}{2} = \dfrac{211}{2}$$

29.

S_5 in the geometric sequence

$$\dfrac{5}{2},\ -\dfrac{25}{4},\ \dfrac{125}{8},\ -\dfrac{625}{16},\ \cdots,$$

can be found from the common ratio

$$r = \dfrac{a_2}{a_1} = -\dfrac{25}{4} \div \dfrac{5}{2} = -\dfrac{25}{4}\cdot\dfrac{2}{5} = -\dfrac{50}{20} = -\dfrac{5}{2}$$

and $a_1 = \dfrac{5}{2}$.

Thus, $\quad S_n = \dfrac{a_1 - a_1 r^n}{1-r}$

$$S_5 = \dfrac{\dfrac{5}{2}-\dfrac{5}{2}\left(-\dfrac{5}{2}\right)^5}{1-\left(-\dfrac{5}{2}\right)}$$
$$= \dfrac{\dfrac{5}{2}-\dfrac{5}{2}\left(-\dfrac{3125}{32}\right)}{\dfrac{7}{2}}$$
$$= \dfrac{5-5\left(-\dfrac{3125}{32}\right)}{7}$$

#29. continued

$$= \frac{5 + \frac{15625}{32}}{7}$$

$$= \frac{5}{7} + \frac{15625}{32}\left(\frac{1}{7}\right)$$

$$= \frac{5 \cdot 32}{7 \cdot 32} + \frac{15625}{7 \cdot 32} = \frac{15785}{224}$$

$$= \frac{2255}{32}$$

31.

S_∞ in the geometric sequence $9, \ 6, \ 4, \ \frac{8}{3}, \ \cdots, $ can

be found from the common ratio $r = \frac{a_2}{a_1} = \frac{6}{9} = \frac{2}{3}$

and $a_1 = 9$. Therefore, since $|r| < 1$ we can use the
infinite geometric sequence formula.

$$S_\infty = \frac{a_1}{1 - r}$$

$$= \frac{9}{1 - \frac{2}{3}}$$

$$= \frac{9}{\frac{1}{3}} = 9(3) = 27$$

33.

S_∞ in the geometric sequence $\frac{1}{3}, -\frac{1}{9}, \frac{1}{27}, -\frac{1}{81}, \cdots,$

can be found from the common ratio

$r = \frac{a_2}{a_1} = -\frac{1}{9} \div \frac{1}{3} = -\frac{1}{9} \cdot \frac{3}{1} = -\frac{1}{3}$ and $a_1 = \frac{1}{3}$.

Thus $\quad S_\infty = \frac{a_1}{1 - r} \qquad$ For $|r| < 1$

$$= \frac{\frac{1}{3}}{1 - \left(-\frac{1}{3}\right)} = \frac{\frac{1}{3}}{\frac{4}{3}} = \frac{1}{4}$$

35.

S_∞ in the geometric sequence $2, \ 4, \ 8, \ 16, \ \cdots,$ can

<u>not</u> be found since the common ratio is $r = \frac{a_2}{a_1} = \frac{4}{2} = 2$.

By definition, $|r|$ must be less than 1.

37.

Expanding the series, we have

$$\sum_{i=1}^{6} 2^i = 2^1 + 2^2 + 2^3 + \cdots + 2^6$$

$$= 2 + 4 + 8 + \cdots + 64$$

Since $r = \frac{a_2}{a_1} = \frac{4}{2} = 2$ and $a_1 = 2$, we can apply the

sum of a geometric sequence formula.

Thus, $\quad S_n = \frac{a_1 - a_1 r^n}{1 - r}$

$$S_6 = \frac{2 - 2(2)^6}{1 - 2}$$

$$= \frac{2 - 2(64)}{1 - 2}$$

$$= \frac{2 - 128}{-1} = \frac{-126}{-1} = 126$$

39.

Expanding the series, we have

$$\sum_{i=1}^{5} 6\left(\frac{2}{3}\right)^{i-1} = 6\left(\frac{2}{3}\right)^0 + 6\left(\frac{2}{3}\right)^1 + 6\left(\frac{2}{3}\right)^2 + \cdots + 6\left(\frac{2}{3}\right)^4$$

$$= 6(1) + 6\left(\frac{2}{3}\right) + 6\left(\frac{4}{9}\right) + \cdots + 6\left(\frac{16}{81}\right)$$

$$= 6 + 4 + \frac{8}{3} + \cdots \frac{32}{27}$$

Since $r = \frac{a_2}{a_1} = \frac{4}{6} = \frac{2}{3}$ and $a_1 = 6$, we can apply the

sum of a geometric sequence formula.

Thus, $\quad S_n = \frac{a_1 - a_1 r^n}{1 - r}$

$$S_5 = \frac{6 - 6\left(\frac{2}{3}\right)^5}{1 - \frac{2}{3}}$$

$$= \frac{6 - 6\left(\frac{32}{243}\right)}{\frac{1}{3}} \cdot \frac{3}{3}$$

$$= 18 - 18\left(\frac{32}{243}\right)$$

$$= 18 - 2\left(\frac{32}{27}\right) = \frac{486}{27} - \frac{64}{27}$$

$$= \frac{422}{27}$$

41.
Expanding the series, we have

$$\sum_{i=1}^{4} \frac{1}{10}\left(\frac{1}{2}\right)^i$$

$$= \frac{1}{10}\left(\frac{1}{2}\right)^1 + \frac{1}{10}\left(\frac{1}{2}\right)^2 + \frac{1}{10}\left(\frac{1}{2}\right)^3 + \frac{1}{10}\left(\frac{1}{2}\right)^4$$

$$= \frac{1}{20} + \frac{1}{40} + \frac{1}{80} + \frac{1}{160}.$$

Since $r = \dfrac{a_2}{a_1} = \dfrac{1}{40} \div \dfrac{1}{20} = \dfrac{1}{40} \cdot \dfrac{20}{1} = \dfrac{1}{2}$ and $a_1 = \dfrac{1}{20}$,

the sum of the geometric sequence is

$$S_n = \frac{a_1 - a_1 r^n}{1 - r}$$

$$S_4 = \frac{\dfrac{1}{20} - \dfrac{1}{20}\left(\dfrac{1}{2}\right)^4}{1 - \dfrac{1}{2}}$$

$$= \frac{\dfrac{1}{20} - \dfrac{1}{20}\left(\dfrac{1}{16}\right)}{\dfrac{1}{2}}$$

$$= 2\left(\frac{1}{20} - \frac{1}{20}\left(\frac{1}{16}\right)\right)$$

$$= 2\left(\frac{1}{20}\right)\left(1 - \frac{1}{16}\right)$$

$$= \frac{1}{10}\left(\frac{15}{16}\right) = \frac{15}{160} = \frac{3}{32}$$

43.
Expanding the series, we have

$$\sum_{i=1}^{\infty} 3^i = 3^1 + 3^2 + 3^3 + \cdots 3^\infty$$

$$= 3 + 9 + 27 + \cdots + 3^\infty$$

Since $r = \dfrac{a_2}{a_1} = \dfrac{9}{3} = 3$, the sum of the infinite sequence

can <u>not</u> be found since $|r|$ must be less than 1.

45.
Expanding the series, we have,

$$\sum_{i=1}^{\infty}\left(\frac{2}{3}\right)^{i+1} = \left(\frac{2}{3}\right)^2 + \left(\frac{2}{3}\right)^3 + \cdots + \left(\frac{2}{3}\right)^\infty$$

$$= \frac{4}{9} + \frac{8}{27} + \cdots + \left(\frac{2}{3}\right)^\infty$$

#45. continued
Since $r = \dfrac{a_2}{a_1} = \dfrac{8}{27} \div \dfrac{4}{9} = \dfrac{8}{27} \cdot \dfrac{9}{4} = \dfrac{2}{3}$ and $a_1 = \dfrac{4}{9}$, the

sum of the infinite sequence is

$$S_\infty = \frac{a_1}{1 - r}$$

$$= \frac{\dfrac{4}{9}}{1 - \dfrac{2}{3}} = \frac{\dfrac{4}{9}}{\dfrac{1}{3}} = 3\left(\frac{4}{9}\right) = \frac{4}{3}$$

47.
Expanding the series, we have,

$$\sum_{i=1}^{\infty} 8\left(-\frac{1}{2}\right)^{i+2} = 8\left(-\frac{1}{2}\right)^3 + 8\left(-\frac{1}{2}\right)^4 + \cdots + 8\left(-\frac{1}{2}\right)^\infty$$

$$= 8\left(-\frac{1}{8}\right) + 8\left(\frac{1}{16}\right) + \cdots + 8\left(-\frac{1}{2}\right)^\infty$$

$$= -1 + \frac{1}{2} + \cdots + 8\left(\frac{1}{2}\right)^\infty$$

Since $r = \dfrac{\dfrac{1}{2}}{-1} = -\dfrac{1}{2}$ and $a_1 = -1$ the sum of the infinite

sequence is

$$S_\infty = \frac{a_1}{1 - r}$$

$$= \frac{-1}{1 - \left(-\dfrac{1}{2}\right)} = \frac{-1}{\dfrac{3}{2}} = -\frac{2}{3}$$

49.
$0.\overline{8} = 0.8888\ldots$

This can be expressed as an infinite geometric,

$$\frac{8}{10} + \frac{8}{100} + \frac{8}{1000} + \cdots,$$

with $r = \dfrac{a_2}{a_1} = \dfrac{8}{100} \div \dfrac{8}{10} = \dfrac{8}{100} \cdot \dfrac{10}{8} = \dfrac{1}{10}$ and $a_1 = \dfrac{8}{10}$.

Thus,

$$0.\overline{8} = S_\infty = \frac{a_1}{1 - r}$$

$$= \frac{\dfrac{8}{10}}{1 - \dfrac{1}{10}} = \frac{\dfrac{8}{10}}{\dfrac{9}{10}} = \frac{8}{9}$$

51.

$0.\overline{56} = 0.565656$

This can be expressed as an infinite geometric series,

$$\frac{56}{100} + \frac{56}{10000} + \frac{56}{1000000} + \cdots,$$

with $r = \dfrac{a_2}{a_1} = \dfrac{56}{10000} + \dfrac{56}{100} = \dfrac{56}{10000} \cdot \dfrac{100}{56} = \dfrac{1}{100}$

and $a_1 = \dfrac{56}{100}$. Thus,

$$0.\overline{56} = S_\infty = \frac{a_1}{1-r}$$

$$= \frac{\dfrac{56}{100}}{1 - \dfrac{1}{100}} = \frac{\dfrac{56}{100}}{\dfrac{99}{100}} = \frac{56}{99}$$

53.

$0.\overline{123} = 0.123123123\ldots$

This can be expressed as an infinite geometric series,

$$\frac{123}{10^3} + \frac{123}{10^6} + \frac{123}{10^9} + \cdots$$

with $r = \dfrac{a_2}{a_1} = \dfrac{123}{10^6} + \dfrac{123}{10^3} = \dfrac{123}{10^6} \cdot \dfrac{10^3}{123} = \dfrac{1}{1000}$ and

$a_1 = \dfrac{123}{1000}$. Thus,

$$0.\overline{123} = S_\infty = \frac{a_1}{1-r}$$

$$= \frac{\dfrac{123}{1000}}{1 - \dfrac{1}{1000}} = \frac{\dfrac{123}{1000}}{\dfrac{999}{1000}} = \frac{123}{999} = \frac{41}{333}$$

55.

$0.5\overline{2} = 0.522222\ldots = 0.5 + 0.02222\ldots$

This can be expressed as an infinite geometric series,

$$0.5 + \underbrace{\left[\frac{2}{100} + \frac{2}{1000} + \frac{2}{10000} + \cdots\right]}_{\text{This is an infinite series to be evaluated}}$$

with $r = \dfrac{a_2}{a_1} = \dfrac{2}{1000} + \dfrac{2}{100} = \dfrac{2}{1000} \cdot \dfrac{100}{2} = \dfrac{1}{10}$ and

$a_1 = \dfrac{2}{100}$. Thus,

#55. continued

$$0.0\overline{2} = S_\infty = \frac{a_1}{1-r}$$

$$= \frac{\dfrac{2}{100}}{1 - \dfrac{1}{10}} = \frac{\dfrac{2}{100}}{\dfrac{9}{10}} = \frac{2}{100} \cdot \frac{10}{9} = \frac{2}{90}$$

Therefore, $0.5\overline{2} = 0.5 + \dfrac{2}{90}$

$$= \frac{1}{2} + \frac{2}{90} = \frac{45}{90} + \frac{2}{90} = \frac{47}{90}$$

57.

$1.\overline{81} = 1 + 0.81818181\ldots$

This can be expressed as an infinite geometric series,

$$1 + \underbrace{\left[\frac{81}{100} + \frac{81}{10000} + \frac{81}{1000000} + \cdots\right]}_{\text{This is an infinite series to be evaluated}}$$

with $r = \dfrac{a_2}{a_1} = \dfrac{81}{10000} + \dfrac{81}{100} = \dfrac{81}{10000} \cdot \dfrac{100}{81} = \dfrac{1}{100}$

and $a_1 = \dfrac{81}{100}$. Thus,

$$0.\overline{81} = S_\infty = \frac{a_1}{1-r}$$

$$= \frac{\dfrac{81}{100}}{1 - \dfrac{1}{100}} = \frac{\dfrac{81}{100}}{\dfrac{99}{100}} = \frac{81}{99} = \frac{9}{11}$$

Therefore, $1.\overline{81} = 1 + \dfrac{9}{11} = \dfrac{20}{11}$

59.

$2.9\overline{34} = 2.9 + 0.0343434\ldots$

This can be expressed as an infinite geometric series,

$$2.9 + \underbrace{\left[\frac{34}{10^3} + \frac{34}{10^5} + \frac{34}{10^7} + \cdots\right]}_{\text{This is an infinite series to be evaluated}}$$

with $r = \dfrac{a_2}{a_1} = \dfrac{34}{10^5} + \dfrac{34}{10^3} = \dfrac{34}{10^5} \cdot \dfrac{10^3}{34} = \dfrac{1}{100}$

and $a_1 = \dfrac{34}{10^3}$. Thus,

#59. continued

$$0.0\overline{34} = S_\infty = \frac{a_1}{1-r}$$

$$= \frac{\frac{34}{10^3}}{1-\frac{1}{100}} = \frac{\frac{34}{1000}}{\frac{99}{100}} = \frac{34}{1000} \div \frac{99}{100}$$

$$= \frac{34}{1000} \cdot \frac{100}{99} = \frac{34}{990}$$

Therefore, $2.9\overline{34} = 2.9 + \frac{34}{990}$

$$= 2\frac{9}{10} + \frac{34}{990}$$

$$= \frac{29}{10} + \frac{34}{990}$$

$$= \frac{2871}{990} + \frac{34}{990} = \frac{2905}{990} = \frac{581}{198}$$

61.

We can describe the motion of the ball pictorially below:

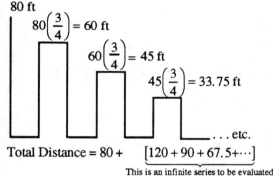

Initial position
80 ft

$80\left(\frac{3}{4}\right) = 60$ ft

$60\left(\frac{3}{4}\right) = 45$ ft

$45\left(\frac{3}{4}\right) = 33.75$ ft

. . . etc.

Total Distance $= 80 + \underbrace{\left[120 + 90 + 67.5 + \cdots\right]}_{\text{This is an infinite series to be evaluated}}$

From the above, $a_1 = 120$ and $r = \frac{a_2}{a_1} = \frac{90}{120} = \frac{3}{4}$.

Evaluating the infinite series portion, we have,

$$S_\infty = \frac{a_1}{1-r}$$

$$= \frac{120}{1-\frac{3}{4}} = \frac{120}{\frac{1}{4}} = 120(4) = 480$$

Thus, $80 + S_\infty = 80 + 480 = 560$

Answer: 560 ft

63.

The depreciation of the home can be described for the first three years as follows:

#63. continued

1st year $= (60000)(0.1) = \$6000 = a_1$

2nd year $= (60000 - 6000)(0.1) = \$5460 = a_2$

3rd year $= (60000 - 6000 - 5460)(0.1) = \$4860 = a_3$

To evaluate the total depreciation after 6 years we must evaluate the geometric series that the above represents for n = 6. Therefore, $r = \frac{a_2}{a_1} = \frac{5400}{6000} = 0.9$

and $a_1 = 6000$, so,

$$S_n = \frac{a_1 - a_1 r^n}{1-r}$$

$$S_6 = \frac{6000 - 6000(0.9)^6}{1 - 0.9}$$

$$= \frac{6000 - 6000(0.531441)}{0.1}$$

$$= \frac{2811.354}{0.1} = \$28113.54$$

Answer: $\$60000 - \$28113.54 = \$31886.46$

65.

The total winnings can be described for the first three holes as follows:

1st hole $= 2^1 = \$2 = a_1$

2nd hole $= 2^2 = \$4 = a_2$

3rd hole $= 2^3 = \$8 = a_3$

To evaluate the total winnings after 10 holes, we evaluate the represented geometric series with

$r = \frac{a_2}{a_1} = \frac{4}{2} = 2$ and $a_1 = 2$. Thus,

$$S_n = \frac{a_1 - a_1 r^n}{1-r}$$

$$S_{10} = \frac{2 - 2(2)^{10}}{1-2}$$

$$= \frac{2 - 2(1024)}{-1} = \frac{2 - 2048}{-1} = 2046$$

Answer: $\$2046$

Exercises 11.4

1.

$(x+2y)^6$

$= x^6 + \_x^5(2y) + \_x^4(2y)^2 + \_x^3(2y)^3$

$\quad + \_x^2(2y)^4 + \_x(2y)^5 + \_(2y)^6$

Using the 6th row of Pascal's Triangle, we have,

$= x^6 + 6x^5(2y) + 15x^4(4y^2) + 20x^3(8y^3)$

$\quad + 15x^2(16y^4) + 6x(32y^5) + 64y^6$

$= x^6 + 12x^5y + 60x^4y^2 + 160x^3y^3 + 240x^2y^4$

$\quad + 192xy^5 + 64y^6$

3.

$(x-3)^5$

$= x^5 + \_x^4(-3) + \_x^3(-3)^2 + \_x^2(-3)^3$

$\quad + \_x(-3)^4 + \_(-3)^5$

Using the 5th row of Pascal's Triangle, we have,

$= x^5 + 5x^4(-3) + 10x^3(9) + 10x^2(-27)$

$\quad + 5x(81) + (-243)$

$= x^5 - 15x^4 + 90x^3 - 270x^2 + 405x - 243$

5.

$\left(\dfrac{5}{2} - 2a\right)^4$

$= \left(\dfrac{5}{2}\right)^4 + \_\left(\dfrac{5}{2}\right)^3(-2a) + \_\left(\dfrac{5}{2}\right)^2(-2a)^2$

$\quad + \_\left(\dfrac{5}{2}\right)(-2a)^3 + \_(-2a)^4$

Using the 4th row of Pascal's Triangle, we have,

$= \dfrac{625}{16} + 4\left(\dfrac{125}{8}\right)(-2a) + 6\left(\dfrac{25}{4}\right)(4a^2) + 4\left(\dfrac{5}{2}\right)(-8a^3)$

$\quad + 16a^4$

$= \dfrac{625}{16} - 125a + 150a^2 - 80a^3 + 16a^4$

7.

$5! = 5 \times 4 \times 3 \times 2 \times 1 = 120$

9.

$\dfrac{4!}{8!} = \dfrac{\cancel{4!}}{8 \times 7 \times 6 \times 5 \times \cancel{4!}} = \dfrac{1}{8 \times 7 \times 6 \times 5} = \dfrac{1}{1680}$

11.

$\dfrac{10!}{4!} = \dfrac{10 \times 9 \times 8 \times 7 \times 6 \times 5 \times \cancel{4!}}{\cancel{4!}}$

$\quad = 10 \times 9 \times 8 \times 7 \times 6 \times 5 = 151200$

13.

$\dfrac{4 \cdot 3 \cdot 2}{3!} = \dfrac{4 \cdot \cancel{3!}}{\cancel{3!}} = 4$

15.

$(a+3)^4$

Applying the Binomial Theorem,

$= a^4 + \dfrac{4}{1!}a^{4-1}(3) + \dfrac{4(4-1)}{2!}a^{4-2}(3)^2$

$\quad + \dfrac{4(4-1)(4-2)}{3!}a^{4-3}(3)^3 + 3^4$

$= a^4 + (4)a^3(3) + \dfrac{4(3)}{2 \cdot 1}a^2(9) + \dfrac{4(3)(2)}{3 \cdot 2 \cdot 1}a(27) + 81$

$= a^4 + 12a^3 + 54a^2 + 108a + 81$

17.

$(3x+4y)^4$

Applying the Binomial Theorem,

$= (3x)^4 + \dfrac{4}{1!}(3x)^{4-1}(4y) + \dfrac{4(4-1)}{2!}(3x)^{4-2}(4y)^2$

$\quad + \dfrac{4(4-1)(4-2)}{3!}(3x)^{4-3}(4y)^3 + (4y)^4$

$= 81x^4 + 4(27)x^3(4y) + \dfrac{4(3)}{2 \cdot 1}(9)x^2(16y^2)$

$\quad + \dfrac{4(3)(2)}{3 \cdot 2 \cdot 1}(3x)(64y^3) + 256y^4$

$= 81x^4 + 432x^3y + 864x^2y^2 + 768xy^3 + 256y^4$

19.

$\left(a - \dfrac{3}{2}b\right)^5$

Applying the Binomial Theorem,

$= a^5 + \dfrac{5}{1!}(a^4)\left(-\dfrac{3}{2}b\right) + \dfrac{5(5-1)}{2!}(a^{5-2})\left(-\dfrac{3}{2}b\right)^2$

$\quad + \dfrac{5(5-1)(5-2)}{3!}(a^{5-3})\left(-\dfrac{3}{2}b\right)^3$

$\quad + \dfrac{5(5-1)(5-2)(5-3)}{4!}(a^{5-4})\left(-\dfrac{3}{2}b\right)^4$

$\quad + \left(-\dfrac{3}{2}b\right)^5$

#19. continued

$$= a^5 + (5)(a^4)\left(-\frac{3}{2}b\right) + \frac{5(4)}{2 \cdot 1}(a^3)\left(\frac{9}{4}b^2\right)$$

$$+ \frac{5(4)(3)}{3 \cdot 2 \cdot 1}(a^2)\left(-\frac{27}{8}b^3\right)$$

$$+ \frac{5(4)(3)(2)}{4 \cdot 3 \cdot 2 \cdot 1}(a)\left(\frac{81}{16}b^4\right) + \left(-\frac{243}{32}b^5\right)$$

$$= a^5 - \frac{15}{2}a^4b + \frac{45}{2}a^3b^2 - \frac{135}{4}a^2b^3$$

$$+ \frac{405}{16}ab^4 - \frac{243}{32}b^5$$

21.

$$(x - 2y)^7$$

Applying the Binomial Theorem,

$$= x^7 + \frac{7}{1!}x^{7-1}(-2y) + \frac{7(7-1)}{2!}x^{7-2}(-2y)^2$$

$$+ \frac{7(7-1)(7-2)}{3!}x^{7-3}(-2y)^3$$

$$+ \frac{7(7-1)(7-2)(7-3)}{4!}x^{7-4}(-2y)^4$$

$$+ \frac{7(7-1)(7-2)(7-3)(7-4)}{5!}x^{7-5}(-2y)^5$$

$$+ \frac{7(7-1)(7-2)(7-3)(7-4)(7-5)}{6!}x^{7-6}(-2y)^6$$

$$+ (-2y)^7$$

$$= x^7 + 7x^6(-2y) + \frac{7(6)}{2 \cdot 1}x^5(4y^2) + \frac{7(6)(5)}{3 \cdot 2 \cdot 1}x^4(-8y^3)$$

$$+ \frac{7(6)(5)(4)}{4 \cdot 3 \cdot 2 \cdot 1}x^3(16y^4) + \frac{7(6)(5)(4)}{5 \cdot 4 \cdot 3 \cdot 2 \cdot 1}x^2(-32y^5)$$

$$+ \frac{7(6)(5)(4)(3)(2)}{6 \cdot 5 \cdot 4 \cdot 3 \cdot 2 \cdot 1}x(64y^6) + (-128y^7)$$

$$= x^7 - 14x^6y + 84x^5y^2 - 280x^4y^3 + 560x^3y^3$$

$$- 672x^2y^5 + 448xy^6 - 128y^7$$

23.

$$\left(\frac{1}{2}x + y\right)^6$$

Applying the Binomial Theorem,

#23. continued

$$= \left(\frac{1}{2}x\right)^6 + \frac{6}{1!}\left(\frac{1}{2}x\right)^{6-1}y + \frac{6(6-1)}{2!}\left(\frac{1}{2}x\right)^{6-2}y^2$$

$$+ \frac{6(6-1)(6-2)}{3!}\left(\frac{1}{2}x\right)^{6-3}y^3$$

$$+ \frac{6(6-1)(6-2)(6-3)}{4!}\left(\frac{1}{2}x\right)^{6-4}y^4$$

$$+ \frac{6(6-1)(6-2)(6-3)(6-4)}{5!}\left(\frac{1}{2}x\right)^{6-5}y^5$$

$$+ y^6$$

$$= \frac{1}{64}x^6 + 6\left(\frac{1}{32}x^5\right)y + \frac{6(5)}{2 \cdot 1}\left(\frac{1}{16}x^4\right)y^2$$

$$+ \frac{6(5)(4)}{3 \cdot 2 \cdot 1}\left(\frac{1}{8}x^3\right)y^3 + \frac{6(5)(4)(3)}{4 \cdot 3 \cdot 2 \cdot 1}\left(\frac{1}{4}x^2\right)y^4$$

$$+ \frac{6(5)(4)(3)(2)}{5 \cdot 4 \cdot 3 \cdot 2 \cdot 1}\left(\frac{1}{2}x\right)y^5 + y^6$$

$$= \frac{1}{64}x^6 + \frac{3}{16}x^5y + \frac{15}{16}x^4y^2 + \frac{5}{2}x^3y^3 + \frac{15}{4}x^2y^4$$

$$+ 3xy^5 + y^6$$

25.

$$(x + 2y^2)^5$$

Applying the Binomial Theorem,

$$= x^5 + \frac{5}{1!}(x^4)(2y^2) + \frac{5(5-1)}{2!}(x^{5-2})(2y^2)^2$$

$$+ \frac{5(5-1)(5-2)}{3!}(x^{5-3})(2y^2)^3$$

$$+ \frac{5(5-1)(5-2)(5-3)}{4!}(x^{5-4})(2y^2)^4$$

$$+ (2y^2)^5$$

$$= x^5 + (5)(x^4)(2y^2) + \frac{5(4)}{2 \cdot 1}(x^3)(4y^4)$$

$$+ \frac{5(4)(3)}{3 \cdot 2 \cdot 1}(x^2)(8y^6)$$

$$+ \frac{5(4)(3)(2)}{4 \cdot 3 \cdot 2 \cdot 1}(x)(16y^8) + 32y^{10}$$

$$= x^5 + 10x^4y^2 + 40x^3y^4 + 80x^2y^6 + 80xy^8 + 32y^{10}$$

Review Exercises

1.

$a_n = 4n + 2$

$a_1 = 4 + 2 = 6$

$a_2 = 8 + 2 = 10$

$a_3 = 12 + 2 = 14$

$a_4 = 16 + 2 = 18$

$a_5 = 20 + 2 = 22$

First five terms are 6, 10, 14, 18, 22.

3.

$a_n = 2n^2 - 3$

$a_1 = 2(1)^2 - 3 = -1$

$a_2 = 2(2)^2 - 3 = 5$

$a_3 = 2(3)^2 - 3 = 15$

$a_4 = 2(4)^2 - 3 = 29$

$a_5 = 2(5)^2 - 3 = 47$

First five terms are -1, 5, 15, 29, 47.

5.

$a_n = (-1)^n (3n + 2)$

$a_1 = (-1)^1 (3(1) + 2) = (-1)(5) = -5$

$a_2 = (-1)^2 (3(2) + 2) = (1)(8) = 8$

$a_3 = (-1)^3 (3(3) + 2) = (-1)(11) = -11$

$a_4 = (-1)^4 (3(4) + 2) = (1)(14) = 14$

$a_5 = (-1)^5 (3(5) + 2) = (-1)(17) = -17$

First five terms are -5, 8, -11, 14, -17.

7.

$a_n = 6n - 5$

$a_4 = 6(4) - 5 = 19$

9.

$a_n = 8 \cdot \left(\dfrac{3}{2}\right)^n$

$a_4 = 8 \cdot \left(\dfrac{3}{2}\right)^4 = 8 \cdot \left(\dfrac{81}{16}\right) = \dfrac{81}{2}$

11.

$a_n = 3^{n-1}$

$a_1 = 1 \quad a_2 = 3 \quad a_3 = 9 \quad a_4 = 27 \quad a_5 = 81$

$S_5 = 1 + 3 + 9 + 27 + 81 = 121$

13.

$a_n = \left(\dfrac{1}{2}\right)^n + 1$

$a_1 = \dfrac{1}{2} + 1 = \dfrac{3}{2}$

$a_2 = \dfrac{1}{4} + 1 = \dfrac{5}{4}$

$a_3 = \dfrac{1}{8} + 1 = \dfrac{9}{8}$

$a_4 = \dfrac{1}{16} + 1 = \dfrac{17}{16}$

$a_5 = \dfrac{1}{32} + 1 = \dfrac{33}{32}$

$S_5 = \dfrac{3}{2} + \dfrac{5}{4} + \dfrac{9}{8} + \dfrac{17}{16} + \dfrac{33}{32}$

$= \dfrac{48}{32} + \dfrac{40}{32} + \dfrac{36}{32} + \dfrac{34}{32} + \dfrac{33}{32} = \dfrac{191}{32}$

15.

$a_n = \dfrac{1}{2n}$

$a_1 = \dfrac{1}{2} \quad a_2 = \dfrac{1}{4} \quad a_3 = \dfrac{1}{6} \quad a_4 = \dfrac{1}{8} \quad a_5 = \dfrac{1}{10}$

$S_5 = \dfrac{1}{2} + \dfrac{1}{4} + \dfrac{1}{6} + \dfrac{1}{8} + \dfrac{1}{10}$

$= \dfrac{60}{120} + \dfrac{30}{120} + \dfrac{20}{120} + \dfrac{15}{120} + \dfrac{12}{120} = \dfrac{137}{120}$

17.

$\displaystyle\sum_{i=1}^{3} \left(\dfrac{2}{5}\right)^i = \left(\dfrac{2}{5}\right)^1 + \left(\dfrac{2}{5}\right)^2 + \left(\dfrac{2}{5}\right)^3$

$= \dfrac{2}{5} + \dfrac{4}{25} + \dfrac{8}{125}$

$= \dfrac{50}{125} + \dfrac{20}{125} + \dfrac{8}{125} = \dfrac{78}{125}$

19.

$$\sum_{i=0}^{6} \left(i^2 + i + 1\right)$$

$$= \left(0^2 + 0 + 1\right) + \left(1^2 + 1 + 1\right) + \left(2^2 + 2 + 1\right) + \left(3^2 + 3 + 1\right)$$

$$+ \left(4^2 + 4 + 1\right) + \left(5^2 + 5 + 1\right) + \left(6^2 + 6 + 1\right)$$

$$= 1 + 3 + 7 + 13 + 21 + 31 + 43 = 119$$

21.

$$\sum_{i=1}^{4} (-1)^i (2i + 1)$$

$$= (-1)^1 (2(1) + 1) + (-1)^2 (2(2) + 1) + (-1)^3 (2(3) + 1)$$

$$+ (-1)^4 (2(4) + 1)$$

$$= -3 + 5 - 7 + 9 = 4$$

23.

Given the series $2 + 4 + 8 + 16 + 32 + 64$ we can see that this is based on a geometric sequence.

Therefore, the common ratio is $r = \dfrac{a_2}{a_1} = \dfrac{4}{2} = 2$

and the first term is $a_1 = 2$ for $n = 6$.

Thus, the general term is

$$a_n = a_1 r^{n-1}$$

$$= 2(2)^{n-1}$$

Expressed as a summation notation, we have,

$$\sum_{i=1}^{6} 2(2)^{i-1} = \sum_{i=1}^{6} 2^i \qquad \text{Replacing n with i}$$

25.

Given the series $\dfrac{3}{2} + \dfrac{4}{3} + \dfrac{5}{4} + \dfrac{6}{5} + \dfrac{7}{6} + \dfrac{8}{7}$ we can see that the numerator and denominator, taken separately, form an arithmetic sequence.

Therefore, for the numerator, $d = 1$ and $a_1 = 3$, and for the denominator, $d = 1$ and $a_1 = 2$. Thus, the arithmetic sequence general terms are:

Numerator: $a_n = a_1 + (n-1)d$

$$= 3 + (n-1) = 2 + n$$

Denominator: $a_n = a_1 + (n-1)d$

$$= 2 + (n-1) = 1 + n$$

#25. continued

Expressed as a summation notation, we have,

$$\sum_{i=1}^{6} \frac{i+2}{i+1} \qquad \text{Replacing n with i}$$

27.

Gus's typing at the end of each week can be expressed with an arithmetic sequence where the first term is $a_1 = 28$ and the the common difference is $d = 12$. Thus, on the 4th week we evaluate the arithmetic sequence general term for $n = 4$.

$$a_n = a_1 + (n-1)d$$

$$a_4 = 28 + (4-1)(12) = 28 + 36 = 64 \text{ wpm}$$

29.

Given the arithmetic sequence, $2, \dfrac{3}{2}, 1, \dfrac{1}{2}, 0, \cdots$ the common difference is

$$a_2 - a_1 = \frac{3}{2} - 2 = \frac{3}{2} - \frac{4}{2} = -\frac{1}{2}$$

31.

Given the arithmetic sequence, $4, 2, 0, -2, -4,$ $\cdots$, the first term is $a_1 = 4$ and the a common difference is $d = (a_2 - a_1) = (2 - 4) = -2$. Therefore,

$$a_n = a_1 + (n-1)d \quad \text{General Term Formula}$$

$$a_{24} = 4 + (24 - 1)(-2)$$

$$= 4 + (-46) = -42$$

33.

Given the arithmetic sequence $5, 8, 11, 14, \cdots$, the first term is $a_1 = 5$ and the common difference is $d = (a_2 - a_1) = (8 - 5) = 3$. Therefore, the general term is:

$$a_n = a_1 + (n-1)d$$

$$= 5 + (n-1)3$$

$$= 5 + 3n - 3 = 3n + 2$$

35.

Given the arithmetic sequence $4, \dfrac{9}{2}, 5, \dfrac{11}{2}, \cdots$, the first term is $a_1 = 4$ and the common difference is $d = \left(a_2 - a_1\right) = \left(\dfrac{9}{2} - 4\right) = \dfrac{1}{2}$. Therefore, the general term is:

$$a_n = a_1 + (n-1)d$$
$$= 4 + (n-1)\dfrac{1}{2}$$
$$= 4 + \dfrac{1}{2}n - \dfrac{1}{2} = \dfrac{1}{2}n + \dfrac{7}{2}$$

37.

If $a_8 = 41$ and $a_{13} = 61$, then a_1 and d can be found by the general term formula:

$$a_8 = a_1 + (8-1)d = a_1 + 7d$$
$$a_{13} = a_1 + (13-1)d = a_1 + 12d$$

Substituting for a_8 and a_{13} yields:

(1) $\quad 41 = a_1 + 7d$

(2) $\quad 61 = a_1 + 12d$

(1a) $\quad -41 = -a_1 - 7d$

(2a) $\quad \underline{61 = a_1 + 12d}$

$\qquad\quad 20 = 5d$

$\qquad\quad\ 4 = d$

Substitute 4 for d in (1).
$$41 = a_1 + 7(4)$$
$$41 = a_1 + 28$$
$$41 - 28 = a_1$$
$$13 = a_1$$

Answer: $\quad a_1 = 13, \quad d = 4$

39.

Since we are given n, a_1, and a_n we apply Equation (3) (see 11.2) where $n = 15$, $a_1 = 8$, and $a_{15} = -14$.

$$S_n = \dfrac{n}{2}\left(a_1 + a_n\right)$$
$$S_{15} = \dfrac{15}{2}\left(8 + (-14)\right) = \dfrac{15}{2}(-6) = -45$$

41.

Since we are given n, a_1, and d we apply Equation (4) (see 11.2) where $n = 12$, $a_1 = \dfrac{5}{3}$, and $d = -\dfrac{2}{3}$.

$$S_n = \dfrac{n}{2}\left(2a_1 + (n-1)d\right)$$
$$S_{12} = \dfrac{12}{2}\left(2\left(\dfrac{5}{3}\right) + (12-1)\left(-\dfrac{2}{3}\right)\right)$$
$$= 6\left(\dfrac{10}{3} + \left(-\dfrac{22}{3}\right)\right)$$
$$= 6\left(-\dfrac{12}{3}\right) = -24$$

43.

Since $\left(\dfrac{3}{8}i - 2\right)$ is first degree i, then $\displaystyle\sum_{i=1}^{24}\left(\dfrac{3}{8}i - 2\right)$ is an arithmetic series. Thus,

$$a_1 = \left(\dfrac{3}{8}(1) - 2\right) = \left(\dfrac{3}{8} - \dfrac{16}{8}\right) = -\dfrac{13}{8}$$

and

$$a_{24} = \left(\dfrac{3}{8}(24) - 2\right) = (9 - 2) = 7$$

From Equation (3) (see 11.2), we have

$$S_n = \dfrac{n}{2}\left(a_1 + a_n\right)$$
$$S_{24} = \dfrac{24}{2}\left(-\dfrac{13}{8} + 7\right)$$
$$= \dfrac{24}{2}\left(-\dfrac{13}{8} + \dfrac{56}{8}\right)$$
$$= \dfrac{3}{2}(-13 + 56) = \dfrac{3}{2}(43) = \dfrac{129}{2}$$

45.

The sum of the even integers between 1 and 201 is an arithmetic series,

$$2 + 4 + 6 + \cdots + 200.$$

Therefore, $a_1 = 2$, $d = (4 - 2) = 2$, and $a_n = 200$.

n by inspection is $\dfrac{200}{2} = 100$.

Applying Equation (3) (see 11.2)

#45. continued

$$S_n = \frac{n}{2}(a_1 + a_n)$$

$$= \frac{100}{2}(2 + 200)$$

$$= 100(1 + 100)$$

$$= 100 + 10000 = 10,100$$

47.

The common ratio of 3, -9, 27, -81, $\cdots$, is

$$r = \frac{a_2}{a_1} = -\frac{9}{3} = -3.$$

49.

Given $a_1 = 7$ and $r = 3$,

$$a_n = a_1 r^{n-1}$$

$$a_5 = 7(3)^{5-1} = 7(3)^4 = 567$$

51.

Given the geometric sequence, 40, -20, 10, -5,

$\cdots$, the common ratio $r = \frac{-20}{40} = -\frac{1}{2}$ and the first

term $a_1 = 40$. The 8th term is:

$$a_n = a_1 r^{n-1}$$

$$a_8 = 40\left(-\frac{1}{2}\right)^{8-1}$$

$$= 40\left(-\frac{1}{2}\right)^7$$

$$= 40\left(-\frac{1}{128}\right) = -\frac{40}{128} = -\frac{5}{16}$$

53.

Given the geometric sequence -15, 45, -135, 405,

$\cdots$, the common ratio is $r = \frac{a_2}{a_1} = \frac{45}{-15} = -3$, and the

first term is $a_1 = -15$. The general term is

$$a_n = a_1 r^{n-1} = -15(-3)^{n-1}$$

55.

Given the geometric sequence $\frac{1}{6}$, $\frac{1}{10}$, $\frac{3}{50}$, $\frac{9}{250}$,

$\cdots$, the common ratio is $r = \frac{a_2}{a_1} = \left(\frac{1}{10} \div \frac{1}{6}\right) = \left(\frac{1}{10} \cdot \frac{6}{1}\right)$

$= \frac{6}{10} = \frac{3}{5}$, and the first term is $a_1 = \frac{1}{6}$.

#55. continued

The general term is

$$a_n = a_1 r^{n-1} = \left(\frac{1}{6}\right)\left(\frac{3}{5}\right)^{n-1}$$

57.

S_5 in the geometric sequence 16, 20, 25, $\frac{125}{4}$, $\cdots$,

can be found from the common ratio $r = \frac{a_2}{a_1} = \frac{20}{16} = \frac{5}{4}$

and $a_1 = 16$. Thus,

$$S_n = \frac{a_1 - a_1 r^n}{1 - r}$$

$$S_5 = \frac{16 - 16\left(\frac{5}{4}\right)^5}{1 - \frac{5}{4}}$$

$$= \frac{16 - 16\left(\frac{3125}{1024}\right)}{-\frac{1}{4}}$$

$$= -4\left(16 - 16\left(\frac{3125}{1024}\right)\right)$$

$$= -4\left(16 - \frac{3125}{64}\right)$$

$$= -64 + \frac{3125}{16}$$

$$= -\frac{1024}{16} + \frac{3125}{16} = \frac{2101}{16}$$

59.

S_∞ in the geometric sequence, 16, 24, 36, 54, $\cdots$,

can <u>not</u> be found since the common ratio is

$r = \frac{a_2}{a_1} = \frac{24}{16} = \frac{3}{2}$. By definition, $|r|$ must be less

than 1.

61.

Expanding the series, we have,

$$\sum_{i=1}^{6} 16\left(-\frac{3}{2}\right)^{i-2} = 16\left(-\frac{3}{2}\right)^{-1} + 16\left(\frac{2}{3}\right)^0 + 16\left(\frac{2}{3}\right)^1$$

$$+ \cdots + 16\left(\frac{2}{3}\right)^4$$

#61. continued

$$= -\frac{32}{3} + 16 + \frac{32}{3} + \cdots + 16\left(\frac{16}{81}\right)$$

$$= -\frac{32}{3} + 16 + \frac{32}{3} + \cdots + \frac{256}{81}$$

Since $r = \frac{a_2}{a_1} = \frac{16}{1} + \left(-\frac{32}{3}\right) = \frac{16}{1} \cdot \left(-\frac{3}{32}\right) = -\frac{3}{2}$ and

$a_1 = -\frac{32}{3}$, we can apply the sum of a geometric

sequence formula. Thus,

$$S_n = \frac{a_1 - a_1 r^n}{1 - r}$$

$$S_6 = \frac{-\frac{32}{3} - \left(-\frac{32}{3}\right)\left(-\frac{3}{2}\right)^6}{1 - \left(-\frac{3}{2}\right)}$$

$$= \frac{-\frac{32}{3} + \frac{32}{3}\left(\frac{729}{64}\right)}{\frac{5}{2}}$$

$$= \frac{2}{5}\left(-\frac{32}{3} + \frac{1}{3}\left(\frac{729}{2}\right)\right)$$

$$= \frac{2}{5}\left(-\frac{64}{6} + \frac{729}{6}\right) = \frac{2}{5}\left(\frac{665}{6}\right) = \frac{133}{3}$$

63.

Expanding the series, we have,

$$\sum_{i=1}^{\infty} 4\left(\frac{1}{2}\right)^i = 4\left(\frac{1}{2}\right)^1 + 4\left(\frac{1}{2}\right)^2 + 4\left(\frac{1}{2}\right)^3 + \cdots + 4\left(\frac{1}{2}\right)^\infty$$

$$= 2 + 1 + \frac{1}{2} + \cdots + 4\left(\frac{1}{2}\right)^\infty$$

Because $r = \frac{a_2}{a_1} = \frac{1}{2}$ and $a_1 = 2$ we can apply the

infinite geometric sequence formula since $|r| < 1$.

Thus,

$$S_\infty = \frac{a_1}{1 - r} = \frac{2}{1 - \frac{1}{2}} = \frac{2}{\frac{1}{2}} = 4$$

65.

$0.6\bar{1} = 0.611111\ldots = 0.6 + 0.0111111\ldots$

This can be expressed as an infinite geometric series,

$$0.6 + \underbrace{\left[\frac{1}{100} + \frac{1}{1000} + \frac{1}{10000} + \cdots\right]}_{\text{This is an infinite series to be evalueted}}$$

#65. continued

with $r = \frac{a_2}{a_1} = \frac{1}{1000} + \frac{1}{100} = \frac{1}{1000} \cdot \frac{100}{1} = \frac{1}{10}$ and

$a_1 = \frac{1}{100}$. Thus,

$$0.0\bar{1} = S_\infty = \frac{a_1}{1 - r}$$

$$= \frac{\frac{1}{100}}{1 - \frac{1}{10}} = \frac{\frac{1}{100}}{\frac{9}{10}} = \frac{1}{100} \cdot \frac{10}{9} = \frac{1}{90}$$

Therefore, $0.6\bar{1} = 0.6 + \frac{1}{90}$

$$= \frac{3}{5} + \frac{1}{90}$$

$$= \frac{54}{90} + \frac{1}{90} = \frac{55}{90} = \frac{11}{18}$$

67.

$0.\overline{285} = 0.285285285\ldots$

This can be expressed as an infinite geometric series,

$$\frac{285}{10^3} + \frac{285}{10^6} + \frac{285}{10^9} + \cdots,$$

with $r = \frac{a_2}{a_1} = \frac{285}{10^6} + \frac{285}{10^3} = \frac{285}{10^6} \cdot \frac{10^3}{285} = \frac{1}{10^3}$ and

$a_1 = \frac{285}{10^3}$. Thus,

$$0.\overline{285} = S_\infty = \frac{a_1}{1 - r}$$

$$= \frac{\frac{285}{1000}}{1 - \frac{1}{1000}}$$

$$= \frac{\frac{285}{1000}}{\frac{999}{1000}} = \frac{285}{999} = \frac{95}{333}$$

69.

We can describe the motion of the ball pictorially below:

Initial position
100 ft

$100\left(\dfrac{1}{4}\right) = 25$ ft

$25\left(\dfrac{1}{4}\right) = 6.25$ ft

$6.25\left(\dfrac{1}{4}\right) = 1.5625$ ft

. . . etc.

Total distance $= 100 + \underbrace{[50 + 12.5 + 3.125 + \cdots]}$

This is an infinite series to be evaluated.

From the above, $a_1 = 50$ and $r = \dfrac{a_2}{a_1} = \dfrac{12.5}{50} = \dfrac{1}{4}$.

Evaluating the infinite series portion, we have:

$$S_\infty = \frac{a}{1-r}$$

$$= \frac{50}{1 - \dfrac{1}{4}} = \frac{50}{\dfrac{3}{4}} = 50 \cdot \frac{4}{3} = \frac{200}{3}$$

Total distance travelled is:

$$100 + S_\infty = 100 + \frac{200}{3} = \frac{300}{3} + \frac{200}{3} = \frac{500}{3} \text{ ft}$$

71.

$(3a + b)^6$

$= (3a)^6 + \underline{}(3a)^5 b + \underline{}(3a)^4 b^2 + \underline{}(3a)^3 b^3$

$ + \underline{}(3a)^2 b^4 + \underline{}(3a)b^5 + \underline{}b^6$

Using the 6th row of Pascal's Triangle, we have,

$= 729a^6 + 6\left(243a^5\right)b + 15\left(81a^4\right)b^2 + 20\left(27a^3\right)b^3$

$ + 15\left(9a^2\right)b^4 + 6(3a)b^5 + b^6$

$= 729a^6 + 1458a^5 b + 1215a^4 b^2 + 540a^3 b^3 + 135a^2 b^4$

$ + 18ab^5 + b^6$

73.

$7! = 7 \cdot 6 \cdot 5 \cdot 4 \cdot 3 \cdot 2 \cdot 1 = 5040$

75.

$(x + 4)^4$

Applying the Binomial Theorem,

$= x^4 + \dfrac{4}{1!}^{4-1} x^{4-1}(4) + \dfrac{4(4-1)}{2!} x^{4-2}(4)^2$

$ + \dfrac{4(4-1)(4-2)}{3!} x^{4-3}(4)^3 + 4^4$

$= x^4 + 4x^3(4) + \dfrac{4(3)}{2 \cdot 1} x^2 16 + \dfrac{4(3)(2)}{3 \cdot 2 \cdot 1} x(64) + 256$

$= x^4 + 16x^3 + 96x^2 + 256x + 256$

77.

$\left(\dfrac{1}{2}x - \dfrac{2}{3}y\right)^5$

Applying the Binomial Theorem,

$= \left(\dfrac{1}{2}x\right)^5 + \dfrac{5}{1!}\left(\dfrac{1}{2}x\right)^4\left(-\dfrac{2}{3}y\right)$

$ + \dfrac{5(5-1)}{2!}\left(\dfrac{1}{2}x\right)^{5-2}\left(-\dfrac{2}{3}y\right)^2$

$ + \dfrac{5(5-1)(5-2)}{3!}\left(\dfrac{1}{2}x\right)^{5-3}\left(-\dfrac{2}{3}y\right)^3$

$ + \dfrac{5(5-1)(5-2)(5-3)}{4!}\left(\dfrac{1}{2}x\right)^{5-4}\left(-\dfrac{2}{3}y\right)^4$

$ + \left(-\dfrac{2}{3}y\right)^5$

$= \dfrac{1}{32}x^5 + (5)\left(\dfrac{1}{16}x^4\right)\left(-\dfrac{2}{3}y\right) + \dfrac{5(4)}{2 \cdot 1}\left(\dfrac{1}{8}x^3\right)\left(\dfrac{4}{9}y^2\right)$

$ + \dfrac{5(4)(3)}{3 \cdot 2 \cdot 1}\left(\dfrac{1}{4}x^2\right)\left(-\dfrac{8}{27}y^3\right)$

$ + \dfrac{5(4)(3)(2)}{4 \cdot 3 \cdot 2 \cdot 1}\left(\dfrac{1}{2}x\right)\left(\dfrac{16}{81}y^4\right) + \left(-\dfrac{32}{243}y^5\right)$

$= \dfrac{1}{32}x^5 - \dfrac{5}{24}x^4 y + \dfrac{5}{9}x^3 y^2 - \dfrac{20}{27}x^2 y^3$

$ + \dfrac{40}{81}xy^4 - \dfrac{32}{243}y^5$

Chapter 11 Test Solutions

1.

Given the arithmetic sequence $4, \frac{11}{2}, 7, \frac{17}{2}, \cdots$,

the common difference is $d = a_2 - a_1 = \frac{11}{2} - \frac{8}{2} = \frac{3}{2}$

and the first term is $a_1 = 4$. Thus,

$$a_n = a_1 + (n-1)d$$
$$= 4 + (n-1)\left(\frac{3}{2}\right)$$
$$= 4 + \frac{3}{2}n - \frac{3}{2}$$
$$= \frac{3}{2}n + \frac{8}{2} - \frac{3}{2} = \frac{3}{2}n + \frac{5}{2}$$

and

$$S_n = \frac{n}{2}(2a_1 + (n-1)d)$$
$$S_{24} = \frac{24}{2}\left(2(4) + (24-1)\left(\frac{3}{2}\right)\right)$$
$$= 12\left(8 + (23)\left(\frac{3}{2}\right)\right)$$
$$= 6(16 + (23)(3)) = 6(85) = 510$$

3.

$$\sum_{i=1}^{5}\left(2i^2 - 3i + 7\right)$$
$$= \left(2(1)^2 - 3(1) + 7\right) + \left(2(2)^2 - 3(2) + 7\right)$$
$$\quad + \left(2(3)^2 - 3(3) + 7\right) + \left(2(4)^2 - 3(4) + 7\right)$$
$$\quad + \left(2(5)^2 - 3(5) + 7\right)$$
$$= 6 + 9 + 16 + 27 + 42 = 100$$

5.

$$\sum_{i=1}^{20}(5i - 7)$$
$$= (5(1) - 7) + (5(2) - 7) + (5(3) - 7) + \cdots + (5(20) - 7)$$
$$= (5 - 7) + (10 - 7) + (15 - 7) + \cdots + (100 - 7)$$
$$= -2 + 3 + 8 + \cdots + 93$$

Since the common difference is $(a_2 - a_1) = 5$, this is an arithmetic series. Therefore, with $a_1 = -2$,

#5. continued

$$S_n = \frac{n}{2}(2a_1 + (n-1)d)$$
$$S_{20} = \frac{20}{2}(2(-2) + (20-1)5)$$
$$= 10(-4 + 95) = 910$$

7.

$$\sum_{i=1}^{\infty}\left(\frac{9}{8}\right)^i = \left(\frac{9}{8}\right)^1 + \left(\frac{9}{8}\right)^2 + \left(\frac{9}{8}\right)^3 + \cdots + \left(\frac{9}{8}\right)^{\infty}$$
$$= \frac{9}{8} + \frac{81}{64} + \frac{729}{512} + \cdots + \left(\frac{9}{8}\right)^{\infty}$$

Since $r = \frac{a_2}{a_1} = \frac{81}{64} \div \frac{9}{8} = \frac{81}{64} \cdot \frac{8}{9} = \frac{9}{8}$, the sum of the

infinite series can <u>not</u> be found since $|r|$ must be less than 1.

9.

$0.\overline{56} = 0.565656\ldots$

This can be expressed as an infinite geometric series,

$$\frac{56}{100} + \frac{56}{10000} + \frac{56}{1000000} + \cdots,$$

with $r = \frac{a_2}{a_1} = \frac{56}{10000} \div \frac{56}{100} = \frac{56}{10000} \cdot \frac{100}{56} = \frac{1}{100}$

and $a_1 = \frac{56}{100}$. Thus,

$$0.\overline{56} = S_{\infty} = \frac{a_1}{1 - r}$$
$$= \frac{\frac{56}{100}}{1 - \frac{1}{100}} = \frac{\frac{56}{100}}{\frac{99}{100}} = \frac{56}{99}$$

11.
$$\frac{12!}{9!} = \frac{12 \cdot 11 \cdot 10 \cdot \cancel{9}!}{\cancel{9}!} = 12 \cdot 11 \cdot 10 = 1320$$

13.

The sale of hamburgers each day can be expressed as the following sequence:

$$a_1 = 64$$
$$a_2 = 64\left(\frac{1}{2}\right) = 32$$
$$a_3 = 64\left(\frac{1}{2}\right)\left(\frac{1}{2}\right) = 16$$

#13. continued

Since $r = \dfrac{a_2}{a_1} = \dfrac{64}{32} = \dfrac{1}{2}$ and $a_1 = 64$, S_n for $n = 7$ is,

$$S_n = \frac{a_1 - a_1 r^n}{1 - r}$$

$$S_7 = \frac{64 - 64\left(\dfrac{1}{2}\right)^7}{1 - \dfrac{1}{2}}$$

$$= \frac{64 - \dfrac{64}{128}}{\dfrac{1}{2}} = 2\left(64 - \frac{1}{2}\right) = 127$$

Answer: 127 hamburgers

Test Your Memory

1.

$$\left(\frac{250x^4 y}{16x^{-2} y^8}\right)^{-\frac{2}{3}} = \left(\frac{2 \cdot 125 x^{4+2}}{2 \cdot 8 y^{8-1}}\right)^{-\frac{2}{3}}$$

$$= \left(\frac{5^3 x^6}{2^3 y^7}\right)^{-\frac{2}{3}}$$

$$= \frac{5^{-2} x^{-4}}{2^{-2} y^{-14/3}} = \frac{4 y^{14/3}}{25 x^4}$$

3.

$$4\sqrt{20x} - \sqrt{180x} - 3\sqrt{45x}$$

$$= 4\sqrt{4 \cdot 5x} - \sqrt{36 \cdot 5x} - 3\sqrt{9 \cdot 5x}$$

$$= 8\sqrt{5} - 6\sqrt{5x} - 9\sqrt{5x} = -7\sqrt{5x}$$

5.

$$\frac{3x^3 + 4x^2 - 12x - 16}{15x^4 - 30x^3 - 120x^2} \cdot \frac{3x^2 - 12x}{x^3 - 8}$$

$$= \frac{(3x+4)(x+2)(x-2)}{15x^2(x-4)(x+2)} \cdot \frac{3x(x-4)}{(x-2)(x^2 + 2x + 4)}$$

$$= \frac{3x(3x+4)}{3x \cdot 5x(x^2 + 2x + 4)} = \frac{3x+4}{5x(x^2 + 2x + 4)}$$

7.

$$g(-3) = 2(-3)^2 - 1 = 18 - 1 = 17$$

9.

$$x = 3f^{-1}(x) + 2$$

$$x - 2 = 3f^{-1}(x)$$

$$\frac{x-2}{3} = f^{-1}(x)$$

11.

$(9,1), (-6,7)$

$$m = \frac{7-1}{-6-9} = \frac{6}{-15} = -\frac{2}{5}$$

$$y - 1 = -\frac{2}{5}(x - 9)$$

$$y - 1 = -\frac{2}{5}x + \frac{18}{5}$$

$$y = -\frac{2}{5}x + \frac{18}{5} + 1$$

$$y = -\frac{2}{5}x + \frac{23}{5}$$

13.-19.

The graphs are in the answer section of the main text.

21.

$$4^0 + 4^{-1} - 4^{\frac{1}{2}} = 1 + \frac{1}{4} - 2 = -\frac{3}{4}$$

23.

$$\frac{10!}{6!} = \frac{10 \cdot 9 \cdot 8 \cdot 7 \cdot 6!}{6!} = 5040$$

25.

$$\sum_{i=1}^{6} 7(3i) = 7(3 \cdot 1) + 7(3 \cdot 2) + 7(3 \cdot 3) + 7(3 \cdot 4)$$

$$+ 7(3 \cdot 5) + 7(3 \cdot 6)$$

$$= 7 \cdot 3 + 7 \cdot 6 + 7 \cdot 9 + 7 \cdot 12$$

$$+ 7 \cdot 15 + 7 \cdot 18$$

$$= 7644$$

27.

$$4x^2 - 4x + 17 = 0$$

$$a = 4, b = -4, c = 17$$

$$x = \frac{4 \pm \sqrt{16 - 272}}{8} = \frac{4 \pm \sqrt{-256}}{8} = \frac{4 \pm 16i}{8} = \frac{1 \pm 4i}{2}$$

$$\left\{\frac{1 \pm 4i}{2}\right\}$$

29.

$$27x^3 + 36x^2 - 3x - 4 = 0$$

$$(3x + 4)(3x + 1)(3x - 1) = 0$$

$$3x + 4 = 0 \quad \text{or} \quad 3x + 1 = 0 \quad \text{or} \quad 3x - 1 = 0$$

$$x = -\frac{4}{3} \qquad x = -\frac{1}{3} \qquad x = \frac{1}{3}$$

$$\left\{ -\frac{4}{3}, -\frac{1}{3}, \frac{1}{3} \right\}$$

31.

$$|5x - 1| = 8$$

$$5x - 1 = 8 \quad \text{or} \quad 5x - 1 = -8$$

$$5x = 9 \qquad\qquad 5x = -7$$

$$x = \frac{9}{5} \qquad\qquad x = -\frac{7}{5}$$

$$\left\{ \frac{9}{5}, -\frac{7}{5} \right\}$$

33.

$$\frac{1}{8} = 16^{3x+1}$$

$$2^{-3} = \left(2^4\right)^{3x+1}$$

$$2^{-3} = 2^{12x+4}$$

Thus, $\quad -3 = 12x + 4$

$$-7 = 12x$$

$$-\frac{7}{12} = x$$

$$\left\{ -\frac{7}{12} \right\}$$

35.

$$\frac{x+6}{x-8} + \frac{2}{x-3} = \frac{3x-19}{x^2-11x+24}$$

$$\text{LCD} = (x-8)(x-3) \text{ or}$$
$$x^2 - 11x + 24$$

$$\frac{(x-8)(x-3)}{1}\left(\frac{x+6}{x-8} + \frac{2}{x-3} \right)$$

$$= \frac{x^2-11x+24}{1} \cdot \frac{3x-19}{x^2-11x+24}$$

$$\frac{\cancel{(x-8)}(x-3)}{1} \cdot \frac{x+6}{\cancel{(x-8)}} + \frac{(x-8)\cancel{(x-3)}}{1} \cdot \frac{2}{\cancel{(x-3)}}$$

$$= \frac{\cancel{\left(x^2-11x+24\right)}}{1} \cdot \frac{3x-19}{\cancel{\left(x^2-11x+24\right)}}$$

#35. continued

$$(x-3)(x+6) + 2(x-8) = 3x - 19$$

$$x^2 + 3x - 18 + 2x - 16 = 3x - 19$$

$$x^2 + 5x - 34 = 3x - 19$$

$$x^2 + 2x - 15 = 0$$

$$(x-3)(x+5) = 0$$

$$x - 3 = 0 \quad \text{or} \quad x + 5 = 0$$

$$x = 3 \qquad\qquad x = -5$$

$$\{-5\} \qquad \text{Note:} \quad x - 3 \neq 0$$

$$\text{So,} \quad x \neq 3$$

37.

$$\left(\sqrt{2x+5}\right)^2 = \left(\sqrt{x+3} + 2\right)^2$$

$$2x + 5 = x + 3 + 4\sqrt{x+3} + 4$$

$$(x-2)^2 = \left(4\sqrt{x+3}\right)^2$$

$$x^2 - 4x + 4 = 16(x+3)$$

$$x^2 - 4x + 4 = 16x + 48$$

$$x^2 - 20x - 44 = 0$$

$$(x-22)(x+2) = 0$$

$$x - 22 = 0 \quad \text{or} \quad x + 2 = 0$$

$$x = 22 \qquad\qquad x = -2$$

$$\{22\} \qquad -2 \text{ does not check.}$$

39.

$$\log_2(2x+1) + \log_2(2x-2) = 2$$

$$\log_2(2x+1) + \log_2(2x-2) = \log_2 2^2$$

$$(2x+1)(2x-2) = 4$$

$$4x^2 - 2x - 2 = 4$$

$$4x^2 - 2x - 6 = 0$$

$$2x^2 - x - 3 = 0 \quad \text{Multiply by } \frac{1}{2}$$

$$(2x-3)(x+1) = 0$$

$$2x - 3 = 0 \quad \text{or} \quad x + 1 = 0$$

$$x = \frac{3}{2} \qquad\qquad x = -1$$

$$\left\{ \frac{3}{2} \right\} \qquad -1 \text{ does not check.}$$

41.

$\log(4x-1) + \log(x-2) = \log(3x+2)$

$$\frac{4x-1}{x-2} = \frac{3x+2}{1}$$

$$4x-1 = (x-2)(3x+2)$$

Cross multiply

$$4x-1 = 3x^2 - 4x - 4$$

$$0 = 3x^2 - 8x - 3$$

$$0 = (3x+1)(x-3)$$

$3x+1=0$ or $x-3=0$

$x = -\dfrac{1}{3}$ $x = 3$

$\{3\}$ $-\dfrac{1}{3}$ does not check.

43.

(1) $x^2 + y^2 = 25$

(2) $x - 7y = -25$

$x = 7y - 25$ Solve for x

Substitute and solve.

(1) $(7y-25)^2 + y^2 = 25$

$49y^2 - 350y + 625 + y^2 = 25$

$50y^2 - 350y + 600 = 0$

$y^2 - 7y + 12 = 0$

$(y-3)(y-4) = 0$

$y-3=0$ or $y-4=0$

$y=3$ $y=4$

$x = 7y-25$ $x = 7y - 25$

$x = 21 - 25$ $x = 28 - 25$

$x = -4$ $x = 3$

$\{(-4,3),(3,4)\}$

45.

Let x = one leg.

Then 3x - 10 = other leg.

Thus, $x^2 + (3x-10)^2 = 10^2$

$x^2 + 9x^2 - 60x + 100 = 100$

$10x^2 - 60x = 0$

$x^2 - 6x = 0$

$x(x-6) = 0$

$x=0$ or $x-6=0$

$x = 6$

$3x - 10 = 3 \cdot 6 - 10 = 8$

#45. continued

 Answer: 6 mm and 8 mm

47.

Let x = width of border.

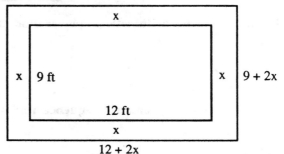

Thus, $(12+2x)(9+2x) - 9 \cdot 12 = 100$

$108 + 42x + 4x^2 - 108 = 100$

$4x^2 + 42x - 100 = 0$

$2x^2 + 21x - 50 = 0$

$(2x+25)(x-2) = 0$

$2x+25=0$ or $x-2=0$

$x = -\dfrac{25}{2}$ $x = 2$

Answer: 2 ft

49.

Let x = number of A's

Thus,

$A = \dfrac{k}{\sqrt{a}}$ $A = \dfrac{24}{\sqrt{9}}$

$12 = \dfrac{k}{\sqrt{4}}$ $A = \dfrac{24}{3}$

$24 = k$ $A = 8$

Answer: 8 A's

331

Chapter 11 Study Guide

$$\boxed{\text{Self-Test Exercises}}$$

I. In the following arithmetic sequence find the common difference, **d**, the general term a_n and S_{24}.

1. $-2, -4, -6, -8, \ldots$

2. $-\dfrac{5}{6}, -\dfrac{7}{6}, -\dfrac{9}{6}, -\dfrac{11}{6}, \ldots$

II. In the following geometric sequence find the common ratio, **r**, the general term, a_n and S_5.

3. $2, -4, 8, -16, \ldots$

4. $6, 3, \dfrac{3}{2}, \dfrac{3}{4}, \ldots$

III. Evaluate the following series, if possible.

5. $\displaystyle\sum_{i=1}^{6} \left(i^2 + 2i - 3\right)$

6. $\displaystyle\sum_{i=0}^{5} \dfrac{(-1)^i}{2^i} - 3$

7. $\displaystyle\sum_{i=1}^{20} (3i + 1)$

8. $\displaystyle\sum_{i=1}^{4} 4^i - 1$

9. $\displaystyle\sum_{i=1}^{\infty} 2 \cdot (-2)^i$

10. $\displaystyle\sum_{i=0}^{\infty} 2 \cdot \left(-\dfrac{1}{2}\right)^{i-1}$

IV. Using an infinite geometric series, write the following repeating decimal as a quotient of two integers.

11. $0.2\overline{2}$

12. $0.\overline{34}$

V. Evaluate the following.

13. $4!$

14. $\dfrac{14!}{8!}$

VI. Use the Binomial Theorem to expand the following binomial.

15. $(x - 2y)^4$

16. $(2x + 3y)^3$

VII. Applications.

17. Melissa found that by dropping a certain ball from a height of 32 feet it always rebounded $\dfrac{1}{2}$ the height it falls. Find the total distance the ball will travel before coming to rest.

18. Jamie saved $500 during her first year of employment for college. In each succeeding year she saves $200 more than she saved the preceding year. How much will she have saved in 5 years?

19. Dennis can bench press 125 pounds. His goal is to increase the weight he can bench press by 15 pounds per week. How long will it take him to press 365 pounds?

20. A giant city center clock is in need of repairs. The clock master finds that the pendulum travels 80% of its previous swing. If the first swing is 12 feet long, what is the total distance traveled by the pendulum by the time it comes to rest?

The worked-out solutions begin on the following page.

1.

$$d = a_2 - a_1$$
$$d = -4 - (-2) = -2$$

The general term is

$$a_n = a_1 + (n-1)d$$
$$= -2 + (n-1)(-2) = -2 - 2n + 2 = -2n$$

$$S_n = \frac{n}{2}[2a_1 + (n-1)d]$$

$$S_{24} = \frac{24}{2}[2(-2) + (24-1)(-2)]$$

$$= 12[-4 + 23(-2)]$$

$$= 12[-4 - 46] = 12(-50) = -600$$

2.

$$d = a_2 - a_1$$

$$d = -\frac{7}{6} - \left(-\frac{5}{6}\right) = -\frac{7}{6} + \frac{5}{6} = -\frac{1}{3}$$

The general term is

$$a_n = a_1 + (n-1)d$$

$$= -\frac{5}{6} + (n-1)\left(-\frac{1}{3}\right)$$

$$= -\frac{5}{6} - \frac{n}{3} + \frac{1}{3} = -\frac{n}{3} - \frac{1}{2}$$

$$S_n = \frac{n}{2}[2a_1 + (n-1)d]$$

$$S_{24} = \frac{24}{2}\left[2\left(-\frac{5}{6}\right) + (24-1)\left(-\frac{1}{3}\right)\right]$$

$$= 12\left[-\frac{5}{3} + 23\left(-\frac{1}{3}\right)\right]$$

$$= 12\left[-\frac{5}{3} - \frac{23}{3}\right]$$

$$= 12\left(-\frac{28}{3}\right) = 4(-28) = -112$$

3.

The common ratio is

$$r = \frac{a_2}{a_1} = \frac{-4}{2} = -2$$

#3. continued

The general term is

$$a_n = a_1 r^{n-1}$$

$$= 2(-2)^{n-1}$$

$$= 2(-2)^{-1}(-2)^n$$

$$= 2\left(-\frac{1}{2}\right)(-2)^n = -(-2)^n$$

$$S_n = \frac{a_1 - a_1 r^n}{1 - r}$$

$$S_5 = \frac{2 - (2)(-2)^5}{1 - (-2)}$$

$$= \frac{2 - (2)(-32)}{3}$$

$$= \frac{2 - 2(-32)}{3}$$

$$= \frac{2 + 64}{3} = \frac{66}{3} = 22$$

4.

The common ratio is

$$r = \frac{a_2}{a_1} = \frac{3}{6} = \frac{1}{2}$$

The general term is

$$a_n = a_1 r^{n-1}$$

$$= 6\left(\frac{1}{2}\right)^{n-1}$$

$$= 6\left(\frac{1}{2}\right)^{-1}\left(\frac{1}{2}\right)^n$$

$$= 6 \cdot 2\left(\frac{1}{2}\right)^n = 12\left(\frac{1}{2}\right)^n$$

$$S_n = \frac{a_1 - a_1 r^n}{1 - r}$$

$$S_5 = \frac{6 - 6\left(\frac{1}{2}\right)^5}{1 - \left(\frac{1}{2}\right)}$$

$$= \frac{6 - 6\left(\frac{1}{32}\right)}{\frac{1}{2}}$$

$$= \frac{6 - \frac{6}{32}}{\frac{1}{2}}$$

#4. continued

$$= \frac{\frac{192}{32} - \frac{6}{32}}{\frac{1}{2}} = \frac{\frac{186}{32}}{\frac{1}{2}} = \frac{186}{32} \cdot \frac{2}{1} = \frac{93}{8}$$

5.

$$\sum_{i=1}^{6} \left(i^2 + 2i - 3 \right)$$

$$= \left(1^2 + 2(1) - 3 \right) + \left(2^2 + 2(2) - 3 \right) + \left(3^2 + 2(3) - 3 \right)$$

$$+ \left(4^2 + 2(4) - 3 \right) + \left(5^2 + 2(5) - 3 \right)$$

$$+ \left(6^2 + 2(6) - 3 \right)$$

$$= 0 + 5 + 12 + 21 + 32 + 45 = 115$$

6.

$$\sum_{i=0}^{5} \frac{(-1)^i}{2^i} - 3$$

$$= \left(\frac{(-1)^0}{2^0} - 3 \right) + \left(\frac{(-1)^1}{2^1} - 3 \right) + \left(\frac{(-1)^2}{2^2} - 3 \right)$$

$$+ \left(\frac{(-1)^3}{2^3} - 3 \right) + \left(\frac{(-1)^4}{2^4} - 3 \right) + \left(\frac{(-1)^5}{2^5} - 3 \right)$$

$$= (1 - 3) + \left(-\frac{1}{2} - 3 \right) + \left(\frac{1}{4} - 3 \right) + \left(-\frac{1}{8} - 3 \right)$$

$$+ \left(\frac{1}{16} - 3 \right) + \left(-\frac{1}{32} - 3 \right)$$

$$= -2 + \left(-\frac{7}{2} \right) + \left(-\frac{11}{4} \right) + \left(-\frac{25}{8} \right) + \left(-\frac{47}{16} \right) + \left(-\frac{97}{32} \right)$$

$$= -\frac{555}{32}$$

7.

$$\sum_{i=1}^{20} (3i + 1)$$

$$= (3(1) + 1) + (3(2) + 1) + (3(3) + 1) + \cdots + (3(20) + 1)$$

$$= (3 + 1) + (6 + 1) + (9 + 1) + \cdots + (60 + 1)$$

$$= 4 + 7 + 10 + \cdots + 61$$

Since the common difference is $(a_2 - a_1) = (a_3 - a_2)$

$= 3$, this is an arithmetic series.

#7. continued

Therefore, with $a_1 = 4$,

$$S_n = \frac{n}{2} \left(2a_1 + (n-1)d \right)$$

$$S_{20} = \frac{20}{2} \left(2(4) + (20 - 1)3 \right)$$

$$= 10(8 + 57) = 650$$

8.

$$\sum_{i=1}^{4} 4^i - 1 = \left(4^1 - 1 \right) + \left(4^2 - 1 \right) + \left(4^3 - 1 \right) + \left(4^4 - 1 \right)$$

$$= 3 + 15 + 63 + 255 = 336$$

9.

$$\sum_{i=1}^{\infty} 2(-2)^i$$

$$= \left(2(-2)^1 \right) + \left(2(-2)^2 \right) + \left(2(-2)^3 \right) + \cdots + \left(2(-2)^\infty \right)$$

$$= -4 + 8 + (-16) + \cdots + \left(2(-2)^\infty \right)$$

Since $r = \dfrac{a_2}{a_1} = \dfrac{8}{-4} = -2$, the sum of the infinite

sequence can <u>not</u> be found since $|r|$ must be less

than 1.

10.

$$\sum_{i=0}^{\infty} -1 \cdot \left(-\frac{1}{2} \right)^i$$

$$= -1 \left(-\frac{1}{2} \right)^0 + (-1) \left(-\frac{1}{2} \right)^1 + (-1) \left(-\frac{1}{2} \right)^2 + \cdots$$

$$+ (-1) \left(-\frac{1}{2} \right)^\infty$$

$$= -1 + \frac{1}{2} - \frac{1}{4} + \cdots + -1 \left(-\frac{1}{2} \right)^\infty$$

Since $r = \dfrac{a_2}{a_1} = \dfrac{1}{2} \div (-1) = -\dfrac{1}{2}$ and $a_1 = -1$, the sum

of the infinite sequence is

$$S_\infty = \frac{a_1}{1 - r}$$

$$= \frac{-1}{1 - \left(-\frac{1}{2} \right)} = \frac{-1}{\frac{3}{2}} = -\frac{2}{3}$$

11.

$0.2\overline{2} = 0.222...$

This can be expressed as an infinite geometric series,

$$\frac{2}{10} + \frac{2}{100} + \frac{2}{1000} + \cdots,$$

with $r = \dfrac{a_2}{a_1} = \dfrac{2}{100} \div \dfrac{2}{10} = \dfrac{2}{100} \cdot \dfrac{10}{2} = \dfrac{1}{10}$ and $a_1 = \dfrac{2}{10}$.

Thus,

$$0.2\overline{2} = S_\infty = \frac{a_1}{1-r}$$

$$= \frac{\dfrac{2}{10}}{1 - \dfrac{1}{10}} = \frac{\dfrac{2}{10}}{\dfrac{9}{10}} = \frac{2}{9}$$

11.

$0.\overline{34} = 0.343434...$

This can be expressed as an infinite geometric series,

$$\frac{34}{100} + \frac{34}{10000} + \frac{34}{1000000} + \cdots,$$

with $r = \dfrac{a_2}{a_1} = \dfrac{34}{10000} \div \dfrac{34}{100} = \dfrac{34}{10000} \cdot \dfrac{100}{34} = \dfrac{1}{100}$

and $a_1 = \dfrac{34}{100}$. Thus,

$$0.\overline{34} = S_\infty = \frac{a_1}{1-r}$$

$$= \frac{\dfrac{34}{100}}{1 - \dfrac{1}{100}} = \frac{\dfrac{34}{100}}{\dfrac{99}{100}} = \frac{34}{99}$$

13.

$4! = 4 \cdot 3 \cdot 2 \cdot 1 = 24$

14.

$\dfrac{14!}{8!} = \dfrac{14 \cdot 13 \cdot 12 \cdot 11 \cdot 10 \cdot 9 \cdot \cancel{8!}}{\cancel{8!}} = 2,162,160$

15.

$(x - 2y)^4$

$= x^4 + \underline{}x^3(-2y) + \underline{}x^2(-2y)^2 + \underline{}x(-2y)^3$

$ + (-2y)^4$

#15. continued

Using the 4th row of Pascal's Triangle,

$= x^4 + 4x^3(-2y) + 6x^2\left(4y^2\right) + 4x\left(-8y^3\right) + \left(16y^4\right)$

$= x^4 - 8x^3y + 24x^2y^2 - 32xy^3 + 16y^4$

16.

$(2x + 3y)^3$

$= (2x)^3 + \dfrac{3}{1!}(2x)^{3-1}(3y) + \dfrac{3(3-1)}{2!}(2x)^{3-2}(3y)^2$

$ + (3y)^3$

$= 8x^3 + 3(2x)^2(3y) + 3(2x)(3y)^2 + 27y^3$

$= 8x^3 + 36x^2y + 54xy^2 + 27y^3$

17.

This is an infinite series where $a_1 = 32$ and

$r = \dfrac{a_2}{a_1} = \dfrac{16}{32} = \dfrac{1}{2}$. Thus,

$$S_\infty = \frac{a_1}{1-r}$$

$$= \frac{32}{1 - \dfrac{1}{2}} = \frac{32}{\dfrac{1}{2}} = 64$$

So, $32 + 64 = 96$

Answer: 96 ft

18.

The sequence of savings is an arithmetic sequence with $a_1 = 500$, $d = 200$, and $n = 5$. Thus,

$$S_n = \frac{n}{2}[2a_1 + (n-1)d]$$

$$S_5 = \frac{5}{2}[2(500) + (5-1)200]$$

$$= \frac{5}{2}[1000 + 800] = \frac{5}{2}(1800) = 4,500$$

Answer: $4,500

19.

This is an arithmetic sequence with $a_1 = 125$, $d = 15$, and $a_n = 365$. Thus,

#19. continued

$$a_n = a_1 + (n-1)d$$

$$365 = 125 + (n-1)15$$

$$365 = 125 + 15n - 15$$

$$365 = 110 + 15n$$

$$255 = 15n$$

$$17 = n$$

Answer: 17 weeks

20.

The problem can be expressed as an infinite geometric series with $a_1 = 12$ and $r = 0.8$. Thus,

$$S_\infty = \frac{a_1}{1-r} = \frac{12}{1-0.8} = \frac{12}{0.2} = 60$$

Answer: 60 ft